Modern Construction Management

Sixth Edition

Frank Harris
Emeritus Professor of Construction Science,
University of Wolverhampton

and

Ronald McCaffer
Professor of Construction Management
Loughborough University

with

Francis Edum-Fotwe
Lecturer, Construction and Project Management,
Loughborough University

Blackwell Publishing editorial offices:
Blackwell Publishing Ltd, 9600 Garsington Road, Oxford OX4 2DQ, UK
Tel: +44 (0)1865 776868
Blackwell Publishing Inc., 350 Main Street, Malden, MA 02148-5020, USA
Tel: +1 781 388 8250
Blackwell Publishing Asia Pty Ltd, 550 Swanston Street, Carlton, Victoria 3053, Australia
Tel: +61 (0)3 8359 1011

First published 2006 by Blackwell Publishing Ltd

ISBN-10: 1-4051-3325-2
ISBN-13: 978-1-4051-3325-8

Library of Congress Cataloging-in-Publication Data
Harris, Frank, 1944–
Modern construction management / Frank Harris and Ronald McCaffer with Francis Edum-Fotwe. – 6th ed.
p. cm.
Includes bibliographical references and index.
ISBN-13: 978-1-4051-3325-8 (pbk. : alk. paper)
ISBN-10: 1-4051-3325-2 (pbk. : alk. paper)
1. Construction industry–Management. I. McCaffer, Ronald. II. Edum-Fotwe, Francis. III. Title.

HD9715.A2H35 2006
624.068–dc22
2006001288

A catalogue record for this title is available from the British Library

Set in Minion 10/12pt
by Best-set Typesetter Ltd., Hong Kong
Printed and bound in Great Britain
by TJ International, Padstow

The publisher's policy is to use permanent paper from mills that operate a sustainable forestry policy, and which has been manufactured from pulp processed using acid-free and elementary chlorine-free practices. Furthermore, the publisher ensures that the text paper and cover board used have met acceptable environmental accreditation standards.

For further information on Blackwell Publishing, visit our website:
www.blackwellpublishing.com

Contents

Preface to the sixth edition

The book is intended for students and graduates of civil engineering, construction management, building and quantity surveying, and is arranged to reflect site, business and corporate responsibilities embraced by the S/NVQ supervisory and management levels of career development. This approach acknowledges that the modern successful construction engineer, builder or quantity surveyor needs to be a competent technologist possessing complementary skills and knowledge in management as well as understanding the business processes. Armed with such expertise the young construction trainee will be better prepared for decision-making and undertaking executive responsibilities.

The new edition has been guided by the drive for improvement in construction industry performance stimulated through the *Rethinking Construction* and *Construction Best Practice* programmes.

Recently restated under the *Accelerating Change* initiative as committed leadership, client satisfaction, integrated processes and teams, quality management and social responsibility, the opening overview sets out the key points which are subsequently interweaved throughout the text. The construction engineer, builder or quantity surveyor is thereby better positioned to understand and implement modern strategies needed in providing value for money for the client and society.

The book begins by emphasising the important role of total quality management and safe working that now pervades every aspect of construction activity. The subsequent sections are: 'project production management' describing the management techniques employed on site; 'business management', which addresses the relevant commercial aspects; and finally 'administration and company management' covering corporate activities including IT systems and international work.

The processes essential in delivering continuous improvement and meeting performance indicators are especially featured, while the principles of lean construction, concurrent engineering, supply networks, re-engineering, value and risk management are given prominence. The latest contractual innovations, notably design and build, PFI/PPP, term, prime, managing agent, early contractor involvement, framework agreements and alliances, are evaluated, with reverse auctions contrasted with negotiated contracts and detailed pre-selection. Issues for business development and business models, business process outsourcing, matrix management, incentives and plant hire are also treated. In addition, topical concerns on construction productivity, relationship marketing, environment and sustainability, the 'Kyoto' protocol, corporate social responsibility, corporate governance, data protection, international construction contracts, investment monitoring and regulation, RFID tagging, health, safety and training are brought to the reader's attention.

Finally, the comprehensive selection of worked examples, designed to help the reader consolidate learning, is augmented in this edition by 25 new tutorial exercises dealing with the Operational Research methods, invaluable for analysing the many challenging facets of construction management featured in the contents.

1
Introduction

Resulting from progressive movements at the operational and business levels in the sector, construction today presents a marked shift from the situation described in earlier editions of this book; for example lean construction, benchmarking, modern forms of contracting and the rapid change in technological possibilities portray significant changes and opportunities. Not least efforts for dealing with environmental, social and economic accountability of construction and prominently displayed under the Constructing Excellence (CE) and government programmes indicate the new challenges and growing competition faced by the industry in a world of intensified global trading.

As a major element in the strategy, education and structured training have come to the fore in the attempt to deal with the pressing demands for a more competent workforce. Notably encapsulated within the Scottish/National Vocational Qualifications (S/NVQ) and Construction Skills Certification Scheme (CSCS), the sixth edition of *Modern Construction Management* has consequently been reorganised to cover the required knowledge for the competencies at each of the three relevant S/NVQ levels covered by the book. These are interwoven throughout the chapters and progressively developed as follows:

(1) Project Production Management generally covers Supervisor/Manager occupations (S/NVQ levels 3 to 4)
(2) Business Management concentrates on the Senior Manager/Contracts Manager/Construction Project Manager/Senior professional functions (S/NVQ levels 4 to 5)
(3) Administration and Company Management mainly treats Company Executive functions (S/NVQ level 5)

Structure of the book

The book is divided into four main sections; in addition, Chapters 1 and 2, which do not form part of the main sections, give specific consideration at the outset to the philosophy of the book as a means of explaining the succeeding chapters. In particular, Chapter 2, which covers quality management in construction, is used to illustrate how quality is intertwined as a thread running through all the subsequent sections. It also explores the emerging strategic role of quality as a driver for competitive advantage in construction.

- Section 1 deals with techniques relating to project production management
- Section 2 treats the business aspects of management at both project and company levels
- Section 3 addresses the executive management responsibilities for overall company control

- Section 4 brings together a selection of self-learning problems complemented with complete worked solutions for use in the classroom environment, tutorial exercises and seminar discussions.

The reasons for this particular presentation are:

(1) Successful construction industry executives have distinct phases in their careers; the initial period is spent on site, followed by middle management duties at the project level, culminating in a career with executive head office activities. The sections are intended to cater for these phases.
(2) The construction industry is inherently uncertain as a result of the nature of the industry itself—the competitive tendering process, the company's turnover, site production rates and the weather are all features that are characterised by variability and a degree of uncertainty. To be able to cope with such uncertainty, construction executives need to be acquainted with the relevant knowledge and tools for addressing these features. The management techniques described in this book help reduce variability and thus provide the basis for sound and effective decisions by aspiring executives. For example, with proper planning the duration of a project is not just an experienced guess. The inevitable residual variability in even the best-run company needs to be controlled by:
    (a) Planning and setting targets
    (b) Choosing methods to achieve such plans and targets
    (c) Monitoring progress
    (d) Taking corrective action when necessary.

This continual monitoring and revision is ultimately the only way to cope with uncertainty and variability.

Objectives and contents

Each chapter deals with a specific topic (which could, if exhaustively treated, form the basis of a whole book; suggestions for further reading appear at the end of chapters).

The level of detail aimed at is that which will provide the reader with a basic working knowledge of the topic, rather than with specialist expertise. For example, the planning section of the book explains the major techniques available for planning both repetitive and non-repetitive works in sufficient detail to allow intelligent engineers to apply them, providing sufficient comprehension for them to converse sensibly with a specialist support group such as a planning department. Engineers and builders need enough knowledge to understand, appreciate and, where necessary, question the work of specialist support staff such as accountants, cost clerks, planners and plant managers. A grasp of the techniques described in the sections should help in achieving this skill. Specialists must not be allowed to hide safely in their own specialisms. Participation in the exercises in Section 4 provides a deeper and better understanding of the implications of the various techniques. Section 4 largely covers the numerical-based aspects of these techniques.

Chapter 2. *Quality management in construction* provides the platform for the succeeding chapters and describes the evolution of quality management from quality control, through quality assurance to total quality management. It also looks at quality from the project perspective, advocates a concerted effort by both client and contractor to make any quality agenda a reality, and explores a systems approach to attaining such an agenda.

The contents of each section are now discussed briefly below.

Section 1

Section 1 relates specifically to project production management, including planning techniques, production process improvement, estimating and tendering, workforce motivation and cost control.

- Chapter 3. *Production process improvement* covers quality management, lean construction, benchmarking, delay surveys, work-study, production sampling and computerised recording techniques.
- Chapter 4. *Planning techniques* deals with the principles of the techniques used in planning repetitive or non-repetitive construction work. The chapter describes bar-charts, linked bar-charts, network analysis and line-of-balance scheduling, PERT, space–time diagrams and The Last Planner. The role and use of computers in planning and the requirements of computer systems in exchanging data are also described.
- Chapter 5. *Workforce motivation* links the use of incentive schemes to motivation theory. It also presents the various payment systems for non-financial, semi-financial and purely financial incentives that can be employed to enhance worker motivation.
- Chapter 6. *Cost control* gives guidance on the various cost control methods available, including profit-related control systems, unit and standard costing approaches and cost monitoring of subcontractors.
- Chapter 7. *Plant management* considers the financing of plant and gives guidance on plant selection. Calculating a hire rate and maintenance procedures are also covered.

Section 2

Section 2 presents business management topics and is intended to assist project-based staff to understand and appreciate the company's attitudes and activities, easing the transition from site to general management. The topics described relate to procurement, bidding, budgets and cash flow, economic assessment and plant management.

- Chapter 8. *Project procurement* introduces the role of project management and reviews various forms of contract. The new developments for procuring construction and engineering such as design and build, PFI and partnering are also explained.
- Chapter 9. *Estimating and tendering* describes the current nature of estimating practised by main and work-package contractors. It describes parties involved in the estimating and tendering process for work packages and outlines the process, including the decisions and calculations involved, and the issues in costing materials and subcontractors. It also addresses the use of computers in estimating and the changing role of the estimator in the face of advances in information technology.
- Chapter 10. *Competitive bidding* examines the effect of estimating accuracy, which implies the need for more resources in the estimating department, reviews how to interpret the

various available items of data relating to competitors' behaviour and comments on improving estimating accuracy.
- Chapter 11. *Budgetary Control* deals with the preparation of budgets and controlling costs for a company or enterprise.
- Chapter 12. *Cash flow and interim valuations* illustrates company cash flow forecasting and provides guidance on how to do this type of forecasting, the use of computers in cash flow calculations, the process of interim valuations and the relationship between interim valuations and cash flow.
- Chapter 13. *Economic assessments* describes the principles employed in economic comparisons and in measuring rates of return, life-cycle costing, cost–benefit analysis and financial modelling. It also provides an introduction to the use of multi-criteria analysis for appraising projects.

Section 3

Section 3 presents the executive management responsibilities largely concerning head office activities, including organisation, business development, global construction, the emerging role of information as a major construction resource and finance.

- Chapter 14. *Company organisation* contains a description and explanation of company structure, organisation and managerial responsibilities.
- Chapter 15. *Business development and market planning* describes a marketing approach to construction and the benefits likely to be derived.
- Chapter 16. *International construction logistics* provides an overview of the problems in raising finance, dealing with unfamiliar conditions of contract and legal systems, transport of goods, payment procedures and local labour, resources and security.
- Chapter 17. *Information resources and IT systems* develops an understanding of the strategic role played by information resources in managing both projects and the business for organisations in the construction industry. It also addresses information systems and its associated technology embracing email, web sites, intranets, on-line information data and transfer, data exchange and integration of systems.
- Chapter 18. *Financial management* describes the sources and means of acquiring capital funds and the use of balance sheets and profit and loss accounts.

Section 4

This section presents 87 tutorial examples with complete worked solutions for students in construction disciplines. It is separated into two chapters, with the first, Chapter 19, covering the worked examples from Chapters 3 to 18. Chapter 20 provides worked examples on operational research techniques.

Students learn by reading texts and attending lectures. However, they need to test their newfound knowledge or skill by attempting to work through example problems and several textbooks are available that offer such examples, either with or without answers. Where an answer is provided, the student's own answer is frequently at variance and he or she is then faced by a dilemma: is the textbook in error or has the author made different, but valid, assumptions? In this book, a complete worked solution to each example is given so that the student has full guidance through the analysis.

The topics covered are those aspects of construction management that may be treated numerically:

- Production analysis
- Planning
- Estimating
- Motivation schemes
- Control of project costs
- Budgetary control
- Cash flow forecasting
- Discounted cash flow
- Investment analysis
- Plant management
- Setting of plant-hire rates
- Financial management
- Development economics
- Operational research

The intention is for the students to test their knowledge by trying the examples and comparing the solutions with those offered in the book. Any differences between the student's solutions and those presented here may be discussed with the tutor, and in this way tutorial discussions may be used advantageously for resolving difficulties rather than for routine learning.

It should be remembered that these are tutorial examples and that each one deals with a limited number of variables and principles, sometimes making simplifying assumptions. Thus, students may test their understanding of the principles and ability to manipulate the variables.

2
Quality management

Summary

Quality control, quality assurance, total quality management and systems quality in construction.

Introduction

Quality management has seen a transition from reacting to the outcome of site production activities to becoming a strategic business function accounting for the raison d'être of construction companies. Unless a construction company can guarantee its clients a quality product, it can now no longer compete effectively in the modern construction market. Crucial to the delivery of such quality products is the quality of processes that produce the product. 'Quality' now stands alongside 'price' as a major factor of differentiation in contractor selection by the client as well as determining the efficiency of processes that the contractor adopts for site operations. To be competitive and to sustain good business prospects, construction companies need a more strategic orientation for the quality systems they deploy.

This chapter focuses on the transitions in quality management for construction companies, culminating in a systems outlook for managing quality in construction. Quality management has to provide the environment within which the tools, techniques and procedures presented in the other chapters can be effectively deployed leading to operational success for the company. The role of the quality management for a construction company is not an isolated activity, but intertwined with all the operational and managerial processes of the company. This chapter reviews various *concepts* associated with quality and then considers the contributions of quality control, quality assurance and total quality management to the quality of construction. It also addresses the growing use of quality management systems for achieving superior performance in construction. It highlights the fact that quality in construction can be achieved only through the direct effort of all stakeholders of the project.

Notions of quality

The management of quality in construction is an area of specialisation that has been growing over the past three to four decades to embrace aspects of the project and company activities that are often seen as remote from the physical product. Figure 2.1 shows various concepts that are considered to have an influence on the quality of the product and which have come to be associated with quality in construction.

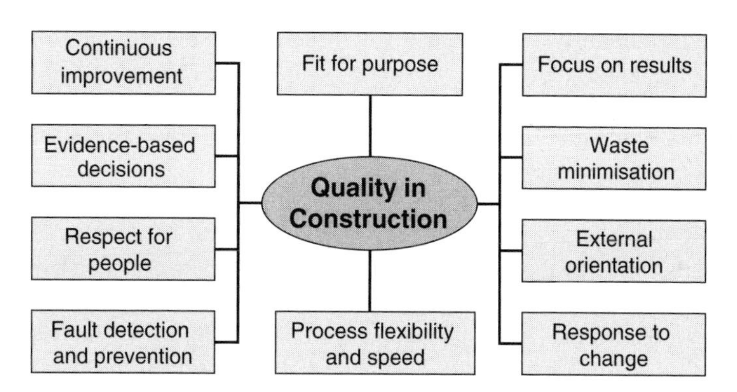

Fig. 2.1 Aspects of construction quality.

The various areas that contribute to quality in construction in Fig. 2.1 reflect the product features, the processes of production and organisation, as well as wider company and industry/business issues. In particular, the management of quality in construction has been embracing considerations that address more of the pre-production processes and organisation/industry issues. For example, a company's quality status is not just seen in isolation, but increasingly from the perspective of industry-wide standards and against that of its competitors.

Quality transition

The modern concept of quality is considered to have evolved through major transition stages over many years. These stages are as follows:

(1) **Quality control and inspection.** Inspection is the process of checking that what is produced is what is required. Quality control introduced inspection to stages in the development of goods and services to ensure that they are undertaken to specified requirements. Usually quality control is done on a sampling basis dictated by statistical methods. Sampling concrete by making cubes is the most common and best-known example in construction.

(2) **Quality assurance.** This has developed to ensure that specifications are consistently met. '*Fit for purpose*' and '*right first time*' are the principles of quality assurance and the frame of reference for quality assurance is the International Quality Standard ISO 9000 family of standards. To be certified as operating to the ISO 9000 standard is now seen as virtually essential in today's construction industry. Many clients simply will not do business with companies not certified to ISO 9000.

(3) **Total Quality Management.** This is based on the philosophy of continuously improving goods or services. A TQM approach is now seen as essential to long-term survival in business, including construction. A key factor is that everyone in the company should be involved and committed from the top to the bottom of the organisation. The successful total quality managed company ensures that their goods and services can meet the following criteria:
- Be fit for purpose on a consistently reliable basis
- Delight the customer with the service which accompanies the supply of goods
- Supply a quality of the product that is so much better than the competition that customers want it regardless of price

(4) **Quality management systems.** A quality management system presents a set of processes that ensure the attainment of defined quality standards for the provision of services and products by the project or a construction company. This can be company- or project-specific or one of several systems that are available on the market. The ISO 9000 is the most commonly used international standard that provides a framework for an effective quality management system.

It is now argued that successful construction companies have to meet at least two of these criteria to stay successful. *The pursuit of total quality is seen as a never-ending journey of continuous improvement.* A fuller description and application of each of the stages is set out below.

Quality control and inspection

The earliest and most basic form of quality management is quality control. This is described under the headings of:

- Definition and objectives of quality control
- Controlling quality
- Quality control implemented in construction

Definition and objectives of quality control

The term quality control is defined by an interpretation of its elements: 'quality' and 'control'.

Quality

The term quality is often used to describe prestige products such as Rolex watches and Mercedes Benz motor cars. However, the term 'quality', although applicable to these items, does not necessarily refer to prestigious products but merely to the fitness of the product to the customers' requirements. *Quality* describes the sum of attributes for a product or service that enables it to meet the requirements or specified need of the *customer*. The concept of quality goes hand-in-hand with *value-for-money* as perceived by the client.

Control

The concept of being 'in control', or having something 'under control', is readily understood. We mean we know what we intend to happen, and are confident that we can ensure that it does. Quality control, however, is primarily concerned with defect detection. The main quality control techniques are inspections and statistical quality control techniques (i.e. sampling). Both are aimed at ensuring that the work produced and the materials used are within the tolerances specified. Some of these limits are left to an inspector's *judgement* and this can be a source of difficulty. The major objectives of quality control can be defined as follows:

- To ensure the completed work meets the specification
- To reduce customers' or clients' complaints
- To improve the reliability of products or work produced

- To increase customers' or clients' confidence
- To reduce production costs

Controlling quality

Quality control involves ensuring that every product or service meets a minimum set of defined criteria for acceptance. The central feature to all quality control systems is that of inspection.

Inspections

To be effective the construction process requires that work items to be inspected must be catalogued in a quality schedule. Inspection in construction takes two forms: that which is objective and quantifiable, e.g. the length of a line, levels of floors or roads, verticality of a wall and volumetric dimensions of a kitchen pod; and that which is open to the inspector's interpretation, for example cleanliness, fit, tolerances and visual checks. The latter method of inspection usually involves simple observation, and relies on the experience of the inspector. Objective inspection requires some form of measurement to support the verification of meeting the quality standard. For example, there are some precise quantified inspections involved in the commissioning of plant and machinery, pressure tests in pipe-work and strength tests on materials such as concrete, each of which involves such physical measurement.

Statistical methods

These methods of quality control are based on the need to sample. In many of the processes of manufacture and construction the scale of the operation is too large to have 100% inspection and therefore sampling techniques are employed.

The main techniques in statistical quality control are:

- *Acceptance sampling*, based on probability theory, which allows the work to continue if the items sampled are within predetermined limits
- *Control charts* that compare the results of the items sampled with the results expected from a 'normal' situation; usually the results are plotted on control charts, which indicate the control limits

In construction it is the quality of materials that is normally controlled by statistical methods, the most common being that of the cube strength of concrete.

Quality control implemented in construction

Traditionally there are two sets of documents that are used to determine the required quality of a construction project. These are the Specifications and the Contract drawings. The contractor uses these two documents during the site operations stage of any project to facilitate 'quality' construction.

The process of actual construction is dissimilar to that of a production line in that there are no fixed physical and time boundaries between each operation of the process; hence the positioning and timing of quality inspection cannot be predetermined. In construction quality checks are undertaken as each operation or sub-operation is completed. The majority of quality checks are undertaken visually. Visual quality checks of each section of construction are undertaken by the contractors' engineers and foremen and then by the resident engineers and inspectors to ensure that it complies with the drawings and specification. Quantifiable quality checks are also made during the construction stage. These include testing the strength of concrete cubes, checking alignment of brickwork and commissioning of services installations. Figure 2.2 provides an example of a quality control sheet for undertaking these inspections. The results of these quality checks are recorded and passed to the resident engineer.

The weakness of quality control is the development of the inspection mentality or culture, whereby the construction contractors' operatives and engineers set their standards to that which they can 'get past the inspector'. In addition to potentially surrendering the standards of workmanship to an inspector, it exposes the contractor to expensive re-work if the standards of workmanship obtained do not meet with the inspector's approval. It would be much better if the contractors' engineers and operatives had a clear understanding of the quality required, were able to recognise it themselves and achieve it first time or regulate it by self-inspection. This concept, which is the basis of quality assurance, potentially reduces the risks of producing unsatisfactory work and becoming involved in expensive re-work. Notwithstanding the existence of quality assurance and the emergence of total quality management most clients still engage inspectors through their resident engineers or architects to reassure themselves. However, the impact and importance of the clients' inspectors is much reduced in a quality assured or total quality managed company.

Quality assurance

Quality assurance emphasises defect prevention, unlike quality control, which focuses on defect detection once the item is produced or constructed. Quality assurance concentrates on the production or construction management methods and procedural approaches to ensure that quality is built into the production system. Quality assurance involves planned and systematic actions necessary both to provide adequate confidence that a product or service will satisfy given requirements or standards and to be able to demonstrate any such compliance to that quality standard.

Quality assurance (QA) is described under the following headings:

- Evolution of QA from quality control
- Definition of quality terms
- QA standards
- Developing and implementing a QA system
- QA in construction

Evolution of QA from quality control

Traditional quality control is the practical implementation of techniques to ensure the quality of work is satisfactory. There are no standard methods for implementing quality control techniques; hence it is unlikely that there is a consistency of quality between companies claiming to use quality control. The variability of quality control results in the loss of the competitive edge that it potentially affords a company in the marketplace, because customers cannot quantify the effectiveness of quality control in any one company. QA was created to remedy this situation. The

Quality Control – RM & Co. Ltd.
Inspection Report

Contract **RMC Installations**
Work Section **Brickwork**
Programme No

Sheet 1 Of 1
Contract No
WS Ref No 0 4 1 1 S 1
Prog. Item No

	SATISFACTORY	UNSATISFACTORY
(1) SETTING OUT		
(a) Check all door and window openings	✓	
(b) Check width of cavity	✓	
(c) Brickwork is plumb (including jambs and corners)	✓	
(d) Check all banding levels and window levels	✓	
(e) Check internal room sizes	✓	
(f) Check all joint sizes	✓	
(2) TECHNICAL ADEQUACY		
(a) Brick ties (number, spacing, insulation clips)	✓	
(b) Insulation (installed and secure)	✓	
(c) DPC (correctly installed)	✓	
(d) Cavity cleaned	✓	
(e) Movement joint (installed correctly)	✓	
(f) Check brickwork support angle	✓	
(g) Check for fire barriers	✓	
(3) APPEARANCE		
(a) Joints and sizes constant	✓	
(b) Faced brick cleaned (free from staining and mortar splashes)	✓	
(c) Bricks level and true	✓	
(d) Overall appearance satisfactory	✓	
(e) Is the work adequately protected	✓	
REMEDIAL ACTION (IF ANY)		
(1)		
(2)		
(3)		
(4)		
(5)		
(6)		
Date of inspection	20/4/99	
In attendance	FCH	
Result of inspection	SATISFACTORY	UNSATISFACTORY

Remedial action (inspected and approval to proceed given)

Signature

Date 20/4/99

Prepared by: RM
Date: 23/04/99

Distribution				
Arch	✓	WS	✓
Const Eng	✓		

Fig. 2.2 Example of an inspection report sheet for undertaking construction quality control.

ultimate objective of QA is to provide the client with the quality of work required without the need for clients to check during the process. A customer for a car does not insist on checking the assembly of the car, for example. This objective is achieved by documenting what processes are performed and how they are accomplished, by self-checking that each process is completed correctly and finally by recording that fact. The policy of recording the processes undertaken, together with the checking and recording of procedures, provides the customer with the assurance that the company is aiming to achieve an acceptable standard of quality. Although 'satisfying the client' is the main object, the essence of QA is primarily to address 'getting it right first time' in order to avoid unnecessary costs to the contractor. Most construction contracts will include a clause requiring contractors to remedy any work that does not meet with the quality requirements of the project. Since this remedial work is undertaken at the contractor's expense, it provides a very strong incentive for the adoption of a QA approach by the contractor. Oakland (1995) defined both quality control and quality assurance; in doing so he also clearly explained their differences.

> *Quality control* is essentially the activities and techniques employed to achieve and maintain the quality of a product, process, or service. It involves a monitoring activity, but also concerns finding and eliminating causes of quality problems so that the requirements of the customer are continually met.
>
> *Quality assurance* (QA) is broadly the prevention of quality problems through planned and systematic activities (including documentation). These will include: the establishment of a good quality management system, the assessment of its adequacy, the audit of the operation of the system, and the review of the system itself.

Definition of quality terms

There are many quality terms in use, for each of which ISO 8402:1994, *Quality management and quality assurance—vocabulary* provides a specific definition. Below are some salient definitions of direct applicability to construction, and for which explanations have been provided. The exact definitions can be obtained from ISO 8402:1994.

Term	Denotation
Conformity	Satisfying pre-defined requirements of a client in a quality system.
Design review	This is a set of activities whose purpose is to evaluate how well a potential design meets all quality requirements. During the course of this review, problems can be identified to which solutions must be developed.
Design verification	This process involves examining design outputs and the use of objective evidence to confirm that outputs meet input requirements.
Preventive actions	Preventive actions are procedures put in place to ensure the elimination of potential causes of non-conformities or to achieve quality improvements.
Quality	Every entity has specific characteristics, some of which are derived from stated or implied needs. The set of these special characteristics makes up the quality of an entity. For example, the need for dependability is met by designing a dependable product. Dependability then becomes a quality of the product.

Term	Denotation
Quality assurance	This is a set of activities whose purpose is to demonstrate that an entity meets all quality requirements, and will do so when the product is finished. QA activities are performed in order to inspire the confidence of both customers and managers, that all quality requirements are being met.
Quality audits	These examine the elements of a quality system in order to establish that they comply with quality system requirements. Elements include responsibilities, authorities, relationships, functions, procedures, processes and resources.
Quality control	This is a set of activities or techniques undertaken to ensure that all quality requirements are being met. In order to achieve this purpose, processes are monitored and performance problems are solved.
Quality management	This includes all the activities that managers perform in an effort to implement their quality policy. These activities include quality planning, quality control, quality assurance and quality improvement.
Quality manual	A quality manual is a document that states the quality policy and describes the quality system of an organisation or a process. It describes the roles, relationships, functions, processes, procedures, systems and resources that affect quality. It can be a paper manual or an electronic manual.
Quality planning	Quality planning is defined as a set of activities whose purpose is to describe quality system policies, objectives and requirements, and to explain how these policies will be applied, how these objectives will be achieved and how these requirements will be met.
Quality policy	A quality policy statement defines the organisation's commitment to quality.
Quality surveillance	Quality surveillance is a set of activities aimed at monitoring an entity and to review its records in order to prove that quality requirements are being met.
Quality system	This is a network of processes made up of responsibilities, authorities, relationships, functions, plans, policies, procedures, practices, processes and resources. Its purpose is to satisfy quality management requirements and to assure customers on the quality of products and services.
Total quality management	This is a management approach that tries to achieve and sustain long-term organisational success by encouraging employee feedback and participation, satisfying customer needs and expectations, respecting societal values and beliefs and obeying governmental statutes and regulations.

Quality standards

The recognised quality standards are the ISO 9000 family of standards, the international standard for quality management and quality assurance (which in this chapter is subsequently referred to as ISO 9000). The ISO 9000 standard was developed to move away from the original 'prescriptiveness' approach of its predecessors and to achieve a more flexible framework, which allows organisations to develop their own policies and procedures. The flexibility of the new standard also allowed its application in the service industries and the IT sector, which impact on the business of construction companies. The ISO 9000 family of standards operates on the assumption that the following factors can have an influence on the quality of a product or a service provided by an organisation.

- Design
- Purchasing
- Management
- Work patterns
- Job descriptions
- Inspection and testing
- Reporting relationships
- Policies and procedures
- Record-keeping systems
- Inventory control
- Training
- Customers
- Technologies
- Resource
- Planning methods
- Production processes
- Transportation services
- Communication patterns
- Service delivery practices
- Employee knowledge and skill

There are three main standards covered by the ISO 9000 family of standards and they are all applicable to any specific industry or sector including the construction industry. Collectively they provide guidance for quality management and the general requirements of quality assurance. ISO 9000 clarifies the principal quality-related concepts and the distinctions and interrelationships among them. It also provides guidance for the selection and use of the ISO 9000 family of International Standards on quality management and quality assurance. The guideline for selection, ISO 9000, is made up of four parts. The issues covered by each of these parts are described in the following sections.

ISO 9000—Part 1 Selection and use

This standard essentially provides the guidelines for the selection and use of the various standards that make up the ISO 9000 family of standards and clarifies the main quality concepts. ISO 9000-1 should be the first reference for any organisation contemplating the development and implementation of a quality system.

ISO 9000—Part 2 Application guidelines

This part provides guidance on the implementation of and application of the quality assurance standards and is applicable to the processes involved at the project and organisation level within the construction industry. ISO 9000-2 is particularly useful at the onset of implementation for any quality system. For example, its specification would apply to design consultants, architects and design-and-build companies.

ISO 9000—Part 3 Software

This part of ISO 9000 deals exclusively with computer software. The standard recognises that the process of development, supply and maintenance of software is different from that of industrial products and construction facilities. The increasing use of IT options for information management and processing during both the design and construction stages makes ISO 9000–3 relevant to the overall quality systems adopted by both contractors and consultants.

ISO 9000—Part 4 Dependability

This deals with reliability, maintainability and availability of input resources for the processes involved in any production or service delivery. It covers the essential features of a comprehensive reliability programme for the planning, organisation, direction and control of resources to produce products that will have to meet the requirements of maintaining consistent and uniform quality.

The ISO 9000 family of standards also has specific application documentation that provides the framework for external quality assurance in which the relationship between parties is essentially contractual. These standards are classed as ISO 9001, ISO 9002 and ISO 9003. In addition, ISO 9004 provides supplementary guidance notes, which allow the principles of quality management to be applied to any organisation irrespective of the nature of the relationships between the parties involved in the process of production or service provision. Each of these application standards is briefly described in the following sections.

ISO 9001

This standard provides a framework for quality assurance in design, development, installation and servicing. ISO 9001 is used when there is a need to demonstrate capability of controlling the processes for design as well as production of a conforming product. The requirements of this standard primarily aim at achieving customer satisfaction by preventing non-conformity at all stages from design through to servicing. A construction company will make use of ISO 9001 if its operations cover all the stages from design to servicing. For example, a design–build–operate contractor will need to utilise this standard to develop their quality system.

ISO 9002

This standard is applicable to manufacturing or installation companies that do not undertake design activities. ISO 9002 requires process control and therefore necessitates inspection during a process. This standard would apply to many construction contractors, where evidence of inspections and tests, during a process, has to be given to the client.

ISO 9003

This standard relies on a quality management system that is based on final inspection and testing. The ISO 9003 standard could be applied to any simple building process including the building of a non-load-bearing brick wall or the laying of roof tiles.

ISO 9004

ISO 9004 provides explanatory clauses to help with the interpretation and clarification of quality system elements, including the three main standards.

Developing and implementing quality systems

The following four stages are common to the development and implementation of any quality assurance system:

- Establish awareness
- Develop quality manuals
- Introduce the system
- Evaluate the system

Establish awareness

To introduce a QA system it is necessary to have the understanding and commitment of top management. Often QA is initiated by top management as they realise that it is necessary for the organisation to remain competitive. The majority of managers are aware of QA and may have prejudices for or against its implementation; it is therefore necessary to gain widespread support for QA by explaining the potential benefits. Just as it is necessary to attain the support and understanding of top management, it is also necessary for senior and middle management to support and understand the QA approach. This is best achieved by means of a short QA training course.

Develop quality manuals

The quality manual is the basis of any QA system. A quality manual usually contains the following:

(1) **Company profile.** The company profile should contain information about the company including the following:
- Date of establishment
- Nature of business
- Annual turnover
- Scope of trade

(2) **Amendments record.** The amendments record lists any modifications to the quality manual so that persons reading the manual can determine the current status of the documents contained within it.

(3) **Policy statement.** The policy statement states that it is the policy of the company to undertake its business activities in accordance with a QA standard.

(4) **Quality standards.** This provides a description of how the criteria set out in the quality standards are addressed by the company quality system. This section describes how the company's quality system equates to the parts of the quality standard with which it is intended to comply.

(5) **Structure of the organisation.** This section should consist of the structure of the functions within the company, the responsibility of each function and the people who undertake each function. This is necessary to allow the identification of the individuals or groups responsible for carrying out specific tasks.

(6) **Procedures.** ISO 9000-1 requires that all quality procedures should be documented for the company's critical functions.

(7) **Work instructions.** The procedures in the above section are concise, it is therefore necessary to support these with work instructions. Work instructions refer to individual tasks and define how each should be completed.

Introduce the QA system

Once the draft quality system has been developed, it should be introduced progressively during a trial period, which lasts between three and six months. The trial period is used to generate understanding and acceptance of QA within the company as well as providing the opportunity to debug the draft quality system information. Problems with the quality system are inevitable, but the majority will be identified by employees during the trial period. The substantial problems will need to be addressed and corrected immediately, while others can be noted and corrected at the end of the trial period.

After the trial period the system can then be introduced formally, by distributing the quality manual to the managers of each section of the company. The section managers then implement the quality procedures within their own department or group.

Evaluate the QA system

The two methods of evaluating a quality system are management review and auditing.

(1) **Management review.** Management review requires managers periodically to review the quality system in their area of responsibility to ensure it is still satisfactory.
(2) **Internal audits.** ISO 9000 requires that a planned sequence of internal audits must be defined to ensure the effectiveness of the quality system. Auditing is a formal procedure undertaken by a trained individual independent of the organisation, who reports directly to senior management. The results of an audit should be documented and the records should detail inadequacies, by issuing non-conformance notices, and suitable corrective action. The person responsible for the corrective action and its timing should also be incorporated in the audit records.

Third party accreditation

Once the quality management system is implemented and operational and internal auditing has helped to refine it, the companies can, if they wish, apply for third party accreditation. Specialist companies such as Lloyd's Register and BSI offer a third party accreditation where they will certify that a company's quality management system meets the requirements of BS EN ISO 9000. To gain third party accreditation the company submits its quality management system and documentation for scrutiny and allows external audits to be conducted. Some large clients of the construction industry undertake their own accreditation inspections for the contractors who wish to work with them.

Quality assurance in construction

The construction process involves three parties: the client; the design consultants; and the contractors. The following are the quality actions required by a contractor in a traditional contract where design is undertaken by an independent designer:

- Receive tender documents; perform tender review; and prepare QA submission
- On award of contract undertake a contract review
- Set up site team

- Prepare Project Quality Plan (PQP) and submit for approval
- Conduct suppliers and subcontractors assessment and appraisal
- Place subcontracts including QA conditions where appropriate to work package. Include requirement for documentation submissions, approvals and records
- Receive Detailed Quality Plans (DQP) from subcontractors for approval prior to start of works
- Prepare DQPs for own work if required
- Place 'hold points' etc. on DQPs to monitor work packages and approve DQPs
- Monitor off-site work against DQPs
- Perform goods inwards inspection to agreed procedure
- Undertake plant inspection to agreed procedure
- Control work on site against PQP, DQPs, inspection checklist, etc
- On-going audits on- and off-site to an agreed audit schedule
- On-going supplier and subcontractor evaluations
- Generate records as construction proceeds
- Mark up drawing to as-built state
- Prepare handover packages and submit

Total Quality Management

This section describes the concepts of Total Quality Management (TQM). The construction relevant aspects are based on the European Construction Institute's (ECI) task force report and the development of their Quality Management measurement matrix. This section addresses TQM in construction under the following headings:

- Definition of TQM and the role of QA in the process
- TQM principles
- Development of TQM in a company
- Tools and techniques

Definition of TQM and the role of QA in the process

Total Quality Management is a process led by senior management to obtain the involvement of all employees in the continuous improvement of the performance of all activities, as part of normal business, and to meet the needs and satisfaction of the customer whether internal or external.

Quality Assurance is all those planned and systematic actions necessary to provide adequate confidence that a product or service will satisfy given requirements for quality. TQM is an umbrella for continuous improvement and incorporates QA. QA is a systematic approach, which controls attitudes and the working environment whereas TQM provides principles, tools and techniques for cultural change and continuous improvement. Table 2.1 sets out the essential role of QA and TQM in achieving a total quality agenda for construction companies.

Table 2.1 Role of QA in the TQM process.

Quality Assurance	Total Quality Management
It is only part of Total Quality Management. It is a systematic approach that provides adequate confidence and satisfies given requirements.	It is a process to obtain continuous improvements of the performance of all activities to provide satisfaction for customers, whether internal or external. It provides principles, tools and techniques.
1 It forms part of a quality improvement process.	It is a process for continuous improvement.
2 It is a systematic approach that influences attitudes and working environment.	It leads to changes in attitudes and the working environment and provides tools, techniques and systems for continuous improvement.
3 It aims to ensure customers' requirements are met every time.	It creates a right first time attitude to delight customers.
4 It provides a base line for measuring the quality.	The cost of quality is recognised as vital, and provides measurement for continuous improvement.
5 It provides confidence to the customer of the quality of the product or service.	The supplier of the product or service is recognised as a quality company by customers and employees.
6 It provides the means to reduce waste.	It seeks to eliminate waste.
7 It enhances publicity and image.	It attracts publicity and the company may be used as a role model for quality.
8 It provides procedures for doing things right.	It provides for doing the right things right.
9 It achieves improvement by eliminating recurring problems.	Improvement is achieved by cultural change based on measurement of performance and elimination of root causes and constraints.
10 It requires a structured organisation and a statement of key responsibilities.	It creates a culture in the organisation that seeks to continuously improve in all its activities.
11 It is directive, and provides procedures for all activities and working practices.	It focuses on a full understanding of the various business processes by the day-to-day involvement of all concerned.
12 It provides quality records of all activities.	It uses quality records for measurement and for continuous improvement.
13 The system relies on regular monitoring and audits to identify and correct non-conformances.	It involves getting ideas and suggestions for improvements from everyone.
14 It relies on regular management reviews of the procedures and working practices to identify options that lead to improvement.	It stresses the importance of products and services delivered to the customer, whether internal or external, meeting requirements whether specified or not.
15 It ensures that people are trained and experienced.	It ensures that everybody in the organisation receives education and training to enable them do their job effectively and achieve personal satisfaction.

TQM principles

Commitment to quality

To successfully promote business efficiency and effectiveness, TQM must be company wide. TQM must be initiated by top management, who must demonstrate that they are serious about quality. Middle management must grasp the principles of TQM and explain them to the people for whom they are responsible and, while doing so, they must also communicate their commitment to quality. It is also the responsibility of middle management to ensure the efforts and achievements of their subordinates obtain the recognition and reward they deserve. If this level of commitment to quality is achieved then TQM will spread effectively throughout the organisation. The successful implementation of TQM results in employees who are committed to quality by taking pride in their work; only then will the full benefits of TQM be realised. TQM is therefore not a set of procedures to achieve quality, but is instead a state of mind, based on pride in the job.

Quality chains

One of the fundamental concepts of TQM is the responsibility of each person to the people they deal with. Every process is formed by the logical progression through a number of operations. If any of the operations in a process is faulty then the effectiveness of the whole process collapses. The people involved in operations that constitute a process therefore form a chain of responsibility, the success of each relying on the success of all of the previous. Each operative can be regarded as an internal customer to the previous operative but also a supplier to the next operative. The concept of a quality chain provides an easily understandable concept to aid the adoption of TQM philosophy.

Development of TQM in a company

There are a number of models used to represent a total quality culture and its development processes. The right model for a company's culture is one that captures the essence of what is trying to be achieved. There are no fixed models for particular types or structure of companies.

Deming (1988) developed a list of points that can be used to aid the development of Total Quality, which is shown in Fig. 2.3. Oakland (1995) developed a pictorial multi-step programme for the development of Total Quality, which is suitable for training purposes. This model is shown in Fig. 2.4.

TQM tools and techniques

The ability of management and employees to control their work processes, to recognise problems, to trace their root causes and to implement effective remedies is the cornerstone of a Continuous Quality Improvement programme.

A wide range of Quality tools and techniques are available to companies and these provide a common language, a consistency of approach to continuous quality improvement. These tools range from simple techniques such as brainstorming to a more sophisticated option including statistical process control techniques.

The most widely used techniques are listed below. While the tools may appear to be little more than applied common sense, they have been proven in many industries and together they form a

(1) Create constancy of purpose for improvement of product and service.
(2) Adopt the new philosophy of refusing to allow defects.
(3) Cease dependence on mass inspection and rely only on statistical control.
(4) Require suppliers to provide statistical evidence of quality.
(5) Constantly and forever improve production and service.
(6) Train all employees.
(7) Give all employees the proper tools to do the job.
(8) Encourage communication and productivity.
(9) Encourage different departments to work together on problem solving.
(10) Eliminate posters and slogans that teach specific improvement methods.
(11) Use statistical methods to continuously improve quality and productivity.
(12) Eliminate all barriers to pride in workmanship.
(13) Provide ongoing retraining to keep pace with changing products, methods, etc.
(14) Clearly define top management's permanent commitment to quality.

Fig. 2.3 Deming's 14 points to achieve TQM (Deming 1988).

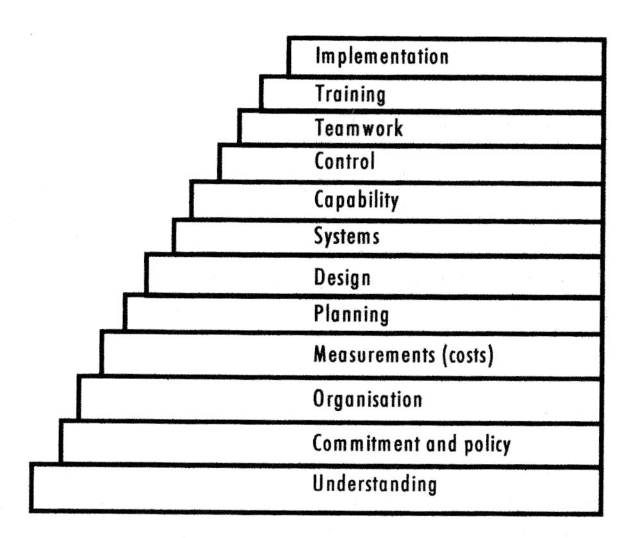

Fig. 2.4 Oakland's steps to TQM (Oakland 1995).

powerful methodology by which individuals or teams are able to continuously improve their work process.

While data collection is the foundation on which a TQM programme is built, it is important that each company selects those tools that work for it and avoids collecting data as an end in itself. The basic principles are summarised as follows:

- Management by fact and not by myth
- No process without data collection
- No data without analysis
- No analysis without a decision
- Avoidance of paralysis by analysis

The main tools and techniques that are used in the development of TQM are described below.

Ways to gather and display data

The systematic collection of data is required to find out the real facts, to prove or disprove what is actually happening, to identify the scale of a problem, to verify the proposed remedy and to control the improvement process. The common methods of collecting and displaying data are:

- Check sheets or tally charts
- Histograms
- Scatter diagrams
- Control charts
- Concentration diagrams

Brainstorming

Brainstorming is a technique for encouraging creative thinking and the generation of ideas from small groups of people using their collective thoughts. It is particularly useful in: generating a list of problems; identifying causes of problems; identifying possible solutions; and developing action plans.

Matrix analysis

Matrix analysis is a procedure for short listing or ranking using a two-dimensional matrix. It is useful in obtaining a group consensus in relation to an agreed set of criteria.

Paired comparisons

Paired comparison is a method of prioritising or ranking a number of alternatives to achieve a specific goal. It is particularly useful for achieving group consensus when prioritising possible causes of problems.

Ranking and rating

Ranking is a structured process of placing an order of preference on a list of options and rating is scoring each option to give a rating on the likelihood of achieving change. Ranking and rating assists with the choice of the best option, makes the choice less emotional and increases commitment to the chosen option. It is particularly useful in deciding which problem to tackle, or which solution to implement.

Pareto analysis

Pareto analysis is a simple technique that helps separate the major causes of problems from the minor ones. It is also known as the 80/20 rule, 80% of the problems are due to 20% of the causes. Plotting 'magnitude of concern' vertically against 'category of problem' horizontally in the form of a vertical bar chart gives a graphic visual display. It is a very effective means of visually representing major causes of a problem. Its main use is to focus attention on the really important issues.

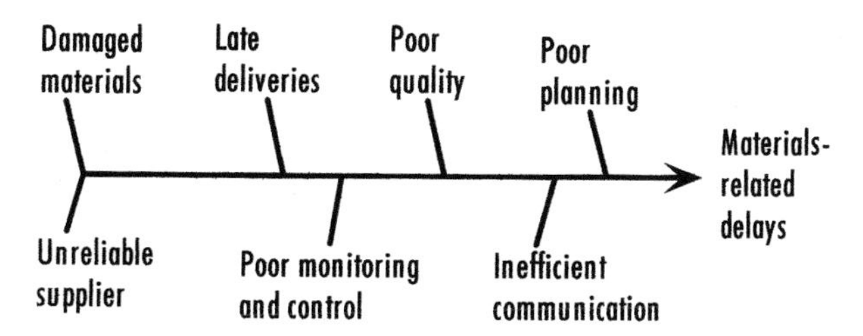

Fig. 2.5 A simple cause and effect diagram for managing materials-related quality.

Cause and effect diagrams

Cause and effect diagrams are a way of displaying potential causes (the cause) of a problem (the effect). These are also known as Ishikawa or Fishbone diagrams. Their main use is in identifying the root cause of a problem and in defining the nature of a problem. Figure 2.5 shows a simple cause and effect diagram for materials-related delays.

Failure prevention analysis

Failure prevention analysis is a technique intended to aid the anticipation of problems before they happen and is designed to promote a move away from reacting to failure towards being proactive in preventing failure. It is particularly useful on new activities or whenever significant changes to a process are planned.

Force field analysis

Force field analysis is a technique to aid the identification of forces that will either help or obstruct a planned change. It is useful in gaining an understanding of what is working for or against any proposal. It is useful in planning to overcome barriers to change.

Process flow chart

A process flow chart, referred to in the chapter on work-study, is a diagram of a process using symbols to represent each element. It displays the sequence of events, activities, stages and decisions in a form that is easily communicated and understood by all. It helps the understanding of the work process and is useful when improving the work process and eliminating waste. To aid the implementation of TQM in the construction industry the International Council for Building Research Studies and Documentation has developed two guides, one for firms of professional consultants and one for building contracting firms. The key elements for a quality system in building contracting firms, as defined in CIB W88 are detailed in Table 2.2.

The European Construction Institute produced a matrix to measure the degree to which a company was operating under TQM and accompanying this matrix were guides to improvement. The 12 elements against which a company can measure its achievement in TQM are listed in Table 2.3. It is interesting to note that the matrix captures additional managerial and social factors that lie beyond the technical requirements of a project.

A systems approach to managing quality

There is recognition that quality in construction involves more than simply focusing on the product. Such recognition provides a strong argument for companies to pay attention to both the internal and external factors and processes within their business environment. Traditionally, the internal factors and processes are addressed by QA and TQM. The external factors, however, hardly receive the same degree of attention in managing for quality. To ensure a more effective

Table 2.2 The key elements for a quality system in a building contracting firm.

1 *Scope of project resource.* A clear understanding by both the firm and its potential customers of the qualifications and expertise of the firm and the scale of work they can realistically undertake.
2 *Resource assessment.* A systematic method to evaluate the resource requirements of a job against the capacity of the firm, before tender or acceptance.
3 *Client brief.* A systematic method to establish the customer's requirements in terms of time, cost and quality; and to identify any consequent need for supplementary specialist subcontractors.
4 *Specialist subcontractor integration.* A systematic method to select supplementary specialist subcontractors on the basis of their expertise, the scale of the work they can undertake, their quality systems and their track record. Its purpose is to ensure that the contribution they make is as vigorously controlled as that of the leading firm and that any design contribution is integrated into the main design at an appropriate time and is well controlled.
5 *Project management plan.* A quality plan relating to the specific requirements of the project: this should include management organisation; quality control; and periodic reviews, involving the customer, to ensure that requirements are being met in terms of time, cost and quality.
6 *Operational information and method statement.* Methods to ensure that site management and workers have information needed for construction easily available to them and understand the construction procedure required. This includes both directly employed and subcontracted employees.
7 *Feedback of system performance.* A method to record problems in the operation of the quality system and to use this feedback to improve the system.
8 *Co-ordinated project information.* A method for organising project information that makes it convenient for the various uses to which it will be put. The UK has, for example, recently begun to adopt a method based on commonly subcontracted packages of work. It has also developed over many years an approach to co-ordinating the work of architects, engineers and quantity surveyors, so that all details relating to the project are in a co-ordinated set of documents. The absence of such a co-ordinated set of documents is frequently the cause of dispute and often of excessive cost in dispute resolution. The technique is known as co-ordinated project information (CPI).

Table 2.3 Key elements to achieve a total quality company within the construction industry.

C—Commitment and leadership by top management at location

At sites the top manager may be the site foreman, agent or construction manager; in offices, the top manager is likely to be the departmental manager, director or managing director. The commitment and leadership to quality of the top-most manager is essential in TQM.

O—Organises process and structure for TQM

The company must structure itself around the TQM process for continuous improvement. How the implementation has been carried out must be monitored.

Table 2.3 *continued*

N—Necessary business performance

Many quality improvement initiatives lose sight of business focus and it is necessary to ensure that it remains a part of the overall company operation. The important measure relates to the unit's performance against its local targets, financial and otherwise.

S—Supplier relationships (internal and external)

Building up a long-term relationship with key internal and external suppliers is an essential feature of TQM.

T—Training, awareness, education and skills

To ensure a high-quality cost-effective operation staff there is a need for training in their job, behavioural skills, safety and environmental awareness, as well as TQM tools and techniques. The hallmark of a TQM organisation is when the staff are asked to assist in the training of others outside their immediate unit of operation. Training is becoming increasingly recognised as essential.

R—Relationships with internal and external customers

Building up a longer-term relationship with key internal and external customers/clients is an essential feature of TQM.

U—Understanding, commitment and satisfaction by employees

The attitude taken by staff in the implementation of the TQM process is important to the success of TQM implementation.

C—Communications

A well-defined system of communication is essential to enable rapid dissemination and feedback of management and staff views, proposals and actions. The system needs both formal and informal channels and needs some co-ordination.

T—Teamwork for improvement

Teamwork is required to implement the TQM process and to solve problems causing barriers to efficient working practices.

I—Independent certification of quality management system

A Quality Management System provides the backbone for a TQM process and full independent certification provides recognition of the company's quality management system.

O—Objective measurement and feedback

Measurements are necessary to judge how far the company has progressed on the TQM process. These must be analysed and the result fed back to monitor improvement.

N—Natural use of TQM tools and techniques

How well each member of staff actually uses the TQM tools and techniques in which they have been trained is a measure of effective TQM. To a TQM Company these are naturally used without consideration.

management for quality, construction companies need to tackle quality through a systems methodology. This will allow them to address the quality of the product or service at the more strategic organisation dimension.

Systems quality management

A systems approach is an interdisciplinary method for the realisation of successful organisational and project systems. While the concept of a systems approach originated from engineering, it has in recent times found application in non-engineering fields in the form of Soft Systems Methodology (SSM).

Definition of a system

A system is a construct or collection of different elements that can be put together to generate outputs that are not obtainable by the elements alone. The elements can include people, technological hardware, software, facilities, policies and documents; that is, all things required to produce systems-level results. A construction project presents a typical example of a system. Viewed from a quality perspective, the output for such a system would be the quality performance of the project. The value added by the system as a whole is usually beyond the simple sum of the contributions of the independent elements because of the synergy that arises from managing the interfaces between the individual elements.

Systems approach

The systems approach integrates all the disciplines and specialty groups into a team effort forming a structured development process that proceeds from concept through design to construction. The systems approach considers both the business and the technical needs of all stakeholders with the goal of providing a quality product that meets the client's needs. When quality is addressed from such a viewpoint, it recognises that every construction project is made up of a series of internal suppliers and customers that create a supply chain and a demand chain respectively. Similarly, each construction company's operations involve a series of such demand and supply chains or *quality chains*. At the same time, the project or company is influenced by several externally driven factors that help to define what constitutes quality and forms the basis of project and company-wide quality improvement. The essence of such an approach for managing quality in construction is that the eventual quality of the project is determined by the quality performance of the *weakest link* in the chain. By using the systems approach, the individual company quality systems at the different stages of the construction process can be aligned to ensure an acceptable overall quality for the project. Figure 2.6 shows how the early stages of the construction project present a better opportunity for setting the project quality regime. Each stage of the process, such as construction, presents a local supply chain, whose quality has to be aligned and managed. The quality requirement and performance of the various stages are then aligned to provide the overall project quality. This is a departure from the situation whereby quality in construction is typically planned and assessed within a stage of the construction process. This gives rise to *silos* of quality performance that are not necessarily aligned at the project level. Aligning the different quality performances would require standardised tools that could be deployed across all the stakeholders that make up the construction supply chain.

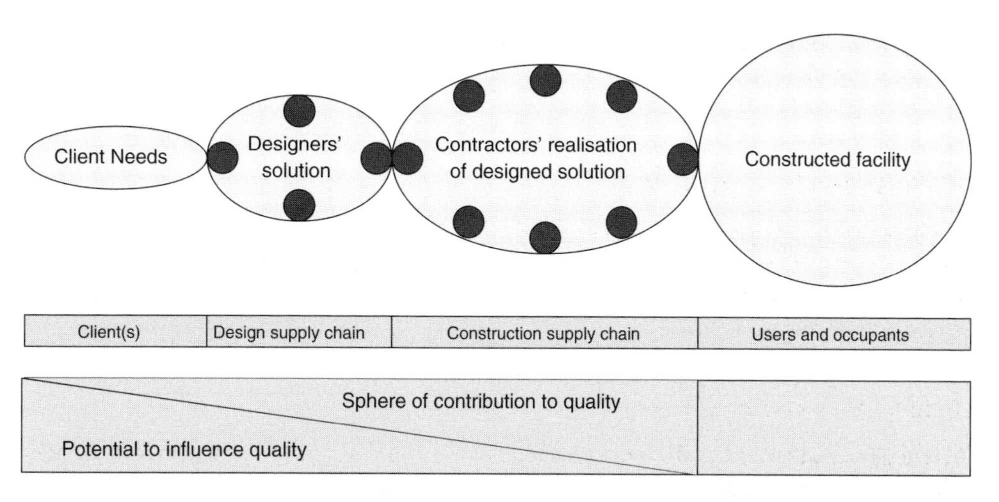

Fig. 2.6 Quality influence and contribution in construction projects.

Quality schemes

There are a number of standard schemes that can be deployed to manage the alignment of the quality performance within and between the stages of the construction process. These include the European Foundation for Quality Management (EFQM), the Malcolm Baldrige National Quality Award (MBNQA) and Six Sigma. All of these schemes recognise product quality as providing a fractional contribution to what is increasingly recognised as project or organisational quality performance.

EFQM Excellence Model

The EFQM Excellence Model was introduced at the beginning of 1992 as the framework for assessing organisations for the European Quality Award. It is now the most widely used organisational framework in Europe and it has become the basis for national and regional Quality Awards in many countries. The Model is a non-prescriptive framework based on nine criteria. Five of these are classified as *enablers* and four as *results*. Figure 2.7 presents the conceptual linkages between the *enablers* and *results* for the model. The *enabler* criteria cover what an organisation or a project actually does. The *results* criteria deal with what an organisation or project achieves. The Model, which recognises that there are many approaches to achieving sustainable excellence in all aspects of performance, is based on the premise that excellent performance results as regards *customers*, *people* and *society* are achieved through *leadership* driving *policy and strategy*, which is achieved through *people*, *partnerships and resources*, and *processes*.

The model emphasises the interdependence between the criteria. It also depicts the continuous learning and innovation that needs to characterise the quality aspirations of any construction company or project. The EFQM Model thus provides a comprehensive basis for assessing the quality performance of construction companies and projects.

Malcolm Baldrige Quality Performance

The Malcolm Baldrige National Quality Award (MBNQA) is an award scheme that was

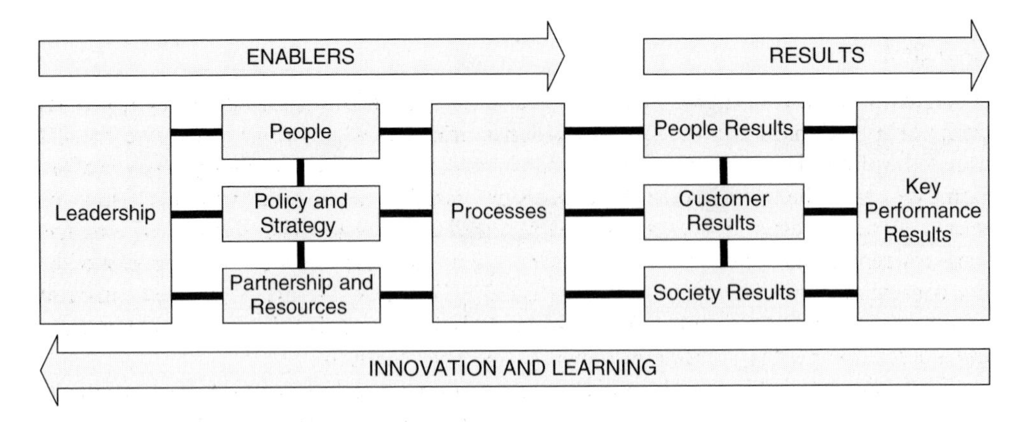

Fig. 2.7 Framework for the EFQM Excellence Model.

originally set up in 1987 to serve as a catalyst for excellence in quality and business by organisations in the USA. The use of the term 'National' provided a guide for deciding what to include in the criteria for judging Performance Excellence for the scheme. This included several factors that are modified versions of the categories, core values and concepts proposed by the Baldrige initiative as drivers or evidence of quality performance.

The Baldrige performance excellence criteria form a framework that any organisation can use to improve overall performance including quality performance. There are seven categories that make up the MBNQA criteria and these are:

(1) The leadership criterion involves how senior executives guide the company and how the company addresses its responsibilities to the public and corporate social responsibility.
(2) The strategic planning criterion focuses on how the company sets strategic directions and how it determines key action plans.
(3) The customer and market focus explores how the company determines requirements and expectations of customers and markets.
(4) Information and analysis delve into the management, effective use and analysis of data and information to support key company processes and the company's performance management system.
(5) Human resource focus deals with how the company enables its workforce to develop their full potential and how the workforce is aligned with the company's objectives.
(6) Process management addresses aspects of how key production, delivery and support processes are designed, managed and improved.
(7) Business results look at the company's performance and improvement in its key business areas: customer satisfaction; financial and marketplace performance; human resources; supplier and partner performance; and operational performance. This category also examines how the company performs relative to competitors.

The seven categories of performance reflected by the MBNQA can be employed as the basis for self-assessment and training, as well as a tool to develop continuous improvement in quality performance and business processes by construction companies.

Six Sigma

Six Sigma is a data-driven and systematic methodology for eliminating defects in any process from site operations to information transactional activities such as design and from construction of the physical product to supporting services in a project. The underlying principle of Six Sigma is to drive the performance of the company towards six standard deviations between the mean and the nearest specification limit to improve the quality and reliability of products. The statistical representation of Six Sigma provides a quantitative assessment of how a process within a construction company or project is performing. To achieve Six Sigma, a process must not produce more than 3.4 defects per million instances for its outputs. In this regard, a Six Sigma defect is defined as anything outside customer or client specifications.

The fundamental objective of the Six Sigma methodology is the implementation of a measurement-based strategy that focuses on process improvement and variation reduction through the application of Six Sigma improvement techniques. This is achieved by the use of one or two of Six Sigma sub-methodologies described as DMAIC and DMADV. The Six Sigma DMAIC process (define, measure, analyse, improve, control) is an improvement system for existing processes falling below specification and looking for incremental improvement. The Six Sigma DMADV process (define, measure, analyse, design, verify) is an improvement system used to develop new processes or products at Six Sigma quality levels. It can also be employed if a current process requires more than just incremental improvement.

References

Deming, W.E. (1988) *Out of the Crisis. Quality Productivity and Competitive Position.* Cambridge University Press, Cambridge.

ISO (2000) *Quality Management Systems—Fundamentals and Vocabulary: ISO 9000.* International Organization for Standardization, Geneva.

Oakland, J.S. (1995) *Total Quality Management: The Route to Improving Performance.* Butterworth-Heinemann, Oxford.

Section one

Project production management

3
Production process improvement

Summary

A process for continuously improving production performance is described and explained with the emphasis on quality management, benchmarking, lean construction, supply chain management, delay surveys, goal setting, work study and the emerging influence of sustainability.

Introduction

An economy comprises many companies and organisations all continually striving to stay profitable or viable. Consequently, the construction sector also must improve relative productivity of resources in order to maintain final prices attractive to potential customers; otherwise demand for its products may stagnate or even fall. However, the propensity of customer taste for bespoke and aesthetically familiar/traditional, but differentiated, building products and architecture, plus burdensome regulation, generally detracts from the application of the mass production process, except for particular manufactured components and specific industrial/prefabricated building systems. The necessity to engage specialists and craft trades further hinders radical change, while rising land values also often generate unavoidable cost increases that have to be passed on. Moreover, other industrial and commercial sectors are commonly able to attract skilled workers more easily, either by providing superior working conditions or in being able to pay higher levels of remuneration arising from more intense application of mechanisation and advanced technology in general.

The primary task in construction is to overcome these difficulties and bring about a climate favourable for management, workforce and resources to combine efficiently (doing things right) and effectively (doing the right things), directed towards raising productivity, quality and value for the client/customer and shareholders, while also providing high rewards and healthy and safe working conditions for those involved. At the same time the activities of construction need to address any adverse impact on the natural environment and meet the changing attitudes of modern society to corporate social responsibility (CSR). Essentially, improvements in construction should address the three axes of sustainability.

Productivity

Productivity is the same as *efficiency*, which is defined as the ratio, output energy divided by input energy. In economic terms Samuelson & Nordhaus (2004) describes the ratio as the Total Factor

Productivity (TFP). In this case, *output* is the value of the Commodities, i.e. goods and services *produced*, and *input* the equivalent value sum of all the Partial Factors of Production (PFP) *consumed*, broadly:

- Natural Resources ('free' finite and renewable resources—land, forests, minerals, fossil fuels, materials, etc.)
- Labour (human resources—mental and physical)
- Capital Goods (production generating resources—infrastructure, facilities, equipment, stock, etc.)
- Entrepreneurship (human capital/skills—financial, managerial, organisational and technical; degree of economic and market/trading flexibility; extent of good governance, institutions, legal, property and social rights, etc.)

The equation may be written as:

$$\mathrm{TFP} = \frac{Output}{Input} = \frac{\text{Commodities}}{\text{Natural Resources} + \text{Labour} + \text{Capital Goods} + \text{Entrepreneurship}} \quad (3.1)$$

The key to raising the productivity level lies in improving the performance and proportional balance of the input factors (each, however, is subject to diminishing marginal returns, only moderated by the constant revolution of techniques and reorganisation of production), especially the application of innovative applied science/engineering and the continuous augmentation and/or replacement of manual (and animal) power, where appropriate, with more (cost-) effective machinery and technology (i.e. capital investment). For example, while the historical ox or horse (1 Horse Power or rate of working) plus driver could haul a mere two tons slowly (say 1/3 HP-hours per mile of work) or a pair intermittently plough an acre a day of light soil (approximately 16 HP-hours per acre), the revolutionary Boulton & Watt steam engine of 1790, consuming relatively small quantities of abundant energy-dense hydrocarbon fuels, worked at the rate of about 950 horses (950 HP).

Gross Domestic Product (GDP)

The productivity improvement process releases redundant labour and other resources (i.e. savings) to meet demands for new activities and specialisations, for the production of extra goods and services, which offer *absolute* and/or *comparative economic advantage*. Such conversions of resources add to total output and thereby generate the country's gross domestic product (GDP, i.e. the total value of the annual output of goods and services produced) and economic growth.

GDP is represented by:

$$\begin{aligned}&[\text{private consumption}] + [\text{government purchases}] + [\text{fixed assets investment}] \\ &\quad + [\text{change in stocks (inventories)}] + [(\text{exports} - \text{imports})].\end{aligned}$$

GDP can be measured in several ways, which would all be the same were statistical collection truly accurate. These are namely:

(1) GDP = income of resident individuals and firms derived from the production of goods and services
(2) GDP = sum of the outputs of the different sectors of the economy at factor cost
(3) GDP = total expenditure on the goods and services produced by residents at market prices after deduction of indirect taxes and before depreciation

To eliminate repeated counting of intermediate stages, only total value added is aggregated. Finally,

GNP (Gross National Product)
= [GDP] + [income earned abroad by residents – corresponding income of foreign investors in the country].

The rate of growth of GDP, commonly stated at constant prices (i.e. market prices adjusted for inflation) is affected by many factors, the significant influences being:

- The steady rise in supplies of energy and the rarer metals (essential for the key alloys and catalysts) reinforced by improved technology, innovation, entrepreneurship, etc., combined with reduced energy usage intensity itself, enables modern technological economies to achieve an annual TFP growth of about 1% (and 2% PFP for labour productivity). This delivers around 2% real GDP growth per year on average when extra inputs such as untapped resources, use of spare cyclical economic capacity and foreign investment, plus favourable trading opportunities and aggressive marketing, which are included to encourage more consumption, are applied. For a rapidly developing economy, mostly driven in its early stages by capital investments in fixed assets, a level of GDP growth up to10% per annum is possible. Notably, as the cost of energy used for an economy rises, the growth rate of GDP will usually be temporarily negatively affected until supply and demand are in better balance. Furthermore, on exceeding about 10% of the total value of economic output, the induced rate of price inflation has also temporarily reached problematic levels as pointed out by Leeb and Leeb. This is then likely to become an intractable problem if determined by continuing energy supply constraints. Current energy expenditures lie in the range of 6–8% of GDP.
- GDP growth may be further regulated through government directives and fiscal measures of tax, borrow and spend augmented with policies attractive to private domestic capital investment and (inward) foreign direct investment in business and technology development. The intention is to generate self-sustaining economic growth, service the associated debt and provide favourable social welfare, health and infrastructure. Ultimately, however, the total of borrowed/investment funds cannot exceed the savings finance available to the economy. Also, conventionally, the more the borrowing, the higher the interest rate demanded by lenders to compensate for the greater risk.
- A little delayed extra GDP may be induced by decreasing the central bank discount interest rate to foster more borrowing for commercial investment, i.e. potential projects yielding a lower nominal rate of return on capital (money) become viable (see profitability measures in Chapter 13 and investment ratios in Chapter 18), and household spending. Conversely, increasing the rate of interest (and/or restricting the amount of credit) assists in dampening down business activity and inflationary pressures (see J.B. Taylor's 1993 rule). However, monetary tactics, if applied excessively (in an open market/commercial system), may trigger counter flows of (international) finance, i.e. savers' and investors' funds, thereby negating the

intended effects and impacting the exchange rate of a floating currency, balance of payments and foreign currency reserves.

Indeed, Government usually develops elaborate partial computer models using economic data and statistical relationships to simulate the workings of the national economy and investigate alternative policy scenarios—partial in that the potential input costs of delivering sustainability, using up non-renewable resources and depreciation of capital stock are conventionally not accounted for. Independently, other governments similarly produce models for their own economies.

Energy consumption

The scale of annual energy consumption driving the global economic system can be appreciated in terms of the $400\,000 \times 10^{15}$ joules or 150×10^{12} HP-hours (111×10^{6} GW-h) primary energy currently used, mostly hydrocarbons, where oil (for example, diesel contains 40×10^{6} joules per litre) is less energy-dense than natural gas equivalent and significantly lower than hydrogen, which possesses no carbon atoms, while coal equivalent has the least energy content and is also the most air-polluting. Energy demand is projected to rise 50% by 2020 (see Fig 3.1); moreover oil production, amounting to over a third of world energy and critically providing 95% of transportation needs, already stands at about 85 million barrels per day (mbd–1 barrel = 159 litres). Illustrated by the US Energy Information Administration data for regional world energy consumption comprising hydrocarbons (0.85), nuclear (0.08) and renewable (0.07) is presently spread approximately as follows:

North America	29%
Asia & Oceania	28%
Europe	21% (UK 2.5%)
Russian Commonwealth area	10%
South & Central America	5%
Middle East	4%
Northern & Southern Africa	3%

Imagine the huge numbers and logistical impracticalities were today's power generated by working animals (1 HP or 746 W) rather than the many hundreds of very large modern power stations world-wide, each with an output rate of about 2 GW (2.7×10^{6} HP).

Sustainability

Although the described approach to the generation of high and directed compound economic growth and industrialisation has proved instrumental in relieving rising populations from poverty, drudgery and creating personal prosperity, to provide a solid base for individual empowerment and proper enfranchisement, disquieting global concerns earlier noted by Packard (1970), Whyte & Nocera (2002), Schumacher (1973) and other writers on the subject, and likely to affect long-term projections, have continued to emerge steadily. Examples of issues for redress which may impact construction industry interests include:

- Progressive removal of the natural habitat and diminishing wilderness
- Gradual degradation of the environment
- Possible waning of oil and gas reserves capacity during the next few decades, perhaps heralding a downward change to the historical pattern of world economic growth, and threatening future world GDP per head as the population steadily expands at about 1% p.a. (see Fig. 3.1); some analysts are projecting peak conventional oil production around 100 mbd and similar for gas equivalent (see Hubbert's Law at www.hubbertpeak.com)
- Global capital investment and trade imbalances marked by regional wealth and welfare variability, speculative investment bubbles (e.g. shares, land, housing, bonds), debt servicing crises for poor countries and little jobs creation for the critical oil exporting states
- Prodigious consumption of goods allied to wasteful planned or built-in obsolescence and burgeoning transport provision
- Intensive farming methods dependent on chemical fertilisers accompanied by soil, wildlife, animal breeding, husbandry, foodstuff and human dietary/health problems
- Much dismal or squalid large-scale urbanisation fostering complex social and cultural tensions, etc.

In attempting partial remedy, the recently approved United Nations 'Kyoto' protocol aims to give impetus for better international environmental conservation co-ordination, starting with a few willing governments of the developed countries. Under the clean development mechanism (CDM) these major emitters are required to reduce atmospheric pollutants during 2008–2012 to a modest average 5.2% below the 1990 level (presumably to be progressively repeated, extended and broadened thereafter).

Launched in the EU as the European Emissions Trading Scheme (ETS), significantly each member state has accepted a voluntary target and agreed to issue emissions allocations (e.g. an emission allowance for a kWh of coal-fired power generated) to every large industrial facility in relation to current polluting amounts produced. Conjunctionally, formal rights are granted to

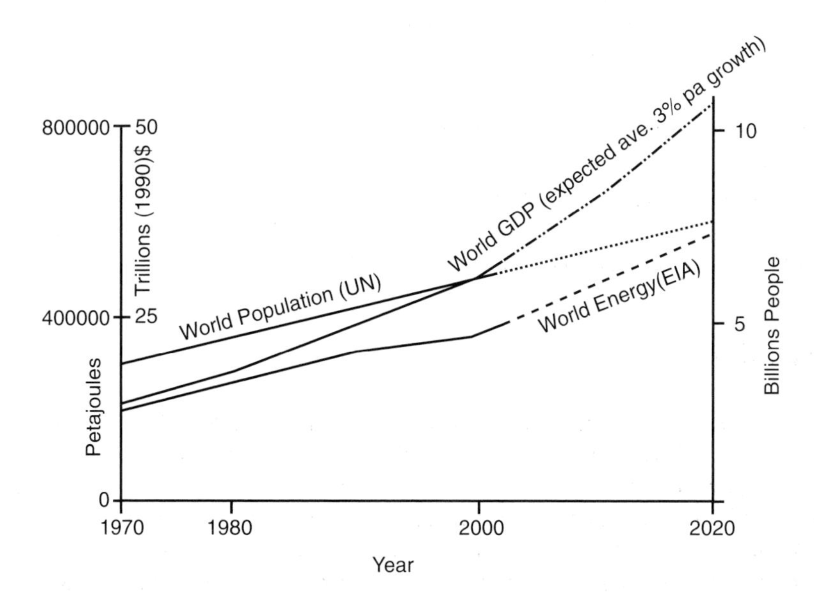

Fig. 3.1 World energy consumption, population and GDP (Based on EIA and UN data).

continue to emit, for example, carbon dioxide from the facility, but the offender may then purchase (trade) credits to cover the extra emissions from others that are able to meet their allowance or organise partnering; otherwise they will face levied financial penalties set by the EU if unable to meet the allocation on re-measurement three years later. Companies may also earn certified emissions reductions (CERs) by cleaning up emissions in developing countries, where a similar monetary market for trading of credits is available.

The trading scheme is intended to encourage heavy polluters to bring in improved technologies to better manage unit-costs of reducing emissions (e.g. converted and stored in other forms, or replaced by 'clean' renewable resources and fuels, etc.). However, any resulting relevant higher commodity prices are likely to be passed up the chain to construction. Hence, as a prodigious consumer of processed materials and fabricated items, it will be imperative for the construction sector to make more carefully considered selection of recycled products, and specifically make less use of energy-intensively produced goods, etc., in order to uphold competitiveness, and not least to help secure the construction sector's own environmental target contribution. Additionally, solutions for reducing the massive 30% of total energy (and associated greenhouse-gas emissions) presently consumed in running buildings may also have to feature more prominently for the construction industry to meet the recently introduced EU Energy Performance of Buildings Directive.

Equally strong remedies, including serious corporate social responsibility (CSR) policies may be required to address some of the other global issues, but would obviously represent foregone, i.e. alternative, economic opportunities and paradoxically contribute to GDP figures.

Productivity improvement

Patently within this context, competitiveness of the 8–10% of GDP generated by the construction industry exercises concern for general economic wellbeing, not least as manifested through the long series of joint government and industry initiatives directed towards improving productivity performance, the most recent being the Accelerating Change/CBPP/Be endeavours, now merged under Constructing Excellence in the Built Environment (CEBE). By focusing on PFP inputs such as integrated systems, processes, supply chains/networks and sustainability, better value for the client, workforce, society and the environment may be attained through application of the principles set out in the productivity improvement model shown in Fig. 3.2.

The objects are to eliminate excess labour and other resources, identify back-end operations more suited to subcontracting/outsourcing and shorten construction times to reduce costs, lift profits and raise the rate of return. Released capital may then be reinvested to generate a favourable cycle of productivity growth and virtuous business performance. Similarly, the core competencies of the business and/or unique products/services have to be identified and any 'added value' realised from them increased, in other words, the value of the output created by the enterprise minus the value of all the inputs bought or received from outside suppliers should have a positive growth – (see the plant acquisition example in Chapter 7).

Systems

The strategy of continuous improvement developed by the manufacturing sector and aimed at achieving 'best in class' or 'world class' standards relies on system concepts such as Total Quality Management, benchmarking and lean production. It also has potential for specific adaptation to

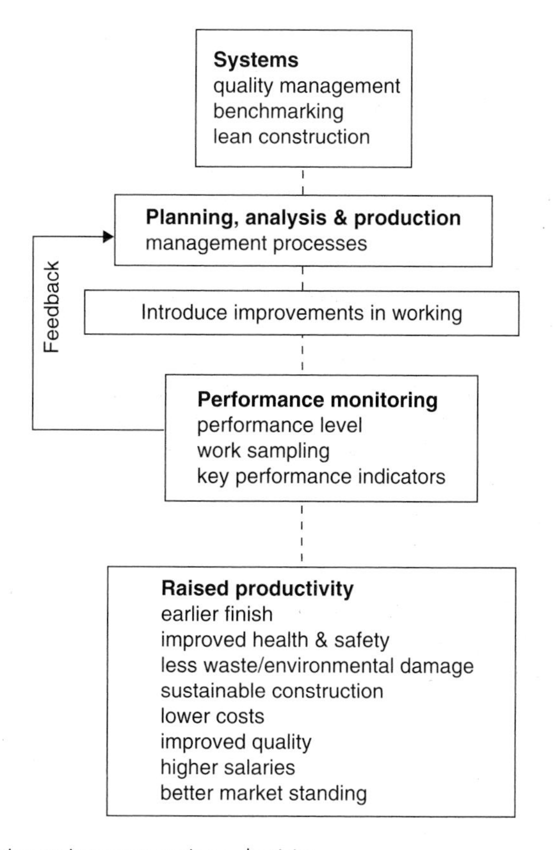

Fig. 3.2 Model for continuous improvement in productivity.

suite construction management and production systems. Founded upon the principles of specialisation and the division of labour, such systems essentially bring economic benefits by encouraging mutual co-operation in work, supply-chain and business relationships based on the assumption of rational calculation of the parties involved, but checked by potential resort to reciprocal reaction in the event of subsequent exploitation and cheating on agreements (see Seabright 2004).

Total Quality Management (TQM) in the production process

A description of TQM has been provided in Chapter 2. From a TQM perspective, every employee in the production chain is regarded as an internal customer to the previous supplier (i.e. adder of value) of a product or service, and who likewise is also a supplier (and adder of further value) to the next point in the chain, i.e. each becomes supplier, processor and customer in turn. All levels of activity must be involved in quality procedures and standards from the individual, work crew, contract site, head office department to the overall corporation itself, including subcontractors and other outside suppliers, all aimed at ensuring that customers receive the products and services *wanted*, and at the standard expected or negotiated.

To be truly effective, the system, in striving for continuous improvement, demands the establishment of clear visions, mission, goals and realistic targets benchmarked against high standards detailed in plans, specifications, approved procedures, work instructions, inspection tests, etc., including supplier involvement backed up by adequate training and communications. Subsequent achievements need to be monitored and amended through regular audits and rigorous acquisition of feedback data, performance indicators and trend ratios such as:

(1) Improved safety statistics, e.g. reportable lost time accident frequency rate—currently 1 per 10^5 mh; or fatal accidents—3.5 per 10^5 workers
(2) More training days provided
(3) Lower absenteeism levels
(4) Fewer defects identified
(5) Reduced staff turnover
(6) Progress on projects
(7) Raised R&D
(8) Higher employee satisfaction
(9) More reuse of waste
(10) Fewer disputes
(11) Improved CSR metrics, etc.

Furthermore, thorough embedment of TQM in the organisation's culture ought to encourage better identification and implementation of appropriate innovations in technology and new management thinking and breakthroughs.

Benchmarking

Benchmarking originated in the manufacturing industry as systematically and continuously measuring the firm's business and management processes and comparing them with those of leaders in the field as a means of identifying areas for potential improvement. Comparison may be with similar internal units in the same organisation or with completely external competitors operating in a different industry even, the primary objective being to achieve 'best practice', principally by introducing a perceived improvement and subsequently measuring its effectiveness against changes in key indicators such as: quality of end product or service; productivity; cost level; safety; delivery time; and sustainability criteria. The approach typically embraces the following procedure:

(1) Establishment of the functions to be benchmarked, e.g. team-building, constructability, accident rate control, pre-project planning, variations management, information technology management, equipment maintenance management, supplies and subcontractor procurement, quality management, disputes resolution, environmental impacts, etc.
(2) Identification of competitor or body for the benchmarking task
(3) Collection and gathering of data
(4) Analysis of the information and comparison with competitor
(5) Implementation of the recommendations for improvement
(6) Monitoring the key indicators and adjustment of the modifications as necessary

Lean construction

The 'Lean' concept also originated in the manufacturing sector and places emphasis on

producing to demand rather than through the batch process, so minimising the c(
work-in-progress and attendant large requirements for buffer stocks. The aim is to eliminate defects while using less input, i.e. less labour, less machinery, less space and less time in design, principally by reducing the number of conversion activities and movement flows in making a product. *Conversion* describes the *operations* performed in adding value to material or information in providing the product, while *flow* conversely represents tasks like inspections, waiting, moving and storing. The end result would be expected to show:

- A reduced proportion of non-value-adding activities
- Less variability in quality
- Fewer steps, stages and linkages in the flow patterns
- Quicker production cycles
- Balance in flows
- Fewer and better quality suppliers and subcontractors

Lean manufacturers improve efficiency by making products only as and when required, rather than in bulk to be held in store waiting for a buyer. Thus, in many ways the construction industry should find this a familiar concept. For this to be truly successful, information needs to be openly shared, with the processes of design, supply, manufacture and construction integrated and with the participants limited to a few trusted partners for competition purposes.

To achieve good results with Lean Construction some of the common tools of production management can be embraced including the following *management processes*:

- *Concurrent engineering*, i.e. the integration of the design and construction phases
- *Just-in-Time* materials management, e.g. ready mixed concrete
- *Re-engineering*, i.e. process redesign
- *Value based and risk management*, i.e. elimination of unnecessary processes, risks and costs
- *Employee participation* in work planning
- *Work task management*, i.e. goal setting, work study, product circles/multi-functional task groups

While difficult to interpret for construction, the lean concept, if adopted as a fundamental philosophy, would be visualised through more accurate pre-planning, an increased use of standard components, prefabrication, modular systems, integrated IT systems and rigorous attention to resources procurement in supply-chain networks where best practices are identified and incorporated, and not least in having a well-trained workforce (see Chapter 14). Notably, in these respects the most successful automobile manufacturers have pared down suppliers to only those capable of designing and producing to high standards, while at the same time reducing costs. In construction the equivalent initiative largely becomes the responsibility of large clients, major contractors and developers able to encourage cultural change, and the application of modernistic forms of contract such as design and build, coupled with partnering attitudes.

The extent for this kind of co-operation can now be seen with manufacturing in proprietary systems for *intranet/extranet* electronic data interchange, established with trusted subcontractors and enabling streamlining of the purchase and design of parts. Impressively, for example, aircraft firms assemble a complete prototype using CAD before building a single plane, with suppliers all integrated into the software ensuring that every nut and bolt fits into place. Furthermore, the *internet's* rapidly improving technology is increasingly being used to establish secure, standing information links between companies, helping firms reach suppliers through a simple connection via a PC and web browsing software. Current examples include creating a *web site* advertising drawings and specifications and enabling suppliers to comply with customised components. At present, construction compares unfavourably with other sectors in the application of IT, occupying a position towards the bottom of the industrial league table of IT spenders per employee (source IDC).

While many of these areas have potential for developing better overall performance, much can also be achieved through regular application of the common work-study principles, especially in reducing management-induced inefficiencies currently responsible for the majority of lost time on projects. Here, techniques like value engineering and worker delay surveys can be revealing, while conventional methods analysis, work measurement and statistical sampling, e.g. Six Sigma techniques, are adept in identifying production problems. Furthermore, when analysed through computer software, the potential ability to simulate different construction methods and evaluate impacts on plans, resources and costs presents opportunities for construction that are already enjoyed in manufacturing. Other novel solutions include bar-coding systems and radio frequency identification tagging (RFID) of items for stock control and materials ordering, including holding drawings registers and record keeping in general. Video-conferencing and closed-circuit televised managerial communication, both on sites and between the site and the outside, are also useful, as has already been demonstrated in other industries. Indeed, the internet facilitates good email and file exchange communications and particularly the world wide web (www), through which voice, pictures, data, tables, diagrams, different operating systems and file storage formats, etc. can be transferred. So, for example, site-based staff can securely log onto a local internet service provider via a broadband connection to access control information located at head office and elsewhere on the web site.

Management processes

Concurrent engineering

The 'lean' concept places emphasis on integration of the design and construction processes to, in effect, form a single entity. For example, in manufacturing design groups and individuals regularly access databases to obtain updates on work that is being done elsewhere, including at suppliers, and vice-versa. Similarly, incorporation of production engineering into the system facilitates early interactive prototyping of the development process, e.g. Early Contractor Involvement (ECI) in the construction context. However, for the construction sector suppliers are generally more transient as a result of the commercial and logistical realities of the industry, and typically a committed network of subcontractors will be more flexibly drawn from previously successful arrangements in contrast to the almost permanent linear chain characteristic of manufacturing industry. Indeed, increasingly the trend, particularly for regular commercial-type building, is towards selection of an experienced lead design-and-build contractor for a project, possessing a demonstrably successful team of partners and subcontractors. The various alternative contract categories are described in more detail under *Project Procurement* in Chapter 8.

Supply-chain management

The objective of supplier management is to provide better value accompanied by steadily reduced costs. In this respect *Just-in-Time* (JIT) delivery, whereby stock holdings and resources supply delays are largely avoided, attempts to offer the optimum solution by immediately and continuously meeting the customers' needs. This kind of approach has been reasonably well demonstrated, for example on the Heathrow Terminal 5 contract with its single materials consolidation centre, from which all supplies are drawn by the various contractors involved on site. Most importantly, for effectiveness all the relevant parties must fully co-operate in planning the schedule, limiting buffer stock/resources to the minimum for uncertain demand.

The whole JIT process is intrinsically bound into concurrent engineering, which, starting with the client, necessitates each supplier in the chain or network having direct and/or indirect presence in the customer's operation, principally manifested by electronic telecommunication involving integrated standard IT, 3D-CAD, MS software and physical co-location as appropriate.

Essentially supply-chain management emphasises the following key aspects:

(1) Rationalisation of existing suppliers and selection of reliable partners for long-term commitment
(2) Establishment of a 'champion' in each party to drive the relationship forward
(3) Involvement of suppliers early in the design process
(4) Developing systems to enable sharing of plans
(5) Managing design integration with supplier(s)
(6) Developing 'open book' procedures and cost information
(7) Establishing performance measures/benchmarks for value, quality, health and safety, and sustainability
(8) Encouraging teamwork and problem-solving with the supply partner(s)
(9) Ensuring procedures for communicating changes speedily
(10) Monitoring progress and addressing issues promptly and regularly
(11) Setting up review mechanisms for feedback, ideas and knowledge exchange

Notably geographical location of production facilities, product range/resource capability, stock levels, transportation provision, communication facilities and currency-exchange rate risk considerations, etc. will vary depending upon the type of supply function involved. For instance, those of a supply manufacturer are likely to differ from a trades subcontractor with a site commitment only and be different again for design services. And the ability to generate acceptable profit will continue to feature strongly in customer/supplier relationships. Indeed, some contractors may attempt to hide from the client transactions details with suppliers through unscrupulous invoicing and similar scams unless vigilance is strongly upheld.

Re-engineering

Developed in the 1990s for the manufacturing and commercial sectors (see Hammer & Champy 2004), re-engineering is described as 'Fundamental rethinking and radical redesign of business processes to achieve dramatic improvements in critical, contemporary measures of performance, such as cost, quality, service, and speed'. A business process in the scheme is defined as a collection of activities that takes one or more kinds of input of value to the customer. The technique seeks to make a marked impact rather than marginal improvements, for ex-ample to significantly reduce time to deliver a product from customer enquiry to despatch and reception. The problem to be investigated, identified for example through benchmarking competitor firms, is selected and the impacting processes associated with this activity evaluated and subsequently redesigned using methods of analysis similar to normal work-study procedures.

A team of five or so headed by a competent leader and selected from people involved in the specific business process, plus outsiders, creatively engage on the problem to discover and synthesise new ways of working until a feasible new solution can be proposed. After agreement and approval by senior management, taking account of the consequences on employees and their concerns, the revised process is thereafter introduced. The subsequent monitoring measurements and results are commonly expressed quantitatively, for example in statistical terms, i.e. the goal is being achieved 99% of the time after introduction of improvements, thereby indicating an acceptable level of variability. The feedback and control loop is thereafter rigorously maintained. In this manner, whole business segments have been reconfigured, even to their complete elimination from the enterprise, most notably functions vulnerable to the introduction of IT such as administration functions, paper-based tasks, manual labour operations, etc. Over-zealous adoption of the technique can lead to excessive downsizing of the organisation; nevertheless insurance, banking, consumer goods and similar organisations have to date seemingly benefited from application of re-engineering. Hence, this kind of radical approach to reorganisation may have a place in some parts of the construction process, for example areas concerned with:

- Concurrent integration of design and construction
- Application of CAD/CAM technology
- Customer bespoke system/modular/prefabricated construction
- Business process outsourcing (BPO) of back-office functions
- Corporate governance, corporate social responsibility and environment
- E-commerce advertising, transactions, payments, online tracking and delivery
- Design for sustainability

Value management

Value management embraces the earlier narrower technique known as Value Engineering, which was originally conceived by the military in the 1940s and later refined by the GE company in the USA as an attempt to meet demand for its products at reduced cost through appropriate selection of design solution, production method, materials, resources and suppliers, while simultaneously maintaining value for money for the customer. Hence, the approach potentially provides a useful evaluation tool for non-profit organisations also.

Essentially the technique comprises a set of procedures directed towards achieving the essential functions of a project, facility, product, system, service, equipment, etc. at the lowest life-cycle cost, consistent with the required form, performance, reliability, quality, delivery and safety.

The three basic stages to value management shown in Fig. 3.3 are:

(1) Value planning
(2) Value engineering
(3) Value review

Stage 1
Value planning is initially carried out at the concept and development stages of the brief and

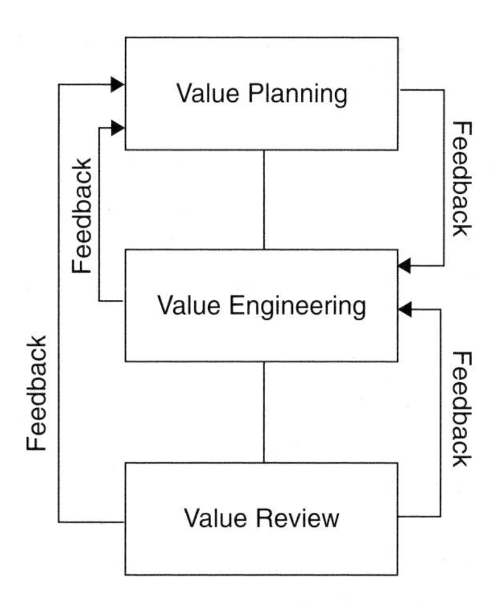

Define brief
Establish functions
Brainstorming/Pareto Analysis etc.
Rank Functions

Confirm project aims
Analyse functions and criteria
Establish alternative solutions
Rate solutions against functions criteria
Weight each solution
Select solution to meet aims and budget

Monitor the value process
Correct defects
Feedback into subsequent work
Set up contingencies

Fig. 3.3 The value management process.

defines the functional requirements of the construction project, product, etc. to meet the client's objectives. Pareto analysis would typically reveal that about 80% represents only 20% of the cost, and these are subsequently represented hierarchically as Primary, Secondary and Non-essential functions.

The ranking process of the functional needs (MUSTS and WANTS) proceeds in a similar fashion to the Kepner & Tregoe (1997) method described in Chapter 7. The process is often executed by means of repeated workshops and brainstorming sessions involving experienced staff including the client and also subcontractors as appropriate.

Stage 2

Value engineering establishes solutions/design alternatives able to satisfy the range of functional needs identified in the previous stage. Ideally achieved by a different review team, preferably basing decisions on reliable past performance evidence and attributes data, the alternative solutions are individually rated against the functions criteria (i.e. MUSTS and WANTS—see Kepner & Tregoe method, Chapter 7) and the total weighting for each option calculated.

Determination of the appropriate rated value for a particular function criterion would likely be assembled from relevant measures of the following attributes:

- Land use, aesthetics, form, function and utility
- Specifications and costs
- Buildability and construction programme
- Energy consumption, conservation and sustainability
- Quality, health and safety aspects
- Life-cycle service and maintenance
- Funding, finance and revenue generation
- Procurement conditions

The selected best value solution delivers the entire Primary and as many of the Secondary needs as possible within the budget acceptable to the client. The chosen solution is subsequently prepared in detail and produced/manufactured/constructed.

Stage 3
Value review is undertaken in conjunction with the contracted parties at each stage of conception, brief development, design, construction (and maintenance) to ensure that the expected enhanced value and functional requirements are being achieved. To this end, progress is continuously monitored in order for sufficient time and opportunity to be given to forestalling any adverse effects before work takes place. The resulting feedback data can then be used to allow modifications to be made and for devising incentives for introducing the improvements, and not least in measuring out-turn results to aid the development of future projects.

Value manager and facilitator

Usually the client's project manager/leader is made explicitly responsible for managing the various stages, particularly scheduling the workshops and reviews, including implementing follow-up actions. In addition, an experienced facilitator is necessary to encourage participants towards achieving the objectives, to tease out background issues, to challenge assumptions, to analyse complex problems, to give leadership and authority, to champion the anticipated savings and to 'sell' the potential improvements in value, etc.

A full value management exercise involving comprehensive stakeholder participation is recommended for high-value/high-risk projects, while a 'paper-based assessment' may prove sufficient for routine construction.

Risk management

Losses and wastage resulting from uncertainty in a project or indeed for the company or firm as a whole can be moderated by rigorous application of risk management techniques using modified work-study procedures. The risks are first identified, control measures allocated and opportunities for innovation exploited, all regularised in a practicable procedure. Defined in the British Standard (ISO 2002, BSI 2002) as:

> the process whereby decisions are made to accept a known or assessed risk and/or the implementation of actions to reduce the consequences or probability of occurrence,

the approach typically embraces the following steps.

Step 1. Risk identification
Risks might include long-term changes in markets, labour, client demands, contract types, legislation, foreign competition, technology, acquisitions and mergers, etc. More long-term impacts might be societal, related to safety, environment, public health, politics and culture.

Specifically for a project, typical aspects of concern are: site accidents/damage to property and the resulting legal, financial and insurance consequences; delays and disputes arising from late working drawings, indecision by the client or advisor and similar, reliability of suppliers and financial sources and the effects of delivery failure; possible strikes and resulting disruptions; potential groundwork and environment hazards and impact on construction methods.

See Chapter 16 *International construction logistics and risks* for other examples.

The identification approach begins with brainstorming sessions involving the manager and team responsible for particular segments of investigation, directed towards identifying and listing potential risks or hazards and likely impacts. However, other techniques such as interviewing experienced internal staff or seeking advice from outside specialists may sometimes be more appropriate.

Step 2. Risk analysis/assessment

Once the potential risks have been tentatively identified, their effects are evaluated qualitatively (e.g. subjectively) and/or quantitatively (e.g. probability, statistical, sensitivity analysis, operations research – OR techniques, etc.), ranked for occurrence and assessed against major/minor impact (see Kepner & Tregoe method—step 6, Chapter 7).

Step 3. Risk response

For each of the considered risk circumstances the appropriate response might include a range of possibilities such as:

- Risk avoidance, e.g. project or construction/temporary works redesign
- Risk transfer, i.e. taking out insurance or passing it on to others in the supply chain
- Risk retention, controllable or otherwise, e.g. instituting appropriate safety measures and checking procedures, duplication of resources, providing outsourced back-up, seeking cash flow from several sources, trading of carbon emissions, etc.

The necessary control measure(s) is accordingly determined, costed and the plan of action ascertained (see Kepner & Tregoe method—step 7, Chapter 7).

Step 4. Install and maintain

Finally, the proposed actions are installed and the subsequent out-turn costs for each of the identified risks, including those that arose unforeseen, evaluated and recorded. Feedback information is redirected into the above stepped loop to help inform and adjust the contingency/continuity plans and estimates as necessary.

The whole process can be prepared as a spreadsheet document to constitute the risk register for each manager, project, department, company, etc., depending upon the organisational level carrying out the evaluation.

Formalisation of the procedure using a suitably designed coding system, analysis and categorisation of the risk data should help evaluation and subsequent identification of 'best practice' measures of control to better inform the enterprise on risk management and thereby improve the production process.

Employee participation

Worker surveys

Research findings by Broomfield *et al.* (1984) indicate that lost production time can regularly account for about 50% of the working day, with even higher levels not uncommon. The causes are usually interconnected and typically arise through unsatisfactory execution of managerial and supervisory functions surrounding short-term planning, daily and weekly scheduling, materials standardisation and control, information flow, constructability of designs, subcontractor performance and suppliers, workforce goals and competency rather than specifically in the methods of working. Indeed, detailed information obtained directly from the workforce in structured questionnaire surveys undertaken by Zakeri *et al.* (1996) shown below illustrate the kind of problems causing serious production difficulties, which wherever identified will need rooting out and remedies devised:

- Lack of materials due to: waste, transport difficulties, improper handling on site, misuse of the specification, lack of a proper work plan, inferior materials, excessive paperwork
- Excessive weather delays
- Equipment breakdowns
- Drawings and/or specification changes
- Variations orders
- Inadequate tools or equipment
- Inspection delays
- Absenteeism
- Poor work planning
- Repeat work due to: poor finishes, negligence, congestion, overcomplicated drawings and/or specification, poor supervision, improper materials, poor design/engineering
- High labour turnover due to: low pay, casual labour, remote site, late pay days, work discontinuities, poor job opportunity, harsh/fatiguing working conditions, poor work facilities, lack of materials
- Work interference
- Poor communication

Supervisor delay surveys

In contrast to the general indicators obtained from a worker survey, a delay survey directly undertaken by each crew supervisor can be very effective as the means of acquiring site-specific data about production problems, especially when repeated regularly over a period. Here the supervisor identifies and records time losses for the crew at the end of the working day in order to reveal areas of inefficiency resulting from factors other than those due to the performance of the workforce. Thus, the supervisor delay survey (SDS) is aimed at determining the administrative items that are outside the supervisor's control, but which nonetheless affect the performance of the work group. A sample survey form is shown in Fig. 3.4 with typical results illustrated in Fig. 3.5. Subsequent diligent application of the recommended method study procedures would enable improvements to be effected and further use of the SDS technique would indicate whether the revisions were proving effective by reducing reported delays. An example illustrating improvement is shown in Fig. 3.6. The SDS technique potentially provides a cost-effective tool that is particularly useful for evaluating managerial related problems and has the following features:

- It provides a link between management and supervisors facilitating discussion of problems and means of solution.
- Delays to particular groups of workers are immediately highlighted.
- Information is provided on particular aspects, such as materials, subcontract interference, drawing information, plant availability, etc.
- The whole project can be readily monitored.
- The survey is inexpensive to carry out, can be done regularly and by untrained personnel.
- The information is current.

Supervisor questionnaire

Problems causing delay	Labour hours lost		
	Number of hours ×	Number in crew =	Labour hours
(1) (a) Waiting for materials (on site)			
(b) Waiting for materials (supplier delivered)			
(2) Waiting for tools and equipment			
(3) Lack of access			
(4) Equipment breakdowns			
(5) Changes/redoing work			
(a) design errors			
(b) site errors			
(6) Moves to other work areas			
(7) Waiting for information			
(8) Lack of continuity			
(9) Overcrowded working areas			
(10) Inclement weather			
(11) Other			
Total			

Comments

Supervisor and crew reference information ______________

Fig. 3.4 Supervisor delay survey.

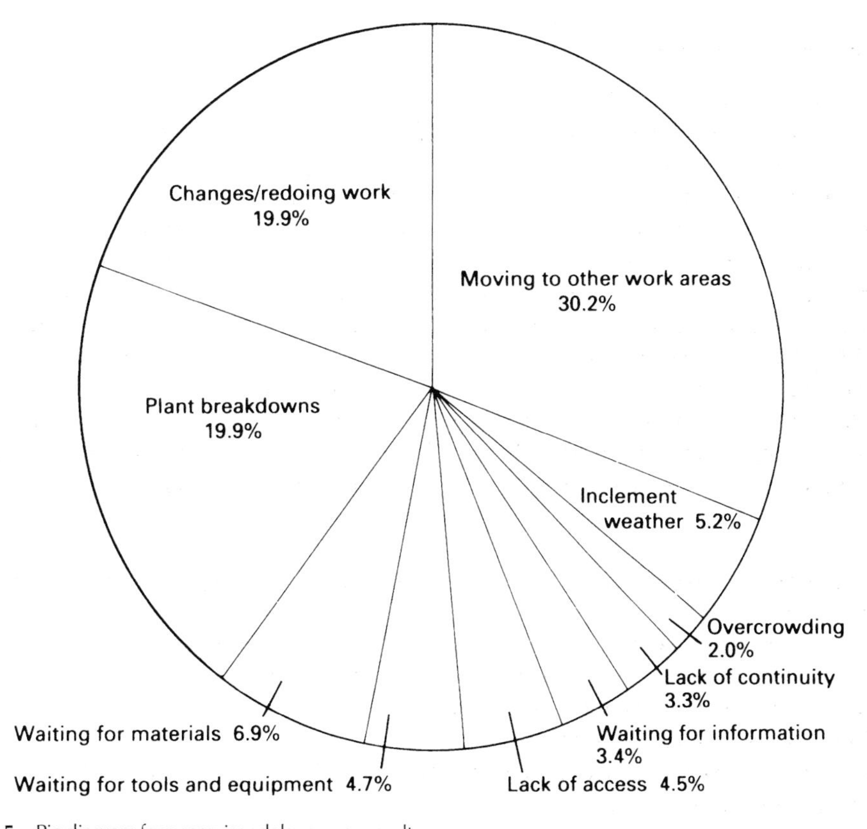

Fig. 3.5 Pie diagram for supervisor delay survey results.

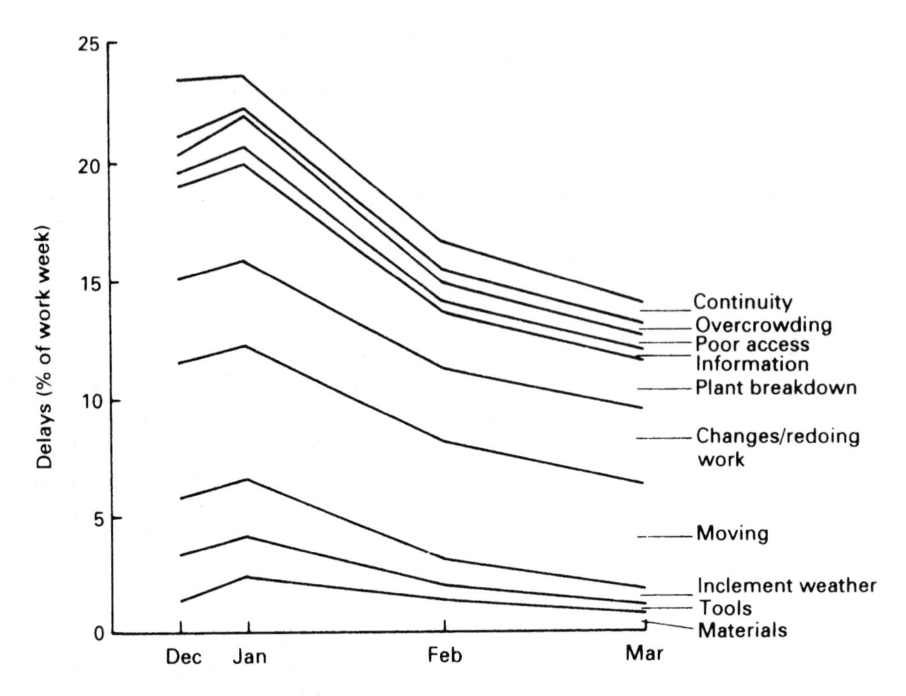

Fig. 3.6 Results of regular supervisor delay survey.

Work task management

Goal/task setting

Management by objectives directed towards raising productivity, first conceived by F.W. Taylor, involves the setting of goals, acquisition of feedback data, monitoring of performance and subsequent adjustment where necessary, with various goal-types having been devised depending on the need for quick or long-term results, personal or organisation needs, etc. Notably, in adopting productivity-type goals the manufacturing sector has demonstrated solid improvements by setting specific targets for individual workers or more commonly work teams; witness the achievements of quality circles in particular. The principles involved also commonly feature, but more weakly, in construction work, particularly crew/gang motivation schemes where close attention is specifically given to targeting the amount and quality of output to be achieved in a set period.

Work task analysis (work study)

Partial factor measures of productivity (PFP) are of special interest in work study. For example, labour-related inputs would be of prime concern to field managers, measured by the ratio of output to labour inputs only. In this respect experience suggests that 5–10% incremental improvements are potentially attainable in production efficiency, diminishing asymptotically towards zero further gains, through regular application of work-study techniques. In 'world class' sectors of industry rigorous application has proved capable of eliminating poor practices and unnecessary manual activities, and helped the introduction of new technology, improved mechanisation and a more cost-effective supply chain. For example, the major automobile manufacturers have standardised the assembly aspect of car production to about 18 to 20 labour hours per unit, with components subcontractors also monitored to similar high standards.

The process of observing, evaluating and improving performance in production operations called *Work Study* is defined (BSI, 1992) as:

> A measurement service based on those techniques, particularly *Method Study* and *Work Measurement*, which are used in the examination of human work in all its contexts, and which lead to the systematic investigation of all the resources and factors which affect the efficiency and economy of the situation being reviewed, in order to effect improvement.

Specifically, *method study* is used to record work procedures, provide systems of analysis and develop improvements. Applications can assist in redesign, detailed planning, site layout evaluation, design of temporary works, equipment selection and other resources, and replanning and progressing of production. In particular, design-and-build contractors engaged with a well-established chain of subcontractors principally working on assembly of standard component/units are best placed to benefit, for example system building processes and prefabricated construction. Notably, variants of the method have been successfully used to reduce inefficiencies and waste on repetitive activities associated with national railtrack network maintenance.

Time study measures the time required to perform a task so that an output standard of production for a worker and/or machine may be established. Such information is required in the estimating process and in setting financial incentives, as part of the data in method study; it can also be used to monitor actual production performance against the standard expected. However, the bespoke, variable and short-term nature of construction projects presents fewer opportunities for its effective application and it is therefore best applied where construction can replicate the mass production process typical of manufacturing.

Method study

Method study is the systematic recording and critical examination of the factors and resources involved in existing and proposed ways of doing work, applied as a means of developing and applying easier and more effective methods and reducing costs, executed procedurally in the following steps:

(1) Define the problem.
(2) Record the work being done, e.g. outline process chart, flow diagram, string diagram or multiple activity chart. Recording is normally carried out manually, preferably aided with time-lapse techniques using lightweight shoulder video equipment, for example, time-lapse 1/80 speed is almost equivalent to one frame per three seconds. The film is best viewed four or five times at about half normal speed to develop a general understanding, followed by viewing at three to four frames per second and finally several times at normal movie speed or using a TV PAL system of about 25 frames per second, looking specifically at each item of equipment, worker or group of workers. The information is thereafter interpreted and reconstructed into charts, flow diagrams and multiple activity charts for subsequent method study analysis.
(3) Analyse the present method and develop alternatives, adopting the following questioning technique:

What . . . ?	Why . . . ?	Could it be . . . ?	Should it be . . . ?
purpose is achieved?	is that necessary?	what else?	what?
place is occupied?	that place?	where else?	where?
sequence is followed?	that time?	when else?	when?
person is involved?	that person?	who else?	who?
means is used?	that way?	how else?	how?

(4) Install and maintain

Situations illustrating applications of the Method Study technique are treated in Chapter 19, specifically Questions and Solutions 1 to 4.

Time study (work measurement)

Work measurement is the application of techniques designed to establish the time for a qualified worker to carry out a specified job at a defined level of performance, indispensable to estimators and planners in contracting organisations in order to establish the 'proper' time for the job.

The procedure involves recording field observations manually or with a video camera and subsequently reinterpreting them as shown in Fig.3.7. Because the objective of the study is to obtain a realistic time for the element, the time study observer must make a judgement on the effective rate of working of the subject under observation since the elapsed time observed for one worker may be different from that for another doing an identical task. The rating thus corresponds to the average rate at which qualified workers will naturally work at a job, provided they adhere to the specified method and are motivated to apply themselves to their work. If the Standard Rating is maintained and the appropriate relaxation is taken, a worker will achieve Standard Performance over the working day or shift.

A rating scale helps provide a suitable description for varying levels of worker performance achieved over a short period of time and is typically divided into 5 point graduations with 100 representing standard rating. Thus:

125:	Very quick; high skill; highly motivated;
100:	Brisk; qualified skill; motivated;
75:	Not fast; average skill; disinterested;
50:	Very slow; unskilled; unmotivated.

As a guide, a worker of average build and a regular walker is said to achieve standard rating when walking unloaded at about 4 mph (6.43 kph) along a straight road and with adequate rest periods.

Calculation of basic time

The basic or normal time for a work task (T_b) is given as the ratio of the assessed rating (R_a) to the standard rating (R_s) multiplied by the observed time (T_o) as reflected by equation 3.2 below.

$$T_b = T_o \times \left(\frac{R_a}{R_s}\right) \tag{3.2}$$

CONTRACT: BOGTOWN SEWAGE WORKS	STUDY NO: 41
OPERATION: Place concrete to walls of Primary Digestion Tank No. 1 (A study of 24 skip placings) OPERATIVES: Concrete gang – four labourers MACHINES: Crane driver, dumper driver OBSERVER: P.L. Sirnam	REFERENCE: TC/WS/704 DATE: 12 September TIME STARTED: 8.00 a.m. TIME FINISHED: 10.18 a.m. ELAPSED TIME: 138 mins TOTAL OT: 700 man mins T.OT ex idle time: 350 man mins
CONDITIONS 15°C, light breeze, cloudy and dull, no rain, dry ground	NOTES Poorly prepared temporary roads between batching plant and digestion tank

R = Rating: OT = Observed Time: CT = Clock Time: BT = Basic Time: IT = Idle Time

Elements	R	OT	CT	BT	Elements	R	OT	CT	BT
A) Crane driver					B) Fills $\frac{1}{2}$ m^3 skip	75	0.30		0.23
B) Dumper driver (Note: Only included when transferring concrete into skip)					B) Attaches hook to crane	75	0.30		0.23
C)									
D) Concrete					A) Waits	IT	(0.6)		
E) Labourers									
F)									
Start time			0800.0		CD) EF) Waits	IT	(2.4)		
B) Fills $\frac{1}{2}$ m^3 skip	110	0.2		0.22				0807.0	
B) Attaches hook to crane	110	0.2		0.22	A) Lifts, travels, slews	75	2.7		2.0
A) Waits	IT	(0.4)			CD) EF) Wait	IT	(10.8)		
					CD) Discharge EF) concrete	85	2.0	0809.7	1.7
CD) EF) Wait	IT	(1.6)			A) Waits	IT	(0.5)		
			0800.4					0810.2	
					A) Returns to B)	60	3.4	0810.2	2.0
A) Lifts, travels, slews	80	2.4		1.9	CD) Vibrate EF) concrete	90	8.8		7.9
CD) EF) Wait	IT	(9.6)							
CD) Discharge EF) concrete	70	2.4	0802.8		CD) EF) Wait	IT	(4.8)		
A) Wait	IT	(0.6)						0813.6	
			0803.4						
A) Returns to B)	70	3.0		2.1	(etc.	–	–	–	
CD) Vibrate EF) concrete	100	8.0		8.0	(etc.	–	–	–	
CD) EF) Wait	IT	(4.0)	0806.4		(etc. up to (24 skips	–	–	– 1018.0	
						Total	OT	CT	BT

Fig. 3.7 Time study observation sheet.

The basic or normal time is therefore the time for a *qualified* worker to carry out a specific job at a defined level of working. In practice, the worker could not be expected to achieve this level of rating over the working day without rest and relaxation.

Factors affecting the rating

(1) The observer should guard against malpractices, whereby workers being timed, for example, to set standards for a bonus incentive scheme, may try to give the impression of working at standard rating while hoping that disguised inefficiencies will go unnoticed by the time study observer.
(2) While the observer must be aware of such practices, they must also try to assess the true effort required for the task. For example, the same rating could be assessed for light and heavy tasks, which would normally be carried out at different speeds even by a qualified and motivated worker.
(3) Factors that can influence the observed time but not necessarily the rating include: quality of tools used; type and quality of material worked on; working conditions; learning period required before task becomes familiar; interruption of supply of materials; quality of working drawings; supervision; quality specification.
(4) Factors attributable to the worker that affect the observed time and therefore should be removed in the rating assessment include: intelligence and education; attitude and motivation; skill and training; personal discipline and organisation; health; level of fatigue.

These factors are difficult to assess and are best included by taking a large number of studies to give a representative sample.

The required number of observations

The observer records the time against the assessed rating and the best straight-line fit through the points is drawn as shown in Fig. 3.8. The basic time is simply read off at 100 rating. The method will also indicate the amount of under- or over-rating included inadvertently by the observer. The true rating line should pass through the origin, since in theory infinite rating should take zero time and infinite time corresponds to zero rating. The method also highlights unrepresentative ratings due either to faulty rating or fraudulent performance by the worker. Such data can be removed from the diagram.

The correct size of sample is difficult to determine but certainly enough observations should be taken to cover the likely changes over a working day, such as starting up and finishing periods, weather, etc. The simplest method is to plot the cumulative average basic time against the number of observations as demonstrated in Fig. 3.9. When the line begins to stabilise, sufficient observations have probably been taken. Unfortunately, as far as construction work is concerned, elements often are insufficiently repetitive for enough observations to be made.

The observer of construction work is frequently faced with variable elements and tasks. For example, the time for placing of concrete by crane will vary with several factors such as the size of the concrete skip, the size of the opening in the shuttering, the quantity of concrete poured, and so on. By plotting the basic times obtained against changes in a variable or combination of variables a line of best fit can quite often be drawn. In this way the basic times, for example of placing concrete per cubic metre, with a range of skips or sizes of pour, etc., could be ascertained, as demonstrated in Fig. 3.10.

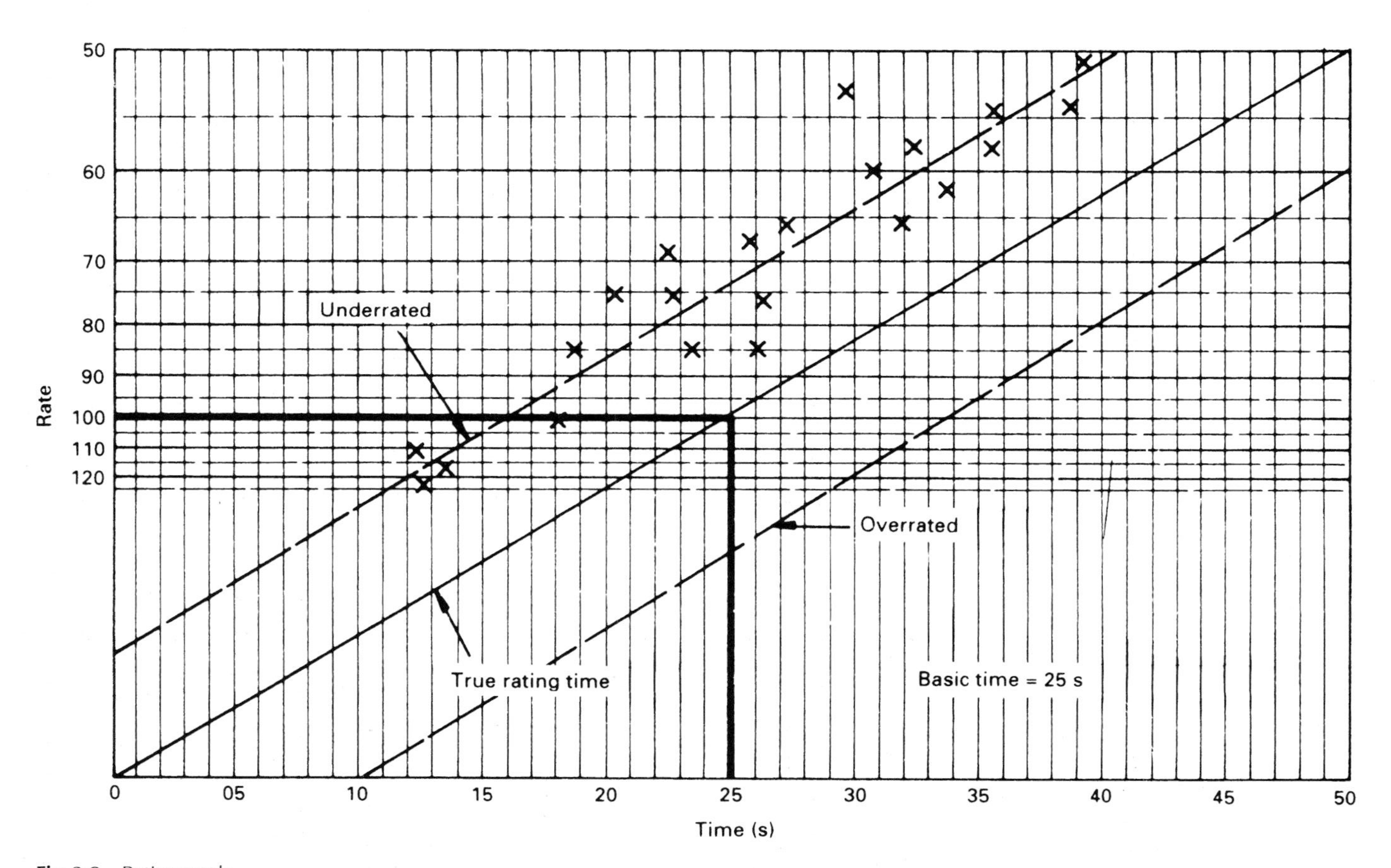

Fig. 3.8 Rating graph.

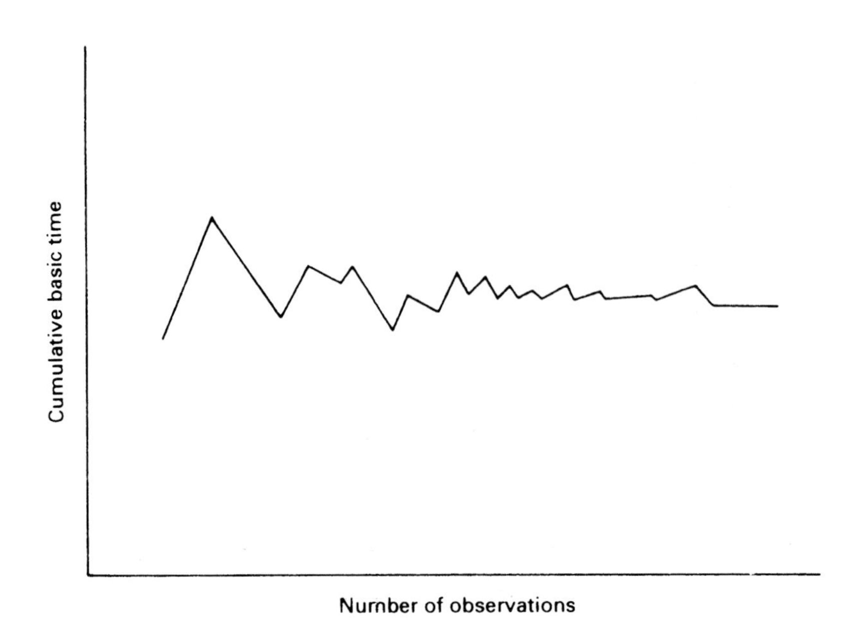

Fig. 3.9 Cumulative mean basic time.

Standard time

The need for workers to have rest periods over a working shift of several hours has been ignored until now in calculating the basic time for an element. To establish the standard time, it is therefore necessary to include some relaxation allowances (A_r) plus a contingency (C). Thus:

$$T_s = T_b + A_r + C \tag{3.3}$$

The units will either be standard hours or standard minutes, and represent the 'proper' time for a qualified worker working at standard rating to execute a task or element of work. If this is obtained, then the worker will achieve standard performance, where standard performance can be defined as the rate of output which qualified workers will naturally achieve without over-exertion as an average over the working day or shift, provided they are motivated to apply themselves to their work.

However, the concept of performance encompasses both the standard rating achieved over short durations and the allowances that must be included if the standard rating is to be maintained over a longer period. Most records or data banks of output times are kept as basic times, with the user applying suitable allowances and contingencies as necessary.

Allowances

Relaxation allowances are given as percentages of basic time and Fig. 3.11 illustrates examples which may be considered.

Contingencies

To obtain the final standard time it is usually prudent to include additional time that cannot be determined accurately but which will almost certainly occur. Indeed, because construction work

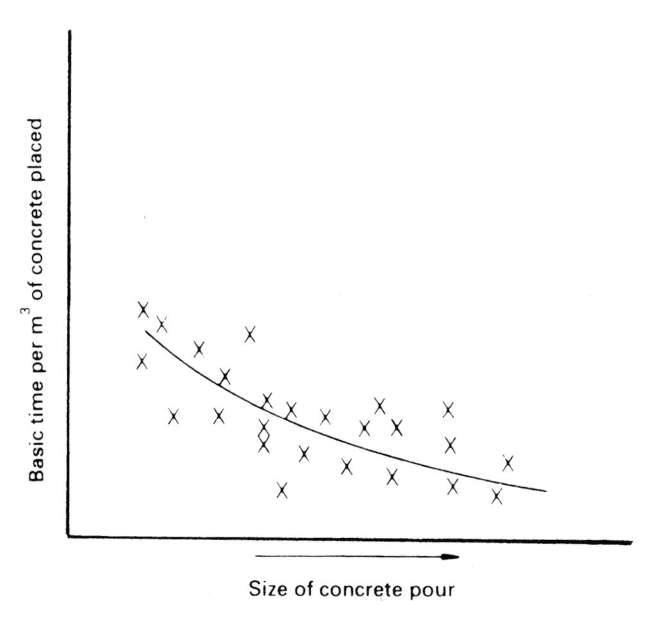

Fig. 3.10 Basic times plotted against volume of concrete.

and physical conditions are very variable the difference between standard time and basic time for a job is generally quite large. Research evidence suggests realistic planning times need to be about double the basic time.

Examples of time study are illustrated in several exercises described in Chapter 19, specifically Questions 1 to 4.

Synthetical estimating

A careful study of work measurement data indicates that there is an underlying base of repetition in many construction activities, particularly building-type work, and thus appropriate for estimating and planning purposes. The computer stored data and files described in Chapter 11 in essence define synthetical estimating, see Fig. 3.12, where small-size elements are used to *build up* the time for the whole series of tasks in an activity. Thus, from experience, the work content of a job, e.g. the foundation formwork mentioned above and the method by which it can be carried out, may be predetermined and broken down into elements. The basic time for each element is selected from established *basic data* and the synthesis of these elements plus allowances gives the standard time for the work at standard performance.

The technique can be incorporated into the unit rate or operational estimating methods currently used by estimators and planners in construction companies.

Feedback evaluation

As indicated in Fig. 3.1 the gathering of feedback information, after installation of the appropriate work-model, is essential in monitoring progress and making subsequent improvements, with the particular approach depending on the management need. These are considered in the following sub-sections.

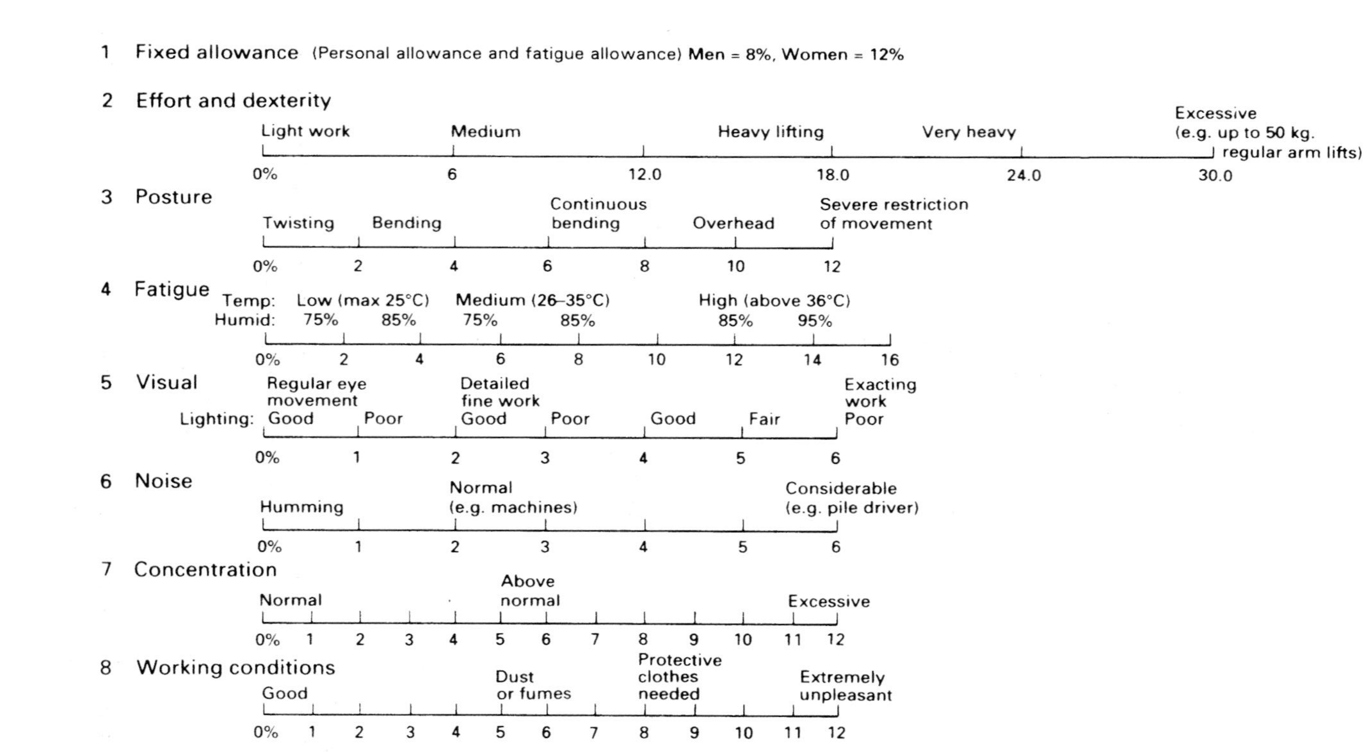

Fig. 3.11 Typical relaxation allowances.

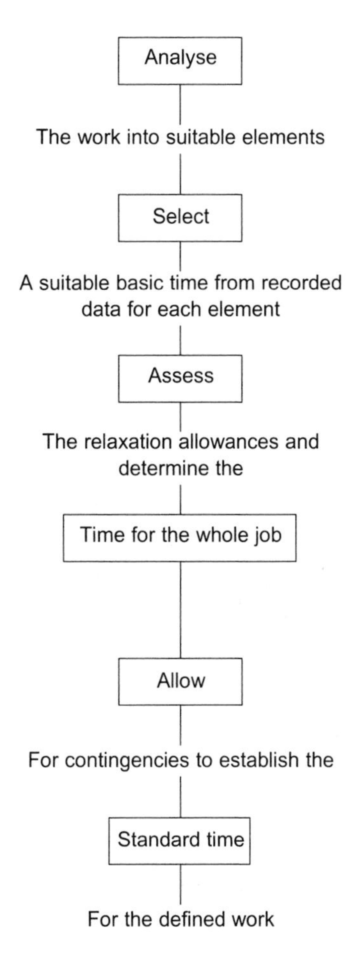

Fig. 3.12 Synthetical estimating.

Performance level

The standard time for an activity when directly compared with the actual time spent on the task indicates the level of performance (P) physically achieved, i.e.

$$P = \frac{\sum T_s}{H} \tag{3.4}$$

where $\sum T_s$ is the total standard time for all measured and estimated work and H is the labour minutes or hours of work time available.

Example

A small project calls for the excavation of a column base foundation. The contractor's programme provides for a 0.5 m^3 backhoe to do the work. Basic time for the task is estimated at 4 minutes per m^3, based on an expected rating of 100 and 50% is to be added for relaxation allowances and contingencies.

At the end of the second day 100 m^3 of excavation had been completed. If an eight-hour day were worked, what level of performance by the machine and its operator would have been achieved?

Solution

$$\text{Standard time} = (1+0.4)\times 4 = 6 \text{ standard minutes}$$

$$\text{Machine performance} = \frac{\text{standard minutes produced}\times 100}{\text{available minutes worked}}$$

$$= \frac{100\times 6}{2\times 8\times 60}\times 100 = \mathbf{62.5P}$$

The result would be expressed as 62.5P, where 100P is standard performance.

The machine was working well below the expected standard level of performance, and either the estimate was misjudged or the driver is inefficient or the allowances were not enough to cover the conditions met on site.

Work (activity) sampling

While the above simple example illustrates the potential function of accurate work-study data as a monitoring tool, the project manager needs to monitor quickly at all times whether work crews are performing as expected and to identify the immediate causes of any problems. Work/activity sampling, by facilitating the determination of the proportion of time spent on effective work, is more suitable for providing this kind of information. Thereafter conventional work-study analysis may be introduced to help devise improvements.

Example

A field count indicates that about 40% of the available work time on a section of construction work is spent on unproductive work. How many observations are required to be sure that the proportion is within 2% accuracy, given that 9.5% confidence is required?

Solution

If the field count is repeated on many occasions and the proportion of unproductive workers plotted graphically, it is quite likely that the data will fit a normal distribution as illustrated in Fig. 3.13. Such data can then be interpreted statistically in the following way.

$$P_x \pm P_u = 2\% \text{ and}$$
$$P_x \pm P_u = \text{standard deviation}\times Z$$
$$Z = 2 \text{ for } 95\% \text{ confidence}$$

Hence

$$0.02 = \text{standard deviation}\times 2 \qquad (3.5)$$

However, the data are also well represented by the binomial distribution, thus

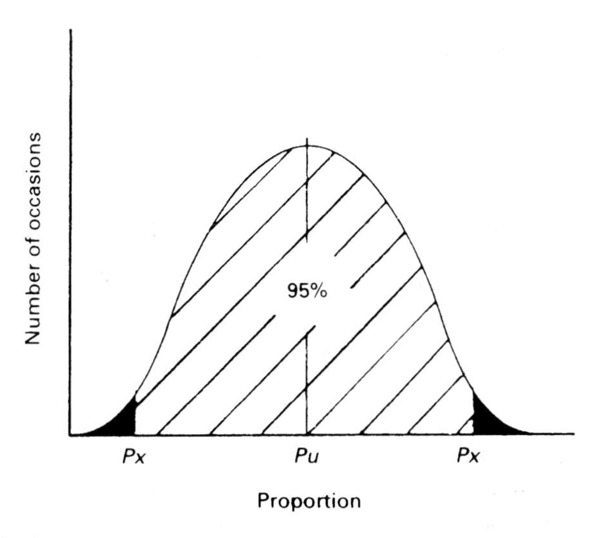

Fig. 3.13 Normal distribution curve.

$$\text{Standard deviation} = \sqrt{\frac{P(1-P)}{n}} \tag{3.6}$$

where P = observed proportion and n = number of observations.

By substituting standard deviation in equations (3.5) and (3.6)

$$0.02 = 2 \times \sqrt{\frac{P(1-P)}{n}}$$

$$n = \frac{4 \times 0.4(1-0.4)}{0.02^2} \qquad n = 2400 \text{ observations}$$

Therefore, by taking 2400 observations, the proportion of unproductive work is determined to within 2% accuracy.

More generally the formula is written as

$$N = \frac{Z^2 \times P(1-P)}{L^2} \tag{3.7}$$

where:

N = number of observations required
P = percentage of activity observed (usually assessed from a pilot study)
L = limit (%) of accuracy required
Z = value obtained from statistical tables depending upon the level of confidence required for the estimate (usually taken as 2, which corresponds to 95% confidence)

Activity sampling procedure

(1) First, carry out a preliminary survey to get a feeling for the problem. The information collected will help in deciding on the size of the section of work to be studied and the number of workers involved.

(2) Identify the workers by name, and list the operations and tasks to be studied. Sometimes this need only be 'Working' or 'Not working', but for a much fuller investigation greater detail will be required, such as 'Fixing formwork', 'Cleaning shutter panels', 'Receiving instructions', etc.

(3) Prepare a suitable observation sheet, Fig. 3.14, for recording the information.

(4) Consult the supervisor of the work and ensure that everyone is fully informed. Failure to do so may cause unrest, which can quickly escalate and feed smouldering grievances.

(5) The number of observations required is normally quite large; therefore, a planned timetable of observation times should now be assembled. In mass production work these times are normally chosen randomly since the work patterns tend to be regular. However, in construction work, activities by their very nature take a random time to complete and thus the observations can be taken at regular intervals. Even so, caution is required with some types of building work, e.g. bricklaying.

(6) Choose a suitable position for taking the observations.

(7) Record each activity that is in operation at the instant it is observed, together with the worker involved.

(8) From the percentages of the activities observed select the activity or activities which show a disproportionate amount of time being spent on them. Corrective action can then be considered.

Observer					Place						Date			
Observation	Time		Worker					Operation						
round	hour	min	A	B	C	D	E	FS	CL	RF	RI	I	A
1.	8	00	✓					✓					
				✓				✓					
					✓					✓			
						✓							✓
							✓				✓		
2.	8	10	✓					✓					
				✓					✓				
					✓						✓		
						✓						✓	
							✓					✓	
3.	8	20	✓									✓	
				✓						✓			
					✓					✓			
						✓							✓

Fig. 3.14 Activity sampling observation sheet.
Key to operations: FS = fixing formwork, CL = cleaning formwork, RF = removing formwork, RI = resting, I = idle, A = absent.

Rated activity sampling
When recording the number of observations for each operation, the performanc operative, or machine, etc. is assessed at each observation against the standard rati scribed previously under *Time study*, and the mean rating for the operation and the ti spent as a proportion of the whole activity are calculated. If the total elapsed time of the activity is also recorded, then the basic times of the individual operations in the task can be calculated.

As in *Time study*, much of the data recording and analysis can be performed on hand-held and/or microcomputers using commercially available software or self-developed programs.

Worked examples are provided in Chapter 19.

Macro-key performance indicators

The CEBE organisation provides the facility to compare and update data for an individual company or project performance against sector averages. Some of the typical indicators are:

(1) Client satisfaction-product rated on a scale 0–10
(2) Client satisfaction-service rated on a scale 0–10
(3) Construction time and cost, i.e. comparison of project this year with similar project in previous year
(4) Predictability construction cost (design), i.e. out-turn cost-cost estimate/cost estimate
(5) Predictability construction cost (construction), i.e. out-turn cost-cost estimate/cost estimate
(6) Predictability-design time (ditto)
(7) Predictability-construction time (ditto)
(8) Safety, i.e. reportable accidents per 100,000 employees as defined by H&S Commission. Defects on a scale 0–10, 10 = defect free
(9) Productivity expressed as company added value (i.e. turnover minus subcontracts) per employee
(10) Profitability, i.e. profit before tax and interest as a percentage of sales

Other indicators suitable for use by construction consultants are also available. A full list is available on website 'constructingexcellence.org.uk'.

References

Broomfield, J.R., Price, A.D.F. & Harris, F.C. (1984) Production analysis applied to work improvement, *Technical Note* 415. *Proceedings of the Institution of Civil Engineers*, **44**, 379–386.
BSI (1992) BS 3138: *Glossary of Terms used in Management Services*. British Standards Institution, London.
BSI (2002) *Guide to Project Management* BS 6079-1:2002. British Standards Institution, London.
Hammer, M. & Champy, J. (2004) *Re-engineering the Corporation*. Harper Business, London.
ISO (2002) *Risk Management: Vocabulary—Guidelines for Use in Standards*, IEC Guide 73:2002. International Organization for Standardization, Geneva.
Kepner, C.H. & Tregoe, B.B. (1997) *The New Rational Manager – an Updated Edition for a New World*. Kepner–Tregoe, Princeton, NJ.
Leeb, S. & Leeb, D. (2004) *The Oil Factor*. Warner Business Books, New York.
Packard, V. (1970) *The Waste Makers*. Penguin.

Samuelson, P.A. & Nordhaus, W.D. (2004) *Economics*. McGraw-Hill Education, Maidenhead.
Schumacher, E.F. (1973) *Small is Beautiful*. Harper Perennial, London.
Seabright, P. (2004) *The Company of Strangers*. Princeton University Press, Princeton, NJ.
Taylor, J.B. (1993). Discretion versus Policy Rules in Practice. *Carnegie–Rochester Conference Series on Public Policy*, **39**, 195–214.
Whyte, W.H. & Nocera, J. (2002) *The Organization Man*. University of Pennsylvania Press, Philadelphia, PA.
Zakeri, M., Olomolaiye, P.O., Holt, G.D. & Harris, F.C. (1996) A survey of constraints on construction operative productivity. *Construction Economics and Management*, **14**, 417–426.

4
Planning techniques

Summary

Bar-charts, linked bar-charts, activity-on-the-arrow networks, precedence diagrams, line of balance schedules, PERT, space-time diagrams and last planner.

Planning in construction

There are two main levels of planning associated with construction projects. These are strategic and operational planning. Strategic planning deals with the high-level selection of overall project objectives, including the scope, procurement routes, timescales and financing options. The different procurement routes and options for financing that are given consideration during planning at this level have been covered in other chapters. Strategic planning for a project results in broad outlines of what the project has to achieve, and how it is to be undertaken. Operational planning, on the other hand, involves establishing a method statement for each activity. Operational planning allows a more detailed look at the project's resource requirements that is not obvious at the strategic level. Examples of operational plans include a Tender plan, Feasibility plan and Construction plan. A method statement is a description of how the work will be executed.

A construction programme of works primarily presents the sequence in which the various activities should occur with their associated durations and resource requirements. Programming of construction works is normally undertaken for a project as a whole and these are called master plans. However, programming of works can also be for only a part of the project, for example the construction of a section for a road project, the procurement of major materials and equipment. Plans are also required for functional aspects of the project such as management of quality, as well as health and safety issues.

Planning at both strategic and operational levels for a construction project employs various tools and techniques to help the planner achieve better optimisation of the decisions involved. The techniques presented in this chapter focus essentially on the operational aspect of planning, but can equally be adapted for planning at the strategic level.

Who plans?

A number of people require plans—proposed programmes of work—of varying complexity for varying purposes. In the construction industry the types of planners can be grouped by the role their organisation plays in the construction process. In this way there are three types of planners: the client organisation; the engineers/architects or designers; and the contractor.

The client

The client organisation is interested in the plan of the overall project from the acquisition of the land to the productive use of the facility. Primarily, the client organisation is interested in determining the times of outflows of cash for which it has to make provision and in the overall strategic decisions of the project's management. The client will operate with activities whose durations are likely to be weeks or months rather than days. Major decisions, such as the type of contract, which may range from a traditional form with design and construction running sequentially to design/construct forms where design and construction overlap, will have an impact on the overall duration of the project and the client's cash flows and will be examined using the overall project plan.

Another example of the use of the project plan will be the client's decision on whether to proceed with procurement of mechanical/electrical plant before construction contracts are let, the procurement times and the overall project duration being the key factors in such decisions.

The client organisation will rely heavily on the project's plan or programme as a decision aid when considering these and other strategic issues. Later the client organisation will use the project's plan to monitor progress.

The designers

The engineers in the case of civil engineering projects or the architects in the case of building projects or the client's project manager, if appointed, will normally carry the responsibility for preparing the plans used by the client in taking the strategic decisions described above.

In addition, the design team has its own resources to manage. It is around a project plan that the design team will determine the order in which various sections of work will be designed, the numbers of designers allocated to each work section and the cost, in terms of designer-hours, of the design tasks. To bring a set of contract documents to a conclusion in order to issue them for tendering purposes is like any process of multiple activities; it is necessary to plan, monitor and control.

The activities to be planned, resourced and controlled include: investigations; design calculations; drawings; statutory planning approvals; preparation of specification; preparation of contract documents, tender programme; and preparation of quality systems. The control of the design process has benefited from the use of planning techniques. The duration of the activities used by designers tends to be in weeks rather than the larger units of time used by the client.

The contractor

The contractor's organisation is the one of the three parties in the construction process that has historically put greatest effort into the planning process because the results of a well-planned, carefully monitored and controlled contract reflect directly in the profitability of the contract and the company. With the benefits of planning clearly visible, it is hardly surprising that the effort is made. The contractor's planning efforts are divided between planning at the estimating stage and production planning. Figure 4.1 shows the interrelationships between the contractor's management functions.

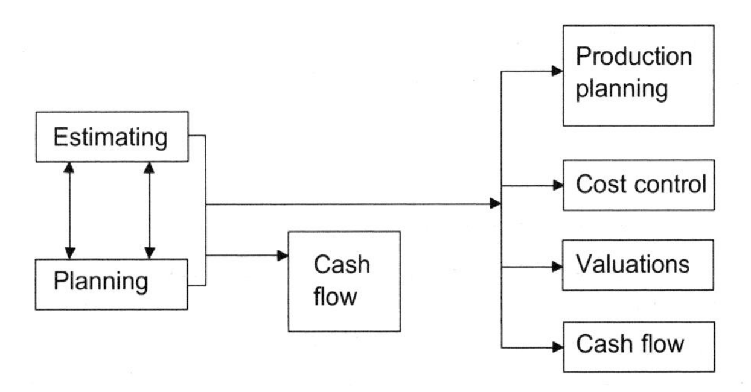

Fig. 4.1 Contractor's management functions.

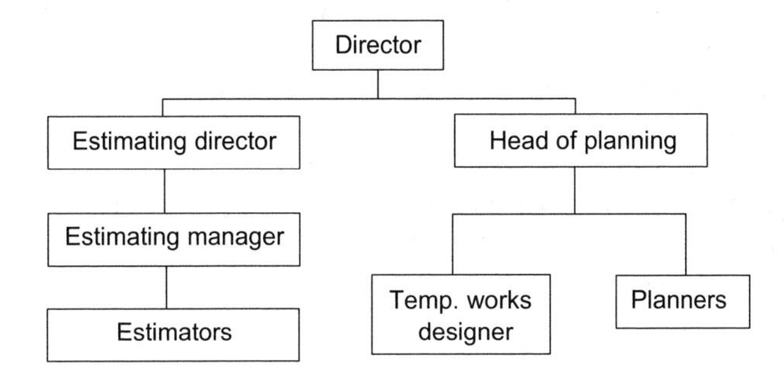

Fig. 4.2 Typical organisation of estimating and tendering in a major construction company.

Here it can be seen that planning is an integral part of the estimating process and again as part of the production control process. Crudely it can be described that *estimating* evaluates the use of resources in terms of *cost* and *planning* evaluates the use of resources in terms of *time*. Adding both together is necessary to obtain the *cash flow*.

In estimating, a project plan of work is required around which to develop proposed construction methods and hence estimates. The units of time used in the activities of the pre-tender programme would normally be weeks or days.

The site manager at the start of the project needs plans or work programmes to determine their resource requirements. During the execution of the project the site manager needs plans to assist in directing those resources, to monitor progress, to evaluate the effect of the changes that may be imposed by varying productivity, by mistakes, by weather or by the client and their designers. In certain forms of contract the site manager needs the project plan to determine payments due at interim stages. At site-manager level the units of time used for activities are usually weeks or days.

The section engineer is usually required to plan the work of their section in great detail week by week and so may use smaller units of time of days or half days.

All the major companies have taken the step of creating a planning department or unit within the company. The purpose of this planning department is to provide a planning service to the estimators and to sites. The control of the planning department usually falls under the same director who is responsible for estimating. Figure 4.2 shows a typical organisation chart for a major construction company. In such a company an estimator and planner would be allocated to each job being tendered for. On larger sites a planning team would be set up on site supported by the central planning department. On the larger sites the section engineers would still have responsibility for planning their section, albeit with support. On smaller sites the site engineers would not have a planning team to call on and would have to undertake their own planning supported from Head Office. The use of computers has now spread to such an extent that they appear on most sites of any reasonable size or are within easy access at Head Office.

In small companies, the existence of a separate planning department is unusual and estimators are responsible for their own planning, as are the site engineers.

It is sometimes considered that centralised planning results in the imposition of a plan of operation, whereas site staffs have a sense of commitment to plans they have drawn up themselves. In all companies work must be planned or the company must suffer the resulting uncertainties.

Planning techniques

The most common and widely used techniques available for planning are bar charts and linked bar charts; network analysis, either activity-on-the-node (often referred to as precedence networks) or activity-on-the-arrow (also known as arrow diagrams); and line of balance, for repetitive construction work.

Bar charts and linked bar charts

Bar charts are the easiest to understand and the most widely used form of planning tool. Even when a more sophisticated technique like network analysis is used, the eventual schedule of work is usually presented in bar-chart form.

The bar chart in Fig 4.3 shows the typical form: a list of activities with the start, duration and finish of each activity shown as a 'bar' plotted to a time scale. The level of detail of the activities

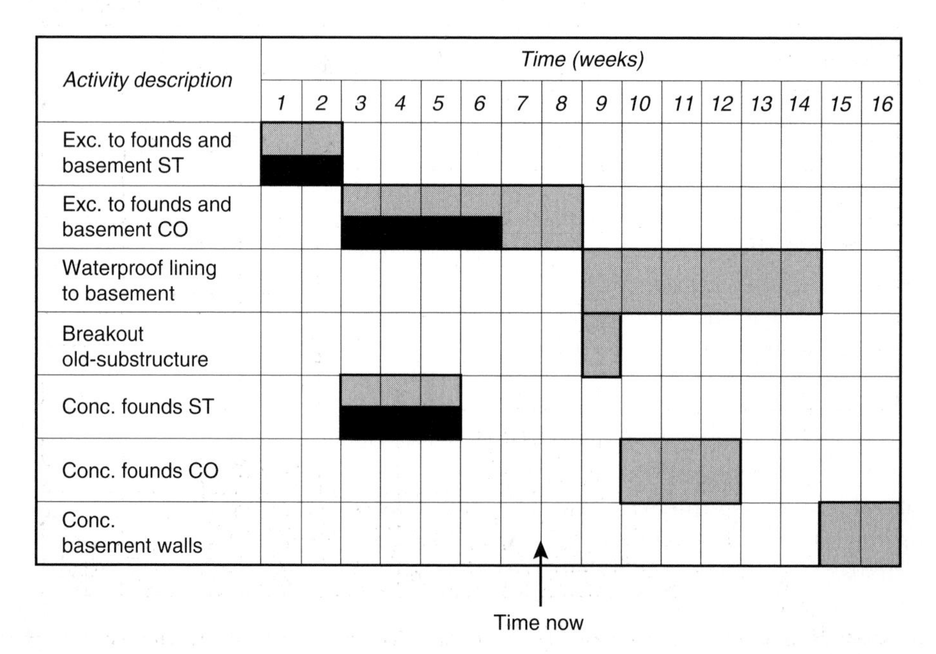

Fig. 4.3 Typical bar chart.

depends on the intended use of the plan. The site manager may be content with an activity as broad as 'Construct foundations'. The section engineer will break this into a finer level of detail, say 'Excavate', 'Blind', 'Fix reinforcement', 'Erect side shutters', 'Concrete', 'Strike shutters', 'Cure' and 'Backfill'. Similarly, the time scale will be chosen to suit the user's purpose. The site manager may use weeks while the section engineer may use days or half days.

Figure 4.3 also shows the bar chart being used as a progress control chart. A second bar is superimposed on the lower section of the main bar showing the planned time, to indicate progress for each activity. At the end of each time period the amount of work done in each activity is recorded and expressed as a percentage of the planned work for that activity. Where such recording is done in a computing environment, this is automatically translated into the equivalent progress bar. Figure 4.3 shows time now as being end of Week 7, the shading on the chart shows activities '*Exc. to founds and basement ST*' and '*Conc. founds ST*' are 100% complete. Activity '*Exc. to founds and basement CO*' is shown as being 67% complete whereas, to be on programme, it should be 83% complete. This means that this activity is unlikely to finish at the end of Week 8 as planned. By working at the achieved rate of progress for that activity, its finishing time will extend by one and a half weeks beyond the planned time. The effect on the other activities would need to be calculated. This effect could be more easily studied by using an extension to the simple bar chart such as the linked bar chart.

The linked bar chart, as in Fig. 4.4, shows the links between an activity and the preceding activities, which have to be complete before this activity can start. Similarly, the links are shown between the activity and the succeeding activities, which are dependent on the activity being completed. This illustration of dependency between activities has the advantage that the effects of delays in any activity are easily seen.

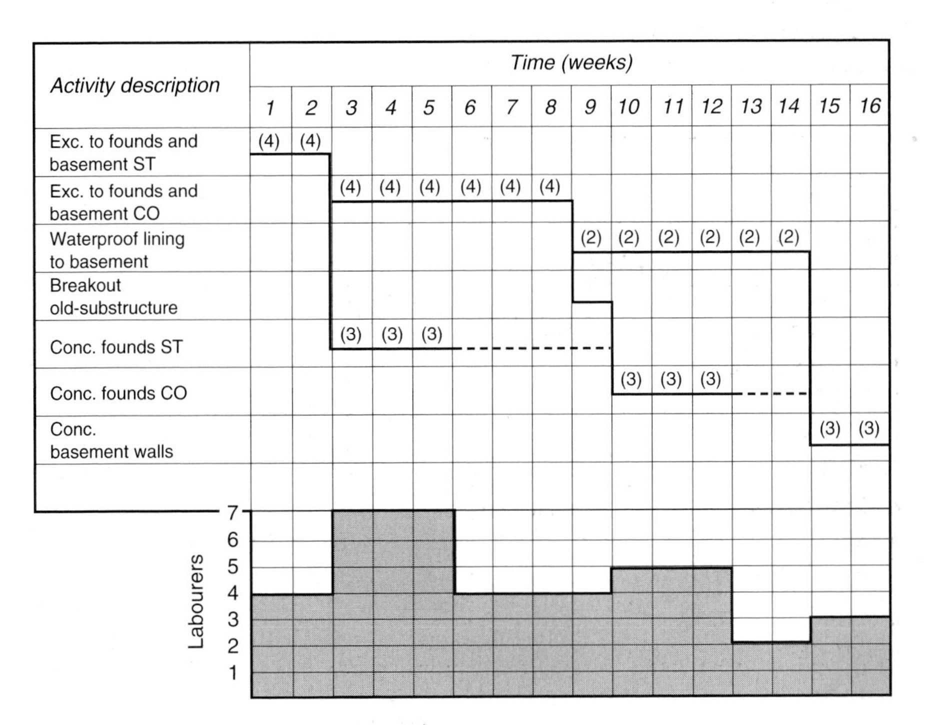

Fig. 4.4 A linked bar chart and resource aggregation chart. The vertical lines indicate dependency between activities, the broken lines indicate 'float'. Labourers required are shown in brackets.

The time available for each activity is also displayed: for example, in Fig. 4.4 activity '*Conc. founds ST*', is shown as starting on Week 3 and ending on Week 5, but it can be seen that the absolute deadline before it interferes with the next succeeding activity, '*Conc. founds CO*', is end of Week 9. This activity has some 'float' or extra time available before any delay affects other activities. Activity '*Exc. to founds and basement CO*', for example, has no float and must be completed by Week 8; this is usually called a critical activity.

The bar chart is also useful for calculating the resources required for the project. To add the resources, say labourers, to each activity and total them as in Fig. 4.4 is called resource aggregation. An aggregation chart similar to the one produced in Fig. 4.4 for labourers can also be done separately for other resource types such as carpenters or steelfixers, or cranes.

The bar chart and the resource aggregation charts are useful for estimating the work content in terms of labour-hours or machine-hours. Similar calculations done on site may be used to check the work content implied by the estimate, so as to determine whether the chosen construction methods will result in a profit or a loss. Cost control is more effectively based on such assessment of construction method than on simple historical cost checks.

Bar charts as means of communication between engineers and foreman are particularly useful and can be improved by colour-coding the activities, for example blue for carpenters, yellow for steelfixers, etc. The main advantage of a bar chart is its *simplicity*. The bar chart is a clear, easily understood document. Thus as a communication document the bar chart is the most successful in conveying the planner's intentions. Using the bar-chart *progress* can be recorded and *schedules* can be extracted, together with *ordering dates* for materials and other requirements.

For the smaller project and the short-term programme for a section of a larger project the bar chart is most useful. The drawbacks to a bar chart relate to its construction. If the bar chart is produced manually, that is, constructed on paper and not as a by-product of a computer system, then the bar chart is:

- Limited in size (say 30 to 100 activities)
- Not easy to update or schedule

Thus, the main limitation in manual bar charts is the inability to manipulate the bar-chart data. This means that there is a tendency not to update and the bar chart fairly soon is out of date, discredited and disregarded.

Network analysis

Network analysis offers all the advantages of being able to manipulate the planning data by holding the data in computer files. The planning data in a network is linked through the logic that defines the relationships between the activities. Thus changes can be made in the data relating to individual activities, i.e. the duration, the resources, etc., or changes can be made in the logical relationships between activities and the consequences re-calculated and re-presented. In addition, the steps to produce and process a network plan are more clearly defined, self-contained and offer a more rigorous approach to planning complex operations. This greater rigour imposed by the logic diagram produces more realistic models of the proposed work.

Finally, only through some form of network analysis is there a possibility of using computers for the calculations. The steps in producing a network are:

(1) Listing the activities
(2) Producing a network showing the logical relationship between activities
(3) Assessing the duration of each activity, producing a schedule, and determining the start and finish times of each activity and the float available
(4) Assessing the resources required

In producing a bar chart (2) and (3) are taken in one step and therefore in complex projects the various alternatives are unlikely to be considered. There are two popular forms of network analysis, activity-on-the-arrow and activity-on-the-node, the latter usually called a precedence diagram. The network technique evolved first with activity-on-the-arrow. Nowadays most planning is undertaken by activity-on-the-node as this is the format adopted in most planning software packages. The logic of planning, however, is more explicit when expressed as activity-on-the-arrow, so this is described first followed by activity-on-the-node.

Activity-on-the-arrow

The preparing of a network follows the steps listed above.

(1) Listing the activities
The comments regarding the level of detail are the same as those given for bar charts.

(2) Producing a logical network of activities
Network logic
In this system of planning, the activity is represented by an arrow. Unless the network is being drawn to a time scale, the length of the arrow is irrelevant. Even if the ultimate network will be drawn to a time scale it is not recommended that it be so drawn in the first instance.

The arrows are joined together in a logical relationship, and as each arrow is included in the network three questions should be asked in order to check that correct logic is being maintained. The questions are:

- Which activities must be complete before this activity starts?
- Which activities cannot start until this activity is complete?
- Which activities have no logical relationship with the activity and can therefore take place at the same time?

Ignoring the restraints that will be placed upon the sequences of activities by resources, either labour or plant, the network that satisfies the above questions will show the logical relationship of all activities. It may be necessary to introduce *dummy* arrows, drawn as broken lines, which do not represent any activity but are simply a logical link. For example, if activity C was dependent on A and B being completed and activity D was dependent only on B being completed, the network would require a *dummy* arrow to represent the logic as shown in Fig. 4.5.

Identifying the activities
The points where arrows start or finish are called *events*. The numbering of these events provides a method of identification. As an example the '*Conc. founds CO*' activity in Fig. 4.6 would be called activity (4)–(5). Except that each event must have a unique number there are no special rules to observe. Most practitioners begin numbering at the start of the network and progress

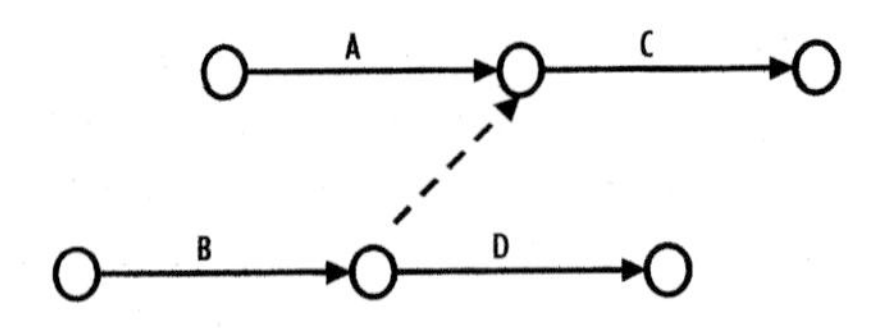

Fig. 4.5 A dummy arrow to maintain correct logic.

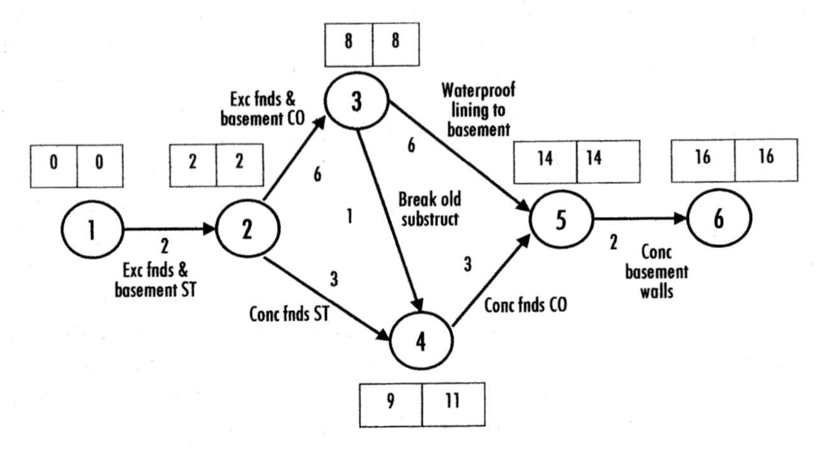

Fig. 4.6 A network showing duration, event numbers and event times.

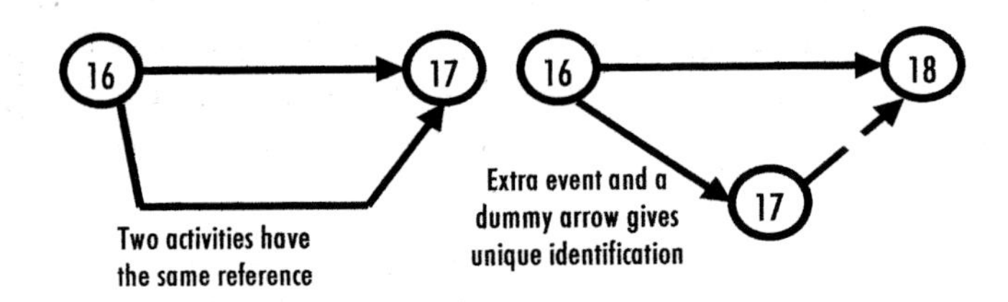

Fig. 4.7 Use of dummies for unique identification of activities.

through the events in numerical order until the end event is reached, ensuring that the number at the tail of an arrow is always smaller than that at the head of the arrow. There may be situations where two arrows leave the same event and arrive together at another event. In this case the activities would have the same identification numbers. To avoid this, a *dummy* is used and an extra event introduced thus ensuring that each activity has a unique identification. Figure 4.7 gives an example.

(3) Producing a schedule

Duration and time analysis

The time required for each activity needs to be estimated; the estimate of duration will be based on knowledge of the work, experience, records and work study. Once estimated, the duration of each activity is marked against the arrow in the logic network. The earliest possible time of each event is then calculated and written in the left-hand square alongside each event. This has determined the earliest possible start time of each activity.

The calculations are shown in Fig. 4.6. For example, the earliest time of event (1) is 0, the earliest time for event (2) is 0 + 2 = 2, and the earliest time of event (3) is 2 + 6 = 8 and so on. The point to watch is that where two paths or chains of activities merge, as for example at event (4), the longest path determines the earliest possible time of the event. At event (4) the path via event (3) produces an earliest time of 8 + 1 = 9 which is greater than the path direct from (2), which produces an earliest time of 2 + 3 = 5. Therefore 9 is the earliest time of event (4). The calculation of the earliest event times is known as the forward pass.

The reverse process, the backward pass, determines the latest possible time for the event, that is the latest possible time for each activity finishing without delaying the completion date of the project. The latest event time is calculated and written in the right-hand square alongside each event. The calculations are shown in Fig. 4.6. The latest time of the finish event, event (6), is taken as 16 weeks; the latest time of event (5) is 16 – 2 = 14; and of event (4) it is 14 – 3 = 11. Event (3) has two activities leaving it, and the latest time of event (3) is determined by the earlier or smaller calculated latest time, i.e. one calculation for event (3) from event (4) is 14 – 6 = 8 and the other is 11 – 1 = 10, therefore the latest time of event (3) is 8. If event (3) were any later than 8 the time to complete activities (3)–(5) and (5)–(6) would extend beyond the project end date of 16.

Having completed the forward and backward passes, the earliest and latest times of each event are known. From this, the 'float' or spare time available for each activity can be calculated. *Critical activities*, which have no float, are those activities whose *earliest* and *latest* times of start event coincide, whose *earliest* and *latest* times of finish event coincide and the time difference between the start event and finish event equals the duration of the activity.

Float

Figure 4.8 shows an activity extracted from the network in Fig. 4.6. The times shown refer to the event times and have the following meanings. The *Earliest time of start event* is the earliest possible time the activity can start. The *Latest time of finish event* is the latest time the activity can finish without delaying the completion of the project. The *Latest time of start event* is the latest time

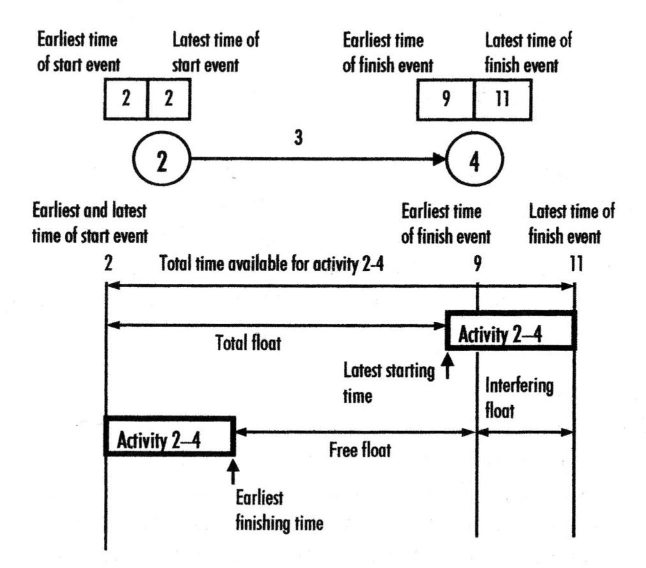

Fig. 4.8 Calculation of float.

a preceding activity may finish, and the *Earliest time of finish event* is the earliest time that a succeeding activity may start. Knowing these times, the float can be calculated; Fig. 4.8 shows the calculation of *total float* and *free float*.

Total float is the total amount of time by which the activity could be extended or delayed and still not interfere with the project end date. *Total float* is the total time available for the activity less the duration, i.e. the latest time of the finish event less the earliest time of the start event less the duration.

If the total float for one activity is completely used up by the activity then some of the total float of the succeeding activity is also used. *Free float*, however, is the amount of time by which an activity could be extended or delayed without interfering with the succeeding activity. *Free float* is calculated by the earliest time of the finish event less the earliest time of the start event less the duration. *Free float* assumes both the preceding and succeeding activities start as early as possible.

Using the examples in Fig 4.8, the total float of activity (2)–(4) is $11 - 2 - 3 = 6$, and the free float is $9 - 2 - 3 = 4$.

The difference between total float and free float is sometimes referred to as *interfering float*, as shown in Fig. 4.8. It is the amount of total float shared by the succeeding activity. *Interfering float* is rarely used in any subsequent calculations.

(4) Assessing resources

In estimating the duration of an activity, the resources required for that activity will have to be considered. The resources can be written alongside each arrow in the network. For example, an activity with a work content of 20 carpenter days requires two carpenters for 10 days. The first and most widely used assessment of resources is the aggregation chart. Figure 4.4 is an example of a resource aggregation chart for labourers. In many cases a more elaborate approach is not justified. The resource aggregation chart is useful in assessing work content for estimates and can be used in conjunction with the linked bar chart (Fig 4.4); a visual assessment of the float available within activities (shown as a broken line) allows the distribution of peaks of resource demand. In many practical situations this is sufficient resource manipulation.

Beyond the use of resource aggregation there are two approaches to assessing resources required. They are the time-limited problem, i.e. the project must be completed by a specific date, and the resource-limited problem, in which a project must be completed within the limited resources available even if this means extending the project deadline.

Time-limited resource considerations

Time analysis will provide the minimum time possible for completing the project. If the minimum is taken as the time limit, adjustments in the timing of any activity that may affect resource requirements must be undertaken within the float available.

The steps in assessing the resources required in a time-limited situation are:

(1) Prepare a list of activities ranked in order of their earliest start dates.
(2) Produce a resource aggregation chart as in Fig. 4.9, based on this list of activities in earliest start order and the resources required as shown in Fig. 4.9. This gives the resources required assuming that all activities start as early as possible.
(3) Produce a list of activities ranked in order of latest start dates. (*Note:* The latest start date of an activity is the latest time of the finish event less the duration.)
(4) Produce a resource aggregation chart as in Fig. 4.9 based on this list of activities in latest start order, and starting at these latest start times. These are the resources required when all activities start as late as possible.

(5) Compare the resource aggregation charts from steps (2) and (4). This provides the two extremes of resource requirements, all activities starting as early and as late as possible. Between these extremes a compromise to produce acceptable resource requirements can be sought by visual inspection and manipulation of activities within the two extremes.

Activities in early start order

Activity	Earliest start	Total float	Duration	Resources (Labourers)
1-2	0	0	2	2
2-3	2	0	6	4
2-4	2	6	3	3
3-5	8	0	6	3
3-4	8	2	1	1
4-5	9	2	3	3
5-6	14	0	2	2

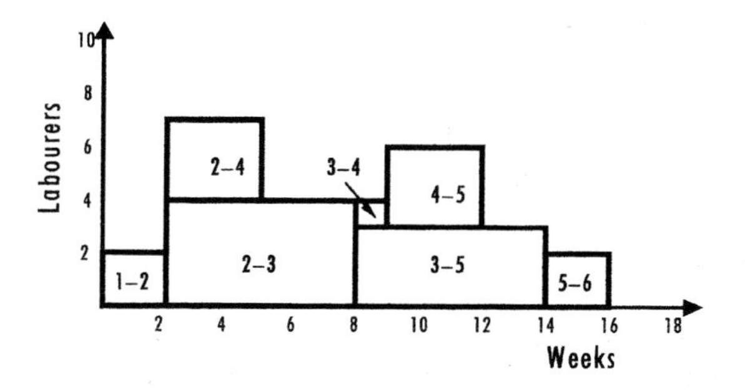

Activities in late start order

Activity	Latest start	Total float	Duration	Resources (Labourers)
1-2	0	0	2	2
2-3	2	0	6	4
2-4	8	0	6	3
3-5	8	6	3	3
3-4	10	2	1	1
4-5	11	2	3	3
5-6	14	0	2	2

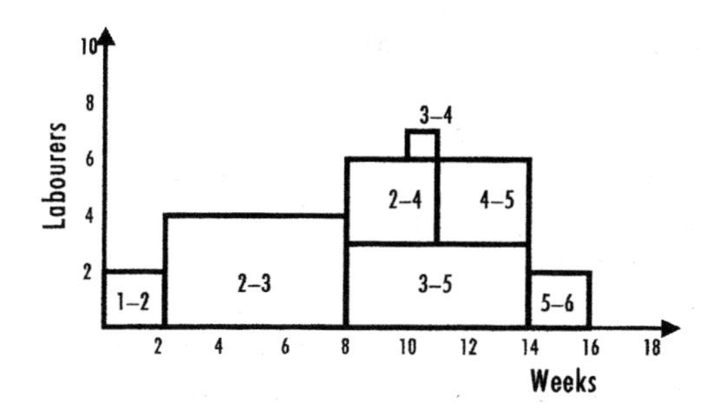

Fig. 4.9 Resource aggregation charts for activities starting as early and as late as possible (based on the network in Fig. 4.6).

Resource-limited resource considerations

The production of a resource-limited aggregation chart is similar to that of the unlimited resource aggregation, except that if the total resource demand of an activity exceeds the specified limit then that activity must be delayed. To produce reasonable results, and so that earlier activities are allocated their resources first, the activities must be arranged according to a system of priorities or 'decision rules'.

The decision rule is a device whereby activities are ranked in the order in which their resource demand is added to the resource aggregation chart. Each activity is thus given its appropriate priority in the queue for resources. It is possible that not all activities can receive the resources they require at the earliest time they require them. Consequently some activities will be delayed until the resources are available. The ordering according to a chosen priority or decision rule ensures that activities high in the priority list receive their resources first. Activities low in the priority list may get delayed.

The ranking in a priority order is known as sorting, and one of the more common sorts or decision rules is to sort in order of 'early start time'. For activities with the same early start time a second sort is required and this could be in order of total float. The upper part of Fig 4.9 gives the activities for the network shown in Fig. 4.6 ranked in order of early start as the first or major sort and in total float order as the second or minor sort. This ranking has been used in preparing the resource aggregation chart shown in Fig. 4.10. This example shows that activity (2)–(4) is the first to exceed the resource limit and is therefore delayed from starting on week 2 to starting on week 8.

When the resources for all the activities have been entered in the resource aggregation chart, a list of scheduled start dates can be extracted as in the table shown in Fig. 4.10; Fig. 4.11 gives similar examples for two types of resource. For demonstration purposes, the decision rule or priority sort used in this case is early start as the major sort and largest duration as the minor sort. The choice of decision rule or sort is left to the user of these techniques. The decision rules in common use are 'early start–total float' and 'late start–total float' and, since each type of decision rule may load the resources into the aggregation chart in a different order, the resulting resource aggregation chart will be different. The priority list guides the user towards an acceptable solution.

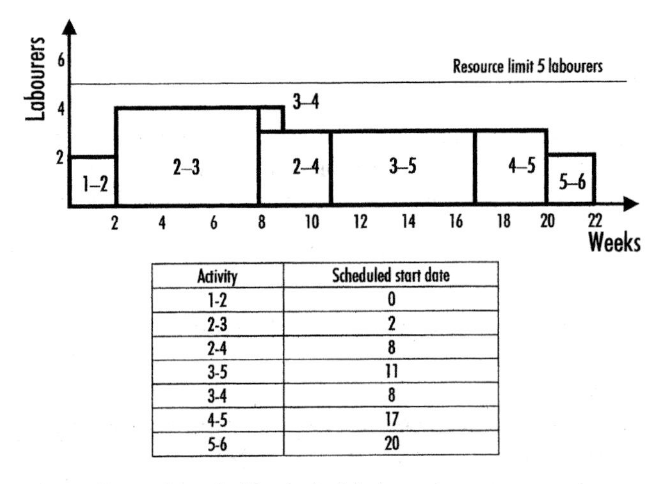

Activity	Scheduled start date
1-2	0
2-3	2
2-4	8
3-5	11
3-4	8
4-5	17
5-6	20

Fig. 4.10 Resource-limited histogram and table of scheduled start dates.

Activity	Earliest start	Duration	Resource	Resource type	Scheduled start date
1-2	0	2	2	C	0
2-3	2	6	4	L	2
2-4	2	3	3	L	8
3-5	8	6	3	L	11
3-4	8	1	1	C	8
4-5	9	3	3	C	11
5-6	14	2	2	C	17

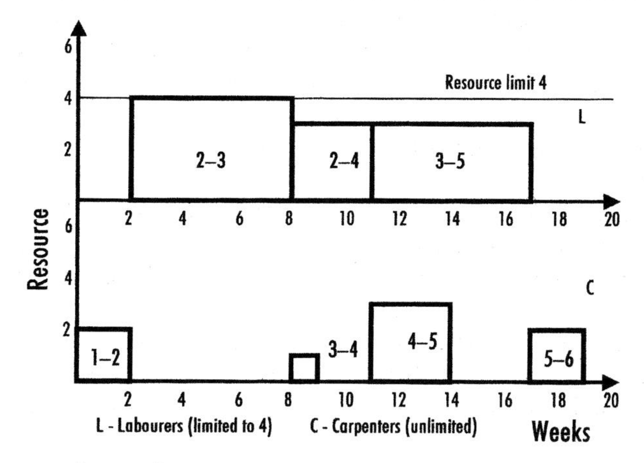

Fig. 4.11 Resource allocation for two resource types.

Activity-on-the node (precedence diagrams)

Network analysis by *activity-on-the-node* follows the same logical steps as the arrow diagrams. The differences in the application are as follows.

(1) Listing the activities

The activity list can be extended to show the dependency between activities, as shown in Fig. 4.12.

(2) Producing a logical network

Network logic

In precedence diagrams the 'node' represents the activity and the link or arrow represents only the logical relationship. Figure 4.12 shows the precedence network and the list of activities on which it is based. No dummies are needed to maintain correct logic or for unique numbering or activities.

Identifying the activities

Each node representing an activity can be given a single unique number.

Durations and time analysis

As for arrows, the duration of each activity is estimated and the forward and backward passes are done and the earliest and latest start and finish times of each activity are entered as shown in Fig. 4.12. These calculated times refer to the activity whereas the calculations for arrows refer

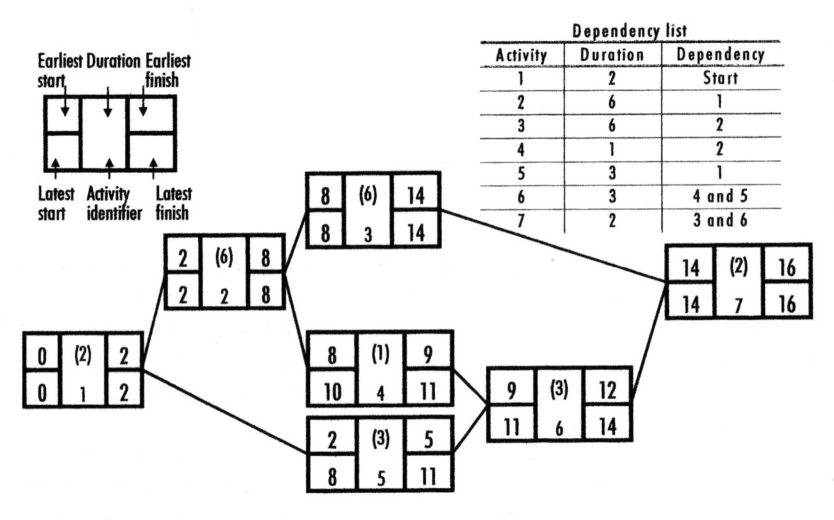

Dependency list

Activity	Duration	Dependency
1	2	Start
2	6	1
3	6	2
4	1	2
5	3	1
6	3	4 and 5
7	2	3 and 6

Fig. 4.12 Precedence list and diagram showing the same network as in Fig. 4.6.

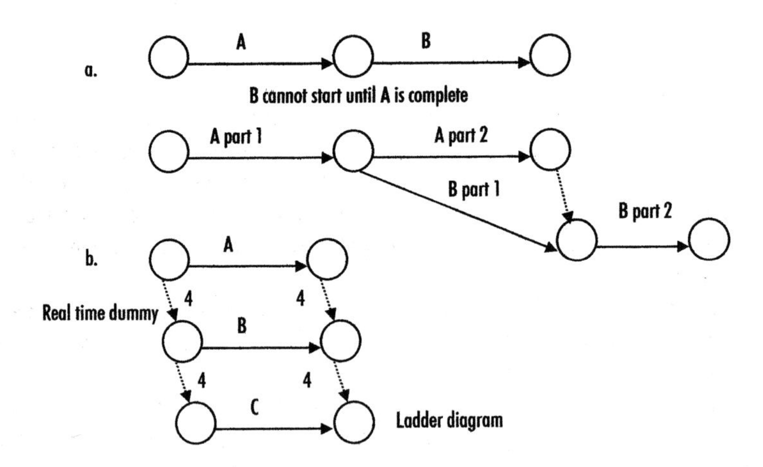

Fig. 4.13 Adjustment to arrow networks to produce overlaps between activities.

to event times. Float can be calculated as before, for example the total float of activity No. 5 is latest finishing time less earliest starting time less duration, that is $11-2-3=6$ (see Fig. 4.8).

Relationships between activities

The major advantage offered by activity-on-the-node diagrams and the available computer packages is that the number of relationships that can exist between activities is more than the simple finish–start relationship offered by arrow networks. The finish–start limitation means that if an activity is to start before the completion of a preceding activity upon which the activity is dependent then the preceding activity must be divided into smaller parts, as in Fig. 4.13(a); otherwise, dummies with a time allowance as in Fig. 4.13(b) must be used to produce overlaps between activities.

In activity-on-the-node diagrams a number of different relationships exist between activities, for example:

- Finish–start
- Finish–finish
- Start–start
- Part complete–start
- Part complete–finish
- Finish–part complete

The particular selection of relationships which exists depends on the computer package chosen.

In practice, the use of many different relationships between activities had proved in some cases to be a disadvantage. The reason is that the user finds the alternatives too complicated. Also, there is now some evidence that when extensive overlapping of activities is used in planning, the project duration tends to be underestimated. This results from a zealous use of the overlap facility in order to meet a predetermined target date.

The main difference between activity-on-the-node diagrams and the arrow equivalent are that node diagrams have no dummies, no change of reference number when additional activities are added and complex relationships. The incorporation of node diagrams in several commercial software packages has contributed to its increased use by planning engineers in construction.

Line of balance

Line of balance is a planning technique for repetitive work; the principles employed are taken from the planning and control of manufacturing processes. The basis of the technique is to find the required resources for each stage or operation so that the following stages are not interfered with and the target output can be achieved. The technique has been applied in construction work mainly to house building and to a lesser extent to jetty work and, in conjunction with networks, to road construction.

Consider a simplified example of the construction of a jetty, which is represented by three operations—say, Drive piles, Construction pile cap, Fix deck—and that this sequence had to be repeated ten times to finish the jetty.

The interrelationship of the three basic operations can be shown in a simple logic diagram, as in Fig. 4.14(a). To provide for a margin of error in the time taken to complete each of these operations, a time buffer is usually placed between each operation, as in Fig. 4.14(b). This construction plan shows that a total time of 50 days is needed to complete one sequence of operations. The target output of the project can be expressed in terms of the completion rate of these sequences. For example, it may be that the jetty comprising 10 sections has to be completed in 100 days. Thus if a target rate of completion of one section per five-day week was chosen, the total time taken for 10 sections would be 95 days, as in Fig. 4.15. The logic diagram or construction plan for section 1 in Fig. 4.14 can be added to Fig. 4.15, as in Fig. 4.16 for the other sections. The sloping lines added to the start and finish for each of the operations allows dates to be established for each section. This is the line of balance schedule if sufficient resources are made available to maintain the production represented in the schedule.

It must be made clear at this point that this is *not* the method by which line of balance schedules are constructed. The schedule shown in Fig. 4.16 is only possible if the required number of teams of operatives are available to start at the scheduled start dates for each section. The method of construction of a proper line of balance schedule is described in the next section. The basic approach is to determine the resources available and calculate the rates of construction that can be achieved. This example is produced only for demonstration purposes.

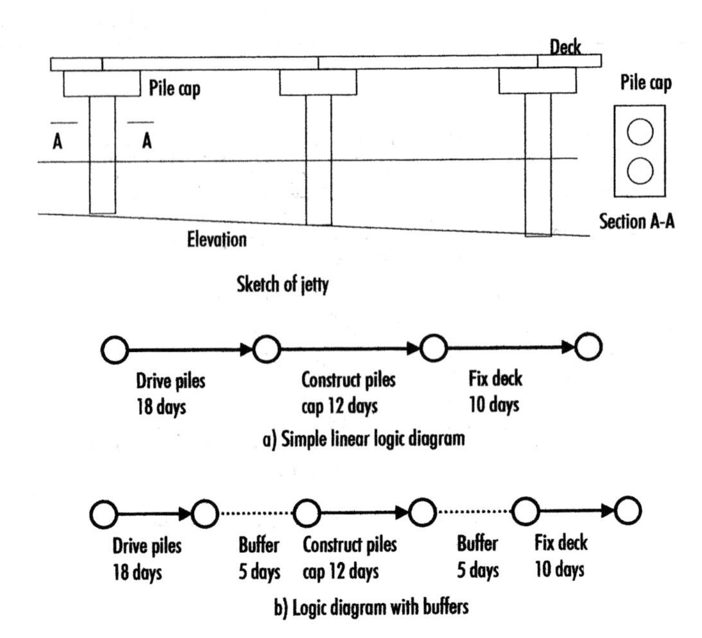

Fig. 4.14 Logic diagram.

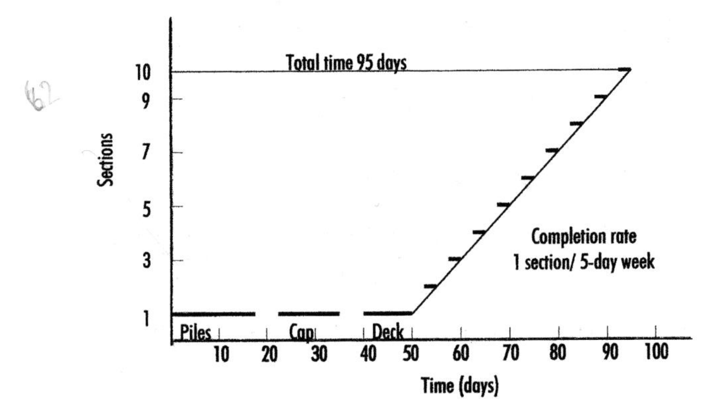

Fig. 4.15 Time for completion of ten repetitive sequences at the rate of one per week.

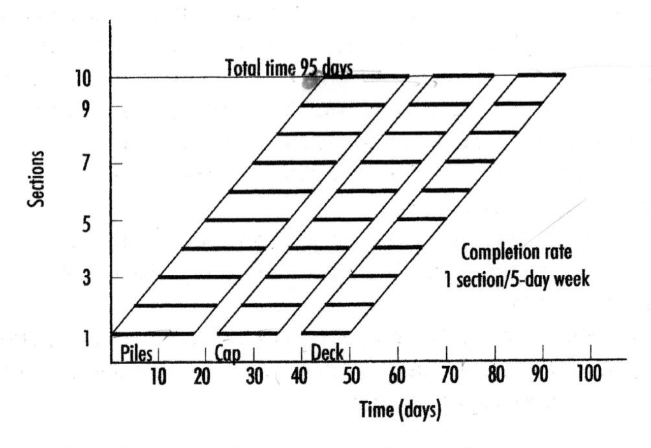

Fig. 4.16 Logic diagram for each section added to the completion schedule.

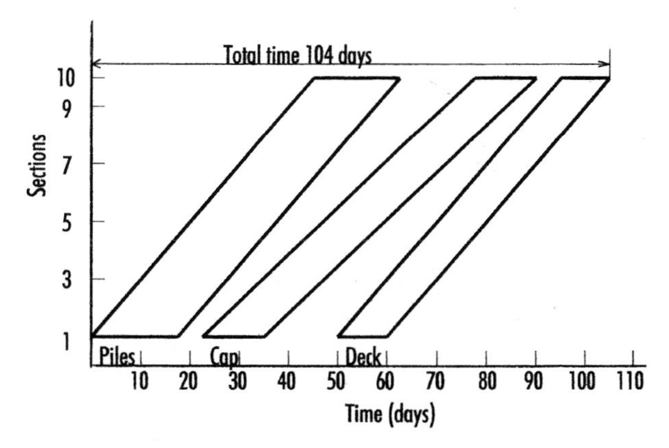

Fig. 4.17 Effect of scheduling operation 'Pile cap' at a completion rate of 0.83 per week.

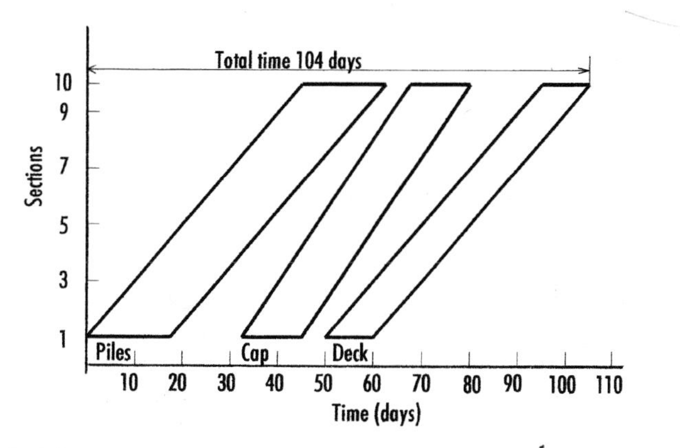

Fig. 4.18 Effect of scheduling operation 'Pile cap' at a completion rate of 1.25 per week.

If, for example, operation 'Pile cap' required a team of six operatives to construct the pile cap in 12 days and two teams of six operatives were employed, the output of the gang of two teams would be 0.83 sections per five-day week and not the one section per five-day week required by the target. This is because, if two teams were employed, team *a* would start on section 1 on day 23 and finish on day 35 (from schedule Fig. 4.16) and team *b* would start on section 2 on day 28 and finish on day 40 (from schedule Fig. 4.16). Section 3, according to the schedule, is due to start on day 33, but team *a* is not available until day 35. Therefore the rate of completion of the operation 'Pile cap' would be less than 1.0 section per five-day week; calculations show this rate to be 0.83 and Fig. 4.17 shows the delay to the project.

If three teams were employed on operation 'Pile cap' the output would be 1.25 sections per week. This is because team *a* completes the first section on day 35 and moves to section 4, which is not due to start until day 38.

Figure 4.18 shows that the project completion date is in fact later than the original, despite the speeding up of operation 'Pile cap'. This is because operation 'Pile cap' is no longer in 'balance' with its preceding and succeeding operations.

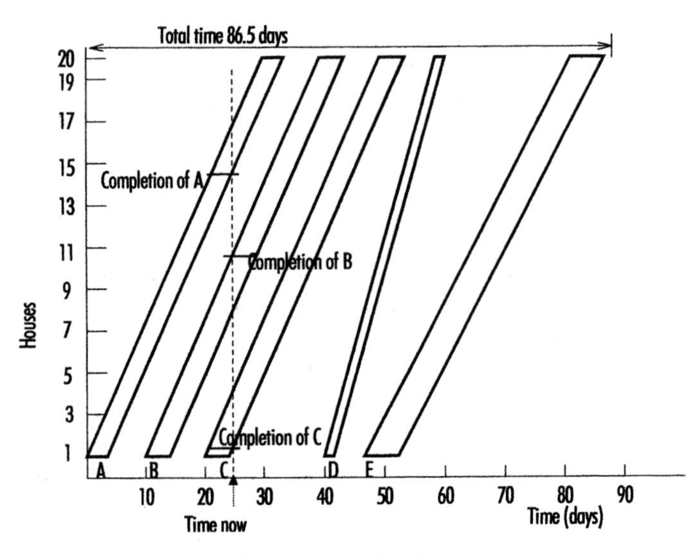

Fig. 4.19 Line of balance schedule drawn from data in Table 4.1.

Preparing a line of balance schedule

(1) Prepare a logic diagram as in Fig. 4.14.
(2) Estimate the labour-hours required to complete each operation.
(3) Choose buffer times that will guard against the risk of interference between operations.
(4) Calculate the required output target in order to meet a given project completion date.
(5) Complete the table as shown in Table 4.1.
(6) Draw the schedule, from the information calculated in Table 4.1. The example is as shown in Fig. 4.19.
(7) Examine the schedule and assess possible alternatives to bring about a more 'balanced' schedule which might include:
   (a) Changing the rate of output of one activity by reducing (or increasing) the gang size partway through the project; an example is given in Fig. 4.20.
   (b) Lay-off and recall one gang; an example is given in Fig. 4.21.
   (c) Overlap some activities, i.e. have a non-sequential logic diagram, which means that the schedules shown as sloping lines on the diagrams would be superimposed. Such schedules become difficult to read and this severely reduces the effectiveness as a means of communication with operational staff such as foremen and gangers.
   (d) Schedule every activity to work at the same rate. This is known as 'parallel scheduling' and involves employing enough resources to ensure that the rate of output required can be achieved. In the example in Table 4.1, if activities C and E had actual gang sizes one team more than shown, then all activities could proceed at a rate of 3.0 per week. However, this means that most gangs of operatives are working at less than their natural rate. The parallel schedule would have a shorter overall duration. Before employing parallel scheduling, examination would be needed as to whether the savings in overall duration and hence indirect overheads offset the extra costs in overmanning.

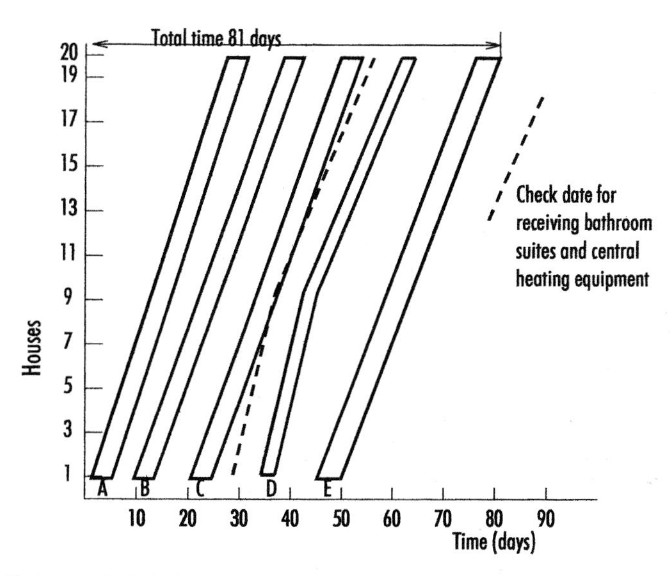

Fig. 4.20 Line of balance schedule drawn from data in Table 4.1.

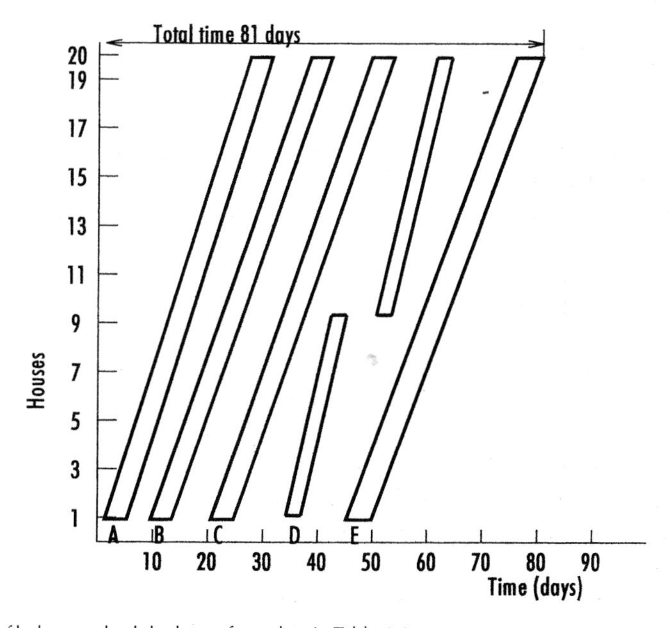

Fig. 4.21 Line of balance schedule drawn from data in Table 4.1.

Using the schedule

The completed schedule is based on chosen resources and therefore the rates of construction calculated have taken account of these resources. This is different from the network calculations, in which logic and resource allocations are kept separate.

Table 4.1 The calculations for a line of balance schedule.

	1 Activity or operation	2 Labour-hours per activity	3 Operatives per activity	4 Theoretical gang size at the chosen output of rate of *R*	5 Actual gang size	6 Actual rate of output	7 Time in days for one activity	8 Time from start on first section to start on last section	9 Minimum buffer time
A	Substructure	110	3	8.25	9	3.27	4.58	29.05	5
B	Superstructure	320	8	24.00	24	3.00	5.00	31.67	5
C	Joinery	365	9	27.38	27	2.96	5.07	32.09	5
D	Plumbing	35	2	2.63	4	4.56	2.19	20.83	5
E	Finishes	210	5	15.75	15	2.86	5.25	33.22	5

Note: Operations are sequential. Total no. of houses = 20. Target rate of build (or completions) $R = 3$ per week (1 week comprises five 8-hour days).

The columns in the table are:

1 Activity description and identification
2 The estimated labour-hours for each activity.
3 The optimum number of operatives for each activity, this is the number of operatives in one team.
4 The theoretical gang size required to maintain the output rate R given by $R \times (\text{Hours per activity})/(\text{Number of working hours per week})$. R would be the number of sections to be completed each week.
5 The actual gang size would be chosen as a number, which would be a multiple of the operatives required for one team (column 3) near to the theoretical gang size. If the actual gang size is greater than the theoretical gang size, the rate of output will be more than the target rate and, if the actual is less than the theoretical, the actual rate of output will be less than the target rate.
6 The actual rate of output is given by $\left(\dfrac{\text{Actual gang size}}{\text{Theoretical gang size}}\right) \times \text{target rate}$.
7 Time taken for one activity in days is given by $\dfrac{\text{Operative-hours for activity}}{\text{No. of operatives in one team} \times \text{no. of hours in a working day}}$.
8 The time in days from start on first section to start on last section is given by $\dfrac{(\text{No. of sections} - 1) \times \text{no. of working days per week}}{\text{Actual rate of build}}$. This is useful in plotting the schedule.
9 Minimum buffer time is assessed from knowledge of the likely variability of the preceding activity.

The schedule and the start and finish dates of the various teams on each operation can be used to monitor progress. Figure 4.19 shows time now as day 25 and the completion of operations A and B marked on their respective lines of balance. The situation portrayed is operation A on schedule and operation B ahead of schedule. The danger from this is that B may run out of work because it is progressing at a faster rate than A. Action is needed either to speed up A or slow down B.

Updating a line of balance schedule once the project has started, particularly if the rates of construction prove to be different from those calculated, can be difficult and quickly becomes unclear.

Another use for the line of balance schedule is checking material orders or deliveries. If a deadline for ordering and/or receiving materials is marked on the schedule, as shown in Fig. 4.20 for plumbing supplies, then a simple check can be made in good time.

Other planning techniques

Programme Evaluation and Review Technique—PERT

The growth in the use of PERT is due in the main to the availability of computing technology. This has made the computations involved in planning with PERT, which until now had been rather cumbersome, much easier to handle. Activity-on-the-arrow and activity-on-the-node assume that information employed in programming is reasonably accurate. In practice, time and cost overruns are not uncommon in construction due in part to the uncertainty of information employed in planning. PERT addresses such uncertainty by introducing three values for an activity's duration to replace the single times adopted for arrow diagrams. These are:

- Optimistic duration (d_o), minimum duration if everything goes well
- Most likely duration (d_m) based on analysis of work from previous projects and the planner's experience and judgement
- Pessimistic duration (d_p), maximum time if everything goes wrong

The technique is based on probability and assumes that the probability of exceeding d_o is about 99% and of exceeding d_p about 1%. The following defined key activity parameters are used for analysis in PERT

Expected duration	d_e	$= (d_o + 4d_m + d_p)/6$
Standard deviation	σd_e	$= (d_p - d_o)/6$
Variance	vd_e	$= [(d_p - d_o)/6]^2$, i.e. $[\sigma d_e]^2$

The following example shows the application of PERT to a small construction project and illustrates the computation of overall project duration and identification of critical activities. The example also illustrates the other uses of PERT such as assessing the risk of timely completion for activities and the overall project in order to facilitate project schedule control.

Table 4.2 outlines the *optimistic*, *pessimistic* and *most likely* time estimates provided by the planning engineer for the project. It presents the calculated *expected duration* and *variance* for each activity. The expected duration for PERT reflects the activity times for simple activity-on-the-arrow and activity-on-the-node techniques. The presentation of a PERT diagram can be as either an arrow or a node diagram, with the additional information on activity variances. For

Table 4.2 Expected duration (weeks) and variances for activities.

Activity	Preceding activity	Optimistic duration (d_o)	Most likely duration (d_m)	Pessimistic duration (d_p)	Expected duration (d_e)	Variance of duration (vd_e)
A	–	1	2	3	2	1/9
B	A	4	5	12	6	16/9
C	B	5	6	7	6	1/9
D	B	0.5	1	1.5	1	1/36
E	A	1	2	9	3	16/9
F	D, E	1.5	3	4.5	3	1/4
G	C, F	2	2	2	2	0

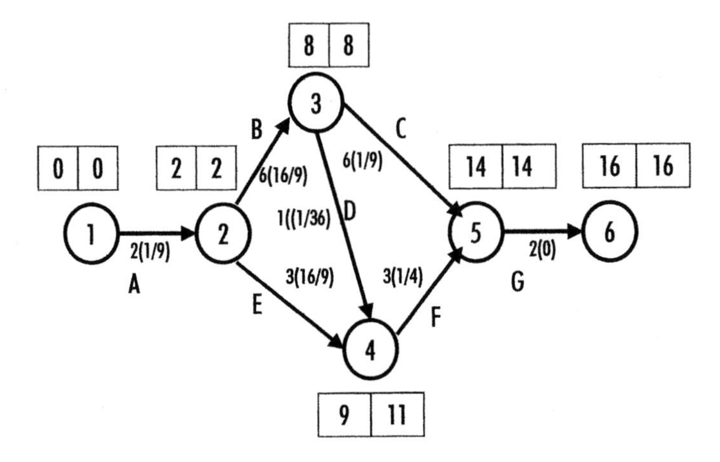

Fig. 4.22 A PERT diagram showing duration, event numbers and event times.

convenience of relating the analysis of the PERT programme to the concepts outlined earlier in this chapter, (forward and backward pass, earliest and latest times) the PERT diagram for our project is presented here as an arrow diagram.

Project duration and critical path

Figure 4.22 shows the PERT programme presented in arrow diagram format. To establish the project duration and critical activities, the activity event times with the expected durations are computed by following the same procedure as in activity-on-the-arrow. Commencing with activity A, a forward and backward pass is run with the expected activity durations. This yields an overall project duration of 16 weeks and activities A–B–C–G as the critical path. The calculation of float times follows the same approach as in the general arrow diagram.

Uncertainty of project completion date

In Fig. 4.22, although the planned project duration is 16 weeks, in practice the completion date for this project will be negotiated with an extra time to cater for any contingencies that might arise during the execution stage. It is essential that there is a clear appreciation of the potential to deliver the project within the negotiated period. This is a measure of the risk associated with the programmed project schedule, and provides useful information for the Contracts Manager, who has to negotiate a delivery date for the project. This is illustrated by using the project in Fig. 4.22 as an example. The Contracts Manager has negotiated an acceptable delivery date of 18 weeks after its start time; then the overall project will have a two-week float to cater for contingencies. An estimate of the risk associated with delivering the project within the negotiated 18 weeks can be assessed utilising the additional information of variances that is provided by PERT.

If it is assumed that the activities of this project are statistically independent of each other, then the probability that it will be completed within the 18 weeks can be established. In order to use the areas under a normal distribution curve to do this, the deviation of the target duration (TD) from the expected project duration (EPD) in units of standard deviation must be calculated. The standard deviation (and variance) represents the degree of spread associated with any distribution. The target duration (TD) is the desired or agreed project completion time. The expected project duration, which is the mean duration for completing the project, is the critical time of the project. This is defined as the sum of expected activity durations (d_e) for those activities on the critical path. The overall variance of the project (v) is the variance of the critical path (the sum of the variances vd_e for all activities on the critical path).

The expected duration of the whole project can be calculated as: EPD $= \sum d_e$ for the critical path, and the variance of the overall project as $v = \sum vd_e$ of the critical activities.

Then Z, the number of standard deviations of a normal distribution, is given by the equation:

$$Z = (\text{target duration} - \text{expected project duration})/\text{standard deviation}$$
$$Z = (\text{TD} - \text{EPD})/(\sqrt{v})$$

The variance (v) of the overall duration for the activities on the critical path (A–B–C–G) is:

$$1/9 + 16/9 + 1/9 + 0 = 18/9 = 2$$

For an 18-week target contract period (TD) based on the programmed expected project duration (EPD) of 16 weeks with a variance (v) of 2 weeks, this yields:

$$Z = (\text{TD} - \text{EPD})/(\sqrt{v})$$
$$Z = (18 - 16)/(\sqrt{2}) = \sqrt{2} = 1.414 \text{ standard deviations.}$$

The equivalent probability can be read off a normal probability distribution table such as Table 4.3:

$$P(Z = 1.414) = 0.92, \text{ which equals } 92\%.$$

This implies that there is a 92% chance that the project will be delivered within the negotiated 18-week period. The higher the percentage, the lower the schedule risk associated with the planned project, and vice versa.

For most practical situations, a working probability of 0.95 or a 95% chance is considered a reasonable level of risk. By adopting this as the level of risk for this project, we can compute the target project duration to be negotiated based on the planned programme. Reading off the corresponding Z value for 95% from the normal probability table gives:

Table 4.3 Normal probability distribution tables. Cumulative probabilities of the normal probability distribution (areas under the normal curve from $-\infty$ to Z).

Z	0	1	2	3	4	5	6	7	8	9	1	2	3	4	5	6	7	8	9
0.0	.5000	.5040	.5080	.5120	.5160	.5199	.5239	.5279	.5319	.5359	4	8	12	16	20	24	28	32	36
0.1	.5398	.5438	.5478	.5517	.5557	.5596	.5636	.5675	.5714	.5753	4	8	12	16	20	24	28	32	35
0.2	.5793	.5832	.5871	.5910	.5948	.5987	.6026	.6064	.6103	.6141	4	8	12	15	19	23	27	31	35
0.3	.6179	.6217	.6255	.6293	.6331	.6368	.6406	.6443	.6480	.6517	4	8	11	15	19	23	26	30	34
0.4	.6554	.6591	.6628	.6664	.6700	.6736	.6772	.6808	.6844	.6879	4	7	11	14	18	22	25	29	32
0.5	.6915	.6950	.6985	.7019	.7054	.7088	.7123	.7157	.7190	.7224	3	7	10	14	17	21	24	27	31
0.6	.7257	.7291	.7324	.7357	.7389	.7422	.7454	.7486	.7517	.7549	3	6	10	13	16	19	23	26	29
0.7	.7580	.7611	.7642	.7673	.7704	.7734	.7764	.7793	.7823	.7852	3	6	9	12	15	18	21	24	27
0.8	.7881	.7910	.7939	.7967	.7995	.8023	.8051	.8078	.8106	.8133	3	6	8	11	14	17	19	22	25
0.9	.8159	.8186	.8212	.8238	.8264	.8289	.8315	.8340	.8365	.8389	3	5	8	10	13	15	18	20	23
1.0	.8413	.8438	.8461	.8485	.8508	.8531	.8554	.8577	.8599	.8621	2	5	7	9	12	14	16	18	21
1.1	.8643	.8665	.8686	.8708	.8729	.8749	.8770	.8790	.8810	.8830	2	4	6	8	10	12	14	16	19
1.2	.8849	.8869	.8888	.8907	.8925	.8944	.8962	.8980	.8997	.9015	2	4	6	7	9	11	13	15	16
1.3	.9032	.9049	.9066	.9082	.9099	.9115	.9131	.9147	.9162	.9177	2	3	5	6	8	10	11	13	14
1.4	.9192	.9207	.9222	.9236	.9251	.9265	.9279	.9292	.9306	.9319	1	3	4	6	7	8	10	11	13
1.5	.9332	.9345	.9357	.9370	.9382	.9394	.9406	.9418	.9429	.9441	1	2	4	5	6	7	8	10	11
1.6	.9452	.9463	.9474	.9484	.9495	.9505	.9515	.9525	.9535	.9545	1	2	3	4	5	6	7	8	9
1.7	.9554	.9564	.9573	.9582	.9591	.9599	.9608	.9616	.9625	.9633	1	2	3	3	4	5	6	7	8
1.8	.9641	.9649	.9656	.9664	.9671	.9678	.9686	.9693	.9699	.9706	1	1	2	3	4	4	5	6	6
1.9	.9713	.9719	.9726	.9732	.9738	.9744	.9750	.9756	.9761	.9767	1	1	2	2	3	4	4	5	5
2.0	.9772	.9778	.9783	.9788	.9793	.9798	.9803	.9808	.9812	.9817	0	1	1	2	2	3	3	4	4
2.1	.9821	.9826	.9830	.9834	.9838	.9842	.9846	.9850	.9854	.9857	0	1	1	2	2	2	3	3	4
2.2	.9861	.9864	.9868	.9871	.9875	.9878	.9881	.9884	.9887	.9890	0	1	1	1	2	2	2	3	3
2.3	.9893	.9896	.9898	.9901	.9904	.9906	.9909	.9911	.9913	.9916	0	1	1	1	1	2	2	2	2
2.4	.9918	.9920	.9922	.9925	.9927	.9929	.9931	.9932	.9934	.9936	0	0	1	1	1	1	1	2	2
2.5	.9938	.9940	.9941	.9943	.9945	.9946	.9948	.9949	.9951	.9952	0	0	0	1	1	1	1	1	1
2.6	.9953	.9955	.9956	.9957	.9959	.9960	.9961	.9962	.9963	.9964		0	0	0	1	1	1	1	1
2.7	.9965	.9966	.9967	.9968	.9969	.9970	.9971	.9972	.9973	.9974	0	0	0	0	0	1	1	1	1
2.8	.9974	.9975	.9976	.9977	.9977	.9978	.9979	.9979	.9980	.9981	0	0	0	0	0	0	0	1	1
2.9	.9981	.9982	.9982	.9983	.9984	.9984	.9985	.9985	.9986	.9986	0	0	0	0	0	0	0	0	0
3.0	.9986	.9987	.9987	.9988	.9988	.9989	.9989	.9989	.9990	.9990	0	0	0	0	0	0	0	0	0
3.1	.9990	.9991	.9991	.9991	.9992	.9992	.9992	.9992	.9993	.9993	0	0	0	0	0	0	0	0	0
3.2	.9993	.9993	.9994	.9994	.9994	.9994	.9994	.9995	.9995	.9995	0	0	0	0	0	0	0	0	0
3.3	.9995	.9995	.9996	.9996	.9996	.9996	.9996	.9996	.9996	.9997	0	0	0	0	0	0	0	0	0
3.4	.9997	.9997	.9997	.9997	.9997	.9997	.9997	.9997	.9997	.9998	0	0	0	0	0	0	0	0	0
3.5	.9998	.9998	.9998	.9998	.9998	.9998	.9998	.9998	.9998	.9998	0	0	0	0	0	0	0	0	0
3.6	.9998	.9998	.9999	.9999	.9999	.9999	.9999	.9999	.9999	.9999	0	0	0	0	0	0	0	0	0
3.7	.9999	.9999	.9999	.9999	.9999	.9999	.9999	.9999	.9999	.9999	0	0	0	0	0	0	0	0	0
3.8	.9999	.9999	.9999	.9999	.9999	.9999	.9999	.9999	.9999	1.000	0	0	0	0	0	0	0	0	0
3.9	1.000	1.000	1.000	1.000	1.000	1.000	1.000	1.000	1.000	1.000	0	0	0	0	0	0	0	0	0

$$Z(P=0.95)=1.65$$

The contract period that will give the project a 95% chance of delivery on time is calculated as follows:

$$\mathrm{TD}_{95}=\mathrm{EPD}+Z(\sqrt{v})$$

where TD_{95} is the target duration to be negotiated for the project to have a 95% chance of completion on time, EPD is the expected project duration based on the critical path and v is the overall project variance.

This gives

$$\mathrm{TD}_{95}=16+1.65\times 1.414=18.33 \text{ weeks or } 18 \text{ weeks } 2 \text{ days.}$$

A similar analysis can be undertaken for individual activities by replacing the target duration (TD) and expected project duration (EPD) with the earliest and latest times of the activity, as shown below. The target duration for the activity is the overall period within which the activity can take place without disrupting the project delivery time. Then:

Period within which activity can take place = Latest Finish Time − Earliest Start Time
Expected activity duration = d_e
Activity variance = vd_e

All activities on the critical path will have a Z value of zero, equivalent to a probability (P value) of 0.50. This indicates that all the activities on the critical path have a 50% chance of slipping out of schedule, and presents a higher risk to the overall project schedule than say activity F, for which there is a lesser schedule risk. By substituting the values of any critical activity with that for activity F, the equivalent probability of completing activity F on time can be obtained as follows:

Period within which activity can take place $\mathrm{F(TD)}=11-2=9$;
Expected activity duration $\mathrm{F}(d_e)=3$;
Activity variance $\mathrm{F}(vd_e)=1/4$; and
Number of standard deviation units associated with activity F is Z_{F}
$Z_{\mathrm{F}}=(9-3)/\sqrt{1/4}=6\times 2=12$ standard deviations

Twelve standard deviations are equivalent to a probability (P value) of 0.9999. This implies that activity F has approximately a 100% chance of being delivered on time.

The introduction of multiple times associated with each activity presents a means of allowing for the uncertainty associated with project delivery. Thus, the main advantage of the use of PERT is in establishing the schedule risk associated with the planned programme. This element is missing from both the simple arrow and node diagrams.

Space–time diagrams

Space–time diagrams for planning construction works were generally used before the advent of computer-based planning packages, which became the norm in industry. They were normally two-dimensional graphs that represented time (elapsed project time) on one axis and a measure for production (chainage or distance, area, volume of works) on the other axis. The resulting plot then presents a pictorial view of how the works should progress on site. Used as a planning tool, space–time diagrams allow the identification of potential constraints in the utilisation of resources, such as plant for excavation, which can then be reorganised. They also serve as a realistic tool for communicating information on the programme of works for a project. Space–time diagrams lend themselves to projects that are essentially linear in nature, for example road-works, pipe laying and overhead pylon construction, where there is considerable repetition. However, they can be applied to projects of a non-linear character where there is considerable repetition of a work package. The main disadvantage associated with space–time diagrams is the difficulty in effecting alterations for control and updating purposes. The use of the space–time diagram as a planning tool is illustrated by the example below.

A new road is required to link a new housing development to the main town, which is 4 km away. The project also includes the construction of a 75 m bridge and two culverts. The symbols that are used in representing a project in a space–time chart are lines and boxes. A line and a box represent respectively a unit or multiples of duration and chainage. The width of a vertical box indicates the chainage that the activity covers and its height represents the time it takes to complete the whole activity. For inclined boxes, the horizontal length and vertical height across the incline are taken to establish the chainage and time it takes to complete a unit of the activity. Figure 4.23 presents a plot of the project in a space–time diagram developed by the planning engineer taking into consideration the associated cut and fill. The plotting of the space–time diagram against the real time calendar enables the identification of potential stops in the project.

Last Planner

The *Last Planner* is a technique that was developed for the management of the site production planning phase of construction projects. The technique is based on the premise that a *planner* should create reliable weekly work plans in order to derive maximum project benefits. Such

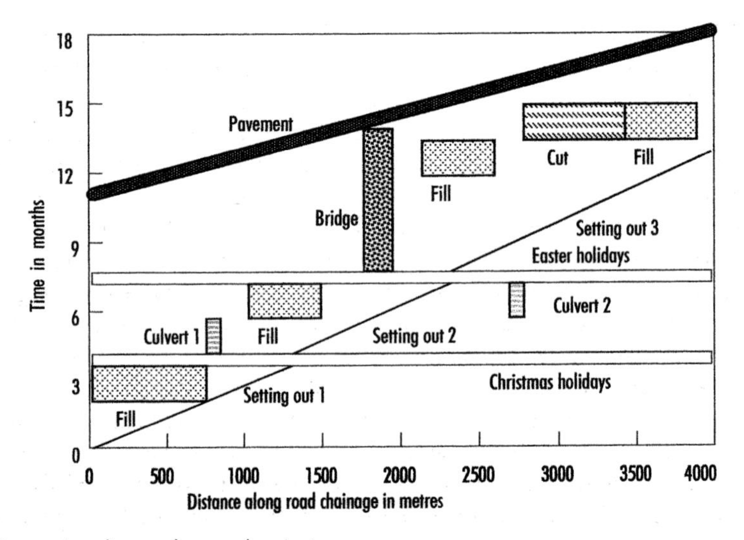

Fig. 4.23 Space–time diagram for a road project.

plans should include only work for which it has previously been determined that it can achieved within the period under consideration.

The need for the Last Planner was influenced by the apparent 'loss of knowledge' of *short-term* and weekly planning for construction sites with the growth in project planning software packages. The traditional short-term planning enabled the programming of site activities in more detail and especially at the operative level. As part of the arguments for the development of the Last Planner system, Ballard (2000) argued that planning of construction just involved scheduling key activities and no resource levelling. The use of *Method Statements* for short-term planning on site provides an opportunity to exercise the same level of rigour and detail advocated by the Last Planner. The extra benefit offered by the Last Planner is the promotion of evidence-based allocation of operative workloads, to minimise the potential for unsuccessful completion of assigned tasks. It also provides a formal devolution of the planning effort to the lowest possible hierarchy within the organisation structure of the project site. This is based on a system whereby the 'Last Planner' is expected to provide firm pledges on what can realistically be delivered for any task under consideration. The *Last Planner* is described as any individual or group that makes decisions on, and assigns, the physical and specific work to be achieved on site on a day-to-day basis.

By recognising the distinction between what *can* be done and what *should* be done, planning at the detailed level on site is able to clearly establish what *will* be achieved on site. By aligning what can be done, any resource bottlenecks and potential delays can be identified and mitigated. According to the Lean Construction Institute, the Last Planner approach can provide project staff with forward information for control because it forces problems to the surface at the planning stage, and it makes it possible to measure planning system performance.

The adoption of the Last Planner system could foster a change in the role of the project planner because a significant proportion of the tasks the project planner performs are devolved to the site.

Modern construction planning

The planning of construction projects without computers is now very rare. The key attraction to construction and project managers is the speed of processing and the manipulative power available. The arrival of microcomputers and interactive systems resolved most of these difficulties and the use of network planning based on a computer system is now much more credible. The introduction of desktop computers for the construction managers and the site planner and, subsequently, laptops that facilitate mobile computing, has generated a new working relationship between the plans developed in the office and the construction site. It is the immediacy of the results that has convinced managers that computer-aided planning is an indispensable aid.

The use of these software planning systems ranges from the simple programming of activities up to a complete time, resource and cost modelling system used in analysis of 'what-if' situations and in making forecasts. The various systems comprise use of spreadsheets (which can be structured to produce a bar chart), off-the-shelf specific software that allows activity scheduling, to a complete information system for the management of the project, such as Primavera. The simpler specific software is normally organised as separate but compatible packages that allow data transfer between packages, with each package focusing on only one aspect of planning. At the other extreme, programming software comes as one integrated package that is able to undertake the whole range of project planning requirements. The choice between an integrated package

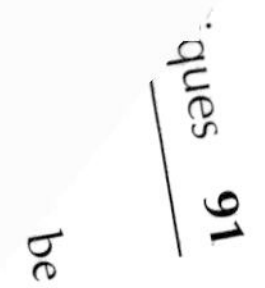

will depend on project complexity and size as well as organisational pref- planning system becomes more complex the data required increases and quired becomes more sophisticated. The following is a review of these

(1) Planning by preparing a schedule of activities
(2) Adding resources to the basic plan and preparing a resource plan
(3) Monitoring progress in terms of time and resources and updating the plan
(4) Costing the resources to produce estimates, including revenues on each activity—producing cost and revenue forecasts and cash flows
(5) Experimenting with the activity schedule and resource allocation to determine likely outcomes to assumed scenarios in terms of time, resources and costs

(1) Preparing a schedule of activities

All systems require project data that describe the overall parameters of the project before accepting data at individual activity level. These project data relate to 'time units', 'start dates', 'days in weeks', 'holidays', etc.

The data required to prepare a schedule of activities will normally comprise for each activity:

- Activity number
- Description
- Management reporting codes
- Duration
- Logic links to other activities

Later, when resources are introduced, this activity data will be extended. Some computer systems will display on the screen the network being constructed. This is helpful in checking the logic of the network.

From this input the most important report that will be available will be a *bar chart*. This will be supported by reports showing the start and end dates of all activities and the float, if required.

The generation of a time-based logic diagram will achieve a basic planning model and will have developed the planner's understanding of the project and the interrelationships between the major activities.

(2) Resources

Allocation of the right amount of resources to the basic activities is essential for the timely execution of the whole project. Most software packages have a facility for creating a list of resources. The list of resources allows the planner to enter all of the resource information for the project in a database, and it can then be assigned to activities. Depending on the sophistication of the system available, the resource requirements of each activity will be identified to cover:

- Individuals or groups or resource types for the duration of the activity
- Resources shared by other activities

The resources will be identified by:

- Resource numbers
- Descriptions
- Management reporting codes

The availability of the resources will be described to the computer system in terms of total availability or restricted data availability.

From this additional data the computer system will calculate the consequences of the resource demands. The report will show the resource requirements, for each resource and for each time period, as a table, histogram or cumulative graph.

This resource report will be available for:

- All activities starting as early as possible
- All activities starting as late as possible
- All activities schedules to suit resource restraints
- Others, depending on the sophistication of the computer system

These reports allow the planner to assess, in terms of resources used, the implications of the basic plan. The logic may be reworked if the resource demands are unrealistic. In practice, the most commonly used form of computer-based planning is to complete the logic diagram and to assess the resource requirements under the assumption that all activities start as early as possible. The reports, produced as *bar charts* and *resource histograms*, are used to take strategic planning decisions. The number of resources analysed in this way is, in practice, usually quite limited and well below the capacity of existing computer systems.

(3) Monitoring and control

Equipped with the basic logic model and the resource requirements of each activity, the planner possesses the tools to monitor the progress of the project. As the project proceeds, many occurrences may not be as planned. Thus the computer system needs to allow updating for each activity by:

- Changing start date
- Changing duration
- Changing end date
- Changing resource requirements

The consequences of these changes can be processed and updated and reports produced. The effect on identified target dates set in the basic plan can be monitored.

The act of checking actual progress and actual resource usage against planned progress and planned resource usage is the act of *monitoring*. The act of taking decisions to alter the likely future outcome and bring the project back on the planned schedule is *control*.

The decisions relating to the rescheduling of activities, the reordering of activities, the altering of resources to change the duration of activities is *control*. Thus the planner must collect the information on the rate of progress to date and current resource usage in order to update the computer model and *monitor* progress. This information can be presented to the construction managers together with suggestions for rescheduling. The rescheduling decisions taken by the *managers* may need processing before the *control* decisions are implemented.

(4) Costs and revenues

Some computer systems allow the cost and revenue to be included. For example, an activity has the following details:

Description: Formwork erection
Duration: 4 weeks
Carpenters: 3
Labourers: 1
Crane: 50%

The costs of resources are:

Carpenters: £345.00 per week
Labourer: £287.50 per week
Crane: £230.00 per week

Thus the cost of the activity is:

			£
Carpenters	£345 × 3 × 4	=	4140
Labourer	£287.50 × 1 × 4	=	1150
Crane	£230 × 0.5 × 4	=	460
	Total		**5750**

All this data entered separately and held on file provides the basis for calculating costs. It also provides a means for exploring the consequences, in cost terms, of any alterations being considered to re-plan the project. For example, the number of carpenters may be increased and the overall duration of this activity reduced; this on its own would be simple enough to calculate but the consequences on the rest of the project would be more tedious. This is where these computer systems show their advantages.

Overheads or on-cost can be held in 'hammocks' that span a number of activities and derive their duration from the interaction of the activities they span. Thus, the manipulation of activity costs can be extended into a trade-off calculation between direct and indirect costs. If provision is also made for the activities to hold details on revenue then income graphs can be produced for the project, for example:

Description: Formwork erection
Duration: 4 weeks
Quantity: 800 m^2
Revenue: £10.40/m^2

With both cost and revenue details it is possible, with additional data relative to payment conditions and credit delays, to calculate the project's cash flow. Thus, with the inclusion of cost and revenues, the simple planning tool has been extended to be a basic *estimating* and *cash flow model*. Thus all the key elements to managing a project are present:

Time – resource – costs – revenues – cash.

(5) 'What-if' modelling

Once the planning model has been developed, it is now possible to use it as a strategic planning tool. This gives the planners the ability to explore envisaged scenarios before they occur. It allows the manipulation of the basic logic network, the resources, resource costs and revenues. As a planning aid, it is this modelling element that is the most powerful and its use has been proved advantageous by clients and their designers, by project managers, by estimators and pre-tender planners and by construction managers on site. Clients and their designers have found such modelling advantageous to study the overall development of their project managers; estimators and pre-tender planners; and by construction managers on site.

Questions such as the effects of procurement delays and their consequences have led to design sequencing being altered to place orders earlier.

Project managers have found the modelling useful to explore the ramifications of the interdependence between the many contractors involved in a major project.

The pre-tender planner finds use in these modelling techniques to explore the effect of different construction methods on time, costs and revenues. The site manager has found such techniques useful in controlling subcontractors and in exploring the consequences of changes of construction methods even after the project has started.

Data exchange

All modern computer systems allow exchange of data between the planning system and other systems. Now that the construction industry is mainly computer literate, there is a reluctance to do much, if anything, by manual calculation. At a minimum, the computer planning systems are able to pass data files to *spreadsheet* packages, where the users can generate their own calculation routines for other purposes such as cash flows. Other examples of data exchange are where the planning system passes on data to estimating or cash flow systems. These are referred to in later chapters. The planning system itself could be the recipient of data from other systems. For example, the *estimating* system may hold details on resources and resource costs on a bill item by bill item basis. By allocating these bill items to activities in the plan the resource details are passed from the *estimating* to the planning system. Thus, an appraisal of the quality of computer systems must now include a review of the system's ability to pass on and receive data from other systems.

As computer systems become more and more 'linked' or 'integrated', the sources of data for the various systems come under scrutiny and it is clear that if contractual systems were being created with computers in mind the contractual system prevalent in the UK involving bills of quantities would not be created. The nature of the data created around the bill of quantities is not ideal for use in other systems. It would arguably be better to construct a contractual system around activities rather than bill items. However, various attempts at new systems have shown how resilient the existing contractual systems are. Forms of contract without a bill of quantities have payment to contractors based, not on bill items, but on completion of activities in a schedule of

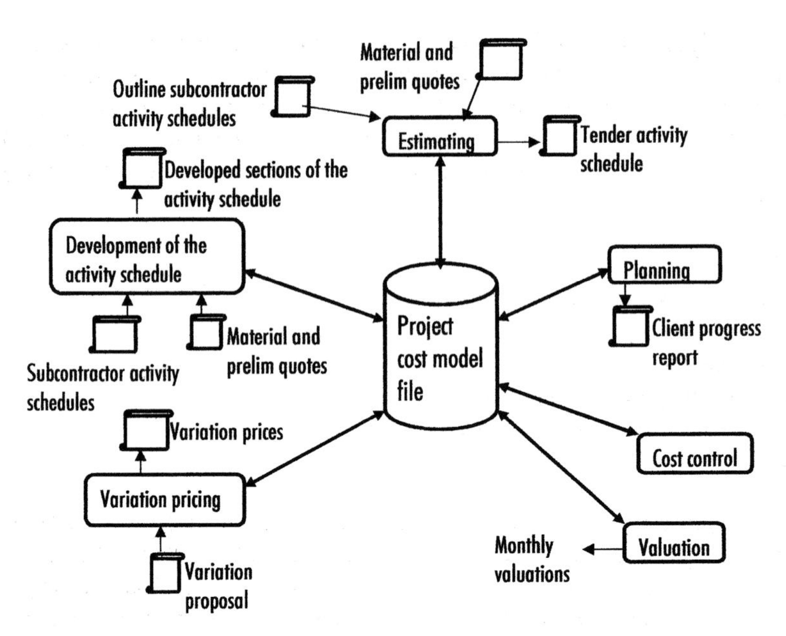

Fig. 4.24 Interrelationships between various management functions using a planning model as a database.

activities. This form of contract, used internationally, offers the contractor the opportunity of using the planning model's database in a number of linked functions such as:

estimating; planning; cost control; valuations; and variation pricing.

Figure 4.24 shows the interrelationships between the various functions on an integrated system. This figure illustrates that the basic planning data was common to all systems. The basic activity schedule had to be developed at the *estimating stage*.

Estimating

The estimating function requires information obtained from the outline tender programme and method statement and the take-off of quantities required for each activity. The resources required for each activity, labour, plant or materials, are entered and costed. Also the subcontractors are identified, allocated to activities and costed. The preliminaries and on-cost calculations are undertaken and allocated to activities or hammocks, as are profit and overheads, the key being that all this assembly of estimator's cost data is being *structured* around the planning model. Estimating by this approach requires that the planning model be developed to a more refined stage than in the traditional process. The value of this is that the data so assembled and structured is available for use in other functions.

Planning

Much of what was described as estimating above could well be described as planning. If the estimate was calculated as described the planner on site would begin their production planning with a well-developed plan. The well-developed plan allows all the efforts of *monitoring* and *control* to begin against a background of knowing how the costs were calculated in terms that were readily understood by site planners and construction managers. Difficulties that arise in merging separately assembled data from estimators and planners are substantially overcome if both the planning and cost monitoring systems have the data they require available from the first day.

Cost control

The elements in a cost control system are:

- Measure progress
- Calculate the budget allowance for that progress
- Compare the budget allowance with actual costs
- Take corrective action

The use of a planning system that has been used for estimating and contains all the planning and estimating data allows the first two items to be executed with speed and efficiency. The monitoring of progress will identify which activities are complete and which are partially complete. The entry of this progress data enables the budget allowances to be calculated. Each activity has allocated to it, at the estimating stage, the resources (labour, plant, materials and subcontractors) and their associated costs. Thus, if the activity is identified as complete, the budget allowance for that activity is the cost of those resources. If the activity is partially complete, the costs can be calculated on a pro-rata basis.

Thus, the use of a planning-based system can considerably ease the calculations of the budget allowance for any measured progress. What such systems cannot do is calculate the actual expenditure. This must come from the accounting systems that control expenditure on labour, plant, materials and subcontractors.

What is of value in any actual vs. allowances comparison is that the planning system holds data on resources, durations and costs. Thus, if actual expenditure is significantly different from the budget allowance, the data are readily available to extend the analysis into whether the resourcing was at planned levels or if time overruns were the cause.

Valuations

In some systems interim valuations are based on completed activities. Since each activity has had its value allocated or calculated at the estimating stage, the calculation of an interim valuation requires only the identification of a completed activity. Thus the process is reduced to identifying the completed activities, updating the planning model and calculating the last valuation.

Variations

One bugbear of all contractual systems is the determination of a fair price for variations. Variation proposals calculated on the same basis as the original estimating calculations, with the implications on overheads because of any disruption, are more clearly demonstrated in a planning model than in other systems.

The above brief review thus demonstrates the advantages to be gained in having systems that are capable of data exchange.

Artificial intelligence

Artificial intelligence or expert systems have not yet had much impact on planning or project management systems. Nevertheless, it is an inevitability that the systems available will become more clever and make suggestions to the user as to how planning might proceed.

Reference

Ballard, H.G. (2000) *The Last Planner System of Production Control.* PhD thesis submitted to the Faculty of Engineering, The University of Birmingham, UK.

5
Workforce motivation

Summary

The main incentive methods adopted in the construction industry are explained in the context of the theories developed by Maslow (1987), McGregor (1985), Vroom (1964) and Herzberg (1975) on workforce motivation. Specific practical examples illustrating the principles for determining suitable schemes for individuals and work crews are demonstrated.

Introduction

Success in the application of work incentives aimed at generating higher levels of performance and production output will largely depend on establishing a careful balance of the many interrelated motivating factors necessary in achieving worker satisfaction, often further complicated by the prevailing nature of the construction environment itself. For example, unique projects, the vagaries of working conditions, industrial relations and historical practices are all especially variable in the construction industry and have tended to hinder precise evaluation of the circumstances surrounding work tasks with inappropriate solutions sometimes evident. Realistic theories based on practical results are, however, available to assist in devising more robust incentives, with perceived shortcomings such as fluctuating earnings, disputes, variable levels of output and indifferent quality moderated.

Motivation theories

Maslow

Maslow suggested that people seek to satisfy needs sequentially as shown in Fig. 5.1, arguing that as each need is gratified then a new set emerges, implying a process of self-motivation.

In a modern society, when the basic needs such as provision of food, clothing, housing, etc., which lie at the root of national economic well-being emphasised through concerns to generate employment, are satisfied, in due course they appear to lead to ever greater demands like better working conditions, shorter hours, etc. Furthermore, the common desire for individuals to try and progress in a job, to gain respect from both the employer and fellow workers and to identify with a particular skill seem to provide additional support to Maslow's theory. Maslow believed that ultimately a level of self-fulfilment can be attained through purely creative work, but would in practice be of limited opportunity given the conditions prevailing in most paid employment.

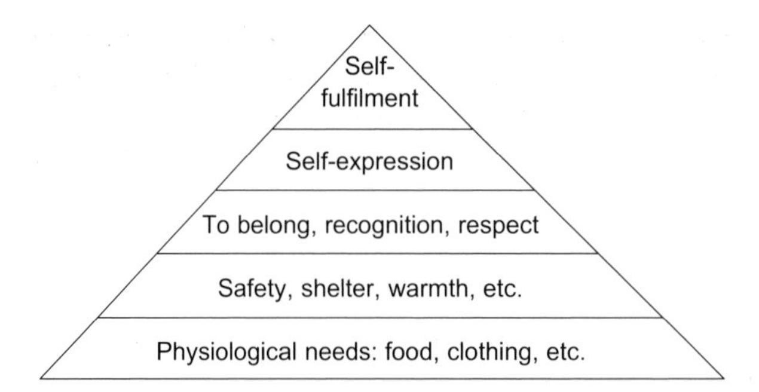

Fig. 5.1 Maslow's hierarchy of needs.

McGregor

McGregor developed the *theory (Y)* suggesting that work is natural, people wish to achieve goals and obtain intrinsic satisfaction when committed to the work objectives and, furthermore, seek responsibility by contributing to problem solving when properly motivated. Conversely, he also proffered the counter *theory (X)* that people are potentially indolent, self-centred, resistant to change, lack ambition, dislike responsibility, and are gullible, etc.

Subsequently Ghoshal & Westney (2005) urged companies in a services-based economy preferably adopt *Theory Y*, arguing that the modern workforce can be better motivated by emphasising 'purpose, process and people' and increasingly being directed through staff appraisals. Contrastingly, mechanised industry tends towards 'strategy, structure and systems' organisation, with incentives and payment by results needed as antidotes to try and treat potential *Theory X* symptoms.

Theory Y assumes:

- The expenditure of physical and mental effort in work is a natural human function.
- A satisfying job leads to worker commitment to the organisation.
- The typical worker seeks responsibility and the opportunity to exercise imagination, creativity and ingenuity.

Theory X assumes:

- The average human being has an inherent dislike of work, to be avoided if possible.
- Because of the dislike for work, workers need to be controlled and threatened to work harder.
- A worker prefers to be directed, dislikes responsibility and desires security.

Process theory

Vroom postulated that individual motivation is influenced by self-judgement, perception and interpretation of comparative treatment in relation to others. The theory measures motivational force behind a task by ascertaining values for:

- Valence (the valued reward)
- Instrumentality (performance)
- Expectancy (perceived connection between effort and performance)

Valence is paternalistic and assumes that the more a worker is rewarded, the greater the motivation to work harder. Instrumentality assumes that motivation to work will be improved if rewards and penalties are tied directly to performance level. Expectancy rewards largely lie intrinsically within the job itself or in the individual's relationship with members of the working team.

Herzberg

Herzberg discovered some interesting pointers to motivation and identified the following inherent factors for job satisfaction:

- Achievement
- Recognition
- The work itself
- Taking responsibility
- The chance to advance

He also concluded that workers could become demotivated when other factors were unsatisfactory these being termed 'hygiene' factors and mainly related to the following factors in job (dis)satisfaction:

- Working conditions
- Salary
- Relations with superiors
- Company policy

Importantly the 'hygiene' factors apparently had little positive effect on job attitudes but served primarily to prevent job dissatisfaction, i.e. if a company fails to provide adequate 'hygiene' factors the worker will become dissatisfied, but no matter how adequate are salaries, working conditions, etc., the worker will remain so unless the job has the intrinsic motivation elements.

Factors affecting construction worker motivation

Survey investigations recently undertaken separately by Kaming et al. (1997) and Zakeri et al. (1996) identified major motivating and 'hygiene' factors typical of those listed below in descending rank order of importance. Surprisingly, the intrinsic job satisfiers described by Herzberg and the other theorists featured only weakly, suggesting that achievement, recognition, etc. opportunities may be naturally inherent in much of construction and that the 'hygiene' elements were therefore necessarily to the fore in the minds of the workers surveyed.

Job satisfaction factors
High to low importance for the worker:

(1) Fair level of pay
(2) Incentive corresponds to financial reward
(3) Pay received on time

(4) Good facilities on site
(5) Safe and healthy working conditions
(6) Good working relations with supervisors
(7) Good working conditions
(8) Favourable promotion prospects
(9) Good working relations with other crewmembers
(10) Job security
(11) Opportunity to select colleagues in the crew
(12) Recognised for doing a good job
(13) Reasonable level of overtime demand
(14) Competent supervisor
(15) Reliable job description/specification
(16) Type of work
(17) Challenging tasks
(18) Responsibility
(19) Good relations with employer
(20) Participation in decision-making process

Job dissatisfaction factors

Reinforcing these findings, other studies by Olomolaiye and Ogunlana (1988) reported the views of production workers on non-financial factors having demotivating effects. Again in descending order of importance, interestingly these were largely negative aspects of the previously mentioned motivators for job satisfaction.

High to low importance for the worker:

(1) Poor treatment by supervisors
(2) Lack of recognition of good effort
(3) Productivity urged with indifference
(4) Reducing work opportunities
(5) Incompetent crew colleagues
(6) Poor communications with management
(7) Under-utilisation of skills
(8) Lack of participation in decision-making
(9) Unsafe working conditions
(10) Poor supervision
(11) Little production

Obviously the specific employer needs to provide a total package embracing at least the most important worker motivators and avoid those elements likely to irritate the workforce and cause dissatisfaction.

Payment systems, remuneration and performance

The achieving of high productivity attended by good quality principally requires careful management of the supply chain (see Chapter 3, *Production process improvement*), with selected participants having previously demonstrated an enlightened approach to motivation theory and

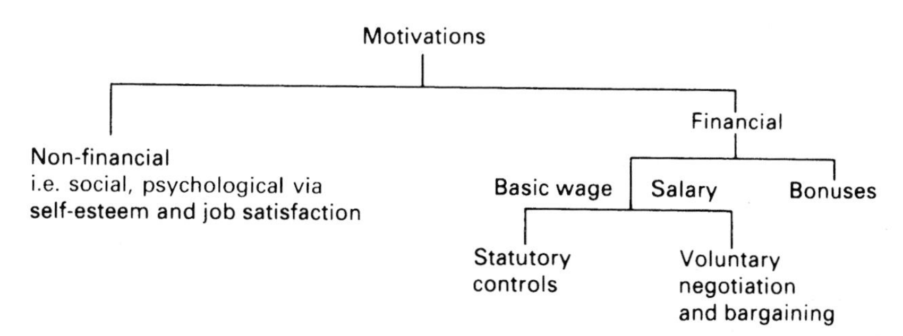

Fig. 5.2 Methods of encouraging motivation to work.

practice, manifested through a proven record of sound TQM results, serious engagement in employee training, progressive skills updating, structured personnel development, attractive remuneration packages, etc. (see Chapter 14 on training and NVQs).

Nevertheless, even if such conditions are reasonably satisfied, some construction work unfortunately remains routine and monotonous, and additional means of inducing better performance for directly employed labour in particular are commonly unavoidable. In this respect the industry has generally concentrated on the Herzberg 'hygiene' factors by offering financial incentives to manual workers and non- or semi-financial incentives to managerial and clerical workers (Fig. 5.2), with some inevitable overlapping as field conditions dictate. Irrespectively, however, attention should be given to all the motivators listed above and below as far as is practicably possible, no matter the shape of the payment-by results scheme.

Non-financial incentives

The incentives involved are fairly intangible, and related to Maslow's higher needs involving in particular the fulfilment of those defined by Herzberg as the '*intrinsic motivators*'. Thus the incentives offered acknowledge the importance of the individual and recognise the need for group participation to provide social satisfaction. For example, there are types of occupation that require little artificial incentive, where much of the work itself is inherently self-motivating, i.e. workers in famine relief, ministers of religion, nurses, etc. Paid or salaried employment seldom offers these kinds of opportunity, although meeting personal career ambitions can sometimes partly suffice.

Semi-financial incentives

Where jobs are difficult to measure in straightforward productive terms, e.g. managerial and supervisory positions, practical experience indicates that, as far as the financially related 'hygiene' motivators are concerned, these apparently function best when introduced through fringe benefits in the form of contributory pension schemes, share options, holidays with pay, restaurant facilities, sports facilities, company cars, paid telephone bills, expense accounts, etc. Good basic salaries and promotion prospects for the employee would also normally be considered fundamental imperatives.

Financial incentive schemes

Since the quantity of construction output achieved by an individual worker or crew is measurable, the 'hygiene' motivators can in part be transformed into financial incentives and used specifically to try and encourage improvements in:

- Productivity
- Methods of working and output quality
- Operative's earnings, but without increasing unit costs

Many schemes are available to suit different situations; those most applicable to the construction industry being demonstrated below and summarised in Table 5.1.

Profit sharing

The company pays out either yearly or half-yearly a lump sum, shares or share options to its employees, based on the profit earned by the company. This system operates best where labour turnover is low and the workforce can meaningfully contribute to the strategies for profit success.

Day work

An hourly rate is paid related to the skill required by the task; the worker is then simply paid for attendance at work. The system is most suitable for craft operations where either there is great complexity involved or a high level of skill is required. High rates of pay coupled with semi-financial incentives are often necessary to attract and keep the worker.

Workers involved in maintenance, inspection, canteens, cleaning, transport and stores, although not of the above category, can often only be paid in such a way—sometimes supplemented with some form of merit-rated bonus.

Piecework

Straight piecework is the payment of a uniform price per unit of production. The principle is expressed graphically in Fig. 5.3 whereby, as a worker improves output, earnings increase proportionately.

Example

An operative paid 10p for each item or unit produced completes 10 units and therefore receives payment of $0.1 \times 10 = £1$.

Such a system would be best installed on repetitive-type activities where the standard time for doing the work can be reasonably accurately determined. Hence, it is popular in the manufacturing industries, but has also gained favour for some of the subcontractor trades particularly construction finishing operations such as flooring, plastering, electrical wiring, etc. The system can sometimes prove problematic in terms of controlling costs from the employer's point of view where a minimum wage is in place, enabling some workers to achieve wages above their productive performance. In practice, if targets are set out of line, morale will quickly fade and industrial unrest is likely to occur until the targets are re-established at an achievable level.

Table 5.1 Summary of incentive schemes.

Incentive scheme	Advantages	Disadvantages
(1) *Daywork* Employee paid a basic wage for attendance	(a) Simple and easy to understand (b) Simple to calculate wages (c) Low clerical requirements (d) Facilitates labour flexibility	(a) No reward for efficiency (b) Slack workers gain at the expense of fast workers (c) Strict supervision required (d) Accurate budget forecasts are difficult
(2) *Piecework* A uniform price is paid for each job or unit completed	(a) Direct incentive to increase output (b) Simple to understand (c) Wage cost per unit of production is constant (only when applied without min. wage)	(a) Changes in wage rate necessitate altering targets (b) Can lead to poor quality
(3) *Straight proportional hours-saved scheme* All time saved on the target set is given to the worker	(a) Incentive is related to effort (b) There is a guaranteed wage (c) Provides cost control data (d) Better quality control than with piecework	(a) Expensive to operate (b) Tends to favour fast workers (c) Requires sound data to establish target rates (d) Causes problem initially when labour is inexperienced
(4) *Geared schemes* Similar to (3) but only a proportion of time saved is given to the worker	(a) Useful to start off new work (b) Provides an incentive for slow and inexperienced workers	(a) Encourages loose rate fixing (b) Fast workers are not fully rewarded
(5) *Group schemes* Similar to (3) and (4), but individuals paid on a shares basis	(a) Assists elimination of slackers (b) Suitable when gangs of workers are necessary, thus particularly suited to construction industry (c) Encourages co-operation	
Taylor two-piece scheme *Rowan* *Halsey (or Weir) scheme* *Graded and measured daywork* *Other piecework schemes*	These schemes are not widely used in the construction industry and are covered adequately in other texts	

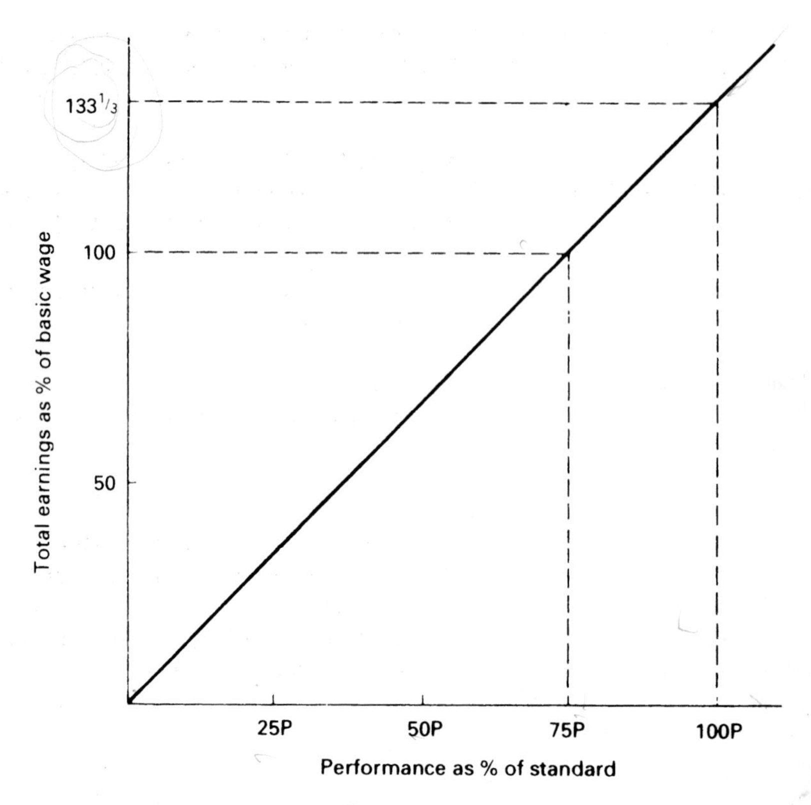

Fig. 5.3 Piecework system expressed graphically.

Standard time, or hour, system

Essentially this is the same as a straight piecework system, but the targets for the worker are expressed in time units rather than money; hence, the system is especially applicable for contractors and subcontractors employing labour directly.

Example
A worker is given a target of 8 hours to complete an activity and completes it in 6 hours. Thus payment = 8 × hourly rate. The next unit can now be commenced 2 hours earlier and thereby the worker's overall earnings are proportionately increased. The advantage of the system is that, if new hourly rates are negotiated, the time standards remain unchanged.

Hours-saved system

If a production bonus scheme is operated at all on a construction project, then the hours-saved system tends to be favoured, and particularly the 75–100 scheme whereby a worker receives a guaranteed wage for any performance up to 75% of standard (100P). When the worker achieves a performance greater than this level, a bonus can be earned, such that at standard performance the earning for a worker reaches 33½ % of the basic wage. Disadvantages relate to abuses and disputes in application, largely stemming from difficulties in establishing reliable output targets.

For example, a misjudged but generous target set for a concrete gang may cause unrest in a formwork gang struggling to earn a bonus on a difficult shuttering operation requiring much skill.

Example

It is considered by a site manager that a bricklayer engaged on building drainage inspection chambers, each containing 720 bricks, should be able to lay 60 bricks per hour at standard performance. If this is achieved, the bricklayer will be entitled to a bonus of 33 ⅓% added to his basic rate of payment.

Calculate the bricklayer's earnings when performance reaches 50P, 75P, 100P and 120P respectively (100P being standard) and the basic rate of payment is £1.00 per hour.

Solution

Actual time required to build the inspection chamber when working at standard performance is:

720/60 = 12 labour hours

If bonus is paid at 33 ⅓% of basic wage for working at standard performance, and in the 75–100 scheme the hours saved are directly paid at the basic rate, then from Fig. 5.3 the gradient for earnings plotted against performance is seen to be unity (or 1).

The following algebraic algorithm can now be formulated with the time allowed for doing the job designated as a and actual time required at standard performance as b. Thus:

$$a = b + \frac{1}{3} \times b \times \frac{1}{\text{Gradient}}$$

and with values substituted:

$$a = 12 + \frac{1}{3} \times 12 \times 1 = 16 \text{ hours}$$

The 16 hour result is the target allowance of time for building the chamber and forms the basis for calculating and comparing actual achievements as illustrated, where Table 5.2 shows the effect of worker performance both on the worker's own earnings and on the cost of the job to the contractor. Clearly, if the performance of the worker falls below 75P the cost of the job increases quite rapidly and readily arises if output targets are set too tightly. The worker in this situation is likely to lose confidence in the arrangement and subsequently perform badly, hence this scheme should only be offered to experienced operatives familiar with the type of work involved.

Table 5.2 Directly proportional incentive scheme calculation (time allowed = 16 hrs at £1 per hour).

Level of performance (P)	Actual time taken (hours)	Hours saved (Number)	Bonus (£)	Hourly earnings (£)	Cost of job (£)
120	10	6	6.00	1.60	16.00
100	12	4	4.00	1.33	16.00
75	16	0	0.00	1.00	16.00
50	24	0	0.00	1.00	24.00

Geared schemes

The direct incentive scheme described above enables only the fast and skilful worker to earn a bonus and they may not always be readily available to an employer, particularly in beginning a new contract, when the work may be unfamiliar to the labour force. Furthermore, in isolated country districts the recruited workers may be unfamiliar with construction and thereby unable to earn a bonus, and consequently an uneconomic level incentive payment for the slower worker may have to be tolerated for a short period during learning. The common solution is to set more generous targets but with payment based only on a proportion of the savings made. To illustrate the method, Table 5.3 shows bonus earnings for payments at 75%, 50% and 25% of the basic rate respectively for hours saved, all interpreted graphically in Fig. 5.4. The results clearly demonstrate that, for example, in a geared scheme of 25% payment, the worker can earn a bonus even when working below 50P performance, although of course the cost of the job is high. Providing the contractor takes this into account in the estimate, advantage can be gained by starting off the project with the low-based scheme, building up gradually over the first few weeks to the full proportional scheme when the workforce has become familiar with its duties.

It will be noticed that in all the schemes the bonus earned at standard performance (33 ⅓%) is the same.

Finally, the 50–100 scheme (Fig. 5.4) tends to be the more popular device used in practice for dealing with the above situation, performance at the 50P level producing a bonus.

Plus rate, or spot, bonus

A minimum bonus, sometimes called a fall back bonus, is paid whatever the output, the incentive simply being used to generate a competitive remuneration rate when the basic rate is low, e.g. where local employment practices may be restrictive.

Job and finish

The worker is offered a lump sum of money to complete an operation, useful on, for example, large concrete pours or similar work, when the gang will complete the work as quickly as possible, but be paid for the full day's work and any bonus earned.

Group scheme

Much construction activity on site performed by work crews/gangs requires the different classes of skilled worker to be acknowledged in earnings levels. The solution normally preferred involves apportioning the bonus on a shares basis, usually with the crew supervisor being paid slightly more than trades persons and so on through the skill levels.

Example

The site manager responsible for the construction of a large concrete pump house sets a crew of five workers a target of 50 labour hours to erect the formwork for the concrete in the base pour. Calculate the bonus earned by each crewmember, given that the work is completed in the times shown below:

Table 5.3 Geared incentive schemes calculation.

Level of performance	Actual time taken (hours)	Target = 17⅓ h		75% payment		Target = 20 h		50% payment		Target = 28 h		25% payment	
		Hours saved	Bonus (£)	Hourly earnings (£)	Cost of job (£)	Hours saved	Bonus (£)	Hourly earnings (£)	Cost of job (£)	Hours saved	Bonus (£)	Hourly earnings (£)	Cost of job (£)
120P	10	$7\frac{1}{3}$	5.50	1.55	15.50	10	5.00	1.50	15.00	18	4.50	1.45	14.50
100P	12	$5\frac{1}{3}$	4.00	1.33	16.00	8	4.00	1.33	16.00	16	4.00	1.33	16.00
75P	16	$1\frac{1}{3}$	1.00	1.07	17.00	4	2.00	1.13	18.00	12	3.00	1.19	19.00
50P	24	0	0	1.00	24.00	0	0	1.00	24.00	4	1.00	1.04	25.00

Payment scheme *Target hours*

$100\% = 12 + (\frac{1}{3} \times 12) = 16\text{h}$ . . . . 75–100 scheme } direct scheme

$75\% = 12 + \left(\frac{1}{3} \times 12 \times \frac{100}{75}\right) = 17\frac{1}{3}\text{h}$

$50\% = 12 + \left(\frac{1}{3} \times 12 \times \frac{100}{50}\right) = 20\text{h}$

$33\frac{1}{3}\% = 12 + \left(\frac{1}{3} \times 12 \times \frac{100}{33\frac{1}{3}}\right) = 24\text{h}$ . . . . 50–100 scheme } geared schemes

$25\% = 12 + \left(\frac{1}{3} \times 12 \times \frac{100}{25}\right) = 28\text{ h}$

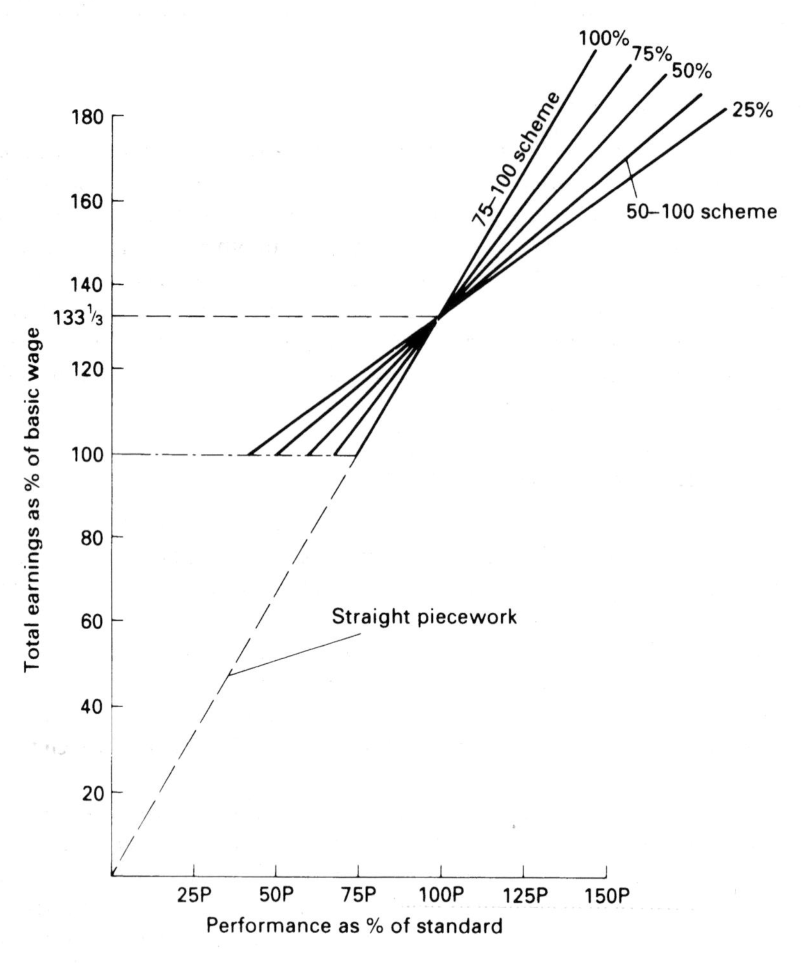

Fig. 5.4 Geared incentive schemes.

Worker Description	Shares entitlement per hour worked	Actual time (hours) worked on activity	Shares
Gang supervisor	1½	8	12
Carpenter No. 1	1¼	8	10
Carpenter No. 2	1¼	8	10
Labourer No. 1	1	8	8
Labourer No. 2	1	4	4
		36	44

Basic rate of payment is £1.00 per hour
Hours saved = 50 − 36 = 14 hours
Bonus = 14 × £1.00 = £14.00
Bonus per share = 14/44 = £0.318 per share

Gang supervisor = 0.318 × 8 × 1½ = £3.83
Carpenter No. 1 = 0.318 × 8 × 1¼ = £3.18
Carpenter No. 2 = 0.318 × 8 × 1¼ = £3.18
Labourer No. 1 = 0.318 × 8 × 1 = £2.54
Labourer No. 2 = 0.318 × 4 × 1 = £1.27
= **£14.00**

Indirect incentive schemes

Certain categories of work cannot be readily measured in physical terms and are therefore difficult to include in normal productive bonus schemes. The tasks might include site fitters, hoist operators, stockyard gang and workers generally providing a service. A common way of dealing with this situation is to pay a bonus related to the number of other workers served, e.g. fitters might be paid a bonus equal to the average bonus for the site as a whole. However, care must be taken when establishing the incentive, since if the number of workers being supplied is increased, those actually providing the service will be doing more but receiving no extra remuneration, and hence, gang ratios demand careful balancing. An alternative approach could be to offer an enhanced rate of pay based on merit.

Principles of a good incentive scheme

The application of suitable motivators, including financial incentives, has proved elusive in practice; nevertheless some of the essential attributes of the various schemes in Table 5.1, together with the following guidelines, may achieve a degree of success in implementation:

- Bonus should be paid to workers in direct proportion to the effort applied.
- The earnings of the worker should not be limited in any way.
- The set targets should be attainable and thereafter remain unaltered. In the construction industry this is not always possible because of the lack of reliable data on which to base the targets.
- Unavoidable hold-ups should be excluded from bonus paid hours and paid at basic rate.
- The scheme must be easily understandable to the worker so that the bonus calculation can be self-determined, otherwise considerable confusion is likely to occur, requiring much time and effort on the part of management to allay any fears of malpractice. It is for this reason that geared and other more complex systems have not found much favour with workers or unions.
- The scheme should comply with local union arrangements.
- Disciplined administrative procedures are essential to ensure that drawings are updated, materials arrive on time, etc.
- The scheme should be integrated with a cost control system.
- Incentive schemes can lead to substandard work; hence the need for the continuous emphasis on achieving—and encouragement to achieve—quality standards, ultimately including penalties for non-compliance as appropriate.
- Importantly, the scheme must be fully compliant with health and safety requirements.

Setting target rates

The information available to the manager for use in setting target outputs for construction work tends to be fairly coarse-grained compared to that used to estimate in manufacturing, for example accurate work-study data. The following sources may therefore provide useful.

Personal experience

Given a wide and varied experience of planning, estimating and site engineering over a period of years, the manager should have gained a reasonable knowledge of gang sizes and expected outputs of plant and labour involved in the major construction operations.

The advantages of using self-knowledge are that it is immediate, inexpensive and readily accepted. It has the disadvantages of being of limited range, not documented, and above all is open to bias. Much of the estimating in the industry is done on this basis and the estimator's build-up sheets generally provide the basis upon which to fix the bonus targets.

Feedback

Often within an organisation there exist records of costs, bonuses and job records for contracts that have been completed. Some of the information may be useful, but frequently cost control and data collection is given low priority on construction contracts, because it is expensive to collect and is historical anyway. Such data offers rough guidelines only, since it normally applies to one-off projects.

Work-study data

Output on construction work can fluctuate wildly depending on the prevailing conditions; hence work-study data would ideally need to be developed by the specific company if it is to be reliable.

Manufacturers' ratings

These provide only a rough guide since they are likely to be optimistic values. Many estimators halve the manufacturer's rating.

Site demonstrations

Site demonstration of equipment offers a guide, but care must be taken to view the exercise in overall terms.

Standard information

Average outputs are available in several published works but usually will only provide approximate guides and not necessarily accurately reflect the achievable value possible in individual firms.

References

Ghoshal, S. & Westney D.E. (eds.) (2005) *Organization Theory and the Multinational Corporation*. Palgrave Macmillan, Basingstoke.

Herzberg, F. (1975) *Work and the Nature of Man*. Crosby Lockwood, London.

Kaming, P.F., Olomolaiye, P.O., Holt, G.D. & Harris, F.C. (1997) Factors influencing construction time and cost overruns on high-rise projects in Indonesia. *Construction Management and Economics*, **15**(1), 83–94.

Maslow, A.H. (1987) *Motivation and Personality*. Longman, New York.

Maslow, A.H., Lowry, R. & Lowry, R.J. (eds.) (1998) *Toward a Psychology of Being*. Wiley, Chichester.

McGregor, D. (1985) *The Human Side of Enterprise*. McGraw-Hill Education, New York.

Olomolaiye, P.O. & Ogunlana, S.O. (1989) A system for monitoring and improving construction operative productivity in Nigeria. *Construction Management and Economics*, **7**(2), 175–86.

Vroom, V.H. (1964) *Work and Motivation*. John Wiley, Chichester.

Vroom, V.H. & Deci, E.L. (eds.) (1989) *Management and Motivation*. Penguin Books, Harmondsworth.

Zakeri, M., Olomolaiye, P.O., Holt, G.D. & Harris, F.C. (1996) A survey of constraints on Iranian construction operatives' productivity. *Construction Management and Economics*, **14**(5) 417–25.

6
Cost control

Summary

The main features of the several cost control methods used by management on construction projects are summarised. The problems of controlling subcontracts and material costs are also described and guidelines enumerated.

A cost control procedure for construction works

Introduction

The construction industry, unlike many manufacturing situations, is concerned mostly with one-off projects. This naturally creates difficulties for effective management control, because each new contract often has:

- A fresh management team
- Labour often recruited *ad hoc*
- Sites dispersed throughout the country, which tends to cause problems in effective communications with other parts of the company
- Frequent use of subcontractors and 'lump-sum' labour items
- Added to all this, the ever-changing weather conditions

Nevertheless, irrespective of the scale of operation from small subcontractor to the multi-faceted project and sophisticated supply chain, production costs need to be monitored and controlled if the anticipated level of profit is to be realised.

Fundamentals

To *control* cost is an obvious objective of most managers, but it should be recognised that no amount of paperwork achieves this control. Ultimately, the decisions of the manager that something should be done differently, and the translation of that decision into practice, are the actions which achieve control. The paperwork can provide guidance on what control actions should be taken and, while we shall continue to call it 'the cost control system', it should more properly be called 'the cost information system'. The elements of any control system are:

- Observation
- Comparison of observation with some desired standard
- Corrective action to take if necessary

The domestic thermostat is a good example of a controller. The instrument measures the temperature, compares it with the desired range and then switches the temperature heating system 'on' or 'off' depending on how the current temperature compares with the desired range.

A cost control system should enable a manager to observe current cost levels, compare them with a standard plan or norm, and institute corrective action to keep cost within acceptable bounds. The system should help to identify where corrective action is necessary and to provide pointers as to what that action should be.

Unlike the humble thermostat, most cost control systems have an inordinately long response time. Even the best current system provides information on what was happening last week or last month. As the work is typically part of a one-off project, it is quite likely that the information is only partly relevant to the work going on now. So the scope for corrective action is limited. For example, the system might indicate on 1 May that the formwork operation in March cost too much. If formwork operations are still continuing the manager will give this work particular attention, but if formwork is complete, nothing can be done to correct the situation.

In the conventional systems described below two fundamental points are important. First, all costs must be allocated even if this is on a very 'coarse-grained' coding arrangement. If only the major items are monitored, you can be sure that the wasted time will be booked against the items which are not being monitored. Thus, the manager will be deluded by the reports on the 'important items' into thinking the whole site is satisfactory; in fact, it might be incurring disastrous hidden losses. Second, there must always be a standard against which to compare recorded costs. In simple projects, this might be the bill of quantities; generally, however, a properly prepared and appropriately updated contract budget forms a better basis.

Systems in current use

The following systems and variants of them are in use in the construction industry. The selection of a system depends in part on the size and complexity of the contract, but more on the attitudes and level of sophistication of top management.

By overall profit or loss

The contractor waits until the contract is complete and then compares the sums of money that have been paid with the monies incurred in purchasing materials, payments for labour, plant and overheads. The figures are normally extracted from the financial accounts compulsorily kept by all companies. Such a system is useful only on minor contracts of short duration involving a small workforce and little construction equipment, typically labour only subcontracting. It scarcely qualifies as a control system as the information it produces can only be used to avoid the recurrence of gross errors in later contracts.

Profit or loss on each contract at valuation dates

The total costs to date are compared with valuations gross of retentions. Care has to be taken to include the cost of materials delivered but not yet invoiced and to exclude materials on site not yet built into the permanent work. If the certificate is not a time reflection of the value of work done, a further adjustment is necessary. This system suffers from the disadvantage that there is no breakdown of the profit figure between types of work; it therefore provides guidance only on which contract requires management attention. It is not suitable for contracts that involve significant set-up costs which are distributed over the unit rates and hence would be more appropriate for a general subcontractor.

Unit costing

In this system, costs of various types of work, such as mixing and placing concrete, are recorded separately. The costs, both cumulatively and on a period basis, are divided by the quantity of work of each type that has been done. This provides unit costs that can be compared with those in the tender. Considerable care must be taken to ensure that all costs are accounted for, as indicated above under 'Fundamentals'. Any miscellaneous costs must be recorded and allowed for in some way, e.g. by proportional distribution over the defined work-types. It is usually best to record site costs only and to compare with bill rates net of contribution for profit and head-office overheads.

Systems based on the principles of standard costing

Standard costing has been used successfully in manufacturing industries, particularly in companies producing a limited range of products or at least a limited range of basic components. Standard minute values are associated with the production of each component and assembly and converted to money values by reference to the hourly rates of the appropriate grades of operatives. Variances are calculated, basically by comparing the value of the output with the cost of producing it. A variance is the amount by which the achieved profit differs from the budgeted profit. With appropriate records, it is possible to analyse the total variance into sub-variances, e.g.:

- Material price
- Material usage
- Labour rate
- Labour efficiency
- Fixed and variable overhead expenditure
- Volume of production
- Sales

Standard costing is seldom directly applicable in construction due to the variety of the product. This makes the use of standard minute values difficult, if not impossible. However, as an alternative, the value of work done can be assessed in relation to the contract budget, which in turn must reflect the amount that the contractor can expect to be paid.

One of the important features is the calculation of a sales variance. This encourages the company to define sales (marketing, public relations, negotiations, estimating and bid strategy) as the responsibility of one department. An adverse variance indicates immediately that the level of acquiring new contracts is inadequate.

Altogether, quite substantial departures from the manufacturing system are necessary and this accounts for the fact that standard costing is not in common use in construction. However, the system is basically very sound and provides comprehensive control of the company from boardroom down to workforce.

Hybrid cost control system

The following example of a cost control system for a multi-storey office building reflects several of the principles discussed above. Strictly it is a hybrid system as one method is used for labour and plant costs and another for materials, and can be usefully applied to general contracting *involving predominantly directly employed labour*. It is illustrative rather than truly representative, as a comprehensive set of figures would occupy a disproportionate part of this book.

Labour, plant and site overheads

In order to reduce booking errors, the contractor sensibly decides to adopt a limited number of cost codes against which to generate variances. These are:

- Cost code 10. All concrete mixing, transporting and placing (labour and plant)
- Cost code 20. All formwork fixing and dismantling (labour and plant)
- Cost code 30. All fixing of reinforcement (labour and plant)
- Cost code 40. All bricklaying (labour and plant)
- Cost code 50. Tower crane
- Cost code 60. All earthworks (labour and plant)
- Cost code 70. Roads and paving (labour and plant)
- Cost code 80. Site overheads and preliminaries

The construction of the works is carefully planned and the budgeted expenditure, excluding head-office overheads, profit, subcontracts and materials for each of the cost codes, is allocated to each activity in the programme. (In order that realistic comparisons of budgeted and actual costs can be made, the budgeted allocations must be derived from accurate quantities, carefully taken from working drawings, and any variations must be included.)

Table 6.1 shows the breakdown of the budget expenditure and Fig. 6.1 shows the first 76 weeks of the programme.

Progress is recorded in the usual way against the programme and this is used as a guide for expediting the work. Costs are recorded against the cost codes. At Week 52, for example, the percentage completion of each activity can be extracted from the progress reports, which are represented in the revised programme shown in Fig. 6.2; the figures are also shown in the last column of Table 6.2. The other columns of Table 6.2 show the budget figures multiplied by the percentage completions. Hence, by adding the columns, the total values of work done in each cost code can be derived. The separately recorded cost figures are also entered in order to calculate the variances. It will be noted that the work to date has cost £2340 more than it should have done and that the reasons for this excess cost are apportioned to the various cost codes. For example, cost code 30, that is steel-fixing, shows a variance of –£1100 and clearly calls for management attention with regard to this trade.

Allocation of costs

The foregoing cost control system in general allocates costs against a gang of workers, e.g. concrete gang. It is therefore the responsibility of the charge-hand of the gang to fill in the daily allocation sheet (see Fig. 6.3), recording a brief description of the operation upon which the gang was employed and the names and hours worked by each member of the gang. Often this recording is not necessary, particularly if the type of work done by the gang is fairly static; the cost control engineer then simply extracts the labour-hours for the gang from the wages sheet. Whenever a bonus system is operated on the project, the gang bonus must be added to the basic costs

Table 6.1 Budgeted costs.

Activity description	Cost codes								
	10	20	30	40	50	60	70	80	Total
Earthworks						900		100	1000
Foundations	5000	6000	7000					2000	20000
Ground floor slab	1000	1000	2500					500	5000
Columns	200	300	300					200	1000
Floor slabs	15000	17000	18000		10000			20000	80000
Lift shaft	2000	3000	4000					1000	10000
Brickwork and blockwork				13000				3000	16000
Windows and glazing				2000	1000			1000	4000
Roofing	100	100	100		100			50	450
Internal finishes				1000	500			3000	4500
Plumbing, heating and ventilating				500	1000			5000	6500
Electrics								4000	4000
External works							1800	200	2000
Clear site							900	100	1000
Budget totals	£23300	£27400	£31900	£16500	£12600	£900	£2700	£40150	£155450

Total value of contract: £800000.
Cost of labour, plant and site overheads: £155450.

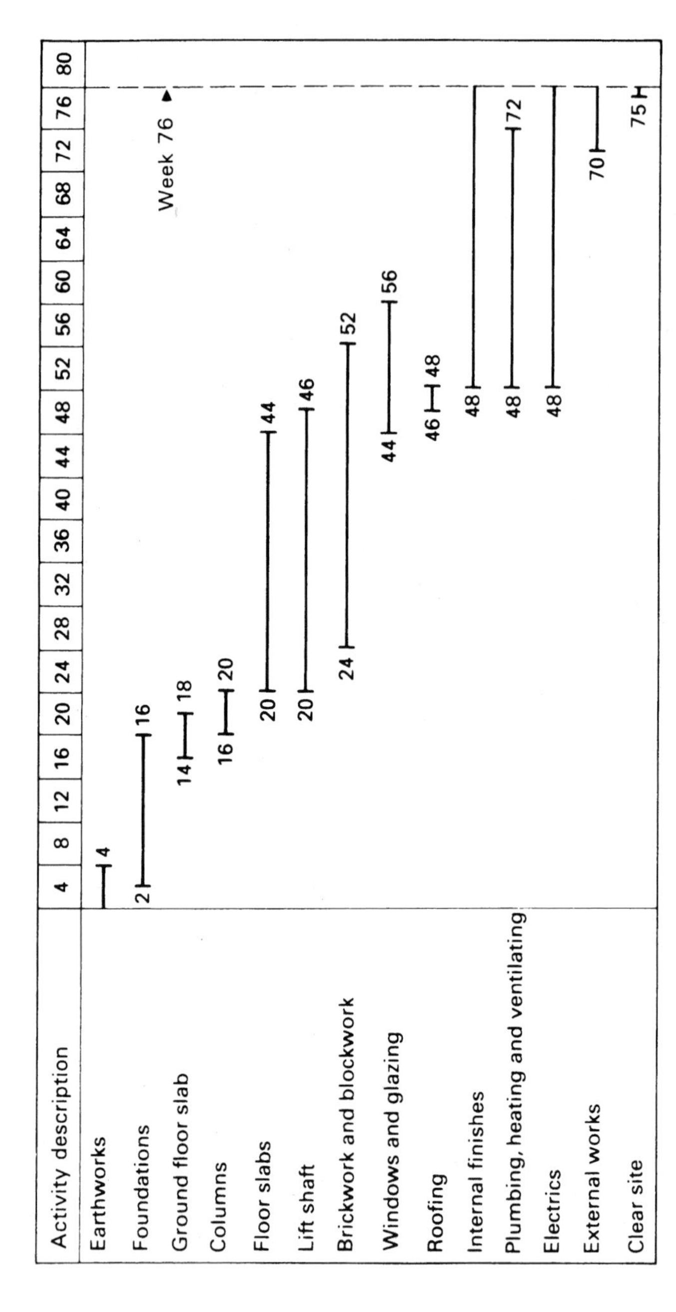

Fig. 6.1 Original programme of works – in weeks.

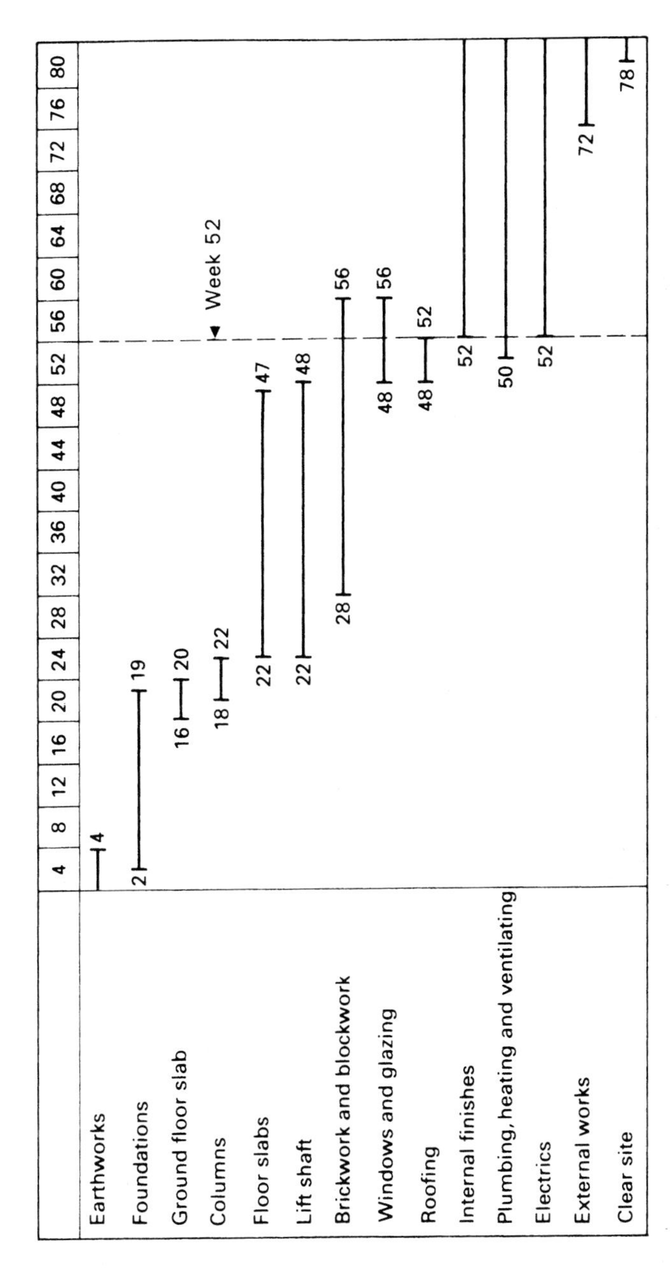

Fig. 6.2 Programme updated at Week 52.

Table 6.2 Updated costs at Week 52.

Activity description	Cost code									Percentage complete
	10	20	30	40	50	60	70	80	Total	
Earthworks							900	100	1 000	100
Foundations	5 000	6 000	7 000					2 000	20 000	100
Ground floor slab	1 000	1 000	2 500					500	5 000	100
Columns	200	300	300					200	1 000	100
Floor slabs	15 000	17 000	18 000		10 000			20 000	80 000	100
Lift shaft	2 000	3 000	4 000					1 000	10 000	100
Brickwork and blockwork				11 200				2 580	13 780	86
Windows and glazing				1 000	500			500	2 000	50
Roofing	100	100	100		100			50	450	100
Internal finishes									–	Nil
Plumbing, heating and ventilating				33	67			330	430	6⅔
Electrics									–	Nil
External works									–	Nil
Clear site									–	Nil
Value to end of Week 52	£23 300	£27 400	£31 900	£12 233	£10 667	£900	–	£27 260	£133 660	
Actual cost to Week 52	£23 000	£28 000	£33 000	£12 500	£11 000	£500	–	£28 000	£136 000	
Variances	+£300	–£600	–£1 100	–£267	–£333	+£400	–	–£740	–£2 340	

Note: In practice the proportions of each code completed for an activity will vary, e.g. 'Construct founds' may involve the concrete, formwork and reinforcement being at different stages of completion.

Company Ltd ______		Form No. ______							
		Daily time allocation sheet							
	Name	Smith	Jones	Roberts	McKay	Baker	Neale	Shaw	Total hours
Cost code	Clock number	041	042	030	061	172	141	124	
20	Formwork to bases	8	8	8					24
20	Formwork to columns				8	8			16
20	Formwork to stairs						8	8	16
		8	8	8	8	8	8	8	56
Contract ______		Date ______							
Chargehand ______		General Foreman ______							

Fig. 6.3 Daily time allocation sheet.

derived from the wages sheet in order to arrive at the actual cost of doing the work. The careful bookkeeping that is usual with the bonus system provides figures that can be used to assess the value of work done, since the measurement engineer has had to take a weekly measurement of progress to enable the bonus to be calculated.

Note that *all costs* of labour and plant should be allocated against the cost codes. If this proves difficult, the device of introducing a code for miscellaneous work is sometimes adopted. This can encourage the inappropriate booking of wasted labour-hours and is not recommended. If, despite this advice, a miscellaneous code is used, the costs in this code should be retrospectively allocated to the other codes.

Materials control

It is commonly more difficult to control material variances than anything else, largely because of the effort necessary to determine accurately what has been already built into the structure.

The factors that add to the difficulty of keeping precise control of material costs can be divided into price and quantity variances.

- Price variances
  - Inflation
  - Changes in the buying situation since the estimate was prepared, e.g. bulk buying, discounts, shortages and changes in quality demanded by the client or available at the time
- Quantity variances
  - Wastage and breakages
  - Theft and loss
  - Short deliveries
  - Remedial work
  - Delays in the recording system
  - Inaccurate site measurement of work done

Very careful, comprehensive records would enable all these variances to be calculated for various material cost codes, but the expense of undertaking this work would probably far exceed the potential savings (see end of chapter for 'points to consider when choosing a cost control system'). In practice, it is probably sufficient to generate an overall materials variance which is expressed in monetary terms (£). Thus:

Materials represented in measured work to end of last period	a
Materials in measured work in this period	b
Materials value to date	$c=a+b$
Cost of materials used to end of last period	d
Cost of materials delivered in this period	e
Total cost of materials purchased to date	$f=d+e$
Materials currently on site	g
Materials used to date	$h=f-g$
Materials variance	$j=c-h$

The value of materials (a, b and c) should be net of contribution. Costs should be based on a careful combination of the information from invoices and delivery notes. (Some relatively sophisticated computer systems have been devised to improve the reliability of this summation.)

When adverse materials variances reach unacceptable levels, the manager will institute *ad hoc* investigations and will take action to prevent recurrences.

There is a tendency for cost control in construction to be directed more towards the control of labour and plant than to the control of materials. There is growing evidence, however, that losses due to materials are often significantly higher than those due to other causes. Thus, greater attention to materials control may pay significant dividends in the form of increased profit. Some of the methods available, in addition to the keeping of reliable records, are:

(1) Keep a neat and well-laid-out site with adequate storage space and room for movement. Use mechanical handling equipment whenever possible.
(2) Employ a reliable storekeeper, possessing clerical experience and well trained in stores control.
(3) Maintain a well-kept bookkeeping system, either manually or using computer software. This often means employing a materials engineer to plan the flow of the right quantity of material at the right time. It will also be their duty to see that all invoices fulfil the original order and to enact reordering when requirements are not met. This duty will be in addition to that of keeping a 'goods received' book from which the invoices are checked. A flow diagram representing a workable procedure is shown in Fig. 6.4.

(4) Double signing of delivery notes, particularly ready-mixed concrete.
(5) Weighbridge spot checks for aggregate deliveries.
(6) Spot checks on moisture content for sand deliveries.
(7) Insistence on palletised delivery of bricks, etc.
(8) A thorough check of all deliveries against the delivery notes as the goods are being unloaded.
(9) Use of bar-codes, electronic and radio frequency identification (RFID) tagging of deliveries in conjunction with (3) above. RFID presents the opportunity to improve effectiveness of the purchase/supply interface, by the incorporation of inexpensive RFID tags on items such as containers, pallets and individual products. Manipulation of scanned information at various points in the supply chain enables the exact location of the product or material to be tracked from the time it leaves the factory until purchased by the consumer. Currently RFID manufacturers are striving to establish and agree a universal global standard for the technology, aimed at offering real-time supply networks driven by consumer needs and behaviour. When fully developed and industrially tested, including practicalities such as cost-effective manual tagging of pallets, reading equipment provision, etc., then having better, more real-time information about products moving through the retail pipeline should enable suppliers to provide customers with a better service, while at the same time reducing inventory levels.

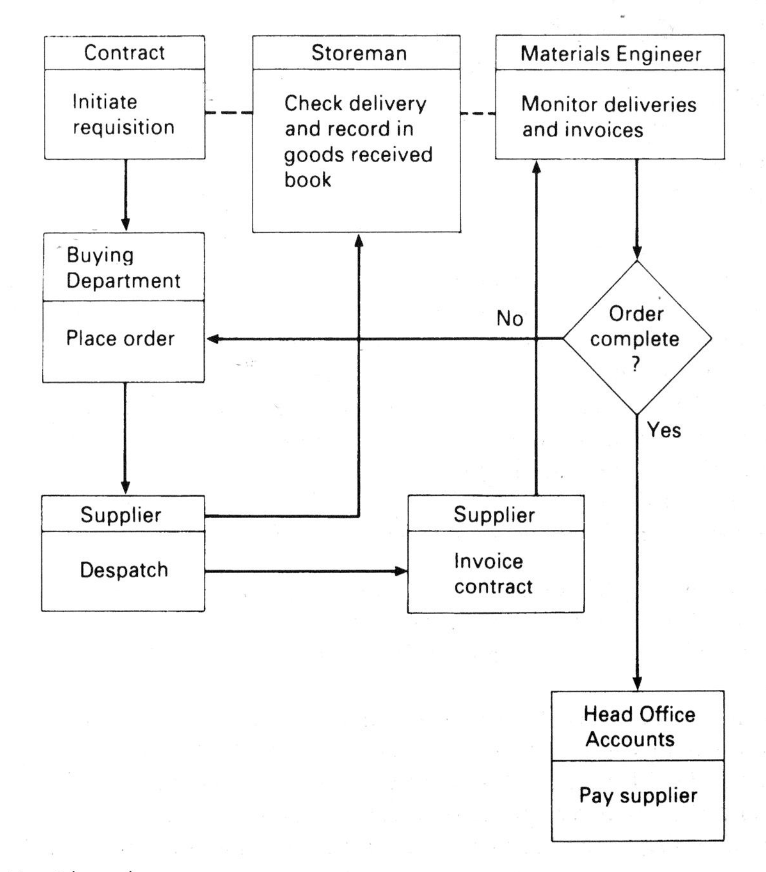

Fig. 6.4 Materials supply.

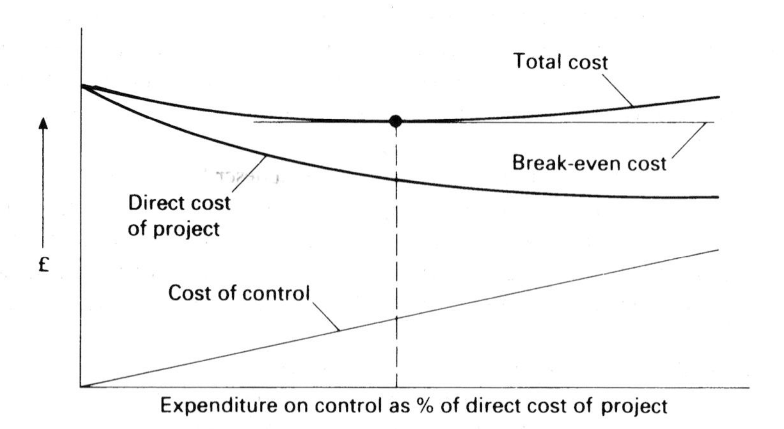

Fig. 6.5 Cost control expenditure graph.

Cost control of modernistic-type contracts

Modernistic-type contracts, such as design and build, managing agent, partnering, alliances, construction management, etc., strongly feature management of the supply chain by a principal contractor. Subcontracts commonly embrace design services, labour, equipment, materials or combinations of all these elements. While prices are usually predetermined and included as fixed or nominal sums in the tender estimates, opportunity is subsequently usually available to re-establish prices that are more realistic after construction commences when drawings and specifications become finalised. Importantly, subcontractors may sometimes also use plant, management and engineering services provided by the employing contractor which will need to be separately accounted for in the latter's cost data, including overheads.

Moreover, the main contract may involve open-book accounting, time and financial targets with incentivised rewards for the contractor(s), which will demand flexible budgetary planning and subsequent variance analysis.

Each subcontract set of activities is formed into a work package, and subsequently networked together with the other subcontracts for cost evaluation purposes. The time update of the network provides the value of work done as a 'by-product' of the calculation, and the value can be divided further by cost code, provided that the work package information is similarly divided. Thus, when incurred payments to subcontractors are recorded against the same codes, variances can be calculated for management information in a similar manner to the variances demonstrated for materials control described under the hybrid system above. Thereby a valuation document is readily provided in a manner that reflects the operations intended to be performed.

Independently, individual subcontractors may control their own costs conventionally as appropriate, valued by a bill of quantities, etc., in relation to completed work.

Points to consider when choosing a cost control system

(1) Any cost control system involves an addition to overheads. Usually the tighter the control, the more expensive will be the system; thus there is a break even point, which is clearly demonstrated in Fig. 6.5. As expenditure on cost control increases, the direct cost of the

project can be expected to decrease. However, there will come a point at which extra effort will secure no further reduction. Thus, the curve is concave upwards as shown in Fig. 6.5. The addition of the cost of control gives the total project cost and this will in most cases come to a minimum and then start rising. This provides some indication of the optimum level of effort.

(2) It is suggested that a coarse-grained system, such as that described involving no more than 15 cost codes, be used. Indeed, recording cost data becomes highly erroneous when large numbers of codes are used and for this reason alone, a coarse-grained system is likely to be the most suitable for construction-type work, but clearly can serve only as a rough check to avoid gross errors. Figures in unpublished work by Fine provide the following guidelines:
   - 30 cost headings. About 2% of items are misallocated.
   - 200 cost headings. About 50% of items are misallocated.
   - 2000 cost headings. About 2% of items are correctly allocated.

(3) Unfortunately the system described, together with almost all others found in the construction industry, is backward-looking. By this is meant that the information provided relates to some past occasion and it is frequently too late to take proper corrective action. It is to be hoped that the two forward-looking systems referred to under 'Fundamentals' will help to improve the position.

(4) Cost control systems are complex, quite expensive to operate and far from perfect. The crucial link in the system is the manager's response to the information provided. For these reasons, any cost information system, if it is to be effective, must be operated by properly motivated cost-conscious staff. Thus, top management will be well advised to institute training courses and seminars which alert those attending to the factors that minimise cost and maximise profit.

(5) Developments in computer hardware and rapid growth of commercially available software have assisted in computerising the cost control process. In principle, the task is relatively straightforward requiring cost data to be recorded and analysed against codes, descriptions, quantities, rates, etc. These developments in computer applications notwithstanding, the difficulty of collecting the data of work done on site, however, is still largely a manual chore and continues to hinder the process. Nevertheless, quite sophisticated computer packages have the ability to integrate data banks of work-study information, estimating, BOQ preparation, valuations, variations adjustments, payroll, materials and cost control and final accounts preparation for construction projects.

7
Plant management

Summary

Plant acquisition options are briefly considered, including hire, purchase, lease and hire purchase. A systematic approach in deciding the best buy for an item of construction equipment is demonstrated. The analysis is then extended to describe the factors involved in setting a realistic hire rate for the machine. Finally, some of the important aspects in the maintenance of plant are highlighted.

Plant acquisition

Generally, a construction company has two options in acquiring plant: it may either own machinery and equipment or hire it. In recent years, the growth of the independent plant hire sector of the construction industry has greatly facilitated this latter option and approximately 50–60% of plant presently used on projects is hired. Many firms, however, prefer to hire only those items of plant that are required to meet peak demand or specialised duties. The alternative decision to purchase will have important financial consequences for the firm, since considerable capital sums will be locked up in the plant, which must then be operated at an economic utilisation level to produce a profitable rate of return on the investment, as can be seen in the following example.

Example
A construction company has an annual turnover of £20 million, of which one-fifth is from equipment operated by its plant department. The expected rate of return on capital employed is 10% .The construction division has a conventional low capital base, which is turned over eight times per year, whereas the plant department derives most of its revenue directly from equipment applications and is therefore relatively capital intensive, typically displaying a turnover ratio of unity or thereabouts:

	Parent company	*Construction division*	*Plant division*
Turnover	£20m	£16m	£4m
Capital employed	£6m	2m	£4m
10% return on capital employed	£0.6m	£0.2m	£0.4m
Profit expressed as a percentage of turnover	$\frac{0.6}{20} \times 100 = 3\%$	$\frac{0.2}{16} \times 100 = 1.25\%$	$\frac{0.4}{4} \times 100 = 10\%$

If the plant division, for instance, had made only £0.2 million profit, then, for the company as a whole, the construction division would have to make £0.4 million profit, or 2.5% on turnover, to achieve 3% overall on turnover. Thus, any shortfall in plant profitability requires a monumental effort of the construction division to redress the balance.

Clearly, therefore, when the need to own plant is examined, the following points must be considered:

(1) Will the item of plant generate sufficient turnover to provide an adequate rate of return on the capital employed?
(2) Is ownership of the plant, rather than obtaining it by some other method, absolutely necessary for the business?
(3) Is outright purchase the only way of acquiring the plant?

If the answer to these questions is not a confident and positive reply, then there should be some other sound commercial reason for making the purchase.

The financing of plant

The firm, having made the decision to purchase rather than hire a piece of plant, has the following methods of arranging the finance.

Cash or outright purchase

The equipment may be paid for immediately at the time of purchase, thereby providing a tangible asset shown on the Balance Sheet. Clearly, this option is only possible if cash is available, and therefore presupposes that profits have been built up from previous trading or that funds are available from investors such as shareholders, bank loans, etc. Also, large or technically unusual contracts sometimes include monies to permit the contractor to purchase the necessary plant at the start of a project.

In some countries, tax legislation with respect to purchase allows capital allowances at a percentage of the purchase price to be set against the firm's profit during the year of acquisition. This facility is a government-inspired policy acting as an investment inducement. Therefore, before the purchase decision is taken, consideration should be given to the level of return expected from the investment to ensure that using the capital in this way is the most profitable method of investing.

Hire purchase

Purchase of plant by this method involves a contract between the purchaser and supplier of finance, in which the purchaser pays specified rentals during the contract period. At the end of this period, the title of the asset may be transferred to the purchaser for a previously agreed sum, which is often nominal in amount. The hire purchase option is particularly useful in that treatment for capital allowances is generally the same as outright purchase. Furthermore, the need for large capital sums is avoided, and the repayments on the borrowed funds can be staged progressively. Hire purchase, however, often involves high levels of interest and is frequently subject to government regulation, but not usually retrospectively.

Leasing

A leasing arrangement is fundamentally different from either outright or hire purchase in that the title theoretically never passes to the lessee (user). A lease may be defined as a contract whereby, in return for payment of specified rentals, the lessee obtains the use of a capital asset owned by another party (the lessor). Within this definition, however, there are several varieties of lease to suit the parties concerned, of which the *finance lease* and *operating lease* are two forms appropriate to plant acquisition.

Finance lease

This type of lease is generally arranged by a financial institution such as a finance house. The rental charges will cover the asset's capital cost, except for its expected residual value at the end of the lease, together with a service charge designed to meet the lessor's overheads, interest charges, servicing costs and an element for profit. The lease usually will be divided into two parts: in the *primary period,* which normally lasts three to five years, rentals are set at a level sufficient to recover all the above-mentioned costs. The *secondary* (or *continuation) period,* which could be a fixed time interval, is extended to suit the lessee's needs and may require only a nominal rental.

At the end of the lease, the asset is sold by the lessor, but not directly to the lessee, this being specifically written into the contract to comply with the terms of the current taxation legislation.

Because the lessor often has no direct interest in the leased asset, it is usual that the contract cannot be cancelled until the lessor has realised the investment at the end of the primary period. Meantime the lessee has full use of the asset as though it were owned. Unlike outright purchase, however, leasing does not afford the lessee company to take advantage of the capital allowances on the asset. But this sometimes may be beneficial to a firm that has little or no corporation tax liability or is even in a tax-deferred situation.

Operating lease

Whereas a finance lease is generally offered by a financial institution, the lessors in the case of an operating lease are likely to be the manufacturers or suppliers of the asset, whose purpose is to assist in the marketing of the item. The primary period lease would be uncancellable and charges are frequently lower than those required by a finance lease, because the type of asset involved either would have a good second-hand value or be tied to a lucrative service agreement or supply of spare parts. Indeed, the profit expected by the lessor might well come from these latter services, which would continue into the secondary period of the lease. Clearly, this type of arrangement is most appropriate for large and/or technically sophisticated plant items where the manufacturers have skilled personnel and are capable of carrying the required servicing and maintenance. Some haulage truck manufacturers have offered this method of trading when a healthy second-hand market existed.

A further advantage to the lessee is that the title remains with the lessor but, unlike finance leases, no entry is required on the lessee's balance sheet. Consequently, the capital gearing would remain unchanged and this could therefore be particularly convenient to a highly geared company, which might otherwise find difficulty in raising loan capital for direct purchase or even hire purchase of the plant item. The lease charges are considered as a business expense and so are included as costs in the profit and loss account.

Summary of plant acquisition options

(1) When capital allowances are available, it may be cheaper to lease than to buy where profits are too low to make full use of capital allowances. The tax allowances are recouped by the lessor and reflected in the lease charges.
(2) For rapidly expanding companies and others with capital raising difficulties or high gearing, the lease allows assets to be acquired and potential turnover increased without altering the firm's existing capital structure.
(3) Both leasing and hire purchase are long-term commitments, and therefore the revenue (turnover) budget and operating costs budget must be able to accommodate the rental charges over the agreed period of the financial contract.
(4) All of the above options imply direct or indirect ownership of the plant, and special facilities will be needed for servicing and maintenance, together with experienced management and skilled staff. Thus, the alternative of hiring in plant should not be overlooked.

Systematic plant selection

The decision to purchase an item of plant should be based on economic considerations because, unless it can be demonstrated that the investment will yield a satisfactory rate of return, there should be no purchase at all. However, in practice several more factors need to be considered before a decision is possible. Many makes of machine are available; sometimes the broad technical details of the products can be closely compared between different manufacturers, but often this is not possible. Intangible areas exist, such as after-sales service, maintenance, delivery and payment arrangements. They are often not quantifiable and frequently take on a disproportionate influence during the decision-making process. Hence, all the important and complex factors involved when deciding on the purchase of an item of plant or equipment should be taken into account. Only after careful consideration of all the facts, involving many separate judgements, can a decision then be reached. Obviously, the final choice is bound to be a compromise between what the manager wants ideally, and what can actually be obtained.

The problem is best tackled in a systematic and disciplined way, whereby the Kepner & Tregoe (1997) decision-making procedure developed for the design process adopted by many manufacturing companies provides a potentially suitable method for application in construction equipment selection.

The essential characteristics of a decision situation

Dixon (1966) described decision as follows:

> Decision making is compromise. The decision maker must weight value judgements that involve economic factors, technical practicalities, scientific necessities, human and social considerations, etc. To make a "correct" decision is to choose the one alternative from among those that are available which best balances or optimises the total value, considering all the various factors.

On the basis of the above, Kepner & Tregoe (1997), among others, established seven essential factors to be considered in a decision situation. This is described as Kepner and Tregoe value judgement procedure:

- Establish the overall objectives that are essential and desirable.
- Classify the objectives according to importance.
- Establish alternative choices.
- Evaluate the outcome for each alternative.
- Choose the best alternative as the preliminary decision.
- Re-evaluate the decision and assess the adverse possibilities of that choice.
- Set up contingency plans to control the effects of the final decision.

To illustrate the advantages to be gained by making decisions in such a logical and disciplined way, the following example is described.

Example: Crane selection

Problem A construction company has just been awarded a contract for the construction of a sewage works. Its policy is to purchase most of its plant requirements, as the managing director believes there is always enough continuity of work involving concrete. The site agent and planning engineer responsible for the project, after careful thought, decide that the most suitable method for placing the concrete is with a crawler-mounted crane fitted with a 1 m^3 capacity concrete skip. At this stage, the plant manager's advice is sought. A simple, but accurate, sketch of the site layout incorporating the crane's position indicates that maximum lift will be 3000 kg at 14 m radius. No such cranes are available in the plant yard, and a decision to purchase one is therefore required.

Investment analysis The crane is to be used on the contract for two years only. Calculations show that, if a ten-year working life is possible, the rate of return of capital using a discounted cash flow (DCF) method is satisfactory. The appraisal is based on the following figures:

Purchase price	£45 000
Working life	Ten years
Resale value	£2000
Running cost	Based on similar machines

The plant manager feels that sufficient work can be obtained for such a crane during a ten-year period, providing that it is possible to convert the machine for dragline and backhoe use, and decides to go ahead with a purchase on that basis having rejected the possibility of hiring a machine. There are many cranes of this type on offer, by British and by foreign manufacturers. The problem is to decide which is the best buy to suit both the short-term needs of the contract and the possible other uses for the crane during its ten-year working life.

Procedure

(1) Choose the overall objectives that are essential and desirable

The individuals involved in selecting the crane, the agent, engineer and plant manager, have so far only loosely established the specification of the machine requirements. When small items of plant are chosen, such information and some experience with a few of the popular manufacturers may be enough evidence on which to base the purchase decision, but too often important purchases are also dealt with in this way, and this has led to some very poor investments. To reduce this possibility, the plant manager decides to adopt a more systematic procedure. This demands that a precise list of requirements be written down, including advice from the foreman or crane driver. These are shown as objectives A–G, 1–23 on the decision analysis sheets (Tables 7.1–7.3). It is recommended that the objectives be classified into two groups,

Table 7.1 Evaluation of craneage selection—Crane no 1.

	Objectives				Alternative			
Ref.	MUSTS				ARB-38	Go/No go		
A	Track-mounted crane				Yes	✓		
B	At least 25 m boom as lift crane				36.6 m max.	✓		
C	Capable of conversion to dragline				Yes	✓		
D	Capable of conversion to backhoe				Yes	✓		
E	Capable of lifting 3 tonnes at 14 m radius				3.6 tonnes	✓		
F	Max. purchase price £45 000				Yes	✓		
G	Conforms to ISO statutory regulations				Yes	✓		
No.	WANTS	E/T	Ranking		Information	Rating	Weighted score	
1	Delivery less than ten weeks	E	10		Six weeks	8	80	
2	High lifting capacity	T		10	30 480 kg max.	8		80
3	Easy dragline conversion	T		9	One day	5		45
4	Easy backhoe conversion	T		9	One day	5		45
5	Good service facilities	E		8		7		56
6	Low maintenance costs	E/T	7	7		9	63	63
7	Low running costs	E	9			8	72	
8	Simple to set up on site	T		6	½ h	9		54
9	Low ground-bearing pressure	T		4	0.55 kg/cm^2	7		28
10	Good safety features	T		8		6		48
11	Simple driver controls/view	T		5		4		20
12	Independent third drum	T		2	500 × 75 mm	10		20
13	Air clutch control	T		4		10		40
14	Fast slewing speed	T		7	3.55 rpm	7		49
15	Fast hoist speed	T		7	45.5 m/min	7		49
16	Good manoeuvrability around site	T		4	1.52 km/h	6		24
17	Good range of spares	T		8		2		16
18	Long working life/good trade-in price	E	7		Ten years – £2 050	8	56	
19	Air control of all movements	T		3		9		27
20	Audible and visible safe load indicator	T		10		2		20
21	Automatic power boom lowering	T		2		10		20
22	Safety crawler brakes	T		6		10		60
23	Low purchase price	E	10		£44 050	7	70	
E	Economic element	27%	43					
T	Technical element	73%		119			341	764
	Totals	100%	162				1105	

Remarks: All MUSTS OK

Table 7.2 Evaluation of craneage selection—Crane no 2.

	Objectives				Alternative			
Ref.	MUSTS				Koppelstein 4B	Go/No go		
A	Track-mounted crane				Yes	–		
B	At least 25 m boom as lift crane				27.4 m max.	–		
C	Capable of conversion to dragline				Yes	–		
D	Capable of conversion to backhoe				Yes	–		
E	Capable of lifting 3 tonnes at 14 m radius				3.4 tonnes	–		
F	Max. purchase price £45 000				Yes	–		
G	Conforms to ISO statutory regulations				Yes	–		
No.	WANTS	E/T	Ranking		Information	Rating	Weighted score	
1	Delivery less than ten weeks	E	10		Nine weeks	7	70	
2	High lifting capacity	T		10	34 860 kg max.	7		70
3	Easy dragline conversion	T		9	4 h	8		72
4	Easy backhoe conversion	T		9	4 h	8		72
5	Good service facilities	E		8		2		16
6	Low maintenance costs	E/T	7	7		2	14	14
7	Low running costs	E	9			4	36	
8	Simple to set up on site	T		6	½ h	8		48
9	Low ground-bearing pressure	T		4	0.5 kg/cm²	8		32
10	Good safety features	T		8		3		24
11	Simple driver controls/view	T		5		4		20
12	Independent third drum	T		2	Not available	0		0
13	Air clutch control	T		4	Mechanical	1		4
14	Fast slewing speed	T		7	4 rpm	8		56
15	Fast hoist speed	T		7	35 m/min	5		35
16	Good manoeuvrability around site	T		4	2 km/h	7		28
17	Good range of spares	T		8		2		16
18	Long working life/good trade-in price	E	7		Eight years – £4 000	8	56	
19	Air control of all movements	T		3	Mechanical	1		3
20	Audible and visible safe load indicator	T		10		7		70
21	Automatic power boom lowering	T		2		10		60
22	Safety crawler brakes	T		6		9	90	
23	Low purchase price	E	10		£40 125			
E	Economic element	27%	43					
T	Technical element	73%		119			266	640
	Totals	100%	162				906	

Remarks: All MUSTS OK

Table 7.3 Evaluation of craneage selection—Crane no 3.

	Objectives	Alternative	
Ref.	MUSTS	Vickerman Lion	Go/No go
A	Track-mounted crane	Yes	✓
B	At least 25 m boom as lift crane	35.4 m max.	✓
C	Capable of conversion to dragline	Yes	✓
D	Capable of conversion to backhoe	No	✗
E	Capable of lifting 3 tonnes at 14 m radius	4.1 tonnes	✓
F	Max. purchase price £45 000	Yes	✓
G	Conforms to ISO statutory regulations	Yes	✓

Remarks: Does not meet MUST requirement D and is therefore eliminated.

economic (E) and technical (T), so that in the final analysis the importance of each is demonstrated.

(2) Classify the objectives according to importance

A careful examination of the listed objectives will reveal that some are essential; some are important but not critical to the machine's function while others are desirable and would be useful if available.

The essential objectives are taken aside and labelled MUSTS. They set the limits that cannot be violated. The possible alternatives are then easily recognised as WANTS, and some WANTS will always be more important than others. The next step therefore is to rank the importance of the WANT objectives (with the inclusion of the MUSTS if appropriate). This is done by attaching a ranking expressed on a suitable scale. In practice, a 1–5 or 1–10 system has been found to be a good compromise, with the importance of the objective increasing with numerical value. The ranking is done on a subjective assessment based on experience.

(3) Establish alternative choices

At this juncture, the plant manager selects those machines on the market that appear to satisfy the MUST objectives. In this case three models only are found:

- ARB-38
- Koppelstein 4B
- Vickerman Lion

This stage should be thorough, but since only the MUSTS are considered the search should not prove to be too tedious. (Much information is available in standard publications, but a good plant department ought to have an up-to-date library of manufacturers' literature.)

(4) Evaluate the outcome for each alternative

Those cranes that do not comply with all the MUST objectives are eliminated. Thus, unsuitable machines are eliminated early on, without the plant manager and his staff devoting too much effort to a detailed analysis. Next, the WANT objectives for each alternative left in the analysis are rated. Again, there arises the problem of choosing the best rating system. The 0–5 or 0–10 method has been widely used. A value of 0 indicates that the alternative does not meet the objective in any way. Both the ranking and rating systems should be used boldly or there will be insufficient 'spread' of the total values in the final result. However, before any analysis of this sort is attempted some of the potential difficulties must be considered.

Economic and financial considerations

In the example, cost is a critical factor, but the advertised prices for the alternatives are not always easy to compare. Table 7.4 demonstrates this problem. The information assembled for each crane varies in this detail and any decision on price can only be a compromise. But a meaningful comparison is often possible within the slight differences in each manufacturer's equipment, given that the plant manager has sufficient experience with the type of plant in question. It must also be remembered that discounts and payment arrangements are an important cost consideration.

Furthermore, the question of a spares holding policy needs to be considered. The importance of the future availability and costs of spares from different suppliers can of course be reflected in the ranking and rating of the desired objectives. However, if the company is already geared up to equipment from a manufacturer and in consequence is carrying stocks of spares, then this situation may override all other considerations. This question of spares must be carefully studied when comparing 'home' and foreign makes, as the cost of foreign-made spares may be affected by changes in the international value of currencies. This aspect should be taken into account in the ranking and rating assessments, when considered necessary.

Technical and performance standards

The plant manager's decision is influenced by factors relating to the technical and performance standards of the equipment. Some may form MUST objectives, while others appear in varying degrees of importance as WANT objectives. Many details can be easily obtained from the manufacturer's literature, such as lifting capacity, machine weight, etc. For other factors, the answers are more intangible, and more subtle sources of information must be found, for example:

Table 7.4 Cost comparison of alternative crane purchases.

	Cost (£)		
Description	ARB-38	Koppelstein 4B	Vickerman Lion
Base machine with:			
* 700 mm wide shoes	31 000	27 000	34 000
* 800 mm wide shoes	31 200	27 150	N/A
Basic boom (1 0.67 m)	Inc.	Inc.	Inc.
Boom inserts:			
* (1.52 m) Varies	(175 ea.)	(225 ea.)	(250 ea.)
* (3.05 m) slightly with	(200 ea.)	(305 ea.)	(325 ea.)
* (4.57 m) each make	(300 ea.)	(450 ea.)	(375 ea.)
Max boom length (27.4m)	1 800*	2700	2 250*
Fly jibs:			
* (4.75 m)	300	310	350
* (6.1 m)	375	400	450
* (7.62 m)	400	475	505
* (9.14 m)	500	550	600
* (Strut)	440	400	375
Counterweight	400	375	500
Hook blocks:			
* 20 tonne	200	180	150
* 10 tonne	175	100	105
* 5 tonne	80	60	75
Telescope boom stops	1500	Inc.	Inc.
Pendant suspension ropes		Inc.	Inc.
Fairlead		Inc.	Inc.
Mast		Inc.	Inc.
Safe load indicator	1800	1000	900
Cab accessories	Inc.	100	110
Parts to convert to 1 m^3 dragline	3000	2900	4000
Parts to convert to 1 m^3 backhoe	4500	5000	N/A
1 m^3 bottom opening concrete skip	150	170	160
Max. total cost	£44 050	£40 125	£42 560

* Boom length can be extended beyond 27.4 m.

- Site demonstrations
- Further discussion with manufacturers
- Company records for similar equipment
- Personal contacts in the plant industry
- Discussions with machine operators

With the details assembled for all the objectives, the task of evaluating each alternative can now go ahead. The MUSTS are assessed on the Go/No go basis and those alternatives violating any objective are eliminated. A comparison of the three cranes proposed reveals that the Vickerman Lion does not have the facility for a backhoe conversion. Hence, although this machine passed the preliminary screening, it must now be marked No go against Objective D and eliminated from further consideration.

The WANTS have previously been considered carefully and ranked on the 1–10 scale, so now each alternative is rated on the 0–10 scale against each objective. By multiplying each individual ranking by its corresponding rating, the weighted score for each objective is determined. This weighted score represents the performance of the alternative against its objective. By adding each of the individual weightings, the total for each alternative gives the relative position in regard to the specified objectives. In this example, the total weighted score for the two remaining alternatives is as follows:

	Percentage of total	Weighted score	
Element		ARB-38	Koppelstein 4B
Cost and service element (E)	27	341 (31%)	266 (29%)
Technical element (T)	73	764 (69%)	640 (71%)
Total	100	1105	906

Note: It must be remembered that the numbers are derived largely by subjective judgement based on careful thought and analysis of the available facts, and do not make the decision in themselves. But this systematic approach does assist in coping with many independent facts which otherwise could not be easily related.

(5) Choose the best alternative as the preliminary decision

The alternative that receives highest weighted score is presumably the best course of action to take, in this case the ARB-38. It is, of course, not a perfect choice but is at least one that draws a reasonable balance between the good and bad features of the machine. Fortunately, both the technical and cost element totals are greater for the ARB-38. The decision would be far more difficult if one of the elements were larger for the Koppelstein 4B crane. Some further balance would then be needed between economic resources and the technical benefits coming to the company.

The preliminary choice is the one that best satisfies the objectives overall. It is thus a compromise, as undoubtedly the alternatives have some superior features. But the method used shows the manager how he arrived at this decision and therefore where the possible pitfalls may lie.

(6) Re-evaluate decision and assess adverse possibilities of that choice

When many alternatives are available, the weighted scores of one or two are sometimes quite close, so, before the manager makes the final decision, any adverse consequences must be considered. Snags, potential shortcomings or anything else that could go wrong must be looked for. The probability (*P*) of the adverse consequences should be assessed and a seriousness weighting (*S*) given to its possible effect. An expression of the total degree of threat may be obtained by multiplying the seriousness weighting by the probability estimate.

In this example, the plant and site managers arrange a meeting to discuss the implications of the proposals.

(a) Possible adverse consequences of the ARB-38
A strike threat is rumoured at the factory, which may delay delivery.
A closer investigation reveals that this type of situation has arisen before with a strike only occurring twice in the past ten years, but strikes have not increased in the past four or five years. The probability of a strike is therefore assessed at a 20% chance of occurring. The duration of the first strike was only one week but the last strike, two years ago, lasted five weeks. However, in view of the existing climate between management and unions, a prolonged strike this time is not envisaged. Should a hire crane be necessary for, say, three weeks, the extra cost to the contract is £250 per week plus transport cost of £50. The degree of threat is therefore $(3 \times 250 + 50) \times 0.2 = £160$.

(b) Possible adverse consequences of the Koppelstein 4B
A serious accident has just revealed a major design fault in the clutch mechanism of the winch.
A temporary modification will be made to all existing models and stocks held. But a completely new part is not available for six months and will require taking the crane out of service for one week at that time. The probability that this service is required is 90% and the cost of hire for another crane is £250 per week plus £50 transport cost of hire for another crane, thereby directly increasing the contract cost at this rate. The degree of threat is $(1 \times 250 \times 50) \times 0.9 = £270$. The choice, therefore, is still the ARB-38. Had the consequences been reversed, then a much closer examination of the advantages and disadvantages would be necessary, making the decision much more difficult.

(7) Set up contingency plans to control effects of the final decisions

The adverse consequences represent potential problems. They must be prevented from causing too much inconvenience. This is done either by taking preventative action to remove the cause or deciding upon a contingency action if the potential problem occurs. In our case, the simple remedy is to ensure that alternative machines are available for hire at suitable times or arrange work on site so that interruption is minimal.

Comments

The decision reached is to buy the ARB-38 and this is the best choice based on the judgement of the plant manager concerned. For someone else with different experience, the assessment of the objectives might be quite different and would possibly produce a different result. But even accepting the obvious weakness, the method at least forces managers to consider most of the facts. It provides a record of the thought processes and will help to sort out points of confusion, for it is almost impossible to memorise all the facts and relationships, except for the very simple choices. In such cases, the decision process can be adequately carried out mentally without the need for the elaborate method described. Finally, the decision process is written down on paper and is available for all to see.

Computer applications

Increasingly, plant owners and hire firms are recording the information needed for high rate calculations, maintenance schedules, purchase prices, dealer facilities, budgeting and cost control, etc. on computer file. Also, equipment specifications are increasingly being made available on computer disk, CD-ROM, etc. Indeed, the selection process is readily achievable with commercially available word processing, spreadsheet and database management software by storing all the relevant information on comparable machines on file in a database, and then subsequently carrying out ranking, rating and weighting manipulations in a spreadsheet.

Setting hire rates

Setting a realistic hire rate is not an easy task, particularly in times of rapid inflation. Factors largely outside the control of the company make it almost impossible to forecast the scrap value, the utilisation factor and running costs over a period of perhaps as long as ten years. Most companies periodically review the high rates to keep them in line with the movements of prices and costs, a policy that will not necessarily generate the replacement cost. To take care of this problem an inflation allowance should ideally be built into the calculations from the outset. This is not easily done when inflation fluctuates over short periods.

Conventional method of tabulating the hire rate

It is usual for companies who own plant to set the hire rate per hour for use in estimating purposes when tendering for new work or for charging an item of plant against a contract. The factors involved are:

- Ownership cost
  - The capital cost of the plant
  - Depreciation
- Operating cost
  - Running and repair costs
  - Licence and insurance cost
  - Overheads

Capital

Quite often, the capital needed to purchase an item of plant is raised through a loan. The interest payments are a charge and therefore should be incorporated into the hire rate. Even when the plant is bought using reserve capital, an equivalent charge should again be allowed since the money, if not used to buy plant, could be earning interest.

Running costs

The cost of maintenance and repairs varies considerably for different types of plant, conditions of work and also with the hours worked. Experience with similar equipment and the keeping of detailed records are the only ways of estimating such costs. The allowance is normally expressed as a percentage of the initial cost, but this is far from an ideal method, since most plant requires more maintenance as it becomes older.

The cost of fuel and lubricants varies with the size, type and age of the equipment. Again, experience is the best guide, but the manufacturer's data does provide figures which, with reasonable judgement, can be used to estimate the machine's consumption of such materials.

Licences and insurances

The amount and type of insurance required is determined by the use of the plant; specifically, whether the plant is to use public roads or not. For plant that will not be used on public roads, the premium is quite small and may only cover fire and theft. Plant using the public highway will naturally have to be covered by the minimum legal requirements, as do any other road users. Again, licences are a small charge when the plant is not used on public roads but become a considerable expense otherwise.

Lifting appliances such as cranes have to be inspected by an insurance company, which adds an additional cost.

Establishment charges

It is usual these days for a contracting company to group its plant either into a separate profit-making division or at least into some sort of self-contained unit to provide a service function. Therefore, an allowance for all the normal administration and other overheads needs to be included.

Depreciation

Depreciation is the loss of value resulting from usage or age. A contractor normally recovers this loss by including a sum of money equivalent to the depreciation cost in his rates for doing the work or hiring out the plant. In normal practice, the revenue is used for many other purposes and not allowed to accumulate idly. When the asset is finally replaced, the capital is either borrowed or withdrawn from the company cash balances. However, several factors can interfere with such arrangements: inflation may mean that the revenue was not enough to replace the old asset, or perhaps the depreciation allowances were misjudged. The only recourse thereafter is extra borrowing.

Methods of depreciation

In the following paragraphs, some fairly common methods of calculating depreciation are described. The reader should note that the choice of method is arbitrary and not of crucial importance. However, depending on tax legislation, the company's profit level may be affected by its depreciation policy, as the amount of depreciation that may be offset against profits in the form of capital allowances is subject to the approval of the Inspector of Taxes.

Straight-line depreciation

Example

It is decided to purchase a mechanical excavator costing £42 000 to work an average 2000 hours per year. The life of the machine is expected to be ten years, after which time the salvage value will be £2000.

Purchase price	£42 000
Residual value	£2 000
Total depreciation	**£40 000**
Annual depreciation	£4 000
Hourly depreciation	£2

Table 7.5 Declining balance depreciation example.

End of year	Depreciation (%)	Dep. for year (£)	Book value (£)
0	26.2	0	42 000
1	"	11 004	30 996
2	"	8 121	22 875
3	"	5 994	16 881
4	"	4 423	12 458
5	"	3 264	9 194
6	"	2 409	6 785
7	"	1 778	5 007
8	"	1 312	3 695
9	"	968	2 727
10	"	727	2 000
Total		£40 000	

Declining balance depreciation

Using the above example, this time the depreciation is made on a fixed percentage basis rather than the fixed sum. Table 7.5 is more easily calculated from the formula

$$d = 1 - \sqrt[n]{\frac{L}{P}} \times 100$$

where L = salvage value, P = purchase price, n = life of asset and d = percentage depreciation.

Sinking fund method of depreciation

A fixed sum is put aside from revenue each year and invested with compound interest throughout the life of the asset. After successive instalments, the sum accumulates to produce the original purchase price less the scrap value.

Taking the given example and assuming 6% interest is earned on savings, the annual amount required is £3034, i.e.:

Sinking fund deposit factor over ten years = 0.07586

Therefore annual payment = 0.07586 × (42 000 − 2 000) = £3034

Table 7.6 gives a detailed breakdown of the analysis.

Sum of digits method

Life = ten years

Digits = 1 + 2 + 3 + 4 + 5 + 6 + 7 + 8 + 9 + 10 = 55

Table 7.7 illustrates the sum of digits depreciation.

Table 7.6 Sinking fund example.

Year	Payment (£)	Interest (£)	Depreciation (£)	Book value (£)
1	3034	0	3034	38 966
2	3034	182	3216	35 750
3	3034	375	3409	32 341
4	3034	581	3615	28 726
5	3034	798	3832	24 894
6	3034	1028	4062	20 832
7	3034	1271	4305	16 527
8	3034	1529	4563	11 964
9	3034	1803	4837	7127
10	3034	2093	5127	2000
Total			£40 000	

Table 7.7 Sum of digits depreciation example.

Year	Factor	Depreciation (£)	Book value (£)
1	10/55	7273	34 727
2	9/55	6545	28 182
3	8/55	5818	22 364
4	7/55	5091	17 273
5	6/55	4364	12 909
6	5/55	3636	9273
7	4/55	2909	6364
8	3/55	2182	4182
9	2/55	1454	2728
10	1/55	728	2000
Total		£40 000	

Free depreciation

The asset is totally depreciated initially.

Graphical comparison of the depreciation methods

Figure 7.1 demonstrates the main features of the declining balance method and the sum of digits method. In the early years the asset is heavily written down, which is particularly helpful in the case of construction plant as the repair and maintenance costs are likely to be low when new, but increase with age and usage.

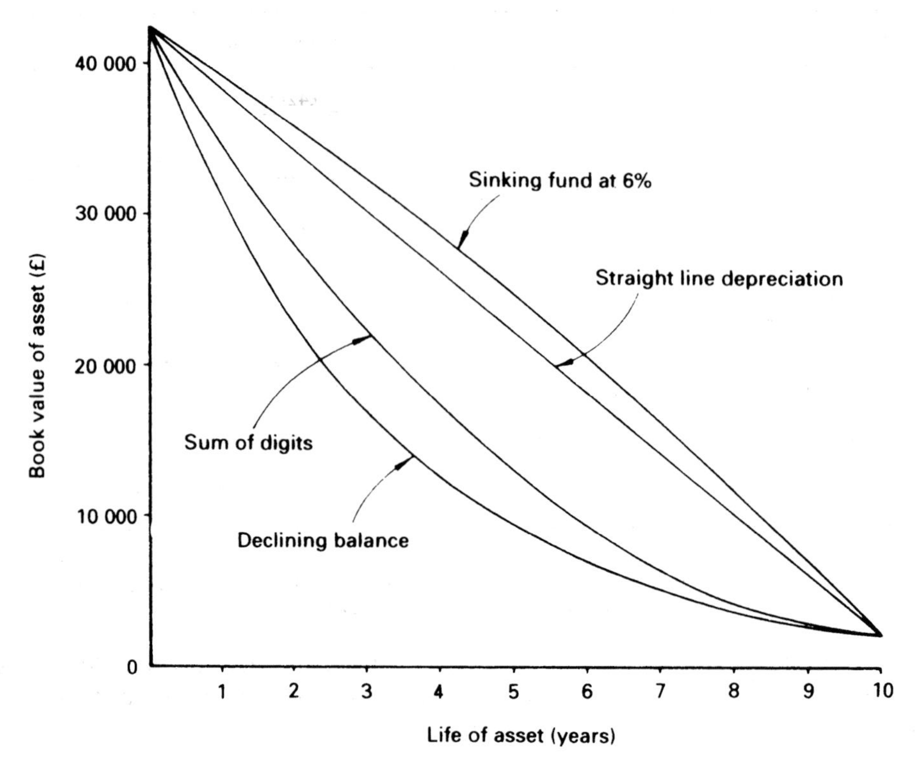

Fig. 7.1 Graphical comparison of depreciation methods.

Example using the crane selection study

Calculate the hourly hire rate for the ARB-38 given the following information:

Initial cost	£44 050
Resale value	£2 050
Average working hours per year	2 000
Years of life of machine	10
Insurance premiums per year	£200
Licences and tax per year	£100
Fuel at 20 litres per hour	£0.10 per litre
Oil and grease	10% of fuel cost
Repairs and maintenance	15% of initial cost per year
Required rate of return on capital	15%

Note: Overheads not included for simplicity.

Item	£ per annum
Depreciation—straight line	$= \frac{£42\,000}{10} = £4200$
Interest on finance, expressed in terms of an annual mortgage type payment	$= \frac{44\,050 \times 0.199 \times 10 - 44\,050}{10} = 4361$
Insurance and tax	= 300
Ownership cost	= **8 861**
Fuel (litres)	= 20 × 0.199 × 2000 = 4000
Oil and grease	= 400
Repairs	= 15% × 44 050 = 6608
Operating cost	= **11 008**
Total cost	= **19 869**
Marginal hire charge	$= \frac{19\,869}{2000} = £9.93$ per hour

Note: When considering other types of equipment, e.g. rubber-tyred excavators, the cost of the tyres should be subtracted from the purchase price and then analysed separately because the rate of depreciation is likely to be much higher on the tyres. This may also apply to tracked machines working in some ground conditions.

The above calculations allow for the generation of sufficient income to replace the asset, to cover operating costs and to provide a return on the initial capital invested. An additional sum must be added to cover company overheads and profit.

Alternative analysis using discounted cash flow

The method described above is widely used and has stood the test of time but no account is taken of the timing of the cash flows. A more logical way of doing the analysis is to express the problem in the way that it actually occurs, i.e. as a series of cash flows. Then an effort is made to balance the outgoings with adequate income to yield a satisfactory return. The problem is thus restricted as shown in Table 7.8. For more information on discounted cash flow, refer to Chapter 12.

Table 7.8 Calculating a crane hire rate using DCF method.

Year	Capital (£)	Scrap value (£)	Operating costs (£)	Tax and insurance (£)	Total cash out (£)	Cash in (£)	Total (£)
0	−44 050		0	−300	−44 350	0	−44 350
1			−11 008	−300	−11 308	×	× −11 308
2			−11 008	−300	−11 308	×	× −11 308
3			−11 008	−300	−11 308	×	× −11 308
4			−11 008	−300	−11 308	×	× −11 308
5			−11 008	−300	−11 308	×	× −11 308
6			−11 008	−300	−11 308	×	× −11 308
7			−11 008	−300	−11 308	×	× −11 308
8			−11 008	−300	−11 308	×	× −11 308
9			−11 008	−300	−11 308	×	× −11 308
10		+2050	−11 008	0	−8 958	×	× −8 958

Clearly, to return 15% on the investment, x must be sufficiently large so that, when cash out and cash in are reduced to net present value, the last column (total) is in balance. Expressed mathematically, the equation is:

$$44\,350 = (x - 11\,308) \times 4.7715 + (x - 8958) \times 0.24718$$
$$44\,350 = 4.47715x - 53956 + 0.2472 - 2214$$
$$100\,520 = 5.02x$$
$$x = £20\,024$$

The annual receipts must be £20 024 or £10.01 per hour hire charge.

Notes: (1) 4.7715 is the present worth factor of a uniform series for nine years at 15% interest: (2) 9.2472 is the present worth factor of a single sum at Year 10 for a 15% interest rate.

The effect of inflation

The traditional method of allowing for inflation in the hire rate is simply to increase periodically the hire rate in line with the movement of prices, but usually only to compensate for the *previous* year's inflation. The effect of this policy is shown in Table 7.9, where all figures are adjusted to take account of a 10% annual rate of inflation including the extra cost of replacing the asset. The results of Table 7.9 indicate that the normal policy of compensating for inflation is totally inadequate if sufficient capital is to be generated to allow for the increase in the price of a new asset (up from £44 050 to £113 811) and still provide the desired rate of return of 15%. The hire rate should be calculated in the context of the whole series of cash flows, which can then be adjusted for any rate of inflation.

Table 7.9 The effect of inflation on the rate calculation.

Year	Capital (£)	Scrap value (£)	Operating cost (£)	Tax and insurance (£)	Total cash out (£)	Cash in (£)	Total cash flow (£)	Adjusted cash flow (£)	Net PW at 10% (£)
0	−44 050		0	−300	−44 350	0	−44 350	−44 350	−44 350
1			−12 108	−330	−12 438	20 024	+7 586	+7 586	+6 890
2			−13 318	−362	−13 680	22 026	+8 346	+8 346	+6 890
3			−14 650	−399	−15 049	24 227	+9 178	+9 178	+6 890
4			−16 115	−439	−16 554	26 664	+10 095	+10 095	+6 890
5			−17 728	−483	−18 211	29 315	+11 104	+11 104	+6 890
6			−19 500	−531	−20 031	32 248	+12 217	+12 217	+6 890
7			−21 451	−584	−22 035	35 472	+13 437	+5 936	+3 040
8			−23 595	−643	−24 238	39 020	+14 782	–	
9			−25 955	−707	−26 662	42 921	+16 259	–	–
10	−69 761*	+5 317	−28 551	0	−23 234	47 214	−45 754	–	–
									Approx. NIL

* Extra cost of replacing the asset at the prices prevailing at Year 10.
Return on capital is 10%. Compare this rate with i_r, the real rate of return given in Chapter 13.

Marginal costing

Marginal costs arise directly from the production process, which for a plant hire company would

Table 7.10

£00s	A	B	C	D	E
Hire revenue obtained over previous week	10	9	5	15	10
Marginal/labour costs	2	1	1	2.5	1
Marginal/material costs	1.5	1	0.5	2.5	1.5
Marginal/expenses	1	1	0.5	1.5	1
Contribution	5.5	6	3	8.5	6.5
Contribution per £1 revenue	45p	67p	60p	57p	65p

be largely connected with maintenance and servicing of the equipment and therefore vary directly with the hiring activity. In contrast, fixed costs generated from the establishment charges fluctuate very little with hiring levels. The purpose of the marginal costing method is to calculate the *contribution* made by each item of equipment on hire towards the fixed costs of overheads, depreciation, interest charges and profit of the business.

This technique can be used to advantage during a short-term period when the market demand is low and hire rates need to be keen to attract custom, the contention being that any hire revenue which exceeds the marginal costs makes a contribution towards the fixed costs. However, such a pricing policy should be considered only during a short and difficult period, since the endeavour must be to realise the budgeted profit for each item over the set financial periods.

Thus, as seen in Table 7.10 for plant item C, for example, a hire rate exceeding £500 per week contributes to the fixed overhead, which may be a better alternative than leaving the machine idle.

Moreover, the method gives a clear indication that the firm should be directing sales effort on plant items B and E, as these machines obtain favourable hire rates and give the best contribution towards fixed costs.

Plant maintenance

A contractor that owns plant must be prepared to provide maintenance and servicing of the equipment if economic levels of utilisation are to be obtained. Effective maintenance is expensive and requires depot facilities, workshops, experienced staff, etc. Many firms have tried to avoid these costs by providing the minimum of maintenance, and the result has often been unexpected breakdowns, lost production and inefficient machinery. Alternatively, the contractor can implement a system of planned maintenance, which can be broadly divided into either preventive or corrective options.

Corrective maintenance

This approach involves both running and breakdown maintenance capabilities sufficient to handle repairs and component changes on site. It is commonly selected by construction plant operators as a cost-effective option by using mobile workshop facilities as an auxiliary service. Typically this would comprise a well-equipped mobile vehicle, such as an 'off-roader', and fitters, with major repairs and overhauls, which are usually outsourced to a specialist maintenance provider.

Fixed time to maintenance (FTTM)

This system requires the implementation of planned regular procedures, aimed at reducing wear, maintaining the plant in good working condition and significantly reducing unforeseen stoppages. The maintenance actions are as follows:

- Daily servicing and superficial inspection performed roughly half an hour before and after working hours
- Regular full maintenance and inspection, including periodic overhaul
- Replacement or repair of component parts within a working period based on the expected duties and conditions

This system falls short of the more comprehensive planned preventive maintenance (PPM) strategy adopted for example by the airlines industry, because failure intervals cannot be accurately established due to the variable working conditions encountered on site; hence minimal breakdowns are expected. FTTM can be properly instituted only by firms with adequate workshops, experienced staff and the ability to substitute other plant whilst major repairs and/or full overhaul is undertaken. Consequently the system requires running, breakdown and shut-down maintenance facilities.

Condition based predictive maintenance (CBPM)

CBPM extends the principles of FTTM by incorporating detailed measurements of wear into the evaluation of component condition. A plot of the actual wear process under field operations enables the appropriate time for replacing the component to be ascertained. Techniques such as used oil analysis, tribology, ferrography, thermography and vibration monitoring provide the relevant information, particularly for engine, transmission and gearing units. Indeed the rapid development of these technologies has led to the adoption of CBPM as the principal maintenance system by the transport and manufacturing sectors of industry.

Statistically based predictive maintenance

All component failures and equipment stoppages are accurately recorded and statistical models for predicting component replacement intervals developed. The approach provides valuable supplementary data for implementation of the full PPM system.

Monitoring of maintenance

In order that the maintenance policy may be executed efficiently, it is necessary to install a recording and costing system, which will include:

(1) An asset register comprising an inventory of each plant item in the fleet, with information on the date of purchase, registration or code number, purchase price, current value, location, hours operated, etc.
(2) A maintenance schedule indicating the type of maintenance and servicing required on each plant item together with the time intervals between each plant maintenance operation.

(3) Job cards filled out by the fitter, commonly recorded on an ongoing basis with an electronic hand-held palm pad each time maintenance work is performed, and which should include the date, description of the work done, materials used, time taken, recurring defects, etc.
(4) History record cards. The information on the job card is transferred for each individual machine to its history card, normally a computer file, together with the hours operated and fuel used. The monthly records are then abstracted to prepare costs for comparison with budgeted values. The variances are subsequently used in controlling maintenance, adjusting the hire rate and ultimately in making decisions with respect to replacement or sale of the item of plant.

References

Dixon, J.R. (1966) *Design Engineering—Inventiveness Analysis and Decision Making*. McGraw-Hill, New York.

Kepner, C.H & Tregoe, B.B. (1997) *The New Rational Manager – An Updated Edition for a New World*, Kepner–Tregoe, Princeton, NJ.

Section two

Business management

8
Project procurement

Synopsis

The latest arrangements for procuring construction work are described, covering project management, management contracting and design and build including PFI/PPP. Traditional contracts are also briefly explained and contrasted with newer systems. Checklists of duties for clients, designers, managing firms and subcontractors are included for guidance.

Introduction

The aim is to achieve the following: quality of the built environment; sustainable and sensitive land use; inspired, sympathetic and aesthetically appealing architecture; affordable functionally well designed and constructed domestic, commercial and industrial buildings, facilities and infrastructure; incorporating high standards of in-built utility/health and safety/energy efficiency/maintainability/environment friendly features.

To do this requires: enlightened and informed clients/developers/promoters/patrons who act in concert with imaginative and able project leaders or advisers, reinforced by competent designers and contractors, all having a common purpose.

Hence, the importance of selecting the appropriate contractual arrangements for the party or parties engaged to undertake the necessary responsibilities and functions, namely:

- Funding and finance
- Aesthetic aspect, form, function and utility
- Design, engineering and cost planning
- Construction
- Operation and maintenance

All of these have to be carried out within the normal bounds of official building controls and regulations. Tam and Harris (1996) identified the following key attributes likely to impact upon the project and parties involved:

- Extent of design complete before construction commences.
- Complexity of the project
- Experience of the contractor's project manager
- Contractor's past performance
- Supervision of quality and progress by client's project leader

The construction process

In providing a building, civil engineering project or engineering construction facility, whether as separate or combined elements of feasibility investigations, design or actual construction, the primary concern is to ensure client satisfaction for the finished product, i.e. fulfilment of conception, function, cost, time, quality, utility and after-sales service objectives.

To this end, the process typically embraces the phases illustrated in Fig. 8.1, as explained in the following sections.

Stage 1. Verification of need

At this initial stage, the terms of reference are established, assisted by an external adviser as required in the case of an inexperienced sponsor. Aspects to be investigated include: determining the need and objectives for the project, identifying who will act as the project sponsor in the client organisation, identification of other stakeholders, prioritising the constraints regarding time, cost, quality, finance, legal issues, impact on the business, etc., including for publicly funded projects, such as local government, fundamental issues concerning demonstration of 'Best Value' procurement. Subsequently, a decision to carry out further study on the potential project options can be implemented.

Stage 2. Assessment of options

The potential project options are evaluated for feasibility and value appraised with techniques such as Value Management (VM), Risk Management (RM) and Cost Benefit Analysis (CBA). A firm business case is subsequently prepared, giving for each option a physical outline as appropriate, base estimate, investment and financial appraisal, risks/life-cycle costs, time plan and technical issues, together with a provisional budget. Thereafter, the decision to go ahead with the project can be made based upon the options information and a Project Manager/Leader appointed to act for the client, which in the case of an experienced client in building procurement might be an in-house function; otherwise, an external specialist professional organisation is sought.

Stage 3. Develop procurement strategy

The project manager/leader in conjunction with the client reviews the business case, and takes on the major task of preparing the project brief covering details such as execution plan, performance criteria, constraints, budget, control and reporting procedures. In particular, VM/RM techniques are further applied to assist in the evaluations; thereafter the decision to proceed with the project can be reconfirmed, the procurement strategy articulated and the perceived best approach selected.

Stage 4. Implement the procurement strategy

The most appropriate contract category (separated, management, integrated or discretionary contract), contractual arrangement or type and form of agreement are chosen, which may have already become evident in the business plan and project brief. The best team/parties to deliver the business solution for the client/sponsor are now procured and contracted. In the case of

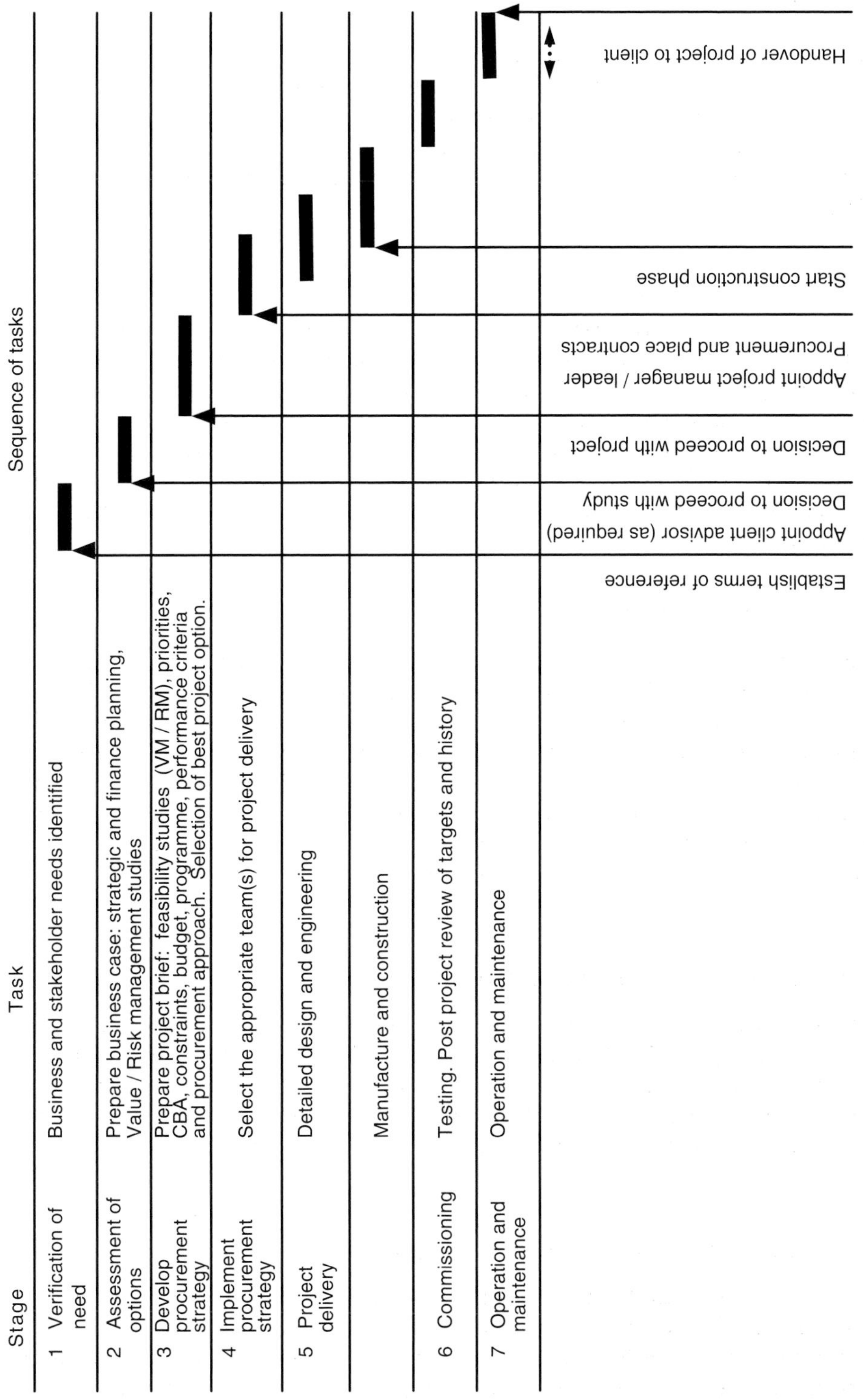

Stage	Task	Sequence of tasks
1 Verification of need	Business and stakeholder needs identified	
2 Assessment of options	Prepare business case: strategic and finance planning, Value / Risk management studies	
3 Develop procurement strategy	Prepare project brief: feasibility studies (VM / RM), priorities, CBA, constraints, budget, programme, performance criteria and procurement approach. Selection of best project option.	
4 Implement procurement strategy	Select the appropriate team(s) for project delivery	
5 Project delivery	Detailed design and engineering	
	Manufacture and construction	
6 Commissioning	Testing. Post project review of targets and history	
7 Operation and maintenance	Operation and maintenance	

Key decisions:

- Establish terms of reference
- Appoint client advisor (as required)
- Decision to proceed with study
- Decision to proceed with project
- Appoint project manager / leader
- Procurement and place contracts
- Start construction phase
- Handover of project to client

Fig. 8.1 Typical stages of a construction project.

public-sector projects, the European Union Procurement Directives require projects exceeding specified values to be advertised across member states in the *Official Journal of the European Union* (OJEU 1998).

Stage 5. Project delivery

Essentially, execution of the project in terms of design and/or construction is now the task of the selected project team or parties. However, the project manager continues to act for the client/sponsor and has specific responsibility for briefing the selected parties/team on all the aspects of the project, inducting new members as and when appropriate, implementing agreed performance, control and modification measures, advising and decision-making relating to the project objectives, requirements, agreeing payments, changes, contractual and legal issues, etc.

Stage 6. Commissioning

The project manager produces a historical review of the project, assesses the asset's performance against the set targets, including any necessary physical testing and minor adjustments to ensure that the completed asset has achieved the objectives and functions satisfactorily, finally accepting handover on behalf of the client.

Stage 7. Operation and maintenance

Some projects such as Private Finance Initiative (PFI), Design–Build–Finance and Operate (DBFO) etc. require the facility to be both operated and maintained profitably by the constructor under a business agreement; hence, the impact of life-cycle costs needs to be carefully considered, particularly materials and technology obsolescence over a long period.

Appointing the team/parties to the contract

Negotiation

Increasingly, clients look to negotiate work with trusted partners, professional service providers and construction contractors where similar projects have been completed and the new prices are unlikely to be much different. This approach is also sometimes appropriate when the project contains many unknowns likely to cause difficulties in detailing, billing and costing.

Notably, selection of the project manager/leader and other members of the professional team often proceeds on a negotiated basis, previous successful relationships with the client or at least having solid recommendations commonly being decisive in the decision. Irrespectively, the references need to be carefully evaluated and investigations attempted to ascertain the outcome on past work, especially team-working relationships with associates and also their past record if possible. Evidence of the following attributes is essential:

- Sufficient capacity of the adviser and team for the project undertaking
- Experience in work of a similar nature to the project
- An ability to demonstrate past results in terms of delivering projects to time, budget and quality requirements
- Possession of well-qualified staff with up-to-date and relevant expertise

- An ability to demonstrate sound procedures in operational management and organisation, combined with financial and economic stability
- An ability to supply written evidence of health and safety knowledge and experience; and to demonstrate expertise in managing design hazards, including subcontractor design responsibilities

Nonetheless, when exclusive appointment of professional service providers such as the project leader and separately contracted design/engineering consultants is undertaken, competition in the professional fees to be charged and setting appropriate incentives continue to influence the client, albeit after full consideration and negotiation of their proposals. Hence, it is important for the interested party to have solid knowledge and good understanding of the client's needs, developed through well-established business relationships.

Open and closed tendering

Construction contractors, in particular, commonly rely on winning work through competitive tendering, as described in Chapter 9, either by responding to advertised opportunities, known as open tendering, or alternatively in the more preferable arrangement of selective tendering. In the latter, the contractor is invited by the client's adviser to submit a bid from a short list of candidates considered suitable for the project. Thereafter, the firm submitting the lowest bid is commonly awarded the contract, unless the price alarmingly exceeds the client's own evaluation, i.e. threatens ability to finance the original budget, when fresh bids might then be called after design adjustments. During tender evaluation, the advisors would normally take into consideration any errors in the submitted bids, abnormal prices for items, contractor's time for project completion, additions required on prime cost items, provisional sums, qualifications, etc. that may have distorted the bids.

Stage tendering

Whatever the procurement approach to prequalification of prospective parties, the adopted procedure needs to be thoroughly executed prior to invitation to tender, in order to eliminate unacceptable candidates. Here, the method developed by Holt *et al.* (1995) comprising a three-stage process is typical of the principles involved. First, potential bidders are quantitatively assessed in general terms covering detailed responses on organisational characteristics, financial strength, management resources, past experience and past performance. After this initial screening, the serious candidates are further analysed against project-specific factors including plant resources, key personnel, staff expertise, experience of similar work, geographical constraints, workload, operational location, health and safety expertise and prior relationships with the client/team. Bids from the successful few are finally manipulated in an algorithm combining the ratings emanating from the three stages, including the bid value itself, to produce a weighted score for each bidder, the winning choice being directed towards the most favourable total score. In practice, however, clients often remain committed to the lowest bid and the method should therefore perhaps be seen as providing additional objective information to aid the evaluation process.

Currently, the Highways Agency's Capability Assessment Toolkit (CAT) provides an example of practical application of these principles, whereby contractors are scored and ranked for entry to preferred bidder lists using measures on past performance such as subcontractor management, management culture, partnering, financial stability and safety record.

Reverse auction

Reverse auctioning is the technique developed for procuring commodities at lowest price by introducing an additional bidding process. A tenderer submits a price and is then matched and ranked against the other participants in an internet-type *Dutch* auction continued over a two-hour session. The bidding price is automatically and gradually reduced until a new bid is received, the final (lowest) bidder being awarded the contract.

Approved as part of the European Union Procurement Directive, public bodies have the authority to use this method for procurement. Although appropriate for highly specified components and similar commodities, adoption by clients for construction projects presents a dilemma for the 'best practice' programme well under way in the UK construction sector, unless tempered by the imposition of rigorous prequalification requirements for bidders.

Public contracts and supplies

Public contracts procured in accordance with the foregoing options are further subject to the EC Public Procurement Directives as follows:

(1) The co-ordination of procurement procedures for the award of public works contracts, public supply contracts and public service contracts 2004, incorporated into UK legislation January 2006
(2) The Public Services Directive, incorporated into UK legislation as the Public Services Contracts Regulations 1993
(3) The Public Works Directive, incorporated into UK legislation as the Public Works Contracts Regulations 1991
(4) The Compliance Directive ensures compliance with the above three Directives

The Directives are intended to ensure that where public funds are to be used for the purchase of supplies (goods), services or building works, there will be full, fair and transparent competition throughout the EU countries.

Project manager/leader

The construction process outlined in Fig. 8.1 may embrace some or all of the contractual and organisational relationships of the parties shown in Fig. 8.2, the project leader/manager/adviser being an essential player in guiding the many separate phases and parties involved towards producing a successfully completed building or facility. Indeed, where clients do not have the necessary internal expertise, an appropriate outside body may be preferred for the purpose.

Standard Forms of Agreement for the appointment of the Project Manager/Leader/Adviser drawn up by various professional bodies, suitably modified for the client's specific needs, may be appropriate for the purpose.

The scope of the project leader may vary depending on the type of contract, ranging from consultant/advisor/representative to overall project management responsibility for dealing with pre-planning, procurement, programming and control, including co-ordination of the design and construction processes from inception to handover, plus managing all of the client's new work portfolio as required. Appropriate fee structure may provide incentives towards

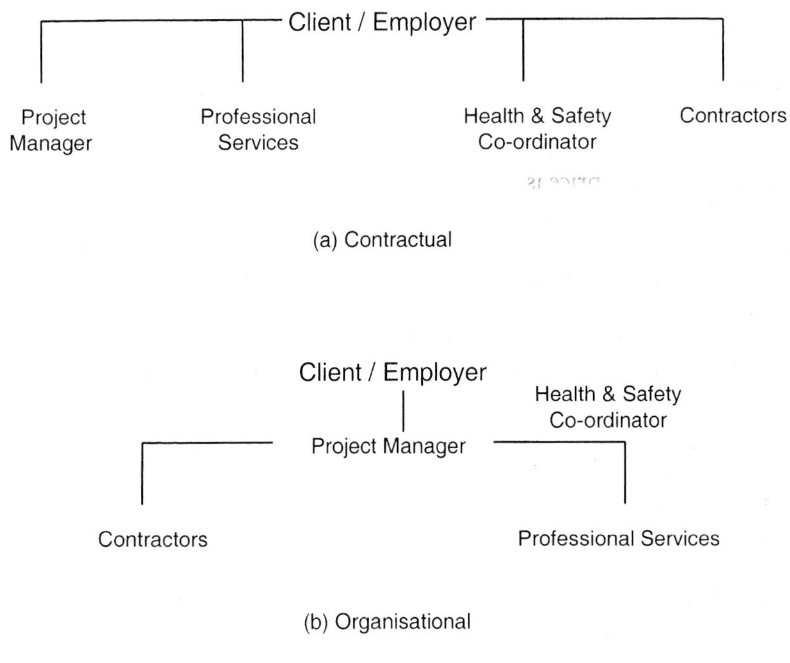

Fig. 8.2 Project management arrangement.

achieving the expected key project stages and effective execution of the relevant elements of the following functions and duties:

Project management functions

General principles	Obtain 'Value For Money' for the client. Delegate sufficient authority and control to those who are responsible for work to enable duties to be properly performed. Ensure provision of a project management and engineering resource compatible with effective and economic project strategies and forward workload.
Verification of client's needs	Identify client's needs, prioritise objectives, and identify other stakeholders and constraints on finance, legalities, time and potential business fluctuations.
Assessment of options	Develop, appraise and review options with the project sponsor and stakeholders. Agree and present the project brief.
Procurement strategy	Establish the procurement brief setting out the appropriate method of selection. Implement the procurement strategy. Ensure compliance with EU Procurement Directives for a public sector contract.
Assess and manage project risks	Specify and implement methods and procedures to manage project risks.
Project definition	Set up at the beginning of each phase of the project the appropriate organisational arrangements. Obtain all consents required for construction of the project.

	Ensure good design definition with the important elements of design work completed before construction commences. Avoid unnecessary differentiation/innovation, and adopt standard designs where practicable. In particular identify the critical or complex architectural or design interfaces, and the key (sub)contractor possession and handover points. Establish health, safety and welfare procedures, and monitor and control outcomes.
Project delivery	Establish and monitor project teams. Establish management structures with good lines of communication and clearly defined responsibilities, including the required IT systems to be used. Place responsibility for the successful completion of the project with a named individual. Establish clear-delegated authorities. Appoint high-calibre participants with relevant skills and experience, having the integrity and capability to lead and motivate. Implement appropriate performance measures.
Programme	Establish a challenging but achievable project plan and drive the project to that programme. Place clear contractual obligations on all parties, including designers, to comply with the programme. Establish viable packages of work arranged for ease of co-ordination and interface management.
Control of design	Apply proven procedures for design control, particularly design interfaces, changes and goodness of fit and function.
Budget and contract control	Establish realistic project budgets, and control expenditure against these budgets. Use experienced contracted parties who have a proven capability. Employ procurement policies that assist contracted parties to perform to programme.
Construction management/ Industrial relations	Employ contract conditions which seek to achieve maximum production on site, for example by employing suitable working hours and payment systems. Place clear obligations on the contracted parties to promote good industrial relations and minimise disruptions.
Quality control	Institute a Quality Assurance scheme encouraging responsible self-inspection by the contributors. Try to ensure that off-site production and inspection are maximised. Ensure that design details and specifications are realistic and practicable.
Project review	Measure and monitor project performance (e.g. Construction Best Practice Programme, now Constructing Excellence Key Performance Indicators, CE–KPI) throughout against targets for improvement and feed information back into the control loop. Manage project completion, commissioning and handover. Conduct post project completion review of delivery against targets.

The contract

A contract does not have to be written or in any special form, but to be valid the agreement must contain various essential requirements, the following being typical:

- There has to be an offer to carry out the work and acceptance by the client so that a contract of agreement can exist binding the contractor to carry out the work and the client to pay.
- The parties to the contract must be acceptable in law by having the capacity to contract.
- The parties must genuinely accept the terms of the contract.
- There has to be an intention to establish legality and the contract must be achievable and legal. In practice, the contract comprises a variety of documents such as:
- *Contract drawings*
- *Specification* (see www.theNBS.com and www.production.information.org)
- *Bill of quantities* (where appropriate)
- *General conditions of contract* setting out the powers, responsibilities, obligations and liabilities of the parties in the contract including payment, extensions of time, alterations, disputes, adjudication, etc.
- *The tender*, i.e. the financial offer submitted by the contractor to carry out the work covered in the above documents together with any letters of explanation qualifying the above matters
- *The legal agreement* itself

Different forms of agreement may be favoured to suit particular types of work such as civil engineering, general building, international work or client preference. Indeed, each country will generally have its own set making foreign contracting a somewhat risky venture from the legal and contractual standpoint without considerable knowledge and experience of the local systems. Also, various alternative contractual arrangements are available described below.

Contractual arrangements

(1) Lump sum
(2) Bill of quantities
(3) Schedule of rates
(4) Fixed or percentage fee
(5) Cost reimbursement
(6) Target cost
(7) Two-stage tender
(8) Serial
(9) Direct labour
(10) Adjudication

(These are explained in the following sections.)

Lump sum contract

The contractor is invited to tender on the basis of drawings and a specification. Usually there is no provision for adjustment to the quoted price unless the client introduces changes subsequent to the letting of the contract. The contractor is responsible for assessing all the costs to be incurred in fulfilling the requirements shown on the drawings and in the specification. The price is often subdivided in relation to phases of work to facilitate stage payments.

Bill of quantities contract

In some countries, clients, or their adviser, prefer to provide contractors with a common document for pricing in the form of a bill of quantities (BOQ). In civil engineering this is based on a detailed bill of approximate quantities, while for a building contract the quantities are prepared by the professional quantity surveyor and not subject to re-measurement for payment purposes; any changes are issued as variation orders.

The bill is derived from the drawings and itemised into elements according to one or other standard method of measurement. Each item contains a brief description of the work to be undertaken and the quantity. The contractor enters a unit rate and price for each, allowing for labour, materials, plant, subcontract work, temporary works, prime-cost items and nominated subcontract work, which constitutes the bid figure. During execution, the actual quantity of work achieved for each item is measured and valued at the quoted price stated in the BOQ. Additional work is either freshly priced or revalued at existing rates for similar work.

Schedule of rates contract

Many projects are so complex that completed designs are impossible to formulate before tendering and piecemeal development of the details has to be produced as construction progresses. In these circumstances, the bidder is requested to submit a unit price or rate against a list of work items typical of those expected for the work. Outline drawings, and sometimes a soils report, are provided together with approximate quantities. Payment to the contractor for the actual work done is made on a similar basis to a BOQ contract with items measured and paid at the quoted rate, new items being renegotiated.

Fixed or percentage fee contract

The contractor and client agree on a fee beforehand to cover head-office overheads and profit. The direct costs of the work are paid by the client as work proceeds based on agreed open-book accounts, i.e. production of receipts, timesheets, etc. prepared under various headings such as salaries, wages, materials, equipment, consumable stores and subcontract payments. The total final cost to the client is therefore the sum of direct costs plus the fee. Sometimes the fee is paid as a percentage of the direct costs, but this leaves the client open to even more risk because the fee will escalate as the direct costs rise.

Cost reimbursement contract

The principal feature is the payment of a fee to the contractor for executing the work, the method being intended to allow work to proceed as the design develops. The disadvantage lies in the opportunity for the contractor to increase the direct costs of the work with little restraint and thereby transferring most of the risk to the client.

Over the past few years, the Institution of Chemical Engineers contract has been used on some specific but high-risk projects, such as tunnelling. Here, the nature of the work is often best negotiated on a cost reimbursable basis using this recognised form of contract.

Target cost contract

To overcome the inherent weaknesses of the ordinary fee contract, clients have tried to encourage contractors to be more cost conscientious by relating the fee to an agreed target estimate based on a set of drawings and a specification or alternatively a bill of quantities. However, to accommodate progression of a design during the construction phase, provision is made for adjusting the target estimate for variations in quantities. The actual fee paid is determined by increasing or decreasing the original fee by an agreed amount or percentage calculated on the saving or excess between the actual cost of the work and the target estimate adjusted for any variations or changes. Nevertheless, unscrupulous contractors continue to enjoy scope to hide from the client details of transactions with suppliers unless vigilance is strongly upheld.

Two-stage tender

Few of the previous methods have proved satisfactory to clients and their advisors for projects where the design is incomplete at the time of tender. The confusion and general lack of discipline by all parties thereafter during the construction phase has often led to cost and time overruns. On particularly complex and large projects, two-stage tendering may permit improved control of costs providing the contractor is prepared to be co-operative. Usually three or four very experienced firms are invited to tender after detailed discussion with the client's advisors on matters relating to the type and scope of the work and the ability of the companies concerned. Each contractor prices an approximate bill of quantities or a schedule of rates organised to reflect the likely elements of the work.

After selection of the contractor, the second phase requires the chosen firm to collaborate with designers by giving advice on construction methods, equipment, subcontractors' work packages, orders of priority, programme and procurement dates, etc. The expected price to the client is then calculated and the contract proceeds on this basis. While the final cost of the work cannot be guaranteed, provided the first stage tender was priced on a carefully prepared elemental cost plan, the rates for the second phase ought to be similar, unless the original concept gets dramatically altered. Two-stage tendering then becomes inappropriate.

Serial contracts

Where a client intends to have a number of similar projects in the future, for example schools, a contractor can be selected at the outset following competitive tendering on a master bill of quantities. This document then acts as a standing offer open to the client to accept on the succeeding projects. The quoted rates are used with a separate bill of quantities for each new contract, updated for inflation or otherwise. The method encourages a more trusting relationship between the two parties.

Direct labour

Some clients have in-house labour, for example government departments, local authorities, nationalised industries, who are employed to carry out construction work designed either internally or by outside consultants. A formal contract therefore does not arise but some competition can be introduced by inviting outside contractors to tender. Evidence of the cost-effectiveness of this approach has been difficult to quantify because of the way labour and resources are entwined in the other activities of the client.

Adjudication—Construction Act

In taking the principle of adjudication beyond a discretionary inclusion in a contract, recent UK legislation (Housing Grants, Construction and Regeneration Act1996 and the sister Scheme for Construction Contracts), i.e. the Construction Act, is mandatory and applies to all parties involved in construction projects including clients, consultants, contractors and subcontractors. As well as dealing with payment clauses between parties to contracts, the legislation provides a swift way of dealing with disagreements, including the right for any party to take a dispute to an independent approved adjudicator for a ruling within a strict time limit, normally not exceeding 28 days. While the result is binding, each party can appeal against the decision by arbitration or through the courts' processes. The anticipated benefit of the Act is a reduction in litigations and in the adversarial manner of conduct prevalent in the construction industry.

Contract risks

Clauses in the general conditions commonly vary depending upon the source of the contract document itself, i.e. sponsored by a specific professional institution, group of employers, international body or even self-developed, and therefore are open to interpretation should disputes arise. In particular, design liability can be problematic; for example, a design carried out by a contractor would need to comply with 'fitness for purpose', whereas an architect or structural engineer is only required to provide 'reasonable skill and care'. The former is almost impossible to fully financially insure. Other areas of ambiguity concern definitions of 'Force Majeure', overall limit of contractor's financial liability, consequential loss and the extent to which the contractor and subcontractors need to insure the work for loss or damage. Hence, clients should carefully consider at the outset the potential consequences of specific forms of agreement.

Health and safety considerations

Health and Safety (H&S) compliance in the UK is governed by EU legislation largely through the Health & Safety at Work Act 1974 and Construction (Design and Management) regulations CDM (2007), pending replacement of CDM (1994), CHSW (1996) and ACoP (1994).

Under the proposed new regulations the role of the Planning Supervisor is to be superseded by the client as statutory H&S duty holder, but can be discharged through an appointed independent and competent (specification pending) H&S co-ordinator for the project. In the case of a notifiable project, (i.e. lasts 30 days or more, or involves at least 500 person-days) no work whatsoever is to commence on the project until the co-ordinator has been appointed and the Health and Safety Executive (HSE) informed. Subsequent submission of stipulated H&S documents and reports then becomes mandatory. Where the client changes during the course of a project's evolution, e.g. PFI, the H&S responsibilities must be fully passed on to the new client.

The client is to ensure the co-ordinator performs the following duties for which they are appointed under the regulations so far as reasonably practicable, broadly:

- The client is advised and updated on its duties under the H&S legislation and the appropriate degree of involvement in the project. *The responsibilities expected of a client inexperienced in commissioning construction works are under review.*
- The client is advised as to the suitability and compatibility of designs, on any need for modification and that H&S requirements are completed at each phase of the project.

- The mobilisation period is specified at the time of tender.
- Parties self-declared as competent for their undertaking are demonstrably so.
- Designers properly assess and incorporate into the design the risks to parties carrying out the construction work, those likely to be affected by construction work, and to parties maintaining, cleaning and using the facility.
- Designers eliminate hazards likely to give rise to risks, and reduce risks from hazards that cannot be removed.
- Designers carry out their responsibilities as far is reasonably practical.
- The principal contractor produces a suitable H&S plan for the construction phase using information provided by the designers as appropriate.
- The co-ordinator should liaise with the contractor on any matters relating to design during the construction phase.
- The co-ordinator should provide an overall structure file containing adequate information to enable evaluation and understanding for potential alterations at a future date.

HSE draft guidance on responsibilities

Client's duties:

- Check competence of all appointees.
- Ensure there are suitable management arrangements for the project.
- Allow sufficient time and resources for all stages.
- Appoint the co-ordinator and ensure the job is done properly.
- Appoint the principal contractor.
- Provide information.
- Make sure that the construction phase does not start unless there are suitable welfare facilities and construction phase plan.
- Retain and provide access to the health and safety file.

Co-ordinator's duties:

- Notify HSE about the project on behalf of the client.
- Co-ordinate design work, planning and other preparation for construction relevant to H&S, but not design checking.
- Manage H&S communication between client, designers and contractors.
- Develop effective management, review and revision arrangements for the project.
- Liaise with the principal contractor regarding ongoing design.
- Prepare/update H&S file.

Designers' duties:

- Prepare designs with adequate regard to H&S.
- Eliminate hazards and reduce risks due to design.
- Provide information about remaining risks.
- Check client is aware of duties and co-ordinator has been appointed.
- Check HSE has been notified.
- Provide any information needed for the H&S file.

Principal contractor's duties:

- Plan, manage and monitor construction phase in liaison with contractors.
- Prepare, develop and implement a written plan and site rules. (Initial plan completed before the construction phase begins.)
- Give other contractors relevant parts of the plan.
- Make sure suitable welfare facilities are provided from the start and maintained throughout the construction phase.
- Check competence of all its appointees.
- Ensure all workers have site inductions and any further information and training needed for the work.
- Consult with the workers.
- Liaise with co-ordinator regarding ongoing design.
- Secure the site.

Other contractors' duties:

- Plan, manage and monitor own work and that of workers.
- Check competence of all appointees and workers.
- Train employees.
- Provide information to workers.
- Comply with requirements in the H&S regulations.
- Ensure there are adequate welfare facilities for workers.
- Check client is aware of duties and a co-ordinator has been appointed and HSE notified before starting work.
- Co-operate with principal contractor in planning and managing work, including reasonable directions and site rules.
- Provide any information needed for the health and safety file.
- Inform principal contractor of problems with the plan.
- Inform principal contractor of reportable accidents and dangerous occurrences.

Everyone's duties:

- Check own competence.
- Co-operate with others involved in the project.
- Report obvious risks.
- Comply with requirements and other regulations for any work under their control.

Construction phase

The principal contractor is responsible for all health and safety provision and management on site and typically requires the following elements:

(1) The principal contractor's policy statement on health and safety
(2) General health and safety arrangements and responsibilities for the project such as how H&S will be managed and communicated during the construction phase, named officers, role of supervisors and individual responsibilities of employees, training regime, personal protection provision for employees, significant risks likely to arise, advice and consultancy support, first aid and fire safety contact points, visitor requirements, etc.

(3) Details of specific hazards and arrangements for health and safety on the project, such as:
- Safety management team
- Safety assessment in appointing subcontractors
- Safety information provision
- Health and welfare facilities
- First aid regulations
- Reporting of accidents and injuries procedure
- Emergency procedures
- Compliance procedures for the statutory construction regulations
- COSSH requirements
- Site safety inspections and meetings
- Site security, power supplies, etc.
- Procedures for maintaining an H&S file of documents
- Method statements
- Permits to work and certificates of compliance
- Requirements for site safety reports, etc.

(4) Relevant sections of the plan provided by the Principal Contractor to other contractors
(5) Relevant sections of the plan provided by every contractor to their employees and self-employed persons under their control
(6) Skills certificate—possession of the appropriate skills card by all workers on site assessed through a written test and interview awarded by the Construction Skills Certification Scheme (CSCS); knowledge and understanding of H&S matters form part of the examination; the card is not presently mandatory but considered good practice

Finally, prosecution for the offence of corporate manslaughter due to negligence, i.e. holding a senior executive(s) responsible for death resulting from an accident, is under consideration in a number of countries.

Categories of contract

The type of contract will be selected by the client or project manager according to need or preference, and will broadly fall into one of the following categories identified by Masterman (1992) and depicted in Fig. 8.3:

- Separated and co-operative, e.g. traditional contracts
- Management-oriented contracts
- Integrated, e.g. design-and-build contracts
- Discretionary, e.g. partnering

Separated and co-operative contracts

Notably, the separated type of contract is preferred for major US defence procurement. Development/design risks are wholly borne by the client via cost-plus contracts, followed by fixed-price contracts for actual production deals. Significantly, the pool of designers and contractors is drawn from a handful of specialist defence industry corporations and their supply chains,

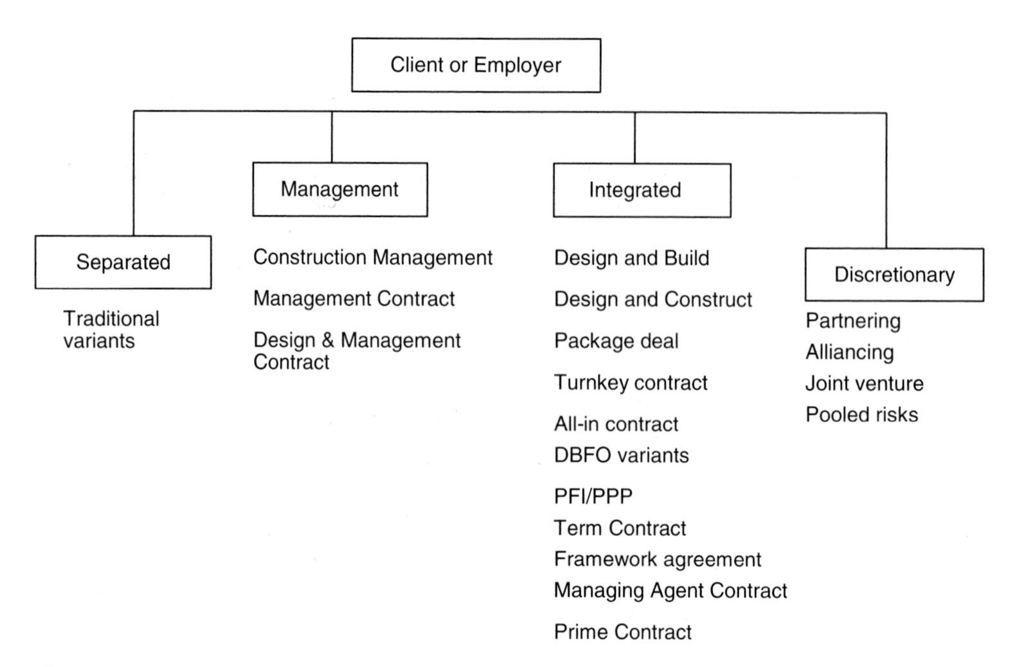

Fig. 8.3 Contract procurement options.

highly experienced and familiar with the advanced technology involved, while the client itself possesses the necessary in-house capability to perform the project management function.

Clients lacking, or with limited, project management expertise more typically prefer to engage an able third party to interpret their needs before proceeding with the project.

Commonly for construction an architect or civil engineer is appointed to act on behalf of the client and also take responsibility for preparation of the design, specification and co-ordination of all of the design input from other specialists, and finally assembling a comprehensive set of contractual documentation, drawings and specification to be put out to tender. Subsequently, the actual builder/construction firm is selected and contracted to produce the facility. Indeed, several contracts and contractors may be appropriate for a multi-faceted project.

Thereafter, the functions may extend to supervising the construction phase invested through the authority to issue instructions, variation orders and sanction payments, including dealing with claims negotiations whenever they arise.

The documentation, while requiring much time, skill and effort to produce, will ultimately prove its worth during construction by clarifying responsibilities and any subsequent arguments with the other concerned parties. Indeed, the significant key to success of the separated contract rests in the setting out of precisely what is to be built and how executed, preferably developed using 3D/VR (Virtual Reality) and IT, leaving as little undecided design as possible to feed through to the construction phase.

Parties to the contract

The parties and contractual relationships in a traditional separated construction project are shown in Fig. 8.4, which illustrates the independent responsibilities, the client being the person or firm requiring the work to be executed through formal contracts. The architect or engineer

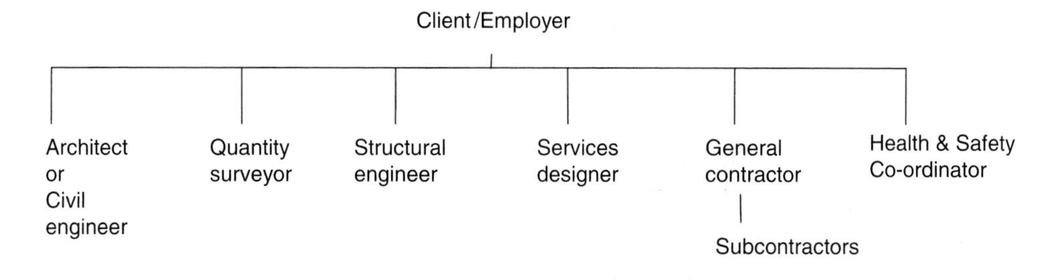

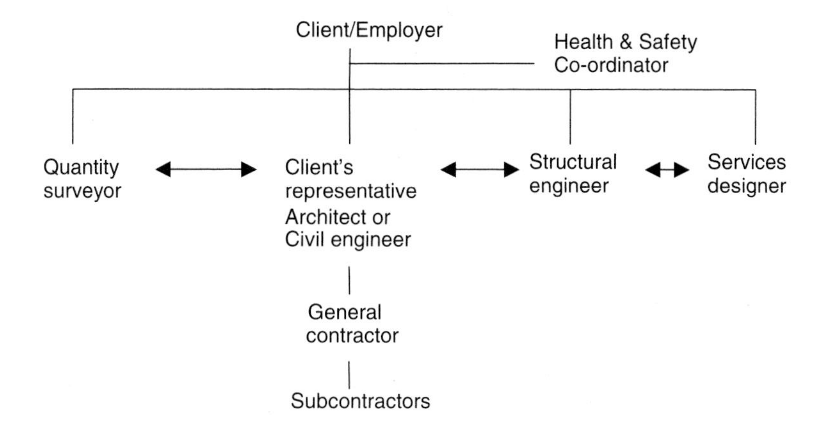

Fig. 8.4 Separated contract.

and other specialists such as structural design, engineering services, etc. undertake agreements describing the services to be provided and the fee to be received. Payment to the architect for site supervision is usually an extra while the contractor executes the work in accordance with the architect's or engineer's drawings and/or instructions. Since no formal contract exists between the architect/engineer and contractor the former acts on behalf of the client, the organisational arrangements are as illustrated in Fig. 8.4.

Ideally, design should be complete before tendering, hence any later variations caused by changes in the quantity of work or site conditions are usually issued as written instructions by the architect or engineer but there may be slight differences in this principle depending upon the particular conditions of contract. In some forms of contract the role of a quantity surveyor is also defined, the duties usually being the agreeing of stage payments, measuring and valuing variations and preparing the final account, as authorised.

Types of separated contract

In the standard arrangement, clients or their project leader (normally an architect or civil engineer) engages designers, consulting engineers and quantity surveyors to prepare schemes, appoints contractors and supervises the work, the latter function usually for an extra fee. The options described under 'Contractual arrangements' are generally acceptable.

Standard forms of agreement in use

Common examples include:

- Joint Contracts Tribunal—JCT 2005 (i) Standard Form of Building Contract (Standard Building Contract); (ii) Intermediate Building Contract; (iii) Minor Works Building Contract
- Conditions of Contract (ICE) for civil engineering work
- Conditions of Contract (FIDIC) for international civil engineering work
- General Conditions (GC/Works) for Government contracts
- NEC3 Engineering and Construction Contract
- Standard Forms of Agreement drawn up by various professional bodies such as RIBA, ACE or NEC3 Professional Services contracts for the appointment of the consultant, architect, safety co-ordinator, etc.

Management oriented contracts

Over recent years, clients of major building and civil engineering work have all too often experienced difficulties in obtaining projects finished on time, to budget and of acceptable quality and serviceability, especially when designs are incomplete before contractors are appointed—intricate projects being particularly affected. Thus for innovative highly complex projects requiring progressive development of the design details as construction proceeds, relatively inexperienced clients occasionally adopt this form of contract, namely:

- Management contract
- Construction management contract
- Design and management contract

In the management-oriented contract, the construction manager or managing contractor joins the project management team at the earliest possible time prior to construction on equal terms to other consultants. Responsibilities cover preparing the overall construction programme and works packages, steering these through the design stage, recommending/appointing the works (sub)contractors and securing their smooth integration. The combined design and supervisory functions of the traditional architect/engineer are removed, being executed by the appointed team. Co-ordination and co-operation is largely achieved by requiring contracted designers and works (sub)contractors to provide detailed programmes with procurement and key dates clearly defined, together with explicit statements on working methods and equipment resources. Considerable experience is essential in judging the viability of the proposals, particularly where works (sub)contractors are expected to bid and execute contracts on an agreed programme and design documents including subsequent co-operation with the other parties involved in the project. Managing all this demands considerable programming expertise and effort, necessitating regular progress meetings and budgets for each work package to be carefully monitored. Indeed, successfully bringing together all the project interfaces and ensuring good fit of others' designs, technical proposals and site work is a major challenge.

Parties to the contract

The well-established choices of contractual arrangement described above generally apply to the management contract, typically with contracts between client and designer, client and contractor, contractor and subcontractors. The contractor, however, has few, if any, obligations to execute actual construction work. A fee is charged for co-ordinating the subcontractors and advising the design team. Similar contractual arrangements are provided for design and management contracts. The construction management contract in contrast is devised for a particular project, with the client having a direct contract with each individual works contractor. As a consequence, the duties and services to be provided by the construction management firm include very detailed schedules of the required responsibilities, akin to those of a project manager. Fees are normally related to the value of the project and the service to be provided and are determined either by negotiation or, increasingly, by competition as more clients and contractors become experienced in this kind of operation.

Standard forms of agreement in use

Common examples include the NEC3 Engineering and Construction Contract–Management contract.

Types of management-oriented contract

Construction management (Fig. 8.5)

The construction manager is appointed *early* to provide a planning, management and co-ordination function. Since all the actual orders with the various works contractors are with the client, the construction manager carries reduced risk. The construction management firm is not allowed to carry out any construction itself, but takes responsibility for advising the designer on buildability, including drawing up suitable work package contracts, arranging procurement contracts and managing the bidding phases of the works contracts. The organisational arrangements are shown in Fig 8.5. Considerable expertise is needed during construction to bring all the elements together, with a view to minimising variations, time delay and general inefficiencies. The main disadvantage to the client is the lack of a firm tender price at the selection stage, other than an approximate overall scheme price. The potential cost then only becomes known as tenders submitted by works contractors are accepted.

The Management Contract (Fig. 8.6)

Many clients are unhappy at bearing the risk of the many works contracts needed on the large and complex projects normally associated with construction management. As a consequence, there has been a trend towards engaging a single experienced contractor early to provide planning, management and co-ordination of construction, who then subcontracts the work in the normal manner. However, like construction management, the principal contractor is barred from carrying out any construction work itself, although on some projects common items for

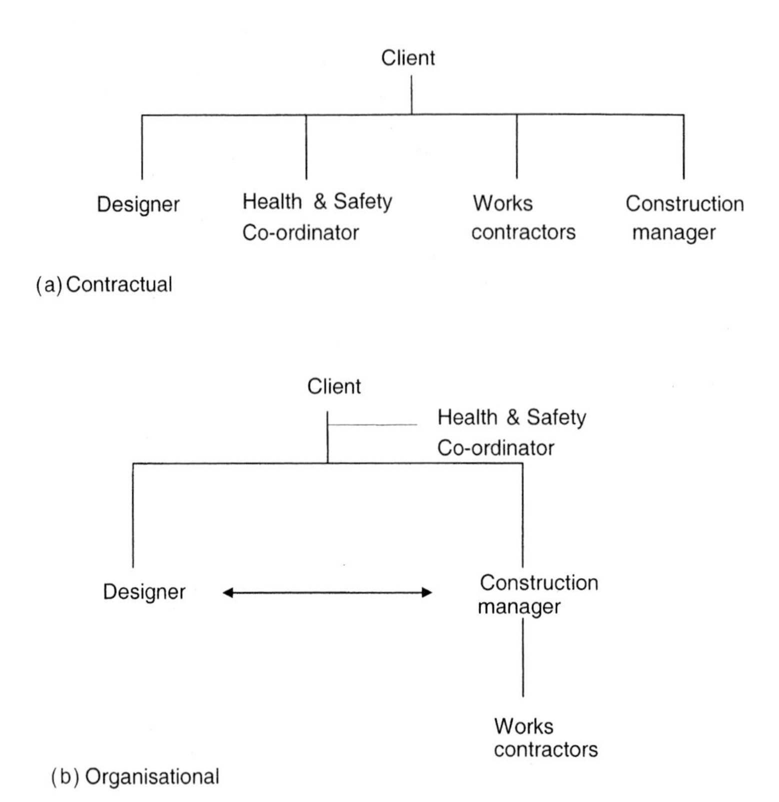

Fig. 8.5 The construction management contract.

subcontractors, e.g. scaffolds, tower crane and access roads, are provided. The management contractor obtains a fee for performing similar duties to the construction manager, but carries more risk and competes for contracts based on reputation of getting work completed on time, to cost and of good quality. The contractual and organisation arrangements are shown in Fig. 8.6.

Design and management contract (Fig. 8.7)

As a natural development from these new systems, clients have invited a principal management contractor to take responsibility for also managing the design phase. The contractual arrangements are shown in Fig.8.7 with all elements sub-let by the contractor. The initial scope design is often executed by the client's own staff or an independent design firm, and forms the basis for inviting tenders. The contractor offering the lowest cost scheme for full design and construction is normally selected but reputation, quality of service and management fee charged are also important considerations. Both design and construction are entirely sub-let; contractual responsibilities are therefore potentially more visible and so less likely to become enmeshed in the commercial operations of the principal management designer/contractor. The organisational arrangements are shown in Fig.8.7.

Quality

In management-oriented contracting the trend in quality control has been away from site

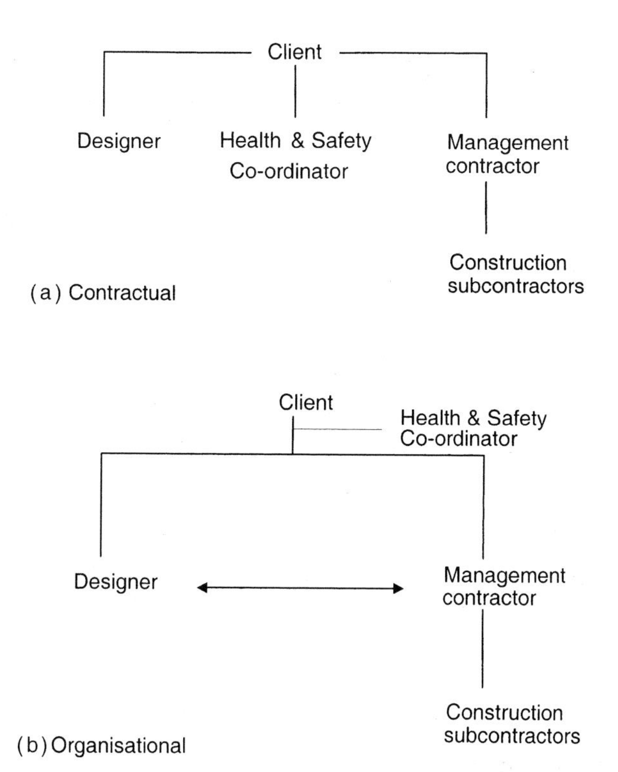

Fig. 8.6 The management contract.

inspection by a resident engineer or clerk of works towards quality management by works (sub)contractors themselves. Some site inspecting and testing will be necessary, but this responsibility is best discharged to an independent team or firm using outside research, development and testing laboratories, especially for off-site inspection of materials and component supplies. Quality management requires the managing firm to encourage and indeed only invite for tender those works (sub)contractors with an ISO 9000 or similar accredited quality system and a proven record of producing good quality work. Procedures for assessing works (sub)contractors' proposals for achieving specified standards should be established at the outset and rigorously upheld. Thus, training in quality management and a conscious determination to ensure high quality are important functions of the managing team. Similarly, with design it is imperative that clients also develop a rigorous attitude to quality. The aim must be to achieve standards reached in the manufacturing sector where companies compete on quality as well as price.

Safety

Like quality management, safety should be seriously treated by all the parties in the project. Regular safety meetings, training and thorough inspection are essential. Works (sub)contractors' proposals must comply with the Construction (Design and Management) regulations (CDM) and additionally meet specified requirements laid down by the managing firm.

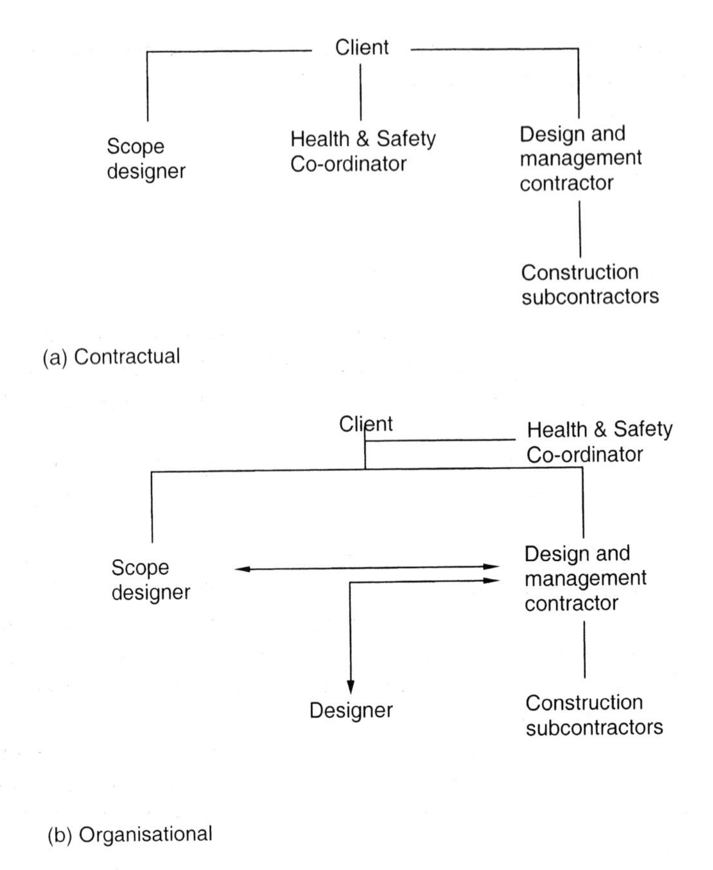

Fig. 8.7 The design and management contract.

Labour

Management-oriented contracting is usually associated with large and complex projects where the design is evolved as construction progresses. Such projects invariably include many works contractors from different industries, for example the civil, mechanical and electrical sectors, where wage rate agreements, conditions of employment, etc. vary. In addition, the individual firms may have different working procedures, payment schemes and so on. Petty jealousies caused by, often small, differentials can lead to unnecessary industrial disputes. Thus, it is imperative that the managing firm lays down a labour relations policy with clear terms and conditions of employment for all works contractors. Procedures for dealing with disputes must be agreed with any unions at both local and national level including a firm policy on the engagement of self-employed labour. Health, safety, welfare and site facilities can also influence the smooth running of a site and therefore particular attention must also be given to these aspects.

Integrated contracts

This form of contract is currently a preferred approach under the *Constructing Excellence Programme*, which seeks better '*value for money*' in reduced whole-life cost and improved quality.

The policy stresses the merits of concurrent engineering so dominant in 'world class' manufacturing, which for the construction context implies integrated design and construction, particularly the key elements of:

- Product development
- Project implementation
- Partnering in the supply chain
- Increased usage of manufactured components, i.e. 'lean construction'

Ideally, the method is best suited to routine construction procured from contractors marketing well-developed standard products, i.e. those which can be readily evaluated by a well-informed client as to what is to be provided, its specification, method of payment and means of monitoring performance and quality.

Regular projects such as office blocks, high-rise buildings, schools, hospitals, housing, factory units, prefabricated/modular system/industrial assemblies, car parks, etc. represent typical examples.

Nevertheless, when the client has developed proven relationships with competent contractors, even complex projects such as engineering-construction or long-term highway and railway maintenance are sometimes considered appropriate. Significantly greater value creation is predominantly sought through an established, well-managed supply chain or network, rigorous measurement of performance against KPI targets and the relentless drive to achieving sustained safety and quality improvements.

Characteristically three or four highly experienced companies are invited to submit proposals for both design and construction, competition being introduced at the design stage through the price offered for the finished product, including commissioning, operation and maintenance, if required.

Inevitably, therefore, the client relinquishes some control over the design, although *early contractor involvement after the feasibility stage* using tendered fees with subsequent as-built sharing of savings on target cost estimates may partly overcome this difficulty—on condition that a thorough procurement process (see Holt *et al.* 1995) is well established in the client organisation.

Notwithstanding this, aspects of design detail may be incomplete when the contract is awarded, and these may subsequently require costly time and effort (to be absorbed by the contractor) in agreeing responsibilities, duties and site coordination with the chain of designers, specialists, suppliers and subcontractors, to achieve the project specification and quality expected by the client for the contracted price.

Parties to the contract

The contractual arrangement shown in Fig. 8.8 is a single-point responsibility and accountability of the contractor to the client for both the design and construction facets, secured either by competitive negotiation or by tendering. Examples include lump sum price, schedule of rates, (target) cost reimbursement or BOQ where a scope design has been provided.

The various functions in the contractor's organisation for a large civil engineering project are demonstrated in Fig. 8.9.

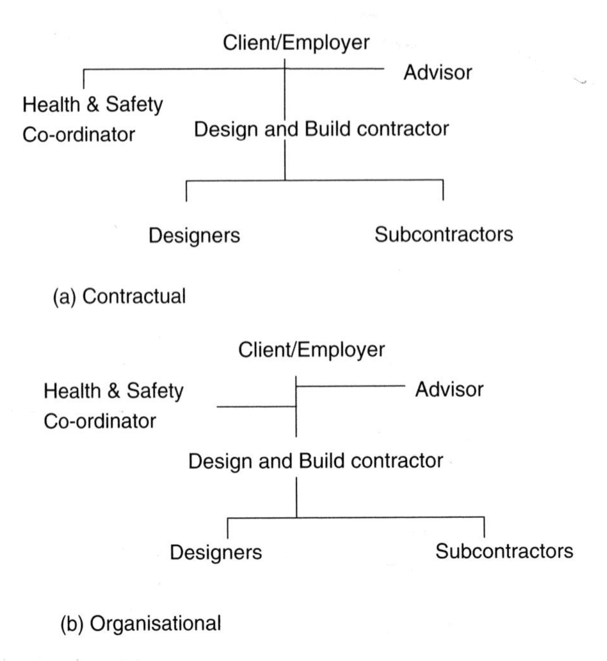

Fig. 8.8 Integrated contract.

Types of integrated contract

Design-and-construct contract

The client separately arranges for an architect, in the case of building contracts, or a civil engineer to produce scope drawings generally relating to particular functional or essential aesthetic details, together with a specification fully describing the design. The contractor(s) thereafter augments the drawings with their own working drawings and secures all statutory approvals including those needed from the advisor.

Design-and-build contract

The contractor is responsible for construction and the full design, including embracing the production of the aesthetic and working drawings, together with obtaining statutory approvals. As there is no sharing of design responsibility, the client needs to exercise caution against utilitarian and mundane design offerings and parsimonious quality specification of the finished product.

Turnkey, all-in, package deal contracts

Typically, in Build Own Operate and Transfer (BOOT) and other similar arrangements such as Design Build Finance and Maintain (DBFM), Design Build Finance and Lease (DBFL), BOO, BOT, DBFO, EPC, etc. the developer is responsible for devising the scheme, raising the finance, operating the facility and maintenance. Ownership is finally transferred to the client after sufficient local experience has been gained in running and managing the facility or when the useful life of the asset has been exhausted. Such projects are usually associated with clear revenue raising opportunities during the period of operation sufficient for a return on the investment to be

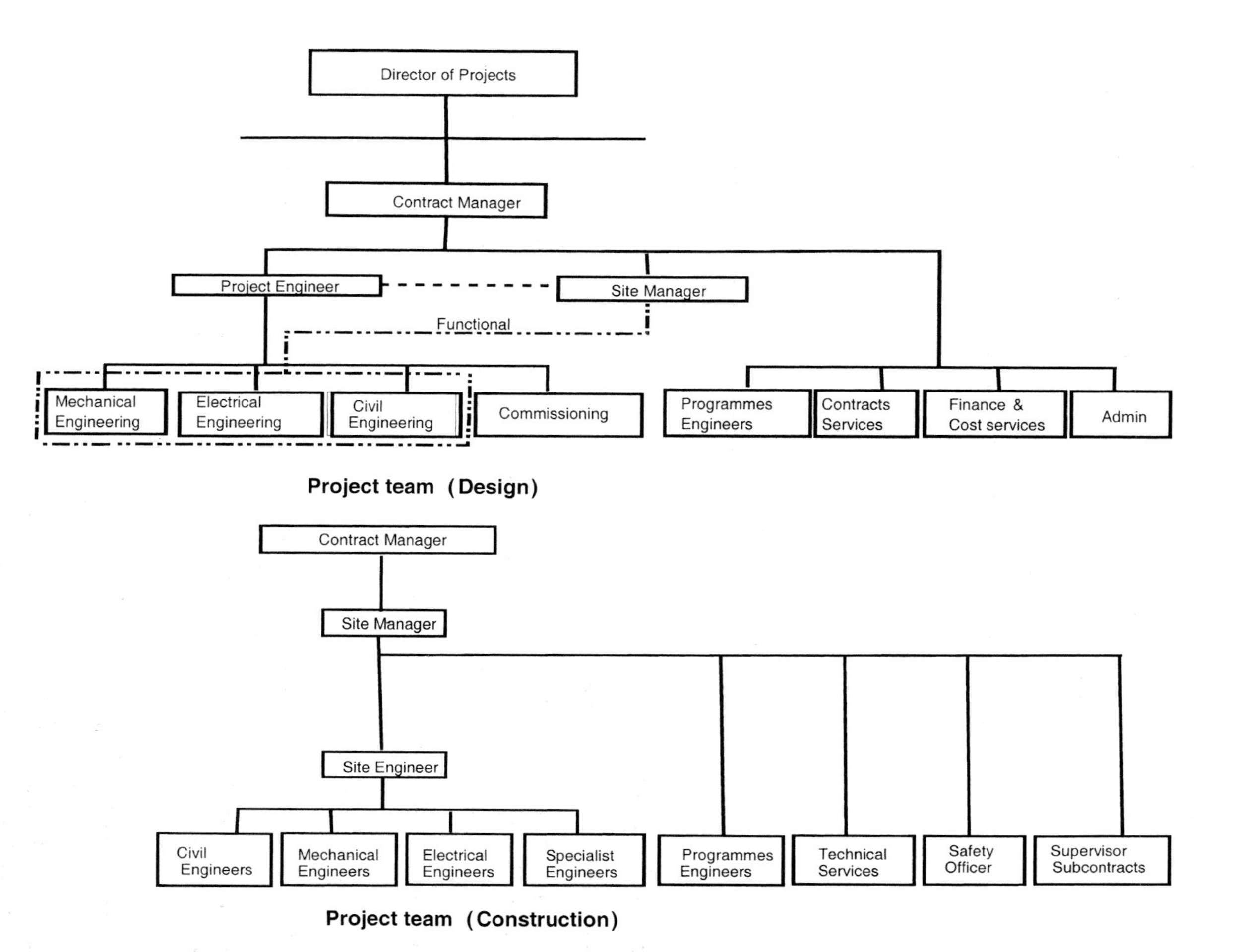

Fig. 8.9 Functions in a large civil engineering project.

generated on the loan capital. Arrangements of this kind are often favoured by clients in countries where problems of raising indigenous capital exist and therefore necessitate experienced firms with access to international finance to take the lead.

However, increasingly such projects can be found in the advanced economies, commonly referred to as the Private Finance Initiative/Public Private Partnerships (PFI/PPP), where private finance is introduced into projects customarily funded wholly by government. The national budget is thus relieved of a large forward capital payment and also subsequent operation of the facility through a leasing arrangement, thereby reducing the impact on short-term demands for public expenditure.

Term contracts and arrangements

Term contracts are the latest attempt at improving the performance of the supply chain, aimed to obtain 'value for money', whereby the client lets a series of service requirements, tasks, projects or contracts to a single provider over a period of perhaps five years or more. To date, this form of arrangement is being adopted and progressively refined in the light of experience.

Framework agreements

A framework agreement involves a partnering-type arrangement linking the contracting authority (the client) and one or more economic operators (suppliers, contractors, service providers) for a fixed period against a set scale of fees. The agreement is commonly, but not exclusively, limited to the provision of design/professional engineering services on a needs basis.

A public sector framework agreement falls under the EC Procurement Directive 2004, which may not exceed four years in duration and can embrace one or more operators selected competitively, where each call-off under the framework is regarded as a purchase, i.e. a contract, and subject to threshold rules. Subsequent call-offs are required to go out to tender. The following types of service, skills and management expertise are commonly provided:

- Civil, structural and traffic engineering design/consultancy
- Review of planning applications
- Feasibility studies
- Modelling
- Site supervision
- Technical support
- Information and physical surveys
- Pollution studies
- Geotechnical advice
- Planning supervision
- Safety auditing
- Preparation of contract documents
- Administrative support
- Procurement advice and support

The main advantage lies in releasing the client's in-house resources, smoothing out workload peaks, rationalising the service supply chain and providing more flexible procurement.

The managing agent contract (MAC)

Typically adopted for large-scale road maintenance schemes for regional government authorities, although a wider range of construction work may be embraced, the contract is secured through a rigorous three-stage tendering process, i.e. expression of interest, pre-qualification, issue of an invitation to tender, and is commonly based on a 70/30 weighting respectively of quality and price assessment.

The contractor is subsequently subject throughout the contract to audit inspection and monitoring using open-book accounting procedures conducted by both internal and external audit groups, but with self-checking and supervision being paramount. Key performance indicators (KPIs) are regularly applied and progressively developed in conjunction with the client, as part of a best-value review process covering each of the main areas of service provision relating to commissioner satisfaction, consultant satisfaction, improvement in facility condition and productivity of the service.

Some clients have preferred to split the MAC into separate 'Stewardship' and 'Works' contracts awarded to individual firms to discourage monopolistic tendencies. The Steward is charged with day-to-day operational functions and has an overall 'eyes and ears' role that requires a proactive approach in supervising the works contractor, which, combined with inspection responsibilities, enables repair or improvement work to be ordered directly and certified after completion, as well as final payment authorised. Work is individually priced by the appointed works contractor, based on the MAC tendered schedule of rates, which may stretch to several thousand items in the case of large area maintenance contracts, with the schedule allowing for variation in volumes of work and percentage increases or decreases, depending on whether the location is more or less inconvenient for the contractor. Bonuses for performance may be included in the scheme, including penalties if work is late in completion, e.g. £/day. Typical KPIs include response time and the number of works orders completed on time.

The most recent innovation of this form of contract embraces PFI principles operating within the terms of EC Procurement Directive 2004, whereby previously successful MAC contractors are invited to bid for new MAC work in the form of DBFM. Payments by the client over a period of 20 years or more are based on contractor performance related to average road condition achieved and maintenance measures for any other ancillary works.

Prime contract

The Prime Contract is principally a feature of the Defence Estates, an Executive Agency of the Ministry of Defence, which has introduced this type of contract in an attempt to achieve better 'Value for Money' in managing both its existing estate portfolio and new build.

Like other integrated-type contracting, a prime contractor has overall contractual responsibility for the management and delivery of a portfolio of maintenance work and new build to meet the required safety standards and quality specification, to cost and to time, all progressively monitored with key performance targets. Responsibilities include subcontractor selection and management, design co-ordination, construction planning and cost control in partnership with the client. The selection process comprises three stages: first, invitation of expressions of interest, followed by pre-qualification and finally the issue of an invitation to tender for the contract. The important assessed factors include project management capability, financial standing and supply chain arrangements.

The Private Finance Initiative (PFI)

While private financing arrangements are fairly familiar to the construction industry, for example in work for foreign governments, where financiers such as the major banks in conjunction with a contractor commonly undertake DBFO, DBFL etc. and similar activities, the major difference in the UK PFI initiative lies in approval only being secured for those 'government' projects having a given share of the risk held by the contractor. The argument is that, unless government is free of the majority of risk, then the project should be counted as an addition to the Public Sector Borrowing (PSB) requirement and hence needs to comply with normal/traditional procurement practices and approvals. For example, in the case of a hospital, the engaged firm might be paid on a leasing basis by government for building and maintaining the facility according to the number of beds provided, regardless of availability; if beds are then subsequently under-utilised, the taxpayer would unfairly foot the bill. Although such shortcomings have now been largely eliminated, the risk-sharing issue and growing volume of such work remain problematic for PSB politics. Meanwhile, other appropriate projects such as roads with tolls or at least scientifically generated shadow tolls, NHS hospitals, HM prisons, local authority (LA) schools and housing schemes, waste recycling services and facilities, road maintenance contracts and similar LA- type projects are being vigorously procured in this manner.

PFI principles

The scheme is usually initiated by central government, one of its agencies or local government. Normally the engaged contractor is required to finance, design, construct, maintain and, where appropriate, operate the facility over a stipulated concessionary period of ownership of perhaps 30 years or more and be paid a rental or lease sum by the client for the services provided. Significantly, participants may sell their stakes in the asset at market values during the concession.

Critically, the client must establish a sound specification for the intended project clearly defining the *scope* of work to be undertaken, *output* needed in amount and quality, and the degree of risk, for example the procedures in the event of the contractor at some stage defaulting. The project's financial viability is evaluated using Net Present Value techniques and the appropriate bidding process and contractor selection criteria carefully ascertained to try and ensure that only suitably competent participants are engaged.

In this respect, the contractor essentially requires a solid balance sheet to meet long-term commitments, so necessitating substantial equity and support from major financial investment banks. Usually the scale of the undertaking also requires the contractor to seek reliable partners either as stakeholders or as independents such as design consultants, works contractors and suppliers, possibly extended to include experienced associates, to operate the facility during its life.

Projects have occasionally failed at the tender stage either because the tendered schemes proved unacceptable when compared to the public sector's own version or because they were overpriced, perhaps resulting in only one tenderer then being left for consideration, so contradicting the main aim of PFI to provide value for taxpayers' money. Indeed, after expending considerable effort, participants may end up with rejection and the scheme finally procured as a normal public sector proposal. Under the new EC Directive 2004 known as 'Competitive dialogue' a short list of perhaps three or four bidders may be selected **early** (selected rigorously) to co-operate in defining the scope of the project, steering it through the statutory processes, agreeing the specification and design. Thereafter the parties are formally invited to tender for construction, maintenance and operation as appropriate, including financing the project, the final DBFO/PFI contract being subsequently awarded to the winning consortium.

Public Private Partnership (PPP)

Similar to the PFI concept, the UK PPP scheme is initiated by central government or a government agency with the consortium appointed in like manner. Some very specific infrastructure projects are currently under way, most notably for the publicly owned and operated London Underground facility. Track, stations, trains, renewal and maintenance, including trains operation, are being carried out and financed by private consortia over 30-year contracts subject to five-year periodic reviews, with payments by the client spread over time. Very complicated and rigorous targets, detailed performance measures, rewards and fines are the key tools used by the London Underground client for monitoring progress, hopefully ensuring that specified high standards are attained. Other similarly novel schemes presently under consideration include the provision of water supply and services to the Ministry of Defence under a contract covering development, management and maintenance of their assets, and the government's Building Schools for the Future programme of rebuilding and refurbishment, etc.

Standard forms of agreement in use

Common examples for integrated contracts include:

- EPC (Engineer Procure Contract)
- BOT (Build Operate Transfer)
- Government Prime Contracting contract
- JCT (i) Design and Build Contract 2005; (ii) Major Project Construction Contract 2005; (iii) Framework Agreement 2005; (iv) Adjudication Agreement
- GC/Works/1 Design and Build contract
- ICE Design and Construct contract
- NEC3 (i) Engineering and Construction Contract; (ii) Professional Services contract; (iii) Term Service contract; (iv) Adjudicator's contract (see Fig. 8.10); (v) Framework contract

Discretionary contracts

Even with the new contract forms, large specialised projects such as power stations, airports, oil refineries and similar complex utilities have still proved difficult to manage satisfactorily from conception through to commissioning and handover. The reasons are many, but in essence have continued to relate to unique single project procurement and the lack of stability in relationships between clients and their designers/constructors, made more difficult by the inability of clients to adequately define their needs at the outset. As a result, whole contracts have been let before designs and control procedures were complete, and all parties concerned, including the different tiers of managers in the client's own organisation, have been able to provide excuses for mistakes, inefficiencies, etc.

The element of discretion attempts to address and improve working relationships between the parties by encouraging teamwork and reducing potential adversarial attitudes. Nevertheless the liability for identifying, managing and taking responsibility for risks and producing the agreed project expected by the client still needs to be properly set out in a formal contract.

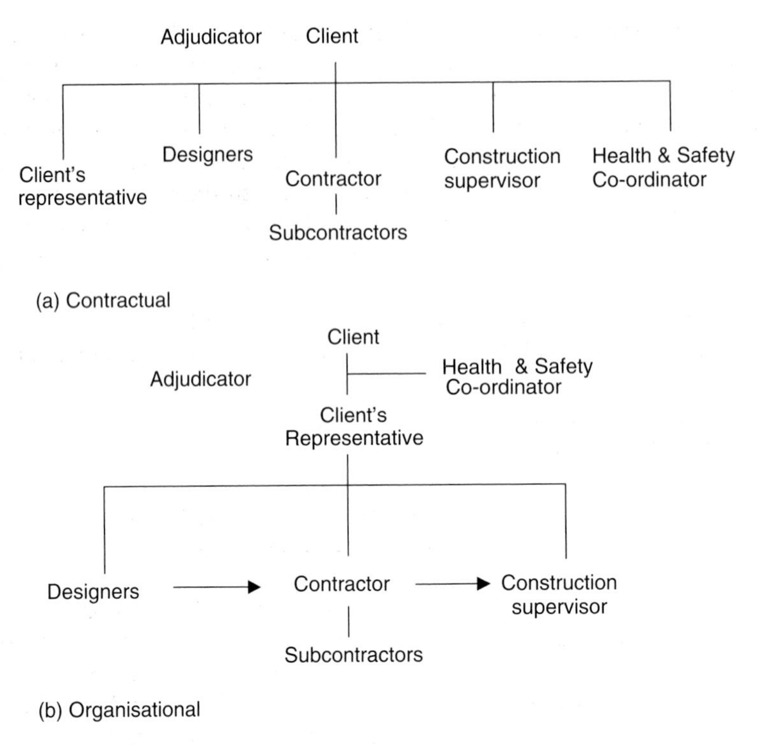

Fig. 8.10 Adjudication contract.

Types of discretionary contract are:

- Partnering
- Alliancing
- Joint venture
- Pooled risks

Partnering

Originally developed for engineering construction-type work, partnering is normally regarded as the strategic long-term arrangement whereby a partner, e.g. a contractor, is selected by a client/advisor for a series of projects aimed at lowering costs and improving efficiency by reducing delays and ensuring completion of projects on time, to budget and quality, although single project arrangements are also increasingly being included in the terminology. Contractors need to be carefully selected based on a whole series of performance measures related to quality and competitiveness, usually conducted through a two-stage tendering process unless already in the position of a favoured contractor, with quality criteria being screened before financial bids are considered.

Notably, partnering involves the client and their PM/Advisor in selecting a particular contractor, negotiating a price for the envisioned project, establishing the programme and agreeing terms and conditions of contract, but with the main objective being to generate a spirit of co-operation. In particular, the contractor is encouraged to put issues to the client at the outset and continue with open attitudes throughout the contract. It is hoped that problems, i.e. claims, disputes, etc., can be foreseen, mutually resolved and eliminated if possible. However, the client may further extend influence by requiring maintenance of key performance levels during the work phase, expecting the contractor take appropriate action if standards begin to stray off target.

For partnering to be meaningful, the client should be well experienced in construction procurement and possess considerable 'in house' skills relating to design and construction of projects in order to have meaningful dialogue with the contractor. Also, the chosen contractor is already likely to have established a proven track record of reliability with respect to time, cost and quality on previous work for the client, where evidence of the following kind would normally be expected:

- Benchmarking in achieving continuous improvement over past contracts
- Settling for a fair reward with the client and other contracted parties through sharing of financial savings
- Trust-building relationships in commercial dealings
- Strategies for meeting clients' needs
- Application of 'best practice' management and engineering methods
- Learning lessons and implementing recognised improvement processes
- All necessary parties are to be involved in the partnering process

Work is generally progressed and paid on a cost-plus-fee basis but with agreements often allowing for sharing the benefits of cost savings between the client and the parties to the contract, achieved largely through open access to plans, estimates, costs and financial accounts. However, additional costs may arise in operating agreements, for example from training, workshop sessions, providing facilitators, extra negotiation tasks, etc. Hence, formulation of contractual conditions reflecting the intentions of the parties concerned presents a challenge compared with conventional practices, especially in expressions of goodwill and co-operation, which, if tested in court should a dispute arise, could prove legally binding. Furthermore, third parties not involved in the partnering agreements, such as suppliers and subcontractors, pose potential points for arrangements to unravel into conventional contractual elements if disputes occur. Indeed, unscrupulous contractors have the scope to hide from the client details of transactions with suppliers unless vigilance is strongly upheld.

Universal 'value for money' is yet to be proven with this kind of arrangement and awaits further trials and reported results from users of the system before firm conclusions can be truly finalised.

Alliancing

Some clients able to offer a series of major long-term construction opportunities, for example the petrochemical industry, have begun to extend partnering to embrace alliancing. Here, successful partners, both design collaborators and contractors, having demonstrated full commitment in terms of previous behavioural attitudes, are invited to co-operate in developing new schemes before even sanction for final approval by the client's main board of directors may have been achieved for the scheme to go ahead.The client's project team, designers and contractors jointly prepare target costs coupled to a risk/award structure based on the final outcome, the implication being that future opportunities will only be forthcoming if the final product meets the client's satisfaction. The management structure requires an executive team with a directorate representative from each alliance member responsible to the client's board for administering the alliance agreement. The project itself is led by the client's project manager heading a team comprising managers from each of the principal parties responsible for implementation of the project, management and administration of the works contracts. Forms of agreement with the principal parties are still evolving either incorporated within the individual contracts for the provision of services or quite separately in an alliance agreement with each.

Joint venture

The unusual step beyond partnering is the joint venture contract between a major client and providers of the facility, perhaps where co-operation is vital, e.g. dealing with a major construction failure. A 51% majority shareholding in the joint company or undertaking provides controlling ownership for one of the partners, but, depending upon circumstances, other proportions are possible. Irrespectively, great care in evaluating the risks and levels of commitments of each, together with establishing sound business and management relations would be essential; otherwise conflicts could eventually surface concerning strategy, decision-making and management style. Even straightforward issues covering accountancy practices, production management, personnel and industrial relations policies could pose potential areas of disagreement and not least confidential business development and research information might prove sensitive issues that could threaten the partnership.

Pooled risks

Conventionally, clients attempt to pass on risks to suppliers by means of the variety of contracts described, with unforeseen circumstances arising during the course of a contract parsimoniously approved as changes and variations. However, when the parties fail to agree on the out-turn work, claims and legal wrangles may result.

As a highly experienced procurer and project manager of construction facilities aiming to reduce time delays and litigation, the BAA client for Terminal 5 at Heathrow airport has piloted a fresh approach by taking responsibility for all of the risk for the project. The T5 Agreement with contracted parties separates out the risk element in tendered prices, the notional amounts from successful supplier quotes being pooled into a reserve.

The individual contracts are subsequently carefully managed on a shared open-book accounting basis, with jobs finished on time and to budget paid a bonus and conversely those leading to legitimate increased costs also drawn from the central reserve. Hence, contractors are encouraged to co-operate since problems caused by one reduce the bonus pot for another.

This complex, large-scale project, financed through bonds issued by the private sector client, claims favourable results compared with construction-industry key performance indicators. However, all aspects were fully designed, 3D-modelled or VR-simulated and tested before contracts were awarded. Moreover, modern Just-in-Time (JIT) scheduling of resources was adopted, and control of supplies for the various contractors involved on site novelly managed through a materials consolidation centre.

Standard forms of agreement in use

Common examples for partnering include:

- Project Partnering Contract 2000(PPC2000) developed by the ACA
- Public Sector Partnering Contract (PSPC) endorsed by the Federation of Property Societies
- BeCollaborative form developed by Collaboration for the Built Environment

- ECC Partnering Agreement
- The proposed joint JCT/Be Partnering Contract

Anticipated performance of different contract categories

Current approximate proportions of total contracts are shown in Table 8.1.

The possible suitability of each is shown in Table 8.2.

Alternatively, subjective assessments can be carried out with, for example, the Office of Government model (Table 8.3).The user is required to decide subjectively on the weight of the important attributes for the project. The scores for the attributes of each separate project category of design and build etc. under consideration are then assessed and allocated. Subsequently

Table 8.1 Current approximate proportions of total contracts.

	By value	By number
Separated	43%	83%
Integrated	43%	15%
M&CM	12%	1%
Discretionary	2%	1%

Table 8.2 Performance expectations of procurement methods.

Project characteristics+	Contract category		
	Separated	Management-oriented	Integrated
Little client experience available	5	4	3
Unique/prestigious design and spec	4	5	3
Small project	3	2	4
Large routine project	3	2	4
Technically complex project	4	5	3*
System/prefabricated construction	2	3	5
Firm price at feasibility stage	4	2	4
Completion on time	2	2	3
Built to tender/budget	2	1	3
Quality, health and safety	3	3	3
Low variations/changes	3	2	4

5 = highly suitable 1 = not well suited

+ Discretionary contract may marginally enhance the rating.

* Higher rating may be justified for a trusted specialist contractor.

Table 8.3 Procurement selection method.

PROJECT No.	Weight (%)	Score for project category (1 to 10)	Weight × Score for project category
Attribute			
Cost-effectiveness			
Certainty of construction cost			
Least disruption			
Flexibility for changes			
Speed of project delivery			
Control over design detail			
Reducing disputes			
Total	100		Points

the total weighted score for each category is calculated. The highest total weight score thereby can be used to inform the final project category selection process.

References

Holt, G.D., Olomolaiye, P.O. & Harris, F.C. (1995) Application of an alternative contractor selection model. *Building Research and Information*, **23**(5), 255–264.

Masterman, J.W.E. (1992) *An Introduction to Building Procurement Systems*. Spon Press, London.

OJEU (1998) Coordination of the laws, regulations and administrative provisions relating to the application of review procedures to the award of public supply and public works contracts. *Official Journal of the European Union*, Directive 89/665/EEC. www.ojec.com.

Tam, C.M. & Harris, F.C. (1996) A model for assessing building contractor project performance. *Engineering Construction and Architectural Management*, **3**(3), 187–203.

9
Estimating and tendering

Summary

The parties involved in the estimating process and the decision to tender; collection and calculation of cost information, all-in rates; project study; preparing the estimate and operational and unit rate estimating; site overheads; estimator's reports; materials and subcontractors; tender adjustments; submitting the tender; computers and estimating.

Introduction

Clients or promoters of the construction industry in the public sector such as central government departments, local authorities and public corporations rely almost exclusively on competitive tendering to justify the awarding of contracts. Clients from private industry tend to follow the practices of the public sector clients and also largely employ competitive tendering procedures. Thus, most construction contracts are awarded after several contractors have submitted a tender and most civil engineering and building contractors derive the major portion of their workload in this way. Previously, most clients based their evaluation of the tenders submitted by contractors essentially on cost. Construction contractors, therefore, based these tenders on an estimate of the cost, to the contractor, of executing the work described in the contract documents. However, there is a growing interest in the use of factors other than cost for evaluating the tenders submitted by construction contractors. Examples of these factors include Quality plans and Health & Safety plans. These additional factors also have an impact on the project costs, which have to be estimated and included in the overall tender price submitted by a contractor. The estimating department is therefore of central importance to the commercial success of the contracting organisation. Most contractors have a director who is responsible for the estimating department and this indicates the importance of the function.

The estimating department is the first department within a contractor's organisation to have any contact with a prospective contract and is required to deal with documents prepared by the client or their representatives. In the course of preparing an estimate and tender, the estimators liaise with the client or their representatives and internally, in the company, with planning staff, buying staff, plant managers, temporary works designers, site management staff and senior management. Thus, the estimators' tasks are not simply ones of calculation but of managing a fairly complex process of assimilation of the contract details, liaising and co-ordinating with all the parties involved, collecting the relevant data, calculating the costs involved and explaining it to senior management.

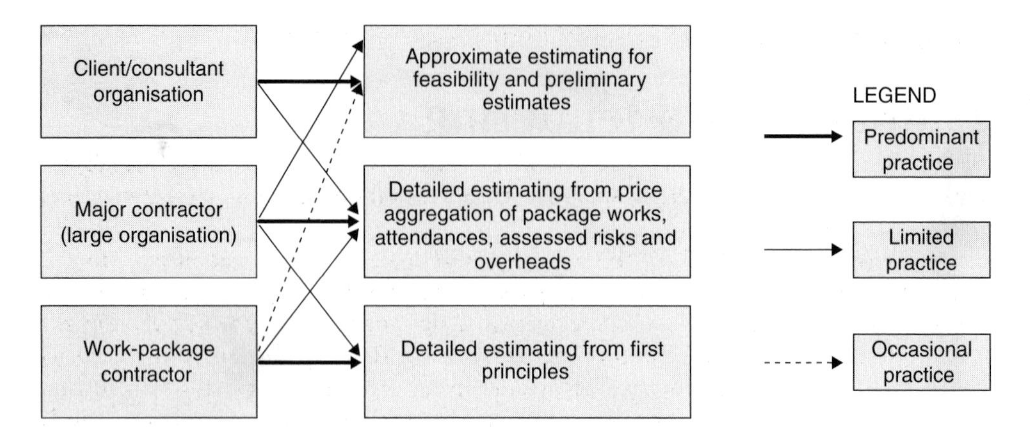

Fig. 9.1 Estimating practices in different types of construction organisations.

Estimating in large and small organisations

Figure 9.1 shows a schematic representation of the nature of the predominant estimating practices that can be associated with different types of construction and related organisations. The client and design consultants often typically rely on approximate estimates to make decisions at the early stages when the scope and scale of the project are being formulated. The gradual progression of the major contractor's role from a traditional works contractor towards the more managerial work-package integrator has resulted in a shift in emphasis on estimating from first principles. Increasingly, such major contracting organisations aggregate price quotes from work-package contractors and provide adjustments for attendances, identified risks from both the quoted prices and other allied project factors and margins.

As such, the type of data collected for the estimating exercise is largely influenced by the nature of the contractor's organisation. The smaller contractors who often deliver the work packages have to compute their estimates from first principles to be able to furnish accurate quotes to the larger contractors. Irrespective of the level at which the estimating is being considered, the important feature is that the estimators have to achieve all this normally within a restricted time period of four to six weeks. The subsequent sections of this chapter describe the parties involved in the production of an estimate and tender, the process of estimating and tendering and the calculations involved. The description of the process of estimating is addressed from the perspective of the work-package contractor since this provides the bedrock of all the other practices of estimating undertaken by other stakeholders in construction.

Parties involved in estimating and tendering

The parties involved in estimating and tendering can be divided into three classes:

- The client's or promoter's staff or their professional representatives
- The construction contractor's personnel including senior management, estimators, planners, buyers, plant managers, temporary works designers and site management staff
- The external organisations, such as material suppliers, plant-hire companies, and subcontractors

The contribution of each of these is described in the following sections.

The client's staff or professional representatives

The person or organisation for whom the building or works is to be constructed is normally referred to as the client when building works are referred to or the promoter when civil engineering works are referred to.

In civil engineering works, the promoter will appoint an engineer or a project manager for the project, and this may be from within their own staff or a consulting engineering practice. The function of the engineer includes the development, design and technical direction of the works as well as the preparation of specifications, bills of quantities, drawings and other contract documents. It is these contract documents that describe the works to the contractor. The drawings, bills of quantities and specifications are the main sources of information for the contractor's estimators, who prepare the cost estimates and tenders.

The method of measurement used in the UK for preparing the bill of quantities is mostly the Civil Engineering Standard Method of Measurement or the Method of Measurement of Road and Bridge Works, or variations of these.

In building contracts, the client appoints an architect, who is responsible for development, design, specification and drawings, and a quantity surveyor, who is responsible for preparation of the bills of quantities and other contract documents and all matters relating to the cost of the project. The architect and quantity surveying staff may be from within the client's own staff or may be from practices engaged by the client. The method of measurement used in the UK for building works usually is the Standard Method of Measurement for Building Works as published by the Royal Institution of Chartered Surveyors (1998).

The construction contractor's personnel

Senior management is an expression used to imply company directors and those who hold responsibilities similar to directors. Senior management are usually involved in the decision whether or not to tender for particular contracts and in the decision on what tender to submit, having considered the estimate of costs and resources involved, as produced by the estimators.

Most companies have a director responsible for the estimating department or an estimating manager who reports to a director. Thus, the day-to-day activities of the estimating department are closely monitored by senior management.

The *estimators* are the personnel employed in the estimating department charged with the responsibility of producing the estimates and managing the process described in this chapter. In both civil engineering and building contractors, the senior estimators are usually professionally qualified staff with extensive experience of the construction industry. These senior estimators are normally supported by junior estimators, who may be aspiring to a career in estimating or may be gaining experience in estimating as part of a wider career development. Other support staff includes estimating clerks and IT support staff.

The *planners* are the personnel employed to produce construction plans or programmes. As far as the estimators' requirements are concerned, this means the pre-tender programme, which may not be as detailed as one produced for site use but will provide the overall duration of the project and the duration of the key activities. It will also provide the sequence of the key activities and approximate resource totals for labour and plant. In some companies, the estimators produce the pre-tender document themselves; in others, the planners produce it, in which case close liaison is required between the estimators and planners. In some companies the planners are part of the estimating department and in others there is a separate planning department.

Buyers are usually responsible for purchasing materials and placing orders with plant hire companies and subcontractors. The service they give estimators is to provide quotations for materials, plant hire and subcontractors. In some companies, the estimators are given the task of sending out the enquiries and receiving the quotations.

Plant managers are responsible for the company's plant department and supply estimators with current internal hire rates and advice on likely availabilities of company-owned plant.

Temporary works designers are responsible for designs of major temporary works such as bridge supports or falsework. The estimators would take their advice on the nature and likely cost of major temporary works in civil engineering contracts.

Site management are the personnel who are employed to take responsibility for the execution of projects on site. The expression covers project managers, agents, works managers, engineers and surveyors. The contribution of site management to estimating is to provide advice to the estimators on methods of construction and to discuss proposed method statements with them.

External organisations

Material suppliers, plant hire companies and subcontractors all get involved in the estimating process in that they receive, and have to respond to, enquiries for quotations from contractors. This also includes consultation with buyers and estimators.

The estimating process

Figure 9.2 shows a generic model of the estimating and tendering process within a construction company. The process involves both the planning and estimating departments. The estimating process commences with an invitation by the client for the contractor to submit a tender for a proposed project.

Estimating practices within construction companies have in the past been based only on the Bill of Quantities (BOQ) type of contract. Although this is still very relevant today, there are a number of variants to the BOQ approach, such as *estimating for work packages*, that are finding increasing popularity. Subsequent sections of this chapter will therefore describe the essential features of the estimating process based on the BOQ contract and highlight any significant differences between the BOQ approach and other contract forms.

BOQ estimating

BOQ estimating is undertaken for contracts that rely on client-prepared contract documents and a bill of quantities. This is still the most common type of contract in the UK. The contractor's role is to estimate how much it will cost to do the project so that it can be priced reasonably.

The process undertaken to produce a cost estimate, which can be used as the basis of a tender, is described in the following steps:

- Decision to tender
- Programming the estimate

- Collection and calculation of cost information
- Project study
- Preparing the estimate
- Site overheads
- Estimator's reports

These steps vary for other forms of contracts such as schedule-of-rates, design-and-construct and all-in contracts.

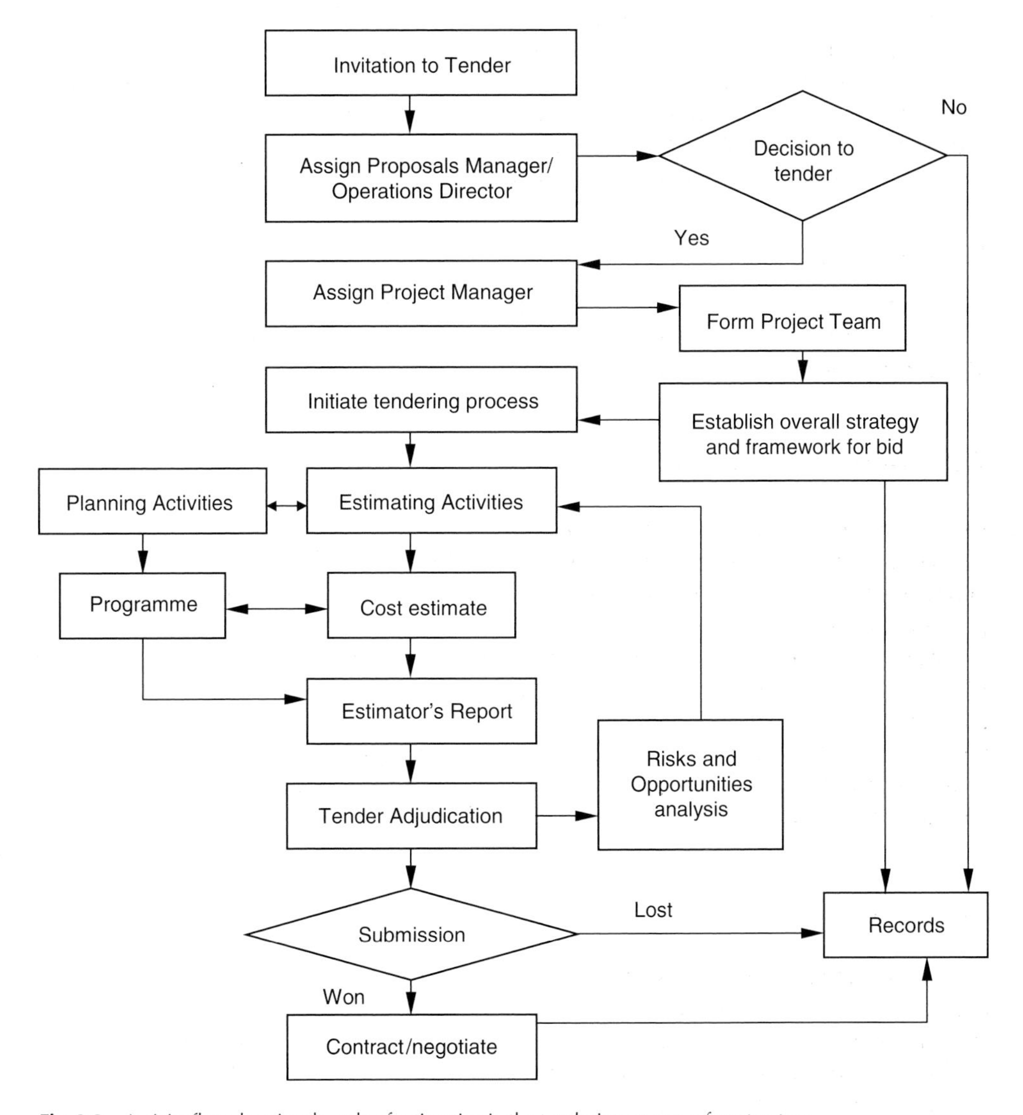

Fig. 9.2 Activity flow showing the role of estimating in the tendering process of contractors.

Decision to tender

The decision to tender for a particular contract is mainly the responsibility of senior management; there are three possible points during the estimating and tendering process where this decision must be made. The first is during the pre-selection stage, if a pre-selection procedure is being used. This decision would be based on pre-selection information provided by the clients' staff or representatives and would include such information as:

- Name of the job
- Name of the client, architect and quantity surveyor or consulting engineers
- Names of any consultants with supervisory duties
- Location of the site
- General description of the work involved
- Approximate cost range of the project
- Details of any nominated subcontractors for major items
- Form of contract to be used
- Procedure to be adopted in examining and correcting priced bill(s)
- Whether the contract is to be under seal or under hand
- Anticipated date for possession of the site
- Period for completion of the works
- Approximate date for the despatch of tender documents
- Duration of the tender period
- Period for which the tender is to remain open
- Anticipated value of liquidated damages (if any)
- Details of bond requirements (if any)
- Any particular conditions relating to the contract

The contractor may have signified their intention to submit a tender at the pre-selection stage, but has a further opportunity after receiving all contract documents to review them and consider, in the light of fuller information, whether they wish to proceed or not. The third and final point of decision whether to submit a tender is after the estimate has been prepared and the contractor is ready to submit. A decision at this stage not to tender is rare but allows the contractor to review the events of the intervening four to six weeks and the changes in workload that have occurred in that time.

The decision to tender is based on such factors:

- Company's current workload, turnover and recovery of overheads
- Company's financial resources
- Availability of resources to undertake the work
- Type of work
- Location of the contract
- Identity of the client or promoter and their representatives
- Detailed examination of the contract documents

Programming the estimate

After the contract documents have been received and a decision made to tender, the tasks required to complete the estimate are programmed and a schedule of key dates established to monitor progress. This is essential as the time to submit a tender is limited. The two main tasks that can now take place are the *collection and calculation of the cost information*, required to prepare the estimate, and a *study of the project* to gain the required appreciation of the work involved in the contract. These two tasks take place in parallel and are interrelated.

Collection and calculation of cost information

The cost information required by the estimator is for:

- Labour
- Plant
- Materials
- Subcontractors

These are discussed below.

Labour

The estimator is required to calculate the 'all-in' rate for each category of labour, which is an hourly rate covering all wages and emoluments paid to the operative, all statutory costs incurred and allowances for holidays and non-productive overtime. As an alternative to the hourly rate, the all-in rate may be calculated as a daily or weekly rate. An example of this calculation is set out below. As the all-in rates are calculated by the estimator, they can be established early and accurately in the estimating process.

The basis of this calculation is to determine the total worked hours in a year and the total cost for the working year and hence the cost per worked hour. The calculation is complicated by the difference between the hours worked and the hours paid, which exceeds the hours worked by such items as overtime allowances, public holidays and inclement weather. The rules for payment are set out in the Working Rule Agreements produced by the Construction Industry Joint Council.

The calculations set out in this example are based on a real example and contain data that are available in the Working Rule Agreements, assumptions made by a contractor's estimator and the data relating to statutory obligations.

Working times

The guaranteed minimum earning period for construction workers is 39 hours. This is based on a 5-day working week for the site works. During the 44-week 'summer' period, the daily hours are 8.00 a.m. to 5.30 p.m. This gives a working day of 8 hours plus 1 hour overtime for Monday to Thursday; and 7 hours plus 2 hours overtime for Friday, making up in all a working week of 39 hours plus 6 hours overtime. During the 8-week 'winter' period, the daily hours are 8.00 a.m. to 4.30 p.m. This gives a working day of 8 hours for Monday to Thursday, and 7 hours plus 1 hour overtime for Friday, and giving a working week of 39 hours plus 1 hour overtime.

Holidays

- Summer holiday is 2 weeks.
- Easter holiday in 'summer' period is 4 days.
- Winter holiday is 7 days.
- Public holidays in 'summer' period total 5 days.
- Public holidays in 'winter' period total 3 days.
- The total holidays are 29 days or 5 weeks, 4 days.

Inclement weather and sick leave

Assume 60 hours of inclement weather and 3 days of sick leave in the 'summer' period and 2 days of sick leave in the 'winter' period.

Weeks worked in a year

- Total weeks in year = 52 weeks.
- Holidays = 5 weeks, 4 days.
- Sick leave = 1 week.
- Weeks worked in year = 45 weeks, 1 day.

Calculation of hours worked

Total hours in year:

Summer period	44 weeks × 45 hours	= 1980 hours
Winter period	8 weeks × 40 hours	= 320 hours
		= 2300 hours

Hours not worked due to holidays:

Total holidays in 'summer' period	= summer holiday + Easter holiday + public holidays	
	= 2 weeks + 4 days + 5 days = 19 days	
Hours not worked due to summer period holidays	= 19 days × 9 hours	= 171 hours
Total holidays in 'winter' period	= winter holiday + public holiday	
	= 7 days + 3 days = 10 days	
Hours not worked due to 'winter' period holiday	= 10 days × 8 hours	= 80 hours
Total hours not worked due to holidays	= 251 hours	
Hours not worked due to inclement weather	= 60 hours	
Hours not worked due to sick leave	= 3 days × 9 hours + 2 days × 8 hours	= 43 hours
Total hours not worked	= 354 hours	
∴ total hours worked in one year	= 2300 hours − 354 hours	= 1946 hours

Calculation of overtime

Possible overtime worked in summer period	= 44 weeks × 6 hours	= 264 hours
Less 19 days holiday and 3 days sick leave		
∴ overtime not worked	= 22 days × 1 hour	= 22 hours
∴ overtime worked		= 242 hours

Costs of employing labour

Basic rates of pay (at the time when this edition of the book was published):

- Labourers £6.52 per hour
- Trades person £8.25 per hour

Overtime is paid at 1½ times basic rate

Assume a 30% bonus including guaranteed minimum.

According to the Construction Industry Joint Council (CIJC) rules, the employer's maximum liability for tools is £400 and for clothing is £30. This requirement takes the place of the tool and clothing allowance, and is covered by the employer's liability.

Plus rates are assumed to average at 3%.

Holiday stamps with death benefit amount to £24.54 per week. Additional requirements under the CIJC rules would be covered by the employer's liability.

Sick pay is £93.60 per week for maximum of 10 weeks per annum. Any incapacity beyond 1 week will be covered by the employer's liability.

National Insurance is 11% on payroll.

The Construction Industry Training Board (CITB) levy is 0.5% for payroll exceeding £64 000.00 of direct employees and 1.5% of payroll exceeding £64 000.00 for labour-only subcontractors (LOSC).

Assume a company employment structure as detailed below:

Employee category (Number)	Wage bill (£)	CITB levy (£)	Levy per employee (£)	Levy per employee per week (£)
25 direct employees	420 500.00	2102.50	84.10	1.79
4 LOSC	80 576.00	1208.64	302.16	6.43

Assume an allowance of 1½ % for severance pay.

Assume an allowance of 2% for employer's liability and public liability insurance.

Calculation of annual costs of labour

Basic cost		Labourer		Trades person	
Item	Factor	Unit	Value (£)	Unit	Value (£)
Annual basic	1946	6.52	12 687.92	8.25	16 054.50
Inclement weather	60	6.52	391.20	8.25	495.00
Basic total			13 079.12		16 549.50
Bonus 30% on basic total	0.30	13079.12	3923.74	16 549.50	4 964.85
Non-productive overtime (basic rate × 0.5 × 242 hours)	121	6.52	788.92	8.25	998.25

Basic cost		Labourer		Trades person	
Public holidays 8 days (basic rate × 8 days × 8 hours)	64	6.52	417.28	8.25	528.00
Sick pay @ £93.60 for 1 week	93.60	1	93.60	1	93.60
Plus rate @ 3% on basic total	0.03	13 079.12	392.37	16 549.50	496.49
Paid total			18 695.03		23 630.69
National Insurance 11% on paid total ('Class 1' NICs)	0.11	18 695.03	2056.45	23 630.69	2599.38
CITB Levy on paid total > £64 000.00 = 0.50% company wage bill of £420 500.00 for direct employment	47	1.76	82.72	1.76	82.72
Annual holiday and death benefits @£24.54 for 47 weeks	47	24.54	1 153.38	24.54	1 153.38
Paid total + statutory costs and benefits			21 987.58		27 466.16
Allowance for severance pay 1.5%	0.015	21 987.58	329.81	27 466.16	411.99
Allowance for employer's liability (under CIJC rules) and public liability insurance 4%	0.040	21 987.58	879.50	27 466.16	1 098.65
Total annual cost			23 196.90		28 976.80

Calculation of all-in rate

The all-in labour rate is defined as total annual cost for labour divided by total worked hours in a year

The total worked hours = 1946 hours

The total annual cost for a labourer = £23 196.90

All-in rate for labourer = £24 978.49 ÷ 1946 hours = **£11.92** per hour

Total annual cost for trades person = £28 976.80

All-in rate for trades person = £35 216.44 ÷ 1946 hours = **£14.89** per hour

The hourly all-in labour rates form the basis of other unit rates such as all-in weekly rates. All-in weekly rates are obtained by substituting the total number of working weeks per year in place of the total hours worked for the calculation. There are 45 weeks and 1 day's working time in a year, yielding 45.2 working weeks per annum.

All-in weekly rate for labourer = £24 978.49 ÷ 45.2 = **£513.21** per week

All-in rate for trades person = £35 216.44 ÷ 45.2 = **£641.08** per week

Plant

The hourly or weekly cost of plant can be either as result of internal calculation or as a result of quotations. Methods of calculating hire rates are given in Chapter 7 *Plant management* within this book, and also in the companion volume by Harris & McCaffer (1991) on *Management of Construction Equipment*. Quotations for hire can be either internal rates from the plant department or the contractor's plant subsidiary, or external hire rates from an independent plant hire company. Calculated rates or internal rates can be established very early in the estimating process. External hire rates may take a little longer to obtain, but it is unusual to suffer serious delays in receiving quotations.

Materials

Materials quotations are more problematic in that materials not only form a significant percentage of the works but, because of the volatility of materials prices, contractors have to send out unique enquiries for almost all materials in every estimate prepared. The materials enquiries include information such as the quantities required, the specification, the approximate delivery dates and the terms and conditions upon which the quotations are being invited.

To enable these enquiries to be sent out, the estimator must go through the bill of quantities and specification, extract the relevant information and prepare a list of required materials. Because of the time taken to send out and receive quotations, this task of preparing the materials lists is undertaken very early in the estimating process. In some companies, a 'buyer' may send out the enquiries, chase the suppliers who fail to respond and check and compare the quotations. In other companies, this task may be undertaken by the estimators.

Enquiries sent to suppliers normally include:

- Specification of the material
- Quantity of the material
- Likely delivery programme including both the period for which supplies would be needed and the daily or weekly requirements
- Address of the site
- Means of access
- Traffic restrictions and conditions affecting delivery
- Period for which the quotation is required to either remain open for acceptance or to be firm
- Date by which the quotation is to be submitted
- Name of the person within the contractor's organisation to whom any reference concerning the enquiry should be made

Buying departments usually have standard letters or forms for issuing enquiries and, where convenient, parts of the contract documents—such as extracts from the specification—are photocopied to accompany the enquiry. As the estimator is responsible for completing the estimate an enquiry index or progress chart is usually kept so that the stage of each enquiry can be readily monitored.

When checking quotations, the estimator is required to ensure that quality and quantity of materials meet the specification of the contract documents, and can be delivered to the site at the times required by the construction programme. In addition, the contractual obligations to be entered into for the supply of the material must be satisfactory.

The determination of the materials prices for inclusion in the direct cost estimate may be considered to be one of the most precise aspects of estimating. The process of obtaining materials prices, as has been described, can be seen to consist solely of contacting suppliers who have the material available and negotiating a suitable rate under satisfactory contractual conditions. In addition, the estimator has to undertake the more difficult task of determining allowances for material wastage, damage, theft, delivery discrepancies and handling insofar as they affect the costs of the works.

Subcontractors

For the traditional BOQ approach, subcontractor enquiries also have to be sent out early and the estimator will prepare lists of the items and work that will be subcontracted. As with materials, these enquiries may be dealt with by a 'buyer'.

Following receipt, the quotations must be compared and the subcontractors selected. The rates for the selected subcontractors will be included in the estimate together with allowances for attendance and other services. The profit to be added to the main contractor's own subcontractors and to the nominated subcontractors may be included at the estimating stage or left until the final additions following the tender meeting. The difference between materials and subcontractors is that in most cases the materials costs are combined with plant and labour to produce costs rates for items of work, whereas the subcontractors' rates in many cases will stand on their own together with an allowance for attendance.

As part of the preliminary study, the work that is to be subcontracted would have been identified by the estimator. The factors that control the decision of which work to subcontract are mainly the specialisation of the work involved and the size of the contract. Most contractors establish by practice the type of work they normally subcontract. In undertaking contracts larger than usual, a company may wish to subcontract some work they would normally undertake themselves, the reason being to offset some of the financial risk. The absence of a direct financial risk in subcontracted work is not a total security because of the indirect risk of losses caused by delay and disruption to the main works if the subcontractors default. For this reason, effective control of subcontracted operations is important and this control begins with the selection of subcontractors. Most companies keep a list of approved subcontractors for various classes of work as a guide to estimators in the comparison of subcontractor quotations.

Since the 1980s, there has been an enormous growth in the use of subcontractors by the main contractor. The reason for this development primarily was to enable the contractor to take on additional workload without the need for increasing internally the level of employment and capacity required to execute the extra workload, thus ensuring greater flexibility. The need for this degree of flexibility by the contractor derived from the volatility in the industry's workload. The use of subcontracting transferred some of the financial risk associated with maintaining the additional capacity to the subcontractor.

The trend towards the use of subcontractors continued through the 1990s. This means that the relative importance of the areas of risk associated with using subcontractors increased. This risk is acknowledged by contractors, who rely on their vetting and selection procedures to reduce it to acceptable levels. The contractor who bases a tender on a subcontractor's quotation merely because it was the lowest would be unwise. Since the beginning of the 21st century, changes in employment laws and taxation have reduced the advantage associated with subcontracting for labour-only arrangements (LOSC). Under these changes, the main contractor is still responsible for ensuring statutory entitlements of all the LOSCs it employs. The following briefly views the risk areas associated with the use of subcontractors.

Attendances

Estimating attendances, such as equipment, scaffolding and tools, when comparing and selecting subcontractors is assuming a greater priority in estimating work. Not all quotations are consistent in the attendance requirement and comparing the simple quotations is inadequate. Very often, a subcontractor will by negotiation take on more onerous attendances within the stated price. Estimating attendances may in itself prove critical in winning a contract or indeed in being able to execute a contract profitably.

Materials

The importance of material costs is that they represent above 50% of the cost profile of the construction industry. A small percentage cut in materials costs could bring a sizeable increase in profits. For example, a 2% cut in materials could increase profits by much more than that which could be achieved by a 2% cut in overheads.

The estimating of material cost in any tender is, therefore, an important element and an inaccuracy in their estimates is likely to affect the outcome of the tender and the profitability of the subsequent contract. Furthermore, it is obvious that any appreciable waste is a significant item of cost and represents a great loss to both the nation's resources and the main contractor.

Many estimators have established within their companies standards or norms for wastage of materials. These standards, however, are largely based on the experience with their own labour force, or where their own labour had a large share of the work. There is potential for wastage rates to rise with the use of subcontractors' labour, whose desire to complete the work as soon as possible is high. Thus, the materials wastage allowance is likely to be a growing factor in the accuracy of an estimate. This may be obviated by tighter materials control, or in some cases passing the responsibility for purchasing materials to subcontractors with agreed limits as to quantity.

Output

The greater use of subcontractors has reduced the importance of one of the estimator's most difficult tasks; that of estimating labour and plant outputs. However, the task is not completely removed; companies do still employ their own labour and even this lessening of the estimator's burden has been partially offset by an increase in other difficulties. These difficulties lie in quantifying the consequences of subcontractors who prove unreliable and fail to complete their work or disrupt the sequence of work. A subcontractor who tackles the straightforward work while failing to complete the smaller but more difficult elements can cause disruption. So can the subcontractor who is slow, starts late and finishes late. The main contractor may not actually suffer any financial loss within the specific subcontract but will be left to face the consequences of the disruption. Evaluating the effects of potential disruption for the purposes of a claim against the subcontractor is difficult. Evaluating the effects of potential disruption at the estimating stage is even more so. As a result, there now seems to be much more effort in the 'back rooms' of companies devoted to trying to determine how this potentially costly element can be evaluated. The use of computer simulations is now being practised. Certainly, potential disruptions are now a major factor in determining the accuracy level of estimators measured by the difference between estimated cost and the actual costs. However, in these competitive times estimators tend to rely on their vetting procedures in assessing the subcontractor's record rather than explicitly allowing for potential disruption.

Management control and efficiency

Site management control and efficiency have always been a major factor affecting the level of estimating accuracy. When a job loses money and the discrepancies between estimates and actuals are large enough for an internal enquiry, the estimators and the site managers can blame each other. That site management got it wrong, or that the estimators got it wrong, have, through the years, been equally valid and equally difficult to prove. The greater use of subcontractors does not make site management less important but it has altered its nature. The control and organisation of the arrival, departure and interactions of many subcontractors has had to be developed in companies that were previously used to making their major efforts in controlling their own labour. Squeezing more productivity out of the company's own labour force and ensuring that subcontractors perform on time, with good quality workmanship, and dovetail into the other activities on site has changed the role of general foremen and the site agents. It has always been true that, all things being equal, the success of a contract is determined by the quality of the site management team. The greater use of subcontractors does not change this but has changed how it must be achieved.

Project study

To gain an appreciation of the project, the estimator and the planner (where this is a separate person) will undertake the following tasks:

- A study of the drawing
- A site visit and meeting with the client's or promoter's representatives
- The preparation of a method statement determining how the project will be constructed

These are discussed below.

Drawings

In civil engineering contracts, the contract drawings are sent to each contractor tendering. In building work, it is now common practice to issue a set of contract drawings. However, in cases where this is not so, estimators and planners may have to visit the architect's offices to inspect a complete set of drawings and to discuss unclear aspects with the architect. Unclear aspects of the drawings are also frequently clarified by telephone consultation.

Site visit

In civil engineering works, it is normal practice for the promoter's engineer to arrange a site visit, which gives the estimators and planners an opportunity to study the site with the design personnel on hand to answer queries. In building work, the site visit may not be arranged so formally but nevertheless most contractors arrange one for themselves. As a result of these site visits, the estimator and planner prepare a report or set of notes which includes:

- Description of the site
- Positions of existing services
- Description of ground conditions
- Assessment of the availability of labour
- Any problems related to the security of the site
- Description of the access to the site
- Topographical details of the site
- Description of the facilities available for the disposal of soil
- Description of any demolition works or temporary works to adjoining buildings

Method statement

Method statements are descriptions of how the work will be executed with details of the type of labour and plant required and a pre-tender programme. It is in the preparation of these method statements that alternative methods of construction are considered, together with alternative sequences of work, differing rates of construction and alternative site layouts. As these evaluations progress, the preferred method of construction is chosen and the pre-tender construction programme illustrating this is prepared.

In preparing the method statements, the estimators and planners work closely and also consult with site staff, plant managers and temporary works designers. The pre-tender programme prepared will show the sequence of all the main activities and their durations, as well as the duration of the overall project. From this pre-tender programme, approximations of the labour and plant resources will be calculated. This pre-tender programme is not an exercise carried out once only and left, but is subject to continual refinement and modification as both the estimator and planner become more and more aware of the implications of the project details. Thus, throughout the preparation of the estimate the estimator and planner remain in close consultation. In many contracts, the pre-tender programme is prepared in the form of a bar chart; in some of the larger and more complex projects some form of network analysis is used. Networks are also used in smaller contract companies, who have become skilled in their preparation and use and aware of their advantages. Bar charts and networks are described in Chapter 4 on *Planning Techniques.*

Preparing the estimate

The estimator's task is to determine the cost to the contractor of executing the work defined in the contract documents. This cost estimate will be modified by senior management in consultation with the estimator to determine the tender or selling price. The estimator is required to establish the direct cost rates for each item in the bill of quantities. A direct cost rate is a rate for the labour, plant, materials and subcontractors, but exclusive of additions for site overheads, head-office overheads and profit. These will be assessed and included later.

Determining a direct cost rate involves selecting the appropriate resources of labour, plant and materials for the item of work (either a single bill item or a group of bill items), selecting the output or usage rates for each resource or determining the elapsed time that each resource (labour and plant) will be employed and combining this with the cost information collected. This combination of the unit cost of resources, together with the usage of resources to produce a direct cost for the work described in the bill item or group of bill items, is illustrated in Fig 9.3.

Types of estimates

There are two main types of estimates employed in construction for which estimators have to go through the full estimating process. These are operational and unit rate estimating.

The method of calculation of a direct cost rate based on the output or usage of the selected resources is known as *unit rate estimating* and the method of calculating direct costs based on the elapsed time of employed resources is known as *operational estimating.* These two methods of estimating are the main methods employed in calculating direct costs for bill items or groups of bill items.

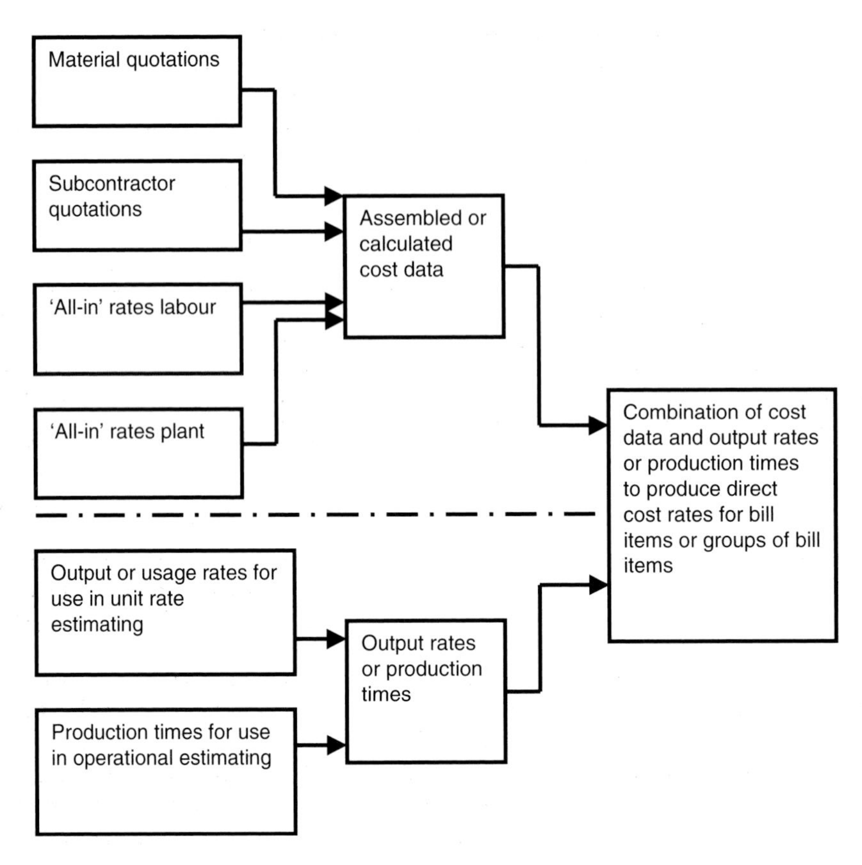

Fig. 9.3 The selection of production rates and cost data and their combination to produce cost rates.

Other methods of producing item rates are *spot* or *gash rate* estimating, whereby the estimator produces estimates of direct cost rates that are based on their experience or use of rates for similar work used in previous contracts. Subcontractor rates, with or without attendances, prime cost and provisional sums are the other major entries against bill items.

Operational estimating

In civil engineering work, there is an extensive use of operational estimating. This is because this method of estimating links well with planning and is effective in allowing for idle time, which is common in most plant-dominated work. For example, an excavator may be on site from the start of excavation operations to the end of excavation operations but may not be working continuously throughout that time period. Operational estimating would base the costs of the excavator on the time that the excavator was on site, not on an assumed output. The time the excavator was on site would be derived from the construction programme.

The link between operational estimating, planning and the bill of quantities is that the estimator would group together the bill items they would wish to estimate as an operation. This group of bill items would be represented in the construction programme as one or more activities and so the construction programme would provide the duration of the operation. The estimator would select the resources to be used in the operation and use the duration of the activities

Table 9.1 Example of an operational build-up for the plant element for placing concrete.

Assumptions				
Total cubic metres of concrete to be placed			$2550\ m^3$	
Duration of concreting operations			38 weeks	
Plant required			1 Mobile crane, 2 Concrete skips, 3 Dumpers, 3 Vibrators	

Calculations Item	a No	b Weekly rate	c No. of weeks	$d = a \times b \times c$ Cost
Mobile crane	1	£845.00	38	£32 110.00
Concrete skip	2	£28.60	38	£2 173.60
Dumper	3	£271.70	38	£30 973.80
Vibrator	3	£41.60	38	£4 742.40
			Total cost =	**£69 999.80**
			Cost per m^3 =	**£27.45**

This table illustrates the operational build-up for the plant element for placing concrete. A similar build-up for labour and for the provision of concrete would be required to satisfy a group of bill items that include providing and placing concrete.

and unit cost of each resource to produce the total cost of the operation. This total cost has to be assigned back to each bill item. The method of assignment is open to the estimator but the simplest method is to apportion the costs pro rata with the item quantity. That is, if a total of $2550\ m^3$ of concrete placing were contained in 40 different bill items of varying quantity and the total cost of the plant element of the operation was £69 999.80 then the rate would be £27.45 per m^3 and this would be set against each item together with the labour costs. In general, conventional bills of quantities are not designed to suit operational estimating and operational bills have not come into common use: thus, estimators are faced with undertaking the grouping of items into operations as described. In CESMM (Civil Engineering Standard Method of Measurement) bills of quantities, there is the opportunity to use 'method-related charges', which suits an operational estimating approach. An example of operational estimating is given in Table 9.1.

Building contractors use operational estimating to a much lesser extent than the civil engineering contractors. This is because their work is less plant-dominated. However, it is used in estimating the costs of crainage and concrete or mortar mixers. Frequently in building contracts these operationally estimated plant costs are not assigned to the bill items but included in the 'Preliminaries'.

Unit rate estimating

Civil engineering contractors' estimators do not use operational estimating exclusively, mainly because of the format of the bill of quantities, and they employ *unit rate estimating* for some of the items or groups of items. The use of unit rate estimating is predominant in building work. The basis of unit rate estimating is the selection of the resources required and the selection of the output or usage rates for those resources. Thus for each of the resources (i.e. labour, plant or materials) the arithmetic calculation is the output or usage rate combined with the unit cost. An output rate is expressed as work quantity per hour (e.g. labour for placing concrete, m^3/h) and a usage rate is the time taken to do a fixed quantity of work (e.g. labour for placing concrete, h/m^3).

Thus, by combining these with the cost of the resource (£/h in the case of labour and plant) the cost per unit (e.g. labour for placing concrete, £/m^3) can be determined.

The sum of the cost per unit for all resources is the estimate of the direct cost rate. An example of a unit rate calculation is given in Table 9.2. Supplementary cost calculations for this example are given in Table 9.3. The selection of resources and their associated output or usage rates for each category of work is:

- Abstracted from previously recorded company manuals
- Taken from estimator's personal manuals or known to the estimators by experience

The use of company manuals for recording standard build-ups giving resources and usage rates is common in larger companies. Estimators are, of course, allowed to modify such standard

Table 9.2 Example of a unit rate build-up.

(a) Example of bill item

Item description	Units	Quantity	Rate	Extension
A. One brick wall in common bricks in 1:3 cement mortar	m^2	100	–	–

(b) Example of unit rate build-up—labour

Labour	Cost/hr	Output/usage	Total cost
Bricklayer—Commons	£20.76	1.8 h/m^2	£37.38
Labour to unload (at 10 minutes/1000 bricks)	£11.27	1.2 min/m^2	£0.22
Total labour cost			£37.60

(c) Example of unit rate build-up—materials

Materials	Quantity/m^2	Net materials cost	Waste	Total cost
Common bricks	120 Nr	£182.00/1000	10%	£24.02
Mortar (3:1)	0.045 m^3	£73.89/m^3	5%	£3.50
		Total material cost		£27.52
		Total cost/m^2		£65.12

Table 9.3 Supplementary cost calculations in unit rate build-up.

(a) Example of calculating the cost of one bricklayer per hour with labourer support

Two bricklayers at £15.12/hour	= £30.24
One labourer at £11.27/hour	= £11.27
Total cost of gang	= **£41.51**
Cost of one bricklayer per hour (÷2)	= **£20.75**

(b) Example of calculating cost of mortar per m^3

Cement 0.44 tonne × £120.35/tonne	= £52.96
Sand 1.83 m^3 × £11.44/m^3	= £20.94
Total cost per m^3	= **£73.90**

build-ups. The growing use of computers in estimating allows the storage in computer files of these standard build-ups and provides for recall and manipulation before inclusion in the estimate.

Combining operational estimating and unit rate estimating

Figure 9.4 shows the build-up for an item where the provision of concrete is calculated on a unit rate basis and the placing of concrete is calculated on an operational basis. The operational rates being calculated in two parts—labour for placing concrete and plant for placing concrete. This illustrates that estimators can use unit rate and operational rates within the same build-up. The choice of use of operational estimating and unit rate estimating varies widely with different estimators and different companies.

Site overheads

The estimator assesses the site overhead based on requirements such as:

- Site staff
- Cleaning site and clearing rubbish
- Mechanical plant not previously included in the item rates
- Scaffolding and gantries
- Site accommodation
- Small plant
- Temporary services
- Welfare, first-aid and safety provisions
- Final clearance and handover
- Defects liability
- Transport of operatives to site
- Abnormal overtime
- Risk

The costs of these site overheads are frequently allocated to the 'preliminary' section of the bill, but may sometimes be allocated, at least partially, to bill item rates.

Estimators' reports

On completion of the estimate, the estimators prepare a set of reports for consideration by the senior management. These reports contain:

- A brief description of the project
- A description of the method of construction
- Notes of any unusual risks that are inherent in the project and which are not adequately covered by the conditions of contract or bills of quantities
- Any unresolved or contractual problems
- An assessment of the state of the design process and the possible financial consequences thereof
- Notes of any major assumptions made in the preparation of the estimate
- Assessment of the profitability of the project
- Any pertinent information concerning market and industrial conditions

DESIGN STRUCT. CONC. 30 MPa 20 mm. AGGR. OPC

Is this the correct item? Y

Section	1	Page	1	Item	3
Library	CESMM 3			Code	F143

DESIGN STRUCT. CONC. 30 MPa 20 mm. AGGR. OPC 158 m^3

Database Resources

Resource	Category	Description	Usage rate	Unit Cost	Total
500	PLT	20/14 MIXER	0.160 hr/m^3	4.80/hr	121.34
501	PLT	2 m^3 READY-CRETE TRUCK	0.140 hr/m^3	8.28/hr	183.15
503	DIR	OPC	0.400 tn/m^3	96.31/tn	6086.79*
505	DIR	SAND	0.350 m^3/m^3	5.95/m^3	329.04*
508	DIR	20 mm AGGR.	1.000 tn/m^3	7.91/tn	1249.78*
511	AUX	DIESEL	1.050 gal/m^3	2.30/gal	381.57
512	PLT	SILO (50 Tonne)	0.160 hr/m^3	1.15/hr	29.07
513	LAB	MIXER DRIVER	0.160 hr/m^3	5.75/hr	145.36
514	LAB	SHOVEL OPERATOR	0.160 hr/m^3	5.75/hr	145.36

*Prices with * indicate Quotes are required*

Operational Rates

Code	Description	Cost	Total
CONCPL	PLANT FOR PLACING CONCRETE	22.02/m^3	
CONCLA	LABOUR FOR PLACING CONCRETE	17.27/m^3	
Item Rate	94.17/m^3		

Do you want reconciliation of database resources? Y

Resource	Category	Description	Amount
500	PLT	20/14 MIXER	29.07 hr
501	PLT	2 m^3 READY-CRETE TRUCK	25.44 hr
503	DIR	OPC	72.68 tn
505	DIR	SAND	63.60 m^3
508	DIR	20 mm AGGR.	181.70 tn
511	AUX	DIESEL	190.79 gal
512	PLT	SILO (50 Tonne)	29.07 hr
513	LAB	MIXER DRIVER	29.07 hr
514	LAB	SHOVEL OPERATOR	29.07 hr

Do you want reconciliation of operational rates? Y

CONCPL PLANT FOR PLACING CONCRETE

Resource	Category	Description	No. Week	Alloc.
P 989	PLT	MOBILE CRANE	38	100
P 991	PLT	CONCRETE SKIP	76	100
P 992	PLT	DUMPER	114	100
P 993	PLT	VIBRATOR	114	100

CONCLA LABOUR FOR PLACING CONCRETE

Resource	Category	Description	No. Week	Alloc.
1	LAB	LABOUR	152	50

Fig. 9.4 An example of unit rate and operation rate estimating combined.

The costs of work included in the estimate are reported to senior management in cost reports that give details of:

- Main contractor's labour
- Main contractor's plant allocated to rates and in preliminaries
- Main contractor's materials
- Main contractor's own subcontractors
- Sums for nominated subcontractors
- Sums for nominated suppliers
- Provisional sums and dayworks
- Contingencies
- Amount included for attendance on domestic and nominated subcontractors
- Amounts included for materials and subcontract cash discounts

In addition, a bill of quantities marked up with the direct cost rates showing the labour, plant, materials and subcontractor breakdown for each rate may form part of the report.

As well as reporting the costs estimated for labour, plant and materials, the estimators also assemble the total hours for each category of labour and the total hours or weeks for each major item of plant and total quantities for materials. These resource totals are compared with the planners' calculated resource totals and any difference reconciled. Figure 9.5 shows an example of such a report showing resource totals and costs.

The estimators may also calculate the cash flow for the contract based on a range of assumed mark-ups, which will assist senior management's judgement as to what is the appropriate mark-up to select. A description of cash flow calculations is given in Chapter 12 on *Cash flow and interim valuations*.

Tendering adjustments

Based on the reports prepared by the estimators, the staff charged with the responsibility of submitting the tender will assess the estimate and decide on the additions to cater for risk, company overheads and profit. The groups of staff concerned, sometimes referred to as the tender adjudicating panel, will comprise representatives of senior management and representatives of the estimating team. It is the responsibility of this panel to satisfy themselves that the estimate is adequate. This is done by studying the reports prepared by the estimator and interrogating the estimator on the underlying assumptions and decisions. It does result on many occasions that the estimate is adjusted, usually in the form of lump-sum additions or subtractions.

The additions for risk, overheads and profit are frequently referred to as the 'mark-up' and are allowances for:

- 'Risk' if the chance of probability of making a loss is assessed as being greater than that of breaking even
- 'Company overheads' to cover the central head-office costs that are involved in administering the contract
- The 'profit' considered to be possible in the existing market conditions

RESOURCE RECONCILIATION					
CONTRACT: A851 BRIDGEWORK				DATE: 5/05/2005	
SUMMARY OF RESOURCES USED IN UNIT RATE ESTIMATING					
Resource Number	Description	Resource			
		Total Quantity	Direct Cost	Contribution	
1	Labour	42.04 hr	£471.38	£471.38	
21	Dump Truck 10.5m3	73.50 hr	£2,772.34	£2,772.34	
136	JCB 3C Excavator	36.75 hr	£890.05	£890.05	
500	20/14 Mixer	57.12 hr	£356.10	£356.10	
501	2m3 Ready-Crete Truck (1 Mile Haul)	49.98 hr	£537.99	£537.99	
503	OPC	142.80 tn	£17,879.46	£17,879.46	Quote
505	Sand	124.95 m3	£4,670.02	£4,670.02	Quote
508	20mm Aggregate	357.00 tn	£3,671.95	£3,671.95	
511	Diesel	374.85 ga	£1,120.81	£1,120.81	
512	Silo (50-Tonne)	57.12 hr	£85.40	£85.40	
513	Mixer Driver	57.12 hr	£426.97	£426.97	
514	Shovel Operator	57.12 hr	£426.97	£426.97	
524	Steel-fixer	952.00 hr	£12,809.16	£12,809.16	
525	Steel-fixer's Labourer	204.00 mh	£2,287.35	£2,287.35	
530	20mm Dia. Mild Steel	74.80 tn	£36,508.95	£36,508.95	
558	Facing Bricks	9.92 tn	£2,224.56	£2,224.56	Quote
560	Mortar	3.76m3	£260.33	£260.33	
561	Scaffolding up to three meters	171.00 m2	£1,227.10	£1,227.10	
562	Bricklayer	328.32 hr	£4,417.54	£4,417.54	
563	Bricklayer's Labourer	328.32 hr	£3,681.29	£3,681.29	

Fig. 9.5 Resource reconciliation report.

The manner in which contractors assess these additions varies enormously from company to company. These additions are incorporated into the tender in a variety of ways, including lump-sum additions and subtractions and *pro rata* adjustments.

The 'discounts' taken on materials and subcontract quotations are sometimes considered as an extra source of profit and it is not unknown for a contractor in very poor markets to submit a tender with a zero or negative profit allowance and rely on such discounts to produce the profit required. Thus, strictly speaking, the 'discounts' should be considered as part of the tender additions.

Submitting the tender

The tender figure arrived at above is entered by the estimators into the contract documents in the manner required by the contract documents, which is either as a priced bill of quantities or submitted on a form of tender with no bill of quantities.

Tendering with a priced bill of quantities

In submission of a tender with a priced bill of quantities the direct cost rates calculated for each item need to be amended to take account of both the estimate adjustments and the mark-up. As previously stated, the estimate adjustments tend to be made as lump-sum additions or subtractions: logically, however, they require the adjustment of the direct cost rates.

In apportioning the mark-up there are a variety of practices such as:

- Marking up all items by a percentage calculated to cover overheads and profit
- Including the mark-up additions as lump sums in the preliminary section of the bill
- Marking up the bill items by different percentages to create some element of rate loading in order to create a favourable cash flow; the effect of rate loading on cash flow is described in Chapter 12 on *Cash flow and interim valuations.*

The preliminary section of the bill is frequently used to include site overheads, estimate adjustments and the balance of the mark-up. In some cases, lump sums are included in the preliminaries and the remainder of the mark-up additions are apportioned over the bill items.

Tendering without a bill of quantities

In cases where tenders require only the submission of the form of tender, the contractor need only submit the global sums as required. If the submitted tender is seriously being considered, then the completed bill of quantities may be called for by the professional quantity surveyor. This practice is frequently used in building contracts.

Estimating in management contracting

Management contracting is a form of procurement strategy in which a contractor is appointed to manage and provide leadership for the project during the pre-construction stage, usually after the preliminary design and project definition have been completed. The main advantage of such an arrangement derives from the experience that the contractor brings to the project to ensure greater buildability and more effective project management. A contractor engaged in such a role is described as a *management contractor*, and performs a *co-ordinating* rather than the traditional *construction* role associated with contractors. Under the NEC: ECC Option F, the management contractor's responsibilities for the construction work are the same as those under priced contract or target contracts. However, other contractors that are appointed in the role of *subcontractors* or *work-package contractors* undertake the construction work as a series of *work packages*. ECC Option F [20.2] requires that all work packages that must be subcontracted should be clearly identified. All other works not to be subcontracted are covered by the fee of the management contractor. With the exception of the managerial role, the management contractor is allowed to subcontract any other work package not identified in the list of work packages. These may include site preparation, the provision of temporary works elements and other general and specific attendances common to all the work-package contractors.

From a management contractor's perspective, the estimating function described for the BOQ type of contracts plays a lesser role in the tendering process. This is because the work-package contractors perform the detailed estimating function. The management contractor's estimate for the actual cost of the works is made up of all the estimates from the work-package contractors. The management contractor tenders a fee for the co-ordinating services in addition to the estimated total of the prices for all work-package contracts. The fee is usually based on the *actual cost* of the work packages defined in Option F [11.2(22, 26)]. The actual cost comprises solely the payments due to work package contractors for those work packages identified in the contract documents. Where additional works not identified in the contract documents are subcontracted, this is covered by the fee of the management contractor.

A typical management contracting tender submission will comprise the following:

- Confirmation of acceptance for the conditions and any specific details defined in the tender documentation of the proposed project
- Fees to cover the management role performed during the pre-construction phase of the project
- Fees to cover the management of the construction phase of the project
- Proposed project team and the management structure
- Project strategy, detailing execution plans, overall programme and method statement
- Project budget estimates covering costs of providing works not identified in the work packages
- Project budget estimates for work packages defined in the tender documents from the work-package estimates submitted by work-package contractors

Confirmation of acceptance of conditions is based on the contractor's decision to tender for the project, which is arrived at after detailed examination of the tender documents, establishing the potential workload and resource requirements for the project and evaluating the conditions of contract for the management service to be provided. In particular, the scope and responsibilities of the management contractor as defined in the conditions of contract will have an impact on the level of fees in order to adequately cover all risks to the contractor. The management decision to tender or decline the offer is taken based on the same factors as in the BOQ approach for tendering.

Estimating the management contractor's fee

The management contractor's fee as defined in Option F [11.2(22)] comprises four main elements. These are:

- Profit
- Corporate overheads
- Other specific items
- Project and business risks

Profit

The level of profit is set to reflect the contractor's long-term business objectives and the conditions prevailing in the industry with respect to common practice and extent of competition. The level of profit is a managerial decision that is taken at the tender adjudication.

Corporate overheads

This covers all overheads not charged directly to the contract. The corporate overheads comprise general head-office expenses such as company administration, market research and business development, tendering, accounting and auditing, head-office support for regional offices, legal services, professional indemnity and public liability insurance. The level of overhead charges is set either as a percentage of the project costs or as lump-sum items to cover each specific overhead by taking account of the size, duration and the extent of support envisaged for the contract.

Specific items

Costs included in the specific items comprise project-related services required for the efficient management of the contract. They include

- Site-based services such as:
  - Inspections
  - Safety advice, planning and supervision
  - Preparation of documentation for cost plan, safety plan and quality plan
  - Industrial relations management
  - Post-contract administration
- Project-related insurances to cover:
  - Physical works
  - Third party
  - Plant and equipment
  - Guarantees
  - Other client interests
- Bonds:
  - Bid bond
  - Performance bond
- Additional works:
  - Any non-management work not defined as part of the project's work packages

Risk

The risk associated with each of the following risk elements will be estimated and included as part of the fee:

- Cost overrun on lump-sum work or management elements to cover any future cost escalation or estimating error
- Fee that is not recovered as a result of variation to the project
- Defective work
- Neglect of duty by work-package contractors
- Bankruptcy of work-package contractors
- Liquidated and ascertained damages not recovered from work-package contractors

Use of estimating software

Changing role of the estimator

The use of computers in estimating by construction companies has grown steadily with the availability of more commercial application software. Early users relied on large computers and developed their own software. Around 1980 the use of computers in estimating became much more widespread and there were two central reasons for this. One was the availability of cheaper microcomputers that removed the capital constraints and the other was the development of interactive computer-aided estimating systems. The use of interactive software whereby the estimator could control each step of the calculations at a computer terminal gave the estimator a high degree of control of the computer system. This resulted in the development of computer-aided systems. Figure 9.6 shows a flow chart for estimating systems.

As a result of these developments, current practice of estimating relies very much on automated and computerised systems that link up databases of output rates and costs. Some estimating software packages also provide dynamic linking of the estimating system directly to CAD and other project-related software as well as enterprise information systems. This enables the automatic measurement of quantities for designed works, a necessary part of estimating without bills of quantities. It also ensures that the data employed for estimating, as well as the output from the estimating effort, are shared with other departments such as finance and planning. The developments in computing have revolutionised not only estimators' attitudes to computer systems, but equally influenced their role in the process.

On-line estimation

The development of the internet has taken the estimating function a step further. On-line estimating utilises databases held outside the company's own computer network to generate the cost estimate for a project. This has potential to overcome the need to maintain large databases in-house, and also ensures that the most current prices are employed for the estimator's work. The companies that stand to benefit greatly from on-line estimating are the small and medium enterprises that mostly undertake subcontract works. This is because the cost of maintaining up-to-date databases is beyond their business capacity. For the larger construction companies, exploitation of on-line estimating is undertaken as an in-house function. All estimating data are

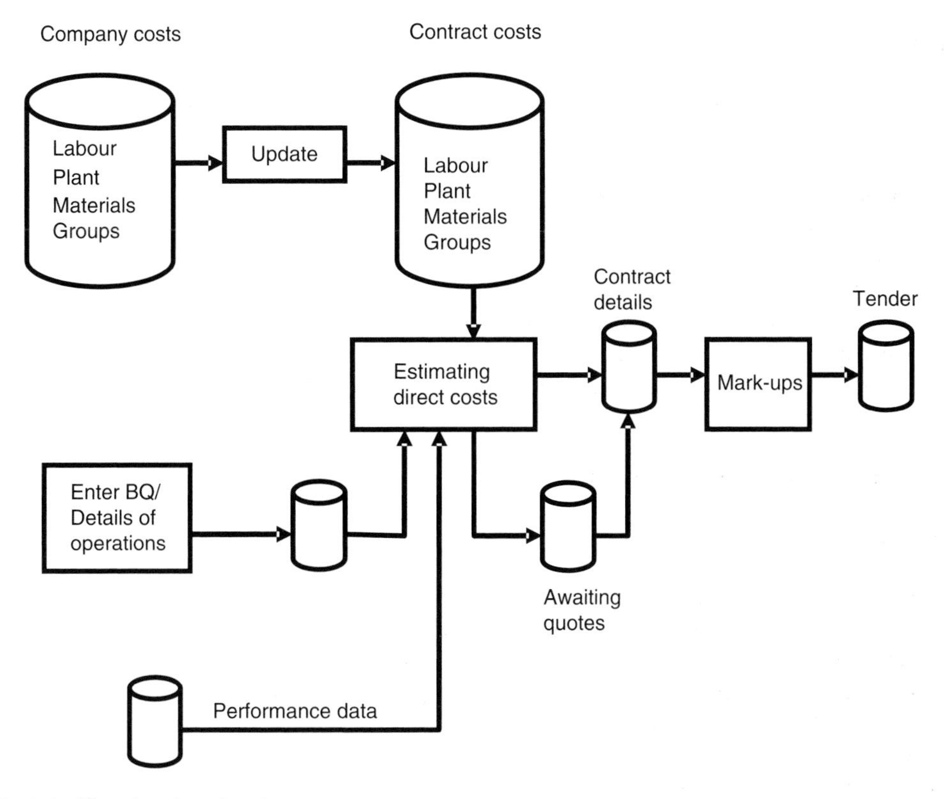

Fig. 9.6 Flow chart for estimating systems.

stored and maintained centrally, and there is access from the different sites on-line to establish the cost of an operation.

Characteristics of estimating systems

Typically, interactive computer-aided estimating systems can be identified with the following essential features:

- Files of supporting information for the estimators' use
- A range of methods of calculating item rates
- Files recording all estimators' calculations
- Facilities for adding mark-ups
- A comprehensive reporting system

Figure 9.6 illustrates how the above features relate to each other in a computer-aided estimating system. Each of the features is discussed below.

Supporting files

The supporting files include:

- Bill details
- Cost files
- Performance data files
- Operational group build-ups

Bill details

The bill details entered by the estimating clerk or estimators' support staff include bill number, section number, page number, bill reference and, if appropriate, a reference code that links the item with the data in the performance data file. This code has been pencilled against the bill item by the estimator. These data are then available on recall for use by the estimator without having to enter them again.

Cost files

The company data cost files contain the all-in cost for different categories of labour, different items of plant and materials prices. Within these cost files, data relevant either to individual resources or to groups of resources may be stored. The contract-specific cost files are the company cost files transferred by the estimator for use in a specific contract. The estimator can either accept the company data, amend it to suit their own requirements for a particular contract or, as is the case with most materials, mark it as *awaiting quotes*. If the awaiting quotes facility is used then any resultant calculation of direct cost rates by the estimator will be held in the awaiting quotes file until the buyer or estimator supplies the quotation to the system.

Performance data files

Performance data files contain build-ups on a unit rate basis with resources and usage rates for commonly recurring items of work. These build-ups are recorded against a set of codes that enable the estimator to identify the build-up required. In some systems, CESMM codes which also identify bill items have been used and with the emergence of SMM8, the codes in SMM8 can also be used as a base for storing estimators' data. However, in the more sophisticated systems, usually developed within large companies for their own use, the codes have also been specially devised by the companies to suit their own estimating and to link to their cost control systems. This ability to link data into other systems has emerged as a major benefit of computerising estimating.

The data contained in these build-ups, which prescribe the configuration of labour, plant and materials together with their associated outputs required to execute a particular item of work, are usually derived from the company's own estimating data. Some proprietary systems supply data as part of the computer system, i.e. an electronic version of the price and estimating data books. For the major companies the data are from their own estimating manuals or their work-study exercises. Figure 9.4 shows an example of the type of standard build-ups that can be stored in these files; in this example, it is combined with two operational resource groups.

Operational group build-ups

The operational resources group files allow for the storage of build-ups for operational resource groups, i.e. groups of resources that can be costed on an elapsed-time basis rather than a usage or output basis. These build-ups can then be used in the calculation of direct rates for bill items. Figure 9.4 shows an example of this: in this example, the operational resource groups are combined with a unit rate calculation taken from the performance data file.

Methods of estimating

The range of methods available to the estimator for calculating the direct cost rates for bill items or groups of bill items are:

- *Unit rate estimating*, which allows the recall of recorded build-ups from the performance data file and permits modifications to the build-up if necessary. The resources and usage rates come from the performance data file, and the costs from the cost files
- *Unit rate estimating*, which provides the facility to build-up the estimate from first principles by inputting resources and usage rates; the costs of the resources come from the cost files
- *Operational estimating*, which enables the build-up of operational resource groups in the operational resource group file and the allocation of costs to bill items
- *A spot or gash rate*, which involves supplying the labour, plant or material rates without calculation
- *Subcontractors*, which involves supplying subcontractor quotations
- *Included*, which provides the facility for indicating that the item has been included in another item
- *Item*, which involves supplying a lump sum to an item

File recording the estimator's calculations

The estimator's build-ups for each item or group of items are recorded in the file of contract details and are available for recall by the estimator for reworking or by senior management for inspection. Any build-up that has used a resource whose cost is not firm but awaiting a quote is flagged until the quotation is supplied.

Add mark-ups

Facilities for adjusting the amount of monies in the cost categories of labour, plant, materials and own subcontractors exist, together with facilities for entering overheads as percentages or lump sums and profits as percentages.

Reports

Figure 9.4 presents a typical layout from a computed operational estimate. Printed reports available include:

Bills of Quantities Listing								
Contract: -A851 ROAD BRIDGE							DATE 5/10/2005	
BILL PAGE		20						
	BILL SECTION		2					
	LAB. rate	PLT. rate	DIR. rate	AUX. rate	S/C. rate	TOT. rate	Quantity	Sum
1	GEN EXC SMALL AREAS MATS REUSE NE 0.25 m CART 100m							
	1.91	11.48	0.00	0.00	0.00	13.39	40.00 m3	£535.60
2	MILD STEEL REINF TO BS 4449 20mm DIA							
	276.39	0.00	488.09	0.00	0.00	764.48	14.00 t	£10,702.69
3	DESIGN STRUT. CONC. 30Mpa 20MM AGGR. OPC							
	23.96	17.52	105.33	2.63	0.00	149.44	78.00 m3	£11,655.93
								£22,894.22
						TOTAL SECTION 2		£22,894.22

Fig. 9.7 Bill of quantities listing with labour, plant, materials and subcontractor breakdowns.

Bill of Quantities Listing							
Contract: -A851 ROAD BRIDGE						DATE 5/10/2005	
DIRECT COST SUMMARY							
PAGE	SECTION	LABOUR	PLANT	DIR. MAT.	AUX. MAT.	SUB. CON	TOTAL
1	1	8151.52	2519.66	21302.42	662.91	0.00	32636.51
20	2	2325.84	912.57	9277.75	104.07	0.00	12620.26
30	3	2475.32	321.28	8646.78	0.00	0.00	11443.38
31	3	2211.98	1059.68	6241.48	322.91	0.00	9836.05
SUM	TOTALS	15164.66	4813.20	45468.44	1089.88	0.00	66536.17

Fig. 9.8 Direct cost summary.

- Bills of quantities giving the labour, plant and materials breakdowns, in directs costs rates and rates including mark-ups; Fig. 9.7 is an example
- The same listing without the labour, plant and materials breakdowns
- Direct cost summaries; Fig. 9.8 is an example
- Resource reconciliation reports; Fig. 9.4 is an example

References

Harris, F. & McCaffer, R. (1991) *Management of Construction Equipment*. Macmillan Education, London.

Royal Institution of Chartered Surveyors (1998) *Standard Method of Measurement of Building Works*, 7th edn. Royal Institution of Chartered Surveyors, Coventry.

10
Competitive bidding

Summary

Review of the historical attempts to develop theories to aid bidding, the accuracy of estimating and its effects, some measures of a company's eagerness to win, and comments on improving estimating accuracy.

Introduction

Competitive bidding based on tender documents prepared by the client's professional advisors is still the most common method of distributing the construction industry's contracts among the contractors willing to undertake the work. Variations such as negotiated contracts, package deals and concession contracts such as private finance initiative (PFI) form only a proportion of the contracts offered to the industry. The acceptance by the majority of clients, mainly central and local government, that competitive bidding is fair and will produce the lowest possible commercially viable tender price in the prevailing market conditions ensures that this form of work distribution will continue for a long time. The random nature of the bidding process also ensures that contracting companies will be unable to plan their company's activities with much certainty, that many contracts will be tendered for with unrealistically low prices and that the pre-occupancy of most contractors with claims will also continue.

From the contractor's viewpoint competitive bidding has the appearance of roulette: sometimes they win when they think their price is high; sometimes they lose when their price is dangerously low, and they have a wry smile for the apparent 'winner'. Often when contractors obtain a contract they resort to claims to ensure that the achieved mark-up is positive because the original tender was based on a low cost estimate. It is not surprising therefore that the subject of 'competitive bidding' has attracted investigations and research by both the contracting companies themselves and a variety of academics. Disappointingly, the results of these investigations and efforts to remove some of the uncertainty from bidding are not conclusive. However, a study of the work will give a better understanding of where the uncertainty arises and provide guidance on how to live with this inherent uncertainty. This chapter is divided into three parts: Part 1 reviews the historical work in bidding and describes earlier attempts to predict the outcome of bidding competition; Part 2 discusses the accuracy of estimating and its effect on a contractor's success in bidding and achieved mark-ups; Part 3 outlines the various ways in which some of the theories are put to use and reports on more recent research.

Part 1. A brief review of bidding strategy

Background

The subject of bidding strategy has interested various researchers in America and Europe since the mid-1950s. The aim of most of these workers has been the development of a 'probabilistic model' that will predict the chances of winning in the type of competitive bidding that is common in the construction industry. These probabilistic models have attempted to give guidance to bidders by producing statements of the type: 'If you bid at a mark-up of 12% you have a 30% chance of winning this contract.' Following on from these calculations of probability, previous workers have attempted to derive a mark-up that purports to represent the 'optimum mark-up', that is the mark-up which in the long term will produce the maximum profit. The optimum mark-up theories so far devised have not taken into account the varying success a company might experience in filling its available capacity or budgeted turnover. Therefore recent work has suggested the use of probability calculations as a means for predicting the overall success ratio (number of jobs won/number of bids submitted) to control work acquired. This is achieved by raising mark-ups when the order book is full and work is plentiful and reducing the mark-up when the market and order book are depressed.

Nature of bidding in construction

Competitive bidding is widely applied in many sectors besides construction. The different forms of bidding reflect one of two types: *open* bidding or *sealed* bidding, or a combination of these two extremes. *Open* bidding employs an iterative negotiation process, whereby each contractor independently negotiates a contract price with the client. Consultation among competing contractors is allowed, and contractors are allowed to revise their bid for as long as the client has not come to a decision on which bid to accept. The open form of bidding is widely used in the commercial sector. *Sealed* bids, on the other hand, are more typical of the construction and civil engineering sector. In *sealed* bids, each contractor is allowed to submit only one bid, and negotiation between the client and competing contractors is barred. Equally, discussion pertaining to the project under bid between the competing contractors is not allowed. Each contractor's bid is submitted by a specified date and once submitted (usually in a sealed envelope) cannot be revised.

Traditionally, bidding in construction and civil engineering is characterised by two main features:

- There is a large input by the client in detailing requirements for the contract, normally established through consultants (in the form of drawings and specification). Thus the essential criterion for evaluating the bids becomes price.
- All the competing contractors bid on the same information.

The above practice reflects a type of sealed bid that is generally described as a Single-Stage/Single-Envelope (SSSE) bidding procedure. Bidders make a submission in a single envelope containing both the *bid price* and any *technical proposals*. The contract is then awarded to the bidder whose bid is determined to be the lowest price evaluated and substantially responsive technical bid. Where it is more appropriate to separate the price from the technical proposals as is the case for design and build, design solutions or consultant selection, the Single-Stage/Double-Envelope (SSDE) bidding procedure becomes more appropriate. Bidders submit simultaneously two sealed envelopes, one containing the technical proposal and the other the bid price, enclosed together in an outer single envelope. Initially, only the technical proposals are opened at the date and time advised in the bidding document. The bid price remains sealed and held in the guardianship of the tender board. The technical proposals are evaluated and are used for an initial pre-selection phase. Bids with technical proposals that do not conform to the specified requirements are rejected as unsatisfactory. The bid prices of the technically compliant bids are then opened in public at a date and time advised by the tender board and the contract is awarded to the bidder with the lowest price evaluated and substantially responsive technical bid.

An alternative to the single-stage bidding is the two-stage bidding procedure. This involves separated submission of bids for technical proposals and price, and is often suitable for large and complex contracts where technical solution is the key criterion for the project. This type of sealed bid is particularly suited to architectural and engineering design competitions. At the first stage, bidders initially submit only technical proposals without prices in response to minimum performance requirements. Each of the submitted unpriced technical bids is reviewed by the tender board in order to agree on an acceptable technical standard for all bids. After the review, a second stage is initiated and bidders are given an opportunity to revise their technical proposals to conform to the agreed standard and to submit price bids together with the revised technical proposals, which shall be evaluated. Skitmore (2002) argues that limitations imposed by the stage selective practice involved in the two-stage bidding has implications for the current models that underpin bidding theories in construction.

Review of conceptual bidding models

The basic assumption of all the bidding calculations is that a relationship exists between the tender sum and the 'probability', or 'chance', of winning the contract. The aim of probabilistic models is to express this numerically. In entering a bidding competition, it is assumed that the contractors first estimate their costs and then add a mark-up to cover profit (or a mark-up to cover contribution, i.e. profit and company overheads). If the contractor is really desperate to win they could submit a bid at something less than cost. If this bid was low enough then it would have a 100% chance of winning. Just as at the lower end there exists the bid with 100% chance of winning, there also exists at the other extreme a bid with no chance of winning (say cost plus 50% mark-up). Between these two extremes there exists a continuum of bids with associated probabilities, which measure the chance of winning. This concept was first introduced in 1956 by Friedman and has been referred to by almost all other researchers since then, including Park (1966), Statham & Sargeant (1969), Gates (1967), Morin & Clough (1969), Whittaker (1981) and King & Mercer (1987).

The method of deriving the relationship between bids and chances of winning depends on the collection and manipulation of historical data as follows:

- We collect data on bids submitted by particular competitors on past contracts in which we have competed with them.
- We divide the competitor's *bid* by our *estimated cost* in each case.
- We group these data and plot a histogram (or frequency distribution) as shown in Fig. 10.1. This histogram now represents a picture of this competitor's historical performance against us.

From this histogram it can be readily seen that with mark-ups of 10% there were 6 out of 31 (i.e. 19.5%) occasions when this competitor would have had bids lower than ours, or conversely 25

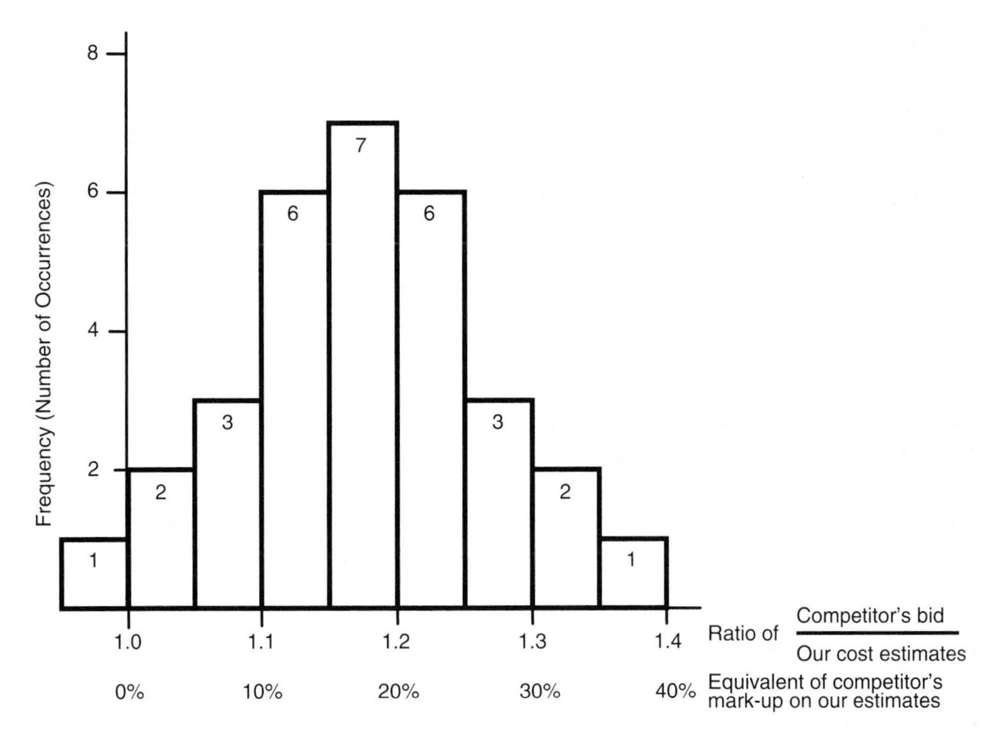

Fig. 10.1 Distribution of 31 competitors' bids compared to our cost estimates.

out of 31 (i.e. 80.5%) occasions where we would have had our bids lower than this competitor. To use this as a guide to the chances of beating this competitor in a future competition assumes that this competitor's future mark-up policy is consistent with previous behaviour. Usually this histogram is converted into a cumulative frequency curve, as shown in Fig. 10.2, with scales that show a direct relationship between our intended mark-up and the chances of beating that particular competitor.

If a record such as Fig. 10.2 can be generated for a particular competitor then the same process can be carried out for all major competitors likely to be met by a single company. Therefore, a collection of such behaviour records for all major companies could exist. In most competitions a contractor faces more than one competitor, and the knowledge that you have a 25% chance of beating 'A', a 40% chance of beating 'B' and a 15% chance of beating 'C', etc. is obviously unsatisfactory since the question that you are trying to answer is, 'What is the chance of winning the contract?'.

Friedman (1956) was the first to suggest a 'model', or expression, which combined these probabilities and predicted the probability of winning a contract, knowing the previous performance of the other competitors. Friedman's model was:

> Probability of winning a contract at a given mark-up competing against a number of known competitors = probability of beating competitor 'A' × probability of beating competitor 'B' × probability of beating competitor 'C' . . . etc.

To deal with the bidding situation of unknown competitors Friedman and later Park (1966) suggested aggregating *all* competitors' bids into a distribution similar to Fig. 10.1. Thus, the

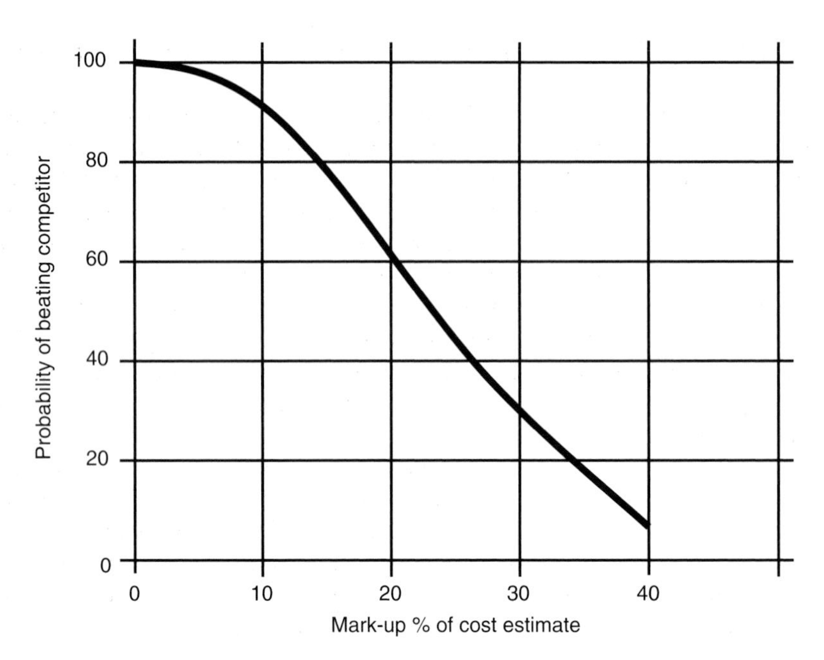

Fig. 10.2 Probability of beating competitor vs. mark-up.

historical behaviour represented by this 'picture' is that of a 'typical bidder' and not of any particular bidder.

To predict the probability of winning a contract against a known number of unknown competitors, Friedman suggested the following:

Probability of winning against n unknown competitors for a given mark-up = (probability of beating one 'typical' competitor)^n

Other writers criticised Friedman's work and suggested alternative models that dealt with the criticisms raised. The most notable was Gates, who in various papers published during the 1960s criticised Friedman's work on theoretical grounds and offered his own solution, which is again based on the collection of historical data and the creation of a distribution of competitors' bids against our estimates. In Gates's model, the probability of winning a contract at a given mark-up against a number of known competitors

$$p = \frac{1}{[(1-p_A)/p_A]+[(1-p_B)/p_B]+[(1-p_C)/p_C]+\ldots+1}$$

where p_A = probability of beating A, p_B = probability of beating B and p_C = probability of beating C.

For the case of unknown competitors, the Gates model becomes;

$$p_n = \frac{1}{n[(1-p_{\text{typ}})/p_{\text{typ}}]+1}$$

where p_n = probability of beating n unknown competitors and p_{typ} = probability of beating a typical competitor.

Friedman's and Gates's models give different answers, and debate over the years has attempted to solve this conflict with several writers offering their own solutions. This debate continued and King & Mercer illustrated that it was still current when they published in 1987. Among the authors was Whittaker (1981), whose model attempts to take account of managerial judgement by including a guess at the general expected level of bids in his calculations. This acknowledges that mathematics is unlikely to supersede judgement entirely. Fine in unpublished papers subsequently suggested a 'low-competitor' model on the strength of the argument that the only competitor one is interested in beating is the lowest competitor. The low-competitor model is based on a collection of historical data of the lowest competitor in each competition entered. Thus a histogram (Fig. 10.3) is produced, this time not for a particular contractor but for the lowest competitor in each competition.

The attraction of this model is that it does not require a complicated expression to combine individual probabilities; instead, the probability of winning is read off the computed distribution (compare with Fig. 10.2). The disadvantage appears to be the large numbers of competitions that need to be entered before a 'stable' distribution is achieved, i.e. a distribution which does not change dramatically with each added item of data. To collect a large set of data requires time, during which the mark-up policy of competitors may change.

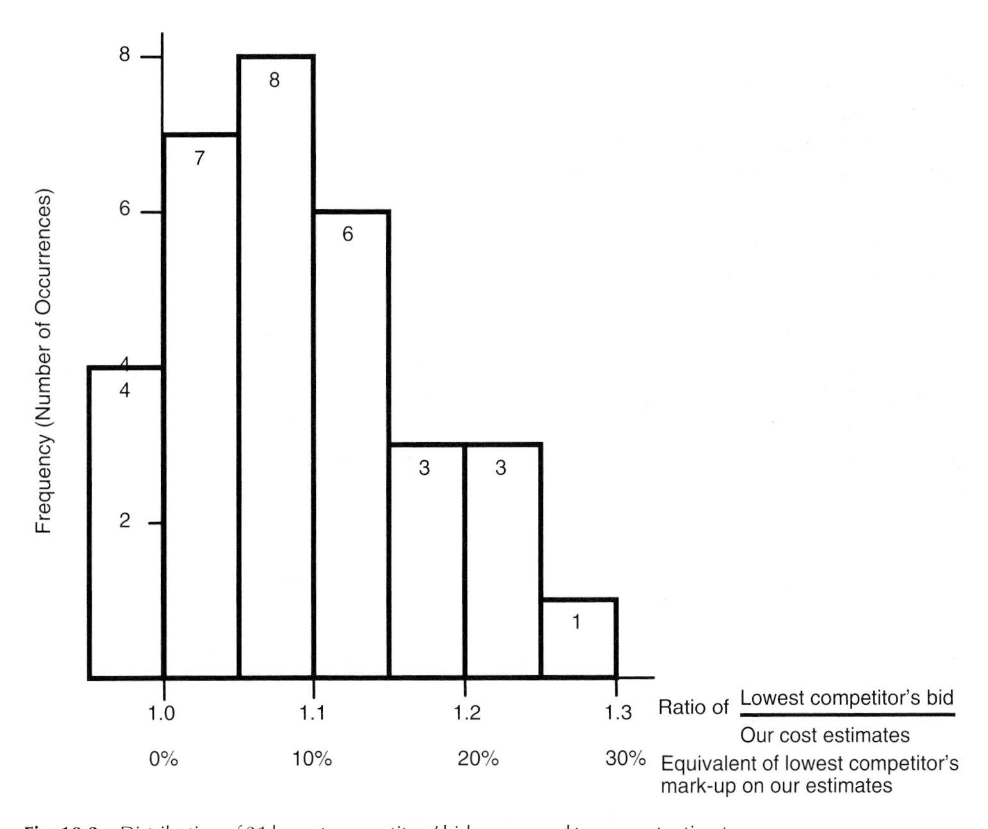

Fig. 10.3 Distribution of 31 lowest competitors' bids compared to our cost estimates.

Among those who have attempted to resolve the Friedman–Gates differences is Rickwood (1972), who concluded that the Friedman model was more correct when the cost estimates used by competitors were the same and the variability in bids was due only to mark-up differences. He also concluded that the Gates model was more correct when the mark-up used by competitors was the same and the variability in bids was due to variations in cost estimate only. The statistical reasons for this are not explained here. Rickwood proposed a compromise, which was no more than a weighted average of the probabilities predicted by Friedman and Gates. The weighting given to each depends on the amount of variability due to estimates and the amount of variability due to mark-ups. However, as yet this has received no practical testing.

Whittaker analysed a number of contracts relating to building projects and produced an 'overall distribution' of bids, which was uniform in shape. Given that you could estimate the mean bid accurately (to within ±2%), the Whittaker distribution, or model, will tell you the probability of any particular bid being the winning bid. Whittaker reports that in tests he has shown a significant improvement when compared to bidding unsupported by his model. Curtis & Maines (1973) emerged as the major critics of Whittaker's work on the grounds of the method of aggregating all contracts, irrespective of the number of bidders, in order to obtain his distribution. They also disputed the claim made by Whittaker that mean bids can be estimated using 'managerial judgement' with the required accuracy of ±2%. Curtis & Maines put forward an alternative to Friedman and Gates, a model based on conditional probabilities, but pointed out that they were of the view that such models were too simple in treating the cost estimate and the mark-up as independent variables.

McCaffer (1975, 1976), sympathising with the approach of Whittaker, undertook a similar analysis. He took account of Curtis & Maines's criticisms and produced distributions of bids for roads and building works which were shown to be virtually normal distributions with standard deviations of 8.4% for road contracts and 6.3% for buildings contracts. The use of these overall distributions, or the distributions for contracts grouped together by the number of bidders, makes it possible to predict the lowest bid from an estimate of the mean bid. Figure 10.4 shows the relationship between mean and lowest bids actually recorded and the predicted values, using 'expected values' from statistical tables, for the variances calculated for the analysis of bidding contracts.

It must be emphasised that such figures would need to be compiled for each type of contract in each area, and that the figures shown are not necessarily universally applicable. McCaffer also attempted to predict the mean tender price, which it is accepted should be an easier task than predicting the less stable lowest bid by means of a price library and inflation indices. The method relied on the compilation of a library of prices for standard items of work, the updating of the prices to time-now using inflation indices, the statistical analyses of selected prices and the use of these statistics to simulate the price of a future tender. The accuracy of this method was shown to have a standard deviation of about 8% for roads contracts, which compares favourably with the accuracies of other methods recorded by McCaffer. This article records accuracies of estimating for regression models with standard deviations ranging from 5% to 20%. The accuracy of traditional estimating by the designer and the client's professionals is recorded in the same article as ranging from a standard deviation of 12% for buildings to 21% for road works. These accuracies of designers and clients were further reviewed by Ogunlana & Thorpe (1987), who quoted estimating accuracies as coefficients of variation of 13%. This they contrasted with contractors' estimating accuracies, which they quoted as having a coefficient of variation of below 6.5%. An accurate estimate of either the mean or the lowest bid could provide the contractor submitting a bid with a reasonable measure of the probability of winning.

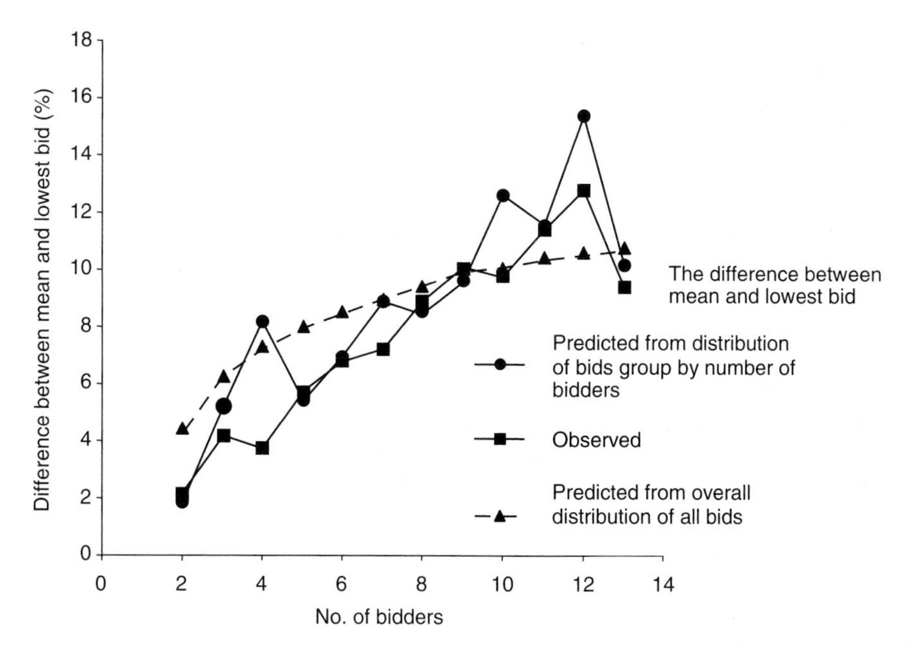

Fig. 10.4 Average difference between mean and lowest bid.

Claims that the accuracy of the estimate is the main controlling variable in determining winning bids have been by Fine (unpublished), Whittaker (1981) and Barnes & Lau (1974). Barnes & Lau noted the accuracy of estimating in the process plant industry and expressed the view that it was impossible to obtain feedback of the effect of different bidding and pricing policies from observing the real situation. The main reason they gave is the level of estimating inaccuracy.

Fine and other investigators have made observations relating the high variability in output of labour and plant, and the resulting question is: *If output varies from 50% to 200% around the mean of 100%, how accurate can any estimator be in forecasting the cost of a contract?* The effects of estimating accuracy are explored in Part 2 of this chapter.

Part 2. The importance of accuracy in estimating

The effect of estimating inaccuracies

In the estimating process the estimating department is required to make several assessments, partly based on recorded data, partly based on experience, partly based on hunches. Typical examples of these assessments are:

- The likely outputs or performance standards of the various trades and of the selected plant (and even the choice of plant for the job)
- An assessment (translated into cost terms) of the ease or difficulty of carrying out the various work items that make up the contract
- The likely trend in material costs over the period of the contract
- The likely trend in wage rates over the period of the contract
- The weather conditions, etc.

Different estimators will obviously assess the effects of the above and other variables differently, and hence a number of estimators are likely to produce a range of estimates.

The cost that the estimator is trying to produce (at least in theory) is the most likely cost to their company of executing the particular contract. This leaves the tendering panel to add profit margins, which represent an adequate return commensurate with the risk involved. In general building where there are a number of contractors of similar efficiencies, especially in the areas where staff and labour move from company to company, a simplification assumes that the 'likely cost' of a contract to each company is similar. Clearly in specialist work, or where the methods of construction vary, the likely cost of a contract will be different from company to company.

Assuming that a contract has a 'likely cost', the range of estimates produced by each company will be 'likely cost' ± $A\%$, where $A\%$ represents the accuracy of the estimator's prediction of likely cost. This simplification can be used to explain partially why contractors' achieved profit/turnover is significantly less than the average of the profit margins added to the cost estimates at the tender stage. The explanation rests on the margin lost in competition, which is described below.

The margin lost in competition

$$\text{Cost estimate range} = \text{likely cost} \pm A\%$$

where $A\%$ is a measure of the accuracy of present estimating methods. To calculate the tender, contractors add a mark-up to their cost estimate, i.e.

$$\text{Tender} = \text{cost estimate} + \text{applied mark-up}$$

In the competitive tendering process the winning tender is usually the lowest. Therefore, the estimator who produces the lowest estimate, say likely cost minus $A\%$, gives their company the best chance of winning the contract. This supports the cliché, 'The estimator who makes the biggest mistake wins the contract.' The attendant cliché, 'What we lose on the swings we make on the roundabouts', is unsupported because the estimator who produces the highest estimate, say likely cost plus $A\%$, gives their company very little chance of winning the contract, and it is probable that most winning tenders will have cost estimates that are low in the range of likely cost $\pm A\%$.

Figure 10.5 shows the range of possible cost estimates as a uniform distribution (a simplification used for demonstration purposes) and all competitors adding the same mark-up (another simplification used for demonstration purposes) and the resulting distribution of winning bids. The difference between the mean of the winning bids and the mean of all the bids is the average margin lost in competition.

Thus, the winning tender based on a cost estimate that is probably less than the likely cost usually results in the mark-up achieved on the contract being less than the mark-up included in the tender. Over a large number of contracts the average difference between mark-up included in tenders and achieved mark-up is the average difference between the likely cost and the estimated costs. The average difference has been called the 'breakeven mark-up'.

If a contractor did not wish to make a profit, but wanted merely to break even and attempted to do so by adding a zero profit mark-up to their tenders, they would, because of estimating inaccuracies, make a net loss over a number of contracts. In order to break even in the long term,

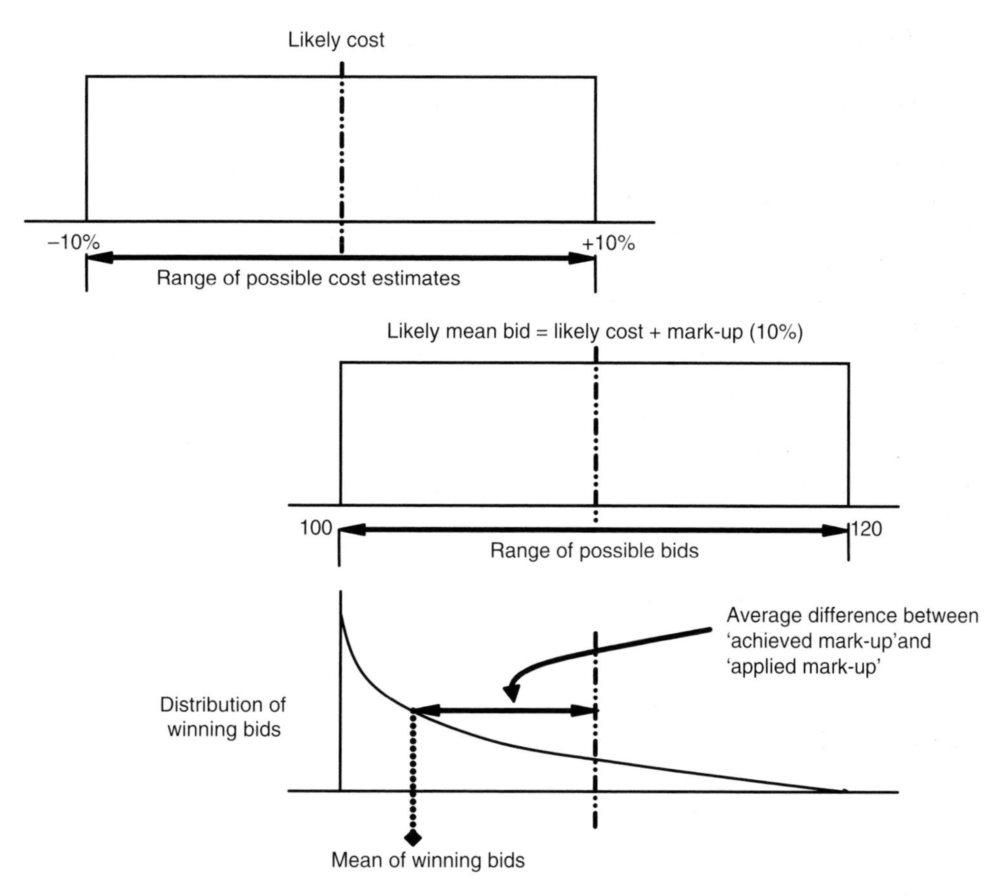

Fig. 10.5 Demonstration of how the estimate of the average amount lost in competition is derived (assuming a uniform distribution of bids).

they would have to apply a profit mark-up greater than zero to compensate for differences between the likely cost and estimated costs of their winning tenders. The mark-up needed to break even depends on (1) the general level of estimating accuracy and (2) the number of competitors.

For example, with estimating inaccuracies of say ±10% and with five competitors, a mark-up of the order of 7% would be needed to break even in the long term. Any long-term achieved profit margins in this situation would be the excess over the 7% breakeven mark-up included in tenders. If a contractor included, say, 10% profit margin in the tenders they submit, the likelihood is that the average achieved profit margin would be of the order of 3%.

In an industry where the achieved mark-up (or profit/turnover ratio) is low, although the applied mark-up at the tender stage is considerably higher, the margin lost in competition offers some explanation of the difference between achieved and applied mark-ups and demonstrates the likely existence of estimating inaccuracies.

To reduce the amount lost in competition and hence increase achieved mark-up, contractors need to improve their estimating accuracy and/or seek tendering situations with fewer competitors. Figure 10.6 gives a graphical representation of the margin lost in competition based on the assumption of uniform distributions and the stated estimating accuracies.

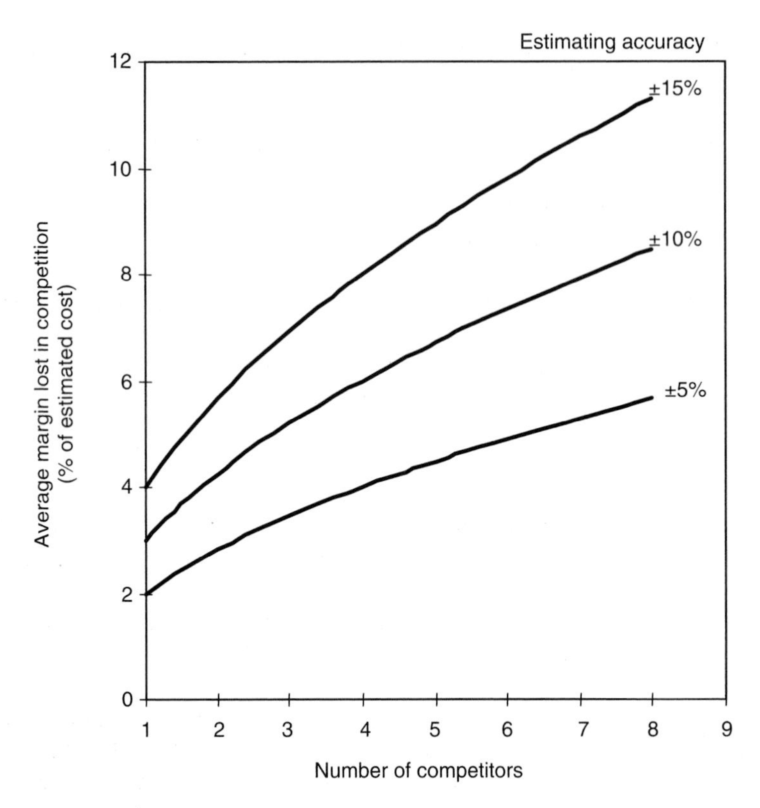

Fig. 10.6 Average margin lost in competition.

The graphs displayed in Fig. 10.6 were generated by McCaffer based on the provision that all competitors add the same mark-up, thus assuming the difference between the applied mark-up of each contractor is zero. This can be revised to assume that there is variability in the mark-ups. This will produce variations in the graphs depending on the assumed variability of the mark-ups. Abdel-Razek & McCaffer (1987) produced sets of alternative graphs.

The effect of improving the accuracy of estimating

If contractors could improve their estimating accuracy they would have to ask themselves several questions, namely:

(1) If I improve the accuracy of my estimating, will I still be successful in obtaining contracts?
(2) If I improve the accuracy of my estimating, what will be the effect on my profits?
(3) If I derive any benefit from improving the accuracy of my estimating, will these benefits remain if the rest of the competitors make similar improvements?

In an attempt to obtain answers to these questions, or at least a guide as to the likely answers, a simulation program was produced and the key results were checked by analytical means. The details of this program are not described here, but briefly it combines the factors that determine a contractor's ability to secure contracts with the resulting achieved profit margins. The factors included were:

- The estimating accuracy of the contractor whose record is being considered (referred to as contractor E)
- The general estimating accuracy of their competitors
- The average applied profit margin of contractor E (the profit margin they included in their tenders)
- The average applied profit margin of their competitors
- The number of competitors they meet when tendering

Using different ranges of values for factors listed, the results obtained gave some insight to the likely effects on achieved profit margins and a contractor's ability to secure contracts.

Results typical of those obtained for the full range of situations simulated are described below. These help to illustrate the effect of improving estimating accuracy.

Situation 1

This situation is the assumed existing set of conditions against which any change will be compared. The levels of estimating accuracy and of average mark-ups are purely assumptions.

Assumptions Contractor E and four others have estimating accuracies of likely cost ±10%. On average, all contractors include a profit margin of 10% in their tenders.

Results All contractors win one contract in every five they tender for. The average achieved profit margin is 3%. The average difference between applied and achieved profit margins is the average difference between likely cost and the estimated cost of the winning tenders.

This means that contractor E's average profit/turnover is 3% and their success ratio (i.e. number of contracts won/number of tenders submitted) is 1:5. The success ratio is a measure of the contractor's ability to obtain contracts.

Situation 2

In this situation contractor E has improved their estimating accuracy and maintains the same average applied profit margin as in situation 1. Their competitors maintain their existing accuracies and profit margins. In this situation the questions being asked are:

(a) What will be the effect on the contractor's ability to obtain contracts (i.e. the effect on their success ratio)?
(b) What will be the effect on their profit margins?

Assumptions Contractor E has an estimating accuracy of likely cost ±5%. The four other contractors have estimating accuracies of likely cost ±10%. On average, all contractors include a profit margin of 10% in their tenders.

Results Contractor E is less successful in securing contracts and now only wins between one in 10 and one in 11 of the contracts they tender. The average achieved profit margin on the contracts they win is approximately 8%. By doubling their estimating accuracy contractor E has vastly improved the profit margins they achieve, but because their ability to secure contracts is reduced, their turnover will also be reduced if they continue to submit about the same number of tenders.

The answers to the two questions are: (a) their ability to secure contracts has been reduced from 1 : 5 to 1 : 11; (b) the average profit margin has increased from 3% to 8%, which, taken together, give a marginal increase in company profits.

Notwithstanding the increase in profits, the decline in ability to secure contracts, which in turn affects contractor E's turnover, is unsatisfactory. However, two solutions are offered: either to compensate by increasing the number of tenders submitted, if a sufficient amount of work is available; or to reduce the applied mark-up. The effect of reducing the mark-up is studied in situation 3.

Situation 3

In this situation contractor E reduces their applied profit margin in order that their ability to secure contracts is the same as in situation 1 (i.e. a success ratio of 1:5). The question is: What effect will this have on company profit?

Assumptions Contractor E has an estimating accuracy of likely cost ±5% and on average includes a profit margin of 7.5% in their tenders. The four other contractors have estimating accuracies of ±10% and on average include a profit margin of 10% in their tenders.

Results Contractor E wins one contract in every five they tender (same as in situation 1). The average achieved profit margin of the contracts they win is in excess of 6% (compared with 3% in situation 1). The conclusions so far and the answers to questions (1) and (2) are: (1) if a contractor unilaterally improves their estimating accuracy and maintains the same average mark-up, their ability to win contracts is substantially reduced. (2) If the contractor reduces the average applied profit margin in order to improve their ability to secure contracts, their average achieved profit margin will still be substantially higher than before. The third question, whether any benefits would remain if contractor E's competitors also made similar improvements, is studied in situations 4 and 5.

Situation 4

In this situation the other four contractors have also improved their estimating accuracies and have also reduced their applied profit margins.

Assumptions All five contractors have estimating accuracies of likely cost ±5% and on average include profit margins of 7.5% in their tenders.

Results Contractor E and the other contractors win one contract in every five they tender. The average achieved profit margin is just over 4%.

For contractor E this is still an improvement when compared with the original situation 1 (i.e. 3%) but is a deterioration from the situation where they alone have improved their estimating accuracy (see situation 3, i.e. 6%).

Situation 5

In this situation all contractors' estimating accuracies remain at the new improved level and their applied profit margins return to the level that existed originally.

Assumptions All five contractors have estimating accuracies of likely cost ±5% and on average include profit margins of 10% in their tenders.

Results Contractor E and the other contractors win one contract in every five tendered. The average achieved profit margin is just in excess of 6.5%.

For contractor E this compares favourably with the original situation 1 (i.e. 3%) and with the situation where they alone improved their estimating accuracy (situation 3, i.e. 6%). (In situation 3, it will be recalled, contractor E had reduced their applied profit margin from 10% to 7.5%.)

The answer to question (3) is that profit improvements do remain even when competitors make similar improvements to estimating accuracy. Finally, the situation concerned was where not all competitors had made similar improvements to estimating accuracies.

Situation 6

In this situation four contractors (including E) have improved their estimating accuracies, the fifth has not.

Assumptions Four contractors (including E) have estimating accuracies of ±5%. The fifth contractor has an estimating accuracy ±10%. All five have on average an applied profit margin of 10%.

Results The ability of contractor E to obtain contracts is slightly reduced from 1 : 5 to 1 : 6. The average achieved profit margin is 5.5%. This compares favourably with situation 5, where all contractors have improved their accuracy. Nevertheless, substantial profit improvements remain.

The other results from the range of situations tested give similar patterns of results. If the original estimating inaccuracy was assumed to be greater, the increase in achieved profit margins on improving the estimating accuracy was also greater. If the original estimating inaccuracy was assumed to be smaller, then the increase in the achieved profit margin on improving the estimating accuracy was also smaller.

Conclusions

The general conclusions drawn from this study are:

(1) The achieved profit margin will be increased if the accuracy of estimating is improved.
(2) If the estimating accuracy is improved and the contractor wishes to maintain the same turnover, they will need to (a) reduce their applied profit margin or (b) increase the number of tenders they submit or (c) make some reduction in their applied profit margin and also increase the number of tenders they submit.
(3) The achieved profit margins will still be greater than the original profit margins when all contractors improve their estimating accuracy. This assumes that contractors fix their mark-up without reference to the current profitability of the company. However, the competitive nature of the industry would probably cause contractors to cut their margins once enhanced profitability had been achieved. It is difficult to assess this effect, but at least one residual benefit would remain, namely the reduction in the number of loss-making contracts.

(4) There are serious consequences for any contractor who allows the accuracy of their estimating to deteriorate.

Part 3. Some ways of using the existing theories

Number of bidders

The theories relating to the margin lost in competition highlights the point that, the greater the number of bidders, the lower the winning bid is likely to be in relation to the 'likely cost'. The advice is simple and is to avoid the bidding competitions with a higher number of bidders. Figure 10.4 shows the average difference between the mean bid and the lowest bid actually recorded. The effect of a higher number of bidders results in a much lower winning bid relative to the mean bid. It is not the mean bid that is increasing; it is the lowest or winning bid that is reducing.

A disturbing piece of evidence established by McCaffer (1976) is presented in Fig. 10.7, which shows the average of *lowest bid/designer's estimate* and the average of *mean bid/designer's estimate*. It was expected that the average of *lowest bid/designer's estimate* would fall off as the number of bidders increased; the designer's estimate took no account of the number of bidders and this fall-off is consistent with the theory relating to the margin lost in competition. However, there is no such explanation for the fall-off of average of *mean bid/designer's estimate* with increasing numbers of bidders; the number of bidders should not have affected this statistic. This fall-off can only be explained as a manifestation of price cutting when competition is fierce. Thus, there seem to be two possible mechanisms forcing down the price of the lowest bid when there are many competitors, one being due to estimating variability and the other being price cutting.

Differences between contractors' average bids

Many of the approaches described in Part 1 of this chapter, such as creating distributions of competitors' bids in comparison with the contractor's own cost estimates, have failed through lack of

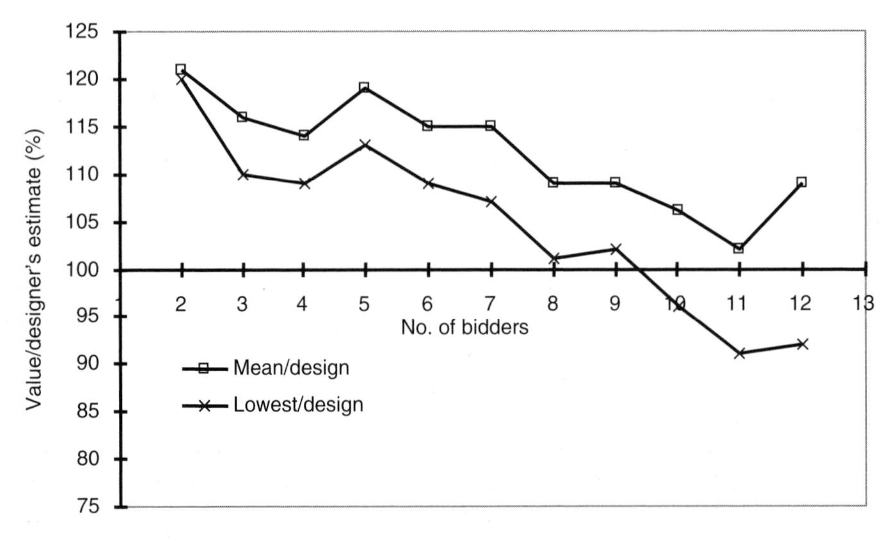

Fig. 10.7 Average of the lowest bid/designer's estimate and average of mean bid/designer's estimate plotted against *N*, the number of bidders (McCaffer 1976).

Table 10.1 Correlation of 'average of mean standardised bid' and 'success rate' (taken from McCaffer 1976).

Company identifier (for anonymity this has been replaced by a letter)	No. of bids submitted by company in the analysis	Average of mean standardised bid	Success rate	
			(No. of winners/No. submitted) (%)	*Statistical test for randomness
A	45	0.954	44.4	Low
B	30	0.971	6.6	Low
C	32	0.978	25.0	Low
D	46	0.983	21.7	R
E	63	0.984	20.6	R
F	21	0.990	23.8	R
G	27	0.991	14.8	R
H	22	0.991	22.7	R
I	66	0.992	13.6	R
J	65	0.997	16.9	R
K	39	0.997	17.9	R
L	27	0.998	11.1	R
M	36	0.998	16.6	R
N	41	1.001	24.3	R
O	25	1.004	20.0	R
P	21	1.005	4.7	R
Q	23	1.006	8.7	R
R	78	1.008	5.1	R
S	26	1.011	7.6	R
T	22	1.018	4.5	R
U	43	1.030	4.6	High
V	29	1.033	6.9	High
W	21	1.050	9.5	High

* Low: A disproportionate number of 'low bids'. High: A disproportionate number of 'high bids'. R: A random mixture of 'low' and 'high' bids.

data. All statistical analysis of competitors' behaviour will fail if insufficient data exist. However, the data requirements of these distributions are specifically restricted to competitions that the contractor enters. This excludes all other competitions that the contractor's competitors enter and so excludes a substantial source of data.

The difficulty of using data from other bidding competitions is in converting this information into some meaningful standard that can then be used for comparison. The most common method of 'standardising' bids from other contracts is to divide each bid by the mean bid to obtain the ratio known as 'mean standardised bid'. There are valid theoretical reasons why this ratio should not be of much value because of the variability of the divisor, the mean bid. Nevertheless, McCaffer (1976) argues that there is also some evidence that demonstrates the usefulness of these mean standardised bids.

If the mean standardised bid is calculated for each contract that a known contractor has entered and the average 'mean standardised bid' is calculated for that known contractor, some measure of the overall performance of this contractor in relation to the others, in particular your own company, can be obtained. Table 10.1 shows the average mean standardised bid for a number of contractors, together with the observed success ratio (number of jobs won/number of jobs bid).

The fifth column in the table also shows the result of a statistical test to determine if each contractor's list of bids could be regarded as a list of random numbers, or if a disproportionate number of low or high bids occurred. This table shows that the average mean standardised bid is a good indicator of success rate and low bidders. This analysis makes use of data from any bidding competition whether your own company is involved or not. The meaningfulness of the results improves as the amount of data available increases.

Differences between contractors' behaviour patterns

Column 5 in Table 10.1 indicates that some contractors have more low bids than is normally expected and some have more high bids. It also indicates that the majority have a reasonable share of high and low bids. A study of observed bids for some contractors by McCaffer revealed that there are occasions when contractors who in the long term have equal shares of high and low bids also have phases of varying length when they display a run of low bids or high bids. These phases when identified help differentiate the less serious competitors from the more serious.

This approach can be improved upon by calculating the cumulative sum of (bid – mean bid)/mean bid and plotting the results, as shown in Fig. 10.8. This value has been given the name *cusum* value. An analysis of this type for some 600 contracts involving almost 400 contractors, also given by McCaffer, showed that there were only 15% of cases when the winning bidder had a rising graph preceding their winning bid. Conversely there were 85% of cases when the winning bidder had a declining graph. Table 10.2 summarises the number of occasions a winning bid was preceded by a drop of one, two, three, four or five steps (a step being a previous bid, each bid being taken in date order). The production of such graphs is useful in identifying work-hungry competitors.

Success-rate sensitivity to change in mark-up

There is obviously a relationship between mark-up and success rate. Increases in the mark-up will reduce success rate. Sometimes the advice given is to increase the applied mark-up and compensate for the reduced success rate by bidding for more contracts. This advice may have a firm theoretical basis, but can only be applied if there are the extra contracts available and should only be applied if the sensitivity of success rate to changes in mark-up is known. The latter can be calculated, and Fig. 10.9 shows the graphs of success rate against changes in mark-up for one contractor. Repeating this for a number of contractors McCaffer (1975) showed that each had a different sensitivity of success rate to change in mark-up. This is thought to reflect the different skills in market judgement.

Changes in the mark-up policy of different companies would clearly lead to different outcomes. The outcome for a particular company should be examined before any unconsidered action is taken. Other variables that should be studied are:

(1) The bidding success rate with different sizes of contract
(2) The bidding success rate with different types of contract
(3) The bidding success rate in different regions
(4) Achieved mark-up versus job size
(5) Achieved mark-up versus job type

Attention to (4) and (5) is recommended as a warning against being too successful in obtaining unprofitable jobs.

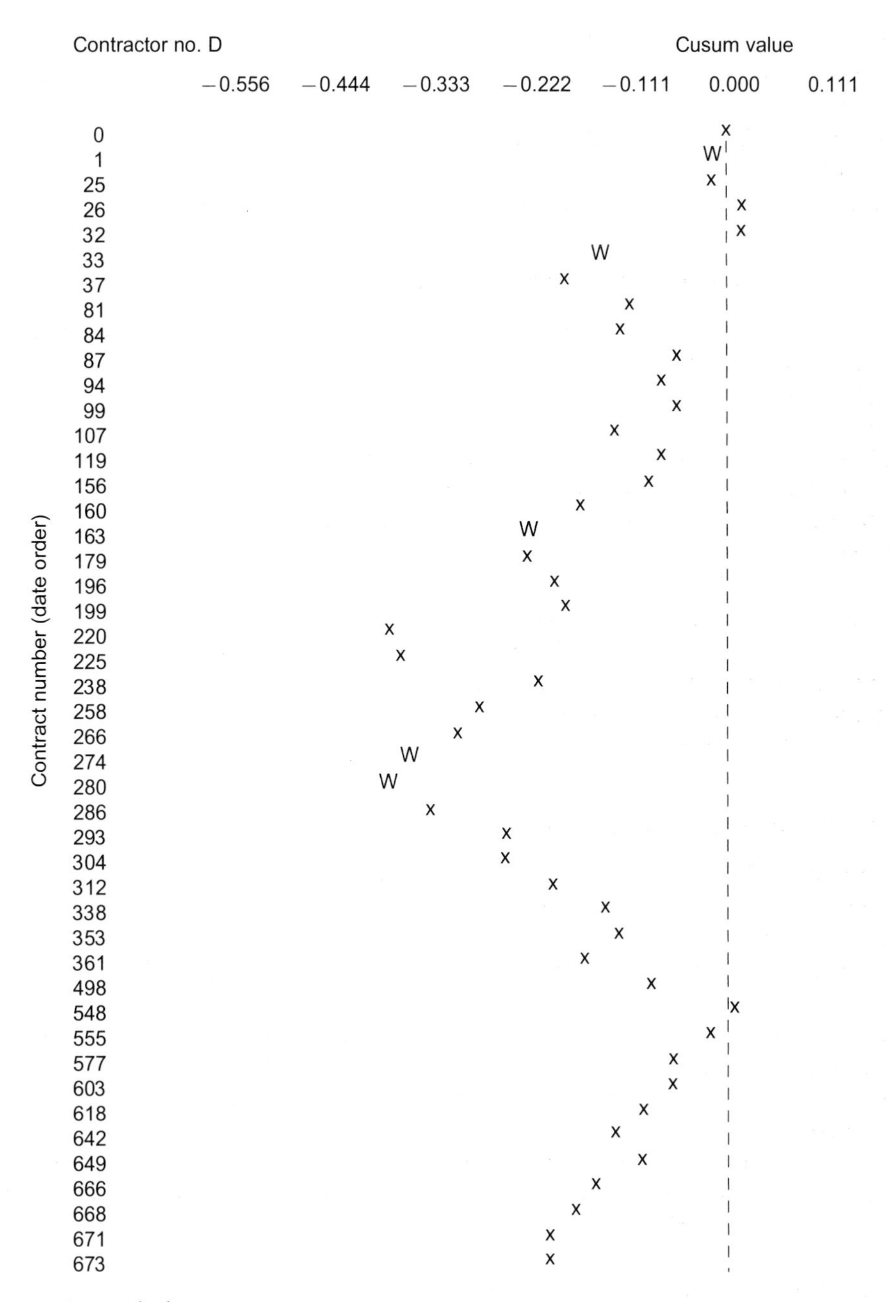

Fig. 10.8 Cusum plot for one contractor.

Table 10.2 Winning bids preceded by 2, 3, 4 or 5 decrements in a contractor's cusum value.

No. of times a winning bid was preceded by M decrements in the contractor's cusum value (%)	No. of decrements (i.e. previous bids)
85	2
75	3
69	4
65	5

```
Contractor CC            Frequency 86

Mark-up  No. of  Percentage                      1
(%)        wins        wins     1 2 3 4 5 6 7 8 9 0

 -44.0      86     100.0   0 0 0 0 0 0 0 0 0 0 0
 -43.0      85      98.8                       X
 -42.0      85      98.8                       X
 -41.0      85      98.8                       X
 -40.0      85      98.8                       X
 -39.0      85      98.8                       X
 -38.0      85      98.8                       X
 -37.0      84      97.7                       X
 -36.0      84      97.7                       X
 -35.0      84      97.7                       X
 -34.0      84      97.7                       X
 -33.0      84      97.7                       X
 -32.0      84      97.7                       X
 -31.0      83      96.5                      X
 -30.0      83      96.5                      X
 -29.0      82      95.3                      X
 -28.0      82      95.3                      X
 -27.0      81      94.2                      X
 -26.0      81      94.2                      X
 -25.0      80      93.0                      X
 -24.0      80      93.0                      X
 -23.0      78      90.7                     X
 -22.0      77      89.5                     X
 -21.0      76      88.4                     X
 -20.0      76      88.4                     X
 -19.0      76      88.4                     X
 -18.0      73      84.9                    X
 -17.0      72      83.7                    X
 -16.0      71      82.6                    X
 -15.0      71      82.6                    X
 -14.0      71      82.6                    X
 -13.0      71      82.6                    X
 -12.0      66      76.7                   X
 -11.0      64      74.4                   X
 -10.0      62      72.1                  X
  -9.0      60      69.8                  X
  -8.0      58      67.4                 X
  -7.0      54      62.8                 X
  -6.0      46      53.5               X
  -5.0      37      43.0             X
  -4.0      29      33.7           X
  -3.0      20      23.3         X
  -2.0      17      19.8        X
  -1.0      13      15.1       X
   0.0      11      12.8       X
```

Fig. 10.9 Sensitivity of success rate to changes in mark-up.

Advice on the data to collect for analysing competitors' bids

If no formalised efforts have been made previously to analyse competitors' behaviour, then the starting point must be the company's own competitions. If the bids are ascertained for all competitors in each competition entered, then the relationship of the competitors' bids to your own estimate can be established, as in Fig. 10.1. Data from competitions that your company does not enter are useful because such details can always be compared with the mean bid, and Table 10.1 indicates that this comparison is also a guide to likely success rate. If sufficient data are gathered relating to your major competitors, then the graph of (bid – mean bid)/mean bid, such as Fig. 10.8, can be drawn and the times when the competitor is looking for work can be identified. However, it is strongly recommended that your own company's performance be examined as listed above and that tight control be exercised on the estimating department to avoid any possibility of deterioration in the accuracy of estimates resulting in a serious loss-making contract.

Improving estimating accuracy

Much has been reported about the accuracy of estimating. The concept of an accurate estimate borders on the philosophic and is difficult to define. The best description might be 'given a particular project, an estimator, by selecting the best method and the most appropriate resources of plant, labour, materials and subcontractors, should be able to calculate the "most likely" cost'. Variability in site performance adds difficulties in checking this calculation. However, assuming that an estimator has produced the calculation above, the question to be asked is 'Will the same estimate be reproduced by another estimator?' This is unlikely because a number of factors intervene that will vary each estimator's approach. These factors include differences in the selection of plant and labour output rates, differences in the calculated costs of labour and plant, differences in assumed wastages and disruption factors. Finally, there exists the possibility of arithmetical errors and errors of omission. Many of these differences are differences in estimators' judgements and differences in determining the cost inputs. If estimators' judgements vary widely, they cannot by definition all be correct. Therefore, the only feasible approach to improving estimating accuracy is to improve the estimators' judgements. This is largely attempted by improving the data that supports the estimators' judgements.

Data supporting estimators

Long ago, most of the larger companies created estimators' manuals, which describe construction methods, the resource inputs to each construction operation and guidance on the selection of the outputs to be used in calculations. These data act as a set of company standards and allow senior estimators to supervise the work of their staff. These manuals will also describe how labour rates are calculated and the assumptions with regard to wet weather allowances, sickness allowances and bonus allowances—all of which vary from project to project. Guidance on wastage of materials and disruption by subcontractors is also usually included.

During the 1980s much of these data were transferred to the companies' computer-aided estimating systems. The advantages to these companies have been to provide well-structured data libraries for use by the estimators. Arithmetical errors and errors of omission are also more easily monitored in such systems. Thus, the advent of computer-aided estimating has greatly improved the 'housekeeping' aspects of estimating and the management of data within the estimating process. Refer to Chapter 9 on *Estimating and tendering* for further descriptions of computers and estimating. Abdel-Razek and McCaffer (1987) undertook studies to determine where any residual variability existed within the estimators' calculations. This work involved constructing computer models to calculate the labour rate, the plant rate and the cost of materials and to explore the sensitivity of the overall cost to variations in the individual factors that make up these costs. For example, in the labour rate calculation the four principal variables studied were:

- Allowance for inclement weather
- Allowance for sick leave
- Allowance for redundancy and sundry costs
- Allowance for supervision

These were chosen as the elements in the calculation that were subject to estimators' and/or company judgement. Other variables such as basic wage rate, national insurance, holiday entitlement, etc. were taken as firm items of data common to all such calculations undertaken by whatever company. Thus, if there were to be variations in the final calculated cost of labour, it would be likely to be from the elements where some judgement had been exercised. Graphs were constructed that showed the sensitivity of the labour rate to variations in these factors. This in turn drew estimators' attention to the importance of each factor and encouraged greater consideration to be given to the more sensitive variables. This exercise was repeated for plant costs, material costs, direct cost rates and subcontractors' disruptions.

Recent developments in bidding

The developments of concepts and models for bidding have been based on certain implicit assumptions. These are summarised below.

- The contractor's sole objective is to maximise profit on any given contract.
- There is no difference in the cost estimate for a particular project for all the competing contractors.
- Competitors will continue to bid in the same way as they have in the past.

In practice, these assumptions, though fair in themselves, do not hold true for all contracting situations. The cost estimate of a project can be different for different contractors as a result of organisational, location and economic factors. Equally, contractors' bidding behaviour can alter as a result of continued success or the lack of it in a particular market. The major contractors that undertake strategic planning often try to attain a balanced portfolio of projects so as to minimise their business risks. This implies that sometimes a contractor may bid for some projects employing other criteria beside profit maximisation.

Changing client requirements

Nowadays, contractors not only find themselves bidding for jobs on the traditional closed system, but are increasingly called upon to deliver a total package under the open bidding option. This evolving trend is reflected in design-and-build procurement contracts and their variants. Contractors are finding that increased participation in the design stage of the project often results in a situation where they compete not only on the basis of price, but also on those features that the client would traditionally define as project specification. Competitive bidding in such a case does not depend only on the price criterion.

Bidding for consultants

Whereas the engagement of consultants by the client in the past has been through a direct approach, there is a trend toward competitive bidding for the selection of these professionals as well. This predominantly reflects the open bidding system, since, in many cases, the consultants are required to help the client define their requirement and provide an appropriate solution for it. The evaluation criterion in this case also goes beyond cost.

Bid/no-bid decision

Bidding by such open systems creates a higher front-end cost for the project, which can lead to increased overheads for an unsuccessful contractor. Since a higher overhead rate can contribute adversely to establishing the optimum bid, contractors carefully select the contracts they bid for from those available to them at an early stage in a bid/no-bid decision. The criteria employed in arriving at this decision have been discussed in Chapter 9 on *Estimating and tendering*. Bidding in an open system naturally will require more information for such a decision. To make this effective a database or expert system can be employed.

Electronic bidding

This involves the electronic transfer of documentation and completed bids between client and contractors. Electronic bidding can either supplement or replace traditional paper-based bid documents. There are several offerings of e-commerce tools that have made electronic bidding a more viable option compared to paper-based options. The use of electronic bidding has also been fostered by the *Official Journal of the European Union* (OJEU) procurement arrangement, where electronic distribution helps to overcome geographical distances across the EU.

Among the advantages presented by electronic bidding are access to bid information and documentation on a 24 hours a day, 7 days a week basis and an increased ease and convenience in submitting bids. Most electronic bid systems (particularly where they are online) provide exclusive access to bid information to the contractor until the date and time of bid opening.

Current options in electronic bidding include offline methods involving the download of bidding documentation for completion in hard-copy or electronic format for submission, or online completion of all the various forms and information to be submitted.

References

Abdel-Razek, R. & McCaffer, R. (1987) Evaluating variability in the estimated all-in plant rate. *International Journal of Construction Technology and Management*, **2**(2).

Barnes, N.M.L. & Lau, K.T. (1974) Bidding strategies and company performance in process plant contracting. *Third International Cost Engineering Symposium*. London, 1974. Association of Cost Engineers, Sandbach.

Curtis, F.W & Maines, P.W. (1973) Closed competitive bidding. *Omega*, **1**(4), 613–619.

Fine, B. Various unpublished papers based on work of the Costain OR Group.

Friedman, L. (1956) A competitive bidding strategy. *Operations Research*, **4**, 104–112.
Gates, M. (1967) Bidding strategies and probabilities. *Journal of the Construction Division, ASCE*, **93**(CO1), proc. paper 5159, 74–107.
King, M. & Mercer, A. (1987) Differences in bidding strategies. *European Journal of Operational Research*, **28**, 22–26.
McCaffer, R. (1975) Some examples of the use of regression analysis as an estimating tool. *The Quantity Surveyor*, **32**(5), 81–86.
McCaffer, R. (1976) *Contractors' Bidding Behaviour and Tender Price Prediction*. PhD thesis. Department of Civil Engineering, Loughborough University of Technology.
Morin, T.L & Clough, R.H. (1969) OPBID: competitive bidding strategy model. *Journal of the Construction Division, ASCE*, **95**(CO1), proc. paper 6690, 85–106.
Ogunlana, O. & Thorpe, A. (1987) Design phase cost estimating: the state of the art. *International Journal of Construction Technology and Management*, **2**(4), 34–47.
Park, W. (1966) *The Strategy of Contracting for Profit*. Prentice-Hall, Englewood Cliffs, NJ.
Rickwood, A.K. (1972) *An Investigation into the Tenability of Bidding Theory and Techniques*. MSc project report, Loughborough University of Technology.
Skitmore, M. (2002) Identifying non-competitive bids in construction contract auctions. *Omega*. **30**(6), 443–449.
Statham, W. & Sargeant, M. Determining an optimum bid. *Building*, **216**(6573), 1969.
Whittaker, J.D. (1981) Implementing a bidding model. *Journal of the Operational Research Society*, **32**, 11–17.

11
Budgetary control

Summary

The emphasis of this chapter is directed towards control of budgets and costs at company level. The preparation of sales, capital, operating and master budgets is explained, together with methods for classifying and allocating costs. An example illustrating the calculations of variances for a portfolio of contracts and other company operations is described.

Introduction

A budget acts as a standard of measure against which the actual performance of a project or company may be compared. The Institute of Cost and Management Accountants defines budgetary control as 'the establishment of budgets, relating the responsibility of executives to the requirements of a policy and the continuous comparison of actual with budgeted results, either to secure by individual action the objective of that policy, or to provide a basis for its revision'. Budgetary control therefore involves:

- Setting targets
- Monitoring progress
- Taking corrective action when necessary

Within a system of budgetary control, budgets are established to relate the financial requirements for the component parts of the firm over the forthcoming period of 12 months to the overall policy of the company. Budgets may be forward estimates of costs or revenues and as such are usually derived from records of past performance adjusted for future expectations.

Preparation of budgets

The budgetary system comprises many individual budgets, which are ultimately integrated into a master budget. The master budget (Table 11.1) is similar in form to a profit and loss account, but, unlike the latter, it is based on forward estimates of costs and revenues and is therefore only a forecast of the anticipated profit to be earned. From such an estimate, other factors related to future expectations may be projected, such as the rate of return on capital employed, dividends to shareholders, capital to be retained in the business for reinvestment in assets and similar items.

Table 11.1 Master budget.

	Year	Month
(1) Budgeted sales (turnover)		
❏ Value of work on future contracts		
❏ Work in progress		
❏ Receipts from claims and other sales		
(2) Budgeted costs		
(a) Contracts		
❏ Wages		
❏ Materials		
❏ Subcontractors		
❏ Rentals on hired plant and equipment		
❏ Salaries and establishment costs		
(b) Head Office		
❏ Salaries		
❏ Rates, rents, insurance, lighting, etc.		
(3) Budgeted trading profit		
❏ Depreciation on plant and equipment		
❏ Expected profit before interest charges and tax		

Unlike a manufacturing company, the construction firm generally has less control over the volume of sales, i.e. turnover, since much of the work tends to be obtained through competitive bidding. Thus, forecasts of future contracts are likely to be inaccurate. However, the company has to plan on obtaining a sufficient workload to cover its commitments, i.e. establishment costs, interest charges on loans, etc. This requires formulating a master budget in conjunction with a sales budget and a corresponding operating budget. By synthesising these latter two budgets with a capital expenditure budget, the cash flow forecast is determined. By a gradual process of refinement, the budgets are adjusted to keep within the financial resources available to the company, culminating finally in the master budget. The procedure is shown logically in Fig. 11.1.

Types of budget (Fig. 11.1)

Sales budget

The *sales budget* is a forecast of the volume of turnover, comprising the expected value of work period by period on the firm's projects, plus any other expected income generated from selling and successful claims on past contracts. It is usual practice to divide the periods into months corresponding to the normal monthly schedule of payments by clients for construction work.

Operating budget

The *operating budget* is obtained from cost estimates of the planned requirements for materials, labour, subcontractors, staff and overheads. The elements of this budget may be updated from past contracts minus the contributions made for head-office overheads and profit. The estimate of the head-office budgeted costs is subsequently included as a total figure.

At this stage, an estimate of the value of depreciation of the firm's assets such as plant and equipment could be included. However, because such costs do not involve cash leaving the business, they are usually omitted and only taken into account as part of the master budget. The

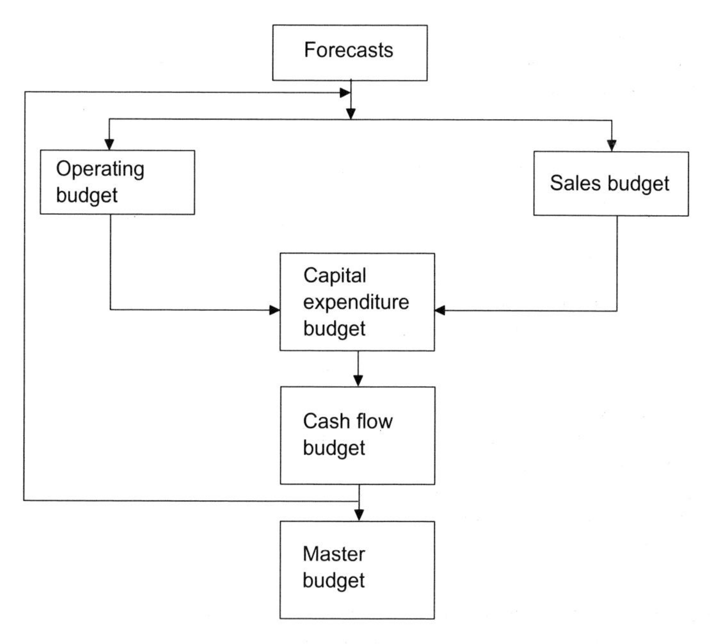

Fig. 11.1 Budget system.

difference in value between the sales and operating budgets is the anticipated profit before deduction of depreciation on assets.

The operating budget may subsequently be divided into separate functions. For example, construction contracts usually have individual budgets, while the head-office functions may include departments for plant, administration, estimating and tendering, planning, contract services, corporate planning, purchasing, legal and insurance services, etc. An example of the form of budget for the administration departments is shown in Table 11.2. By providing each department with a separate financial budget, a target is available against which subsequent performance may be monitored.

Capital expenditure budget

The *capital expenditure budget* very much depends upon the nature of the volume of work to be expected during this coming year. For a construction firm that has a policy of purchasing most of its plant, equipment must be scheduled to meet the requirement of the contracts. The budget therefore sets out the capital needs to meet this schedule and may include other significant items of expenditure such as new premises, costly materials and special items for contracts. In addition, old plant will have to be replaced or planned for.

Cash flow forecast

The *cash flow forecast* is produced by integrating the sales, operating and capital expenditure budgets, taking into account payment delays and retentions, interest charges on loans, corporation tax and capital allowances. In this way, the period by period cash needs are determined.

Table 11.2 Administration budget.

Code (A)	Item	Annual (£)	Month
Direct employment costs			
300.0	Staff salaries	9 500	
Direct material costs			
310.0	Stationery	200	
Direct expenses			
320.0	Photocopying	100	
Indirect costs			
330.1	Telephone	1 000	
330.2	Postage	700	
330.3	Electricity	1 000	
340.1	Rates	500	
340.2	Rent	1 000	
340.3	Office equipment	400	
340.4	Insurances	600	
Total cost		15 000	

Where the results of this synthesis indicate that the firm will not have the ability to finance this level of business, then the budgets must be refined to meet the constraints.

Classification of costs

Budgets consist of cost forecasts of the requirements for materials, labour and expenses. A coding system is used to allocate the amount to particular departments or functions concerned and like items are collected under the same alphanumeric code. It is usual to present both an annual budget and a weekly budget, as short-term fluctuations may be the more usual pattern of expected performance.

The costs of materials, labour and expenses that can be clearly allocated to a specific cost centre are called direct costs, and usually vary with the volume of production, i.e. turnover. Indirect costs are those materials, labour and expenses that cannot be directly identified to the cost centre, but which provide some functions or service, such as a computer or the rent of the firm's offices and works. Indirect costs are mostly fixed costs such as staff salaries, rent and rates, insurances, office equipment, maintenance tools and machines, which remain constant irrespective of the volume of work done. A direct or variable overhead is one which varies in cost with the volume of production, such as electricity.

Costing

While budgets are prepared from predetermined costs, because of short-term changes in company performance, it is essential that the actual costs incurred are continuously monitored and compared with budgeted costs in order that changes may be implemented. The difference between the actual and predetermined cost is called a variance. A costing system should be updated regularly on a weekly basis and the variances calculated for each function, department or cost centre. The procedure should also include analysis of variances incurred by the individual construction contracts as described in Chapter 6 on *Cost control.*

A note of caution with regard to the budgets, particularly departmental budgets, is advised. It

Table 11.3 Budgeted sum for overheads and profit.

Contract	HO Overheads (£)	Profit (£)	Total (£)
1	4000	5000	
2	10000	14000	(Represents 10% of sales)
3	3000	3000	
4	3000	6000	
Total	20000	28000	48000

is useful to compare the actual results with the value for the same month or week of the previous year. Astute managers can be adept at 'hiding' behind a 'stuffed' budget.

A budget may be 'stuffed' due to changed circumstances, e.g. reduction in the assumed rate of inflation, manipulation of the figures, etc. Managers' performance should therefore be measured against both the budget and previous year results.

Example of budgetary control

A construction company engaged in heavy civil engineering work prepares at the beginning of each year a financial budget for its contract department. The situation existing at 1 January is as follows:

Contract	Fixed tender price	Period of work
1	£90 000	1 Jan to 31 Dec
2	£240 000	1 Jan to 31 Dec
3	£60 000	1 Jan to 31 Dec

However, the company is at present negotiating a further contract valued at £90 000 to commence on 1 April with a duration of 12 months. Each of these contracts includes an allowance to cover head-office overheads and profit as indicated in Table 11.3.

The budget is reviewed six months later, to reveal the following position at 30 June:

Contract	Value of work	Direct cost to date
1	45 000	42 000
2	120 000	110 000
3	25 000	20 000
4	Contract did not materialise	

The actual cost of head office overheads to date on each contract is £2500.

(1) Assuming that the value of work in each project can be spread uniformly throughout its duration, prepare a budget for the year.
(2) Contract 4 did not materialise, thus HO overheads and profit will be under-recovered as a result of this shortfall in obtaining contracts.

Table 11.4 Budgeted receipts.

Period	Contract 1 (£)	Contract 2 (£)	Contract 3 (£)	Contract 4 (£)	Total (£)
1 Jan	7 500	20 000	5 000	–	32 500
Feb	7 500	20 000	5 000	–	32 500
Mar	7 500	20 000	5 000	–	32 500
Apr	7 500	20 000	5 000	7 500	40 000
May	7 500	20 000	5 000	7 500	40 000
Jun	7 500	20 000	5 000	7 500	40 000
					217 500
Jul	7 500	20 000	5 000	7 500	40 000
Aug	7 500	20 000	5 000	7 500	40 000
Sept	7 500	20 000	5 000	7 500	40 000
Oct	7 500	20 000	5 000	7 500	40 000
Nov	7 500	20 000	5 000	7 500	40 000
Dec	7 500	20 000	5 000	7 500	40 000
Total	90 000	240 000	60 000	67 500	457 000
1 Jan (one year on)				7 500	
Feb				7 500	
Mar				7 500	
Total				90 000	

Budgeted profit = £5000 + £14 000 + £3000 + £4500
= £26 500
Budgeted HO overheads = £4000 + £10 000 + £3000 + £2250
= £19 250
= £45 750 at end of December

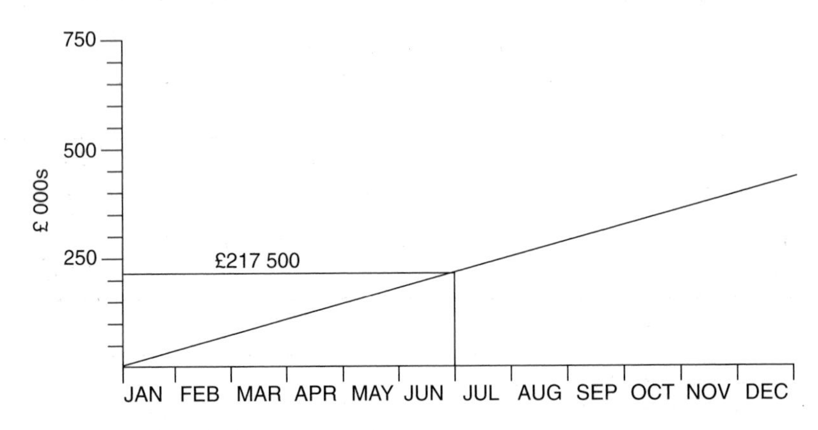

Fig. 11.2 Monthly cumulative budgeted turnover in year.

Calculate the company sales and overhead variances, and the total variance on each contract at 30 June and thereby obtain the actual profit and compare it with that budgeted.

Solution

(1) At the beginning of the year, the budget for receipts is as shown in Table 11.4.
(2) Position at June on each contract. This is shown in Table 11.5.

The budgeted sales figure to the end of June is £217 500 (from Fig. 11.2). However, contract 4

Table 11.5 Table of variances.

		Budgeted for this date (£)	Value of work done (£)	Actual cost (£)	Variance (£)
Contract 1					
❑ Direct		40 500	40 500	42 000	−1500
❑ HO overhead		2 000	2 000	2 000*	0
❑ Profit		2 500	2 500	2 500*	0
	Total	45 000	45 000	46 500	−1500
Contract 2					
❑ Direct		108 000	108 000	110 000	−2000
❑ HO overhead		5 000	5 000	5 000*	0
❑ Profit		7 000	7 000	7 000*	0
	Total	120 000	120 000	122 000	−2000
Contract 3					
❑ Direct		27 000	22 500	20 000	+2500
❑ HO overhead		1 500	2 250	1 500*	−250
❑ Profit		1 500	1 250	1 500*	−250
	Total	30 000	25 000	23 000	+2000

* Fixed costs allocated by Head Office.

did not materialise, and therefore the budget that can be realised by this date is only £195 000, i.e. £45 000 + £120 000 + £30 000.

Shortfall = £22 500.

This shortfall will mean that head office overheads and profits will be under-recovered by:

Sales variance = 10% of £22 500 = £2250.

Total costs to date (£)		
Direct costs	= 42 000 + 110 000 + 20 000	= 172 000
Head-office overheads	= 2 500 + 2 500 + 2 500	= 7 500
		179 500
Total receipts	= 45 000 + 120 000 + 25 000	= 190 000
Company profit to date		10 500

Reconciliation	
Budgeted profit to date from Table 11.3	(£)
(i.e. 5 000 × 1/2 + 14 000 × 1/2 + 3 000 × 1/2 + 6 000 × 1/4)	+12 500
Contract 1 =	−1 500
Contract 2 =	−2 000
Contract 3 =	+2 000
	−1 500
Total variance for all contracts	= −1 500
Sales variance	= −2 250
Budgeted head-office overheads (2 000 + 5 000 + 1 500 + 750)	= £9 250
Actual head-office overheads	= £7 500
Head-office overheads variance	= +1 750
Company profit at 30 June	= +10 500

12
Cash flow and interim valuations

Summary

The need for cash flow forecasting, the requirements of a cash flow forecasting system, some practical suggestions to simplify cash flow forecasting and measuring interest charged on capital lock-up, the role of computers in cash flow calculations and interim valuations.

The need for cash flow forecasting by contractors

Each year the construction industry usually experiences a proportionally greater number of bankruptcies than do other industries. One of the final causes of bankruptcy is inadequate cash resources and failure to convince creditors and possible lenders of money that this inadequacy is only temporary. The need to forecast cash requirements is important in order to make provision for these difficult times before they arrive. In times of high interest rates the need for cash flow forecasting is even more important. Many of the large contractors operate internationally and undertake projects in countries that are characterised by high levels of inflation. The need to forecast cash flows in terms of inflation and thus avoid an embarrassing cash deficit when replacing old equipment at new inflated costs becomes an essential aspect for profitable business by the contractor.

There is evidence that some smaller companies confuse profit flows with cash flows and make misleading calculations. A cash flow is the transfer of money into or out of the company. The timing of a cash flow is important. There will be a time lag between the entitlement to receive a cash payment and actually receiving it. There will be a time lag between being committed to making a payment and actually paying it. These time lags are the credit arrangements that contractors have with their creditors and debtors. It is these credit arrangements, stock levels and depreciation that make cash and profit different. The following simple example highlights this difference due to credit arrangements at the time of the transactions.

A company buys goods at £6 each and sells for £8 each; there are a number of different credit arrangements as follows:

Credit Arrangements	Profit Flow (£)	Cash flow (£)	
		Out	In
Buy ten for cash Sell ten for cash	20		20
Buy ten on credit Sell ten for cash	20		80
Buy ten for cash Sell ten for credit	20	60	

The credit arrangements cause very different cash flows. There is a need therefore for separate cash flow forecasting. Some companies forecast their cash flows for only a few months ahead, say three or six months. This is not adequate and does not look far enough ahead. The factors that affect cash flow are the duration of new projects, the profit margin on these projects, the retention conditions, the delay in receiving payment from the client, the credit arrangements with suppliers, plant hirers and subcontractors and the phasing of the projects in the company's workload and the late settlement of outstanding claims. As many projects last one, two or three years, the cash flow forecasts should also look this far ahead.

Some companies quickly point out that forecasts are guesses and therefore are probably wrong and useless and not worth the effort. However, this argument is equally applicable to estimating, which is essential to a company's operations. Cash flow forecasting is, like any forecasting, the result of calculations based on the information available at the time and a few assumptions as to what will happen. If, as is likely, the data contained in the information changes or the assumptions alter, the forecast will be in error and a new forecast is required. When a company equips itself to forecast its cash flows efficiently and without great expense, it usually forecasts every quarter and in some cases every month. This is sufficient to monitor the ever-changing situation.

Contractors who undertake cash flow forecasting do so at two levels. One is at the estimating and tendering stage, when the forecast is just for the single project being estimated. The other level is the calculation of a cash flow forecast for the company, division or area; this involves aggregating cash flows for all active projects and is done regularly every month or quarter.

These two types of forecasts require different treatments. The estimator has all the project details at the estimating stage and, because the forecast applies only to one project, the estimator can produce a carefully calculated forecast based on these details by allocating bill items to 'activities' on the pre-tender 'bar-chart' or 'network'. This creates a direct link between the estimator's build-ups for each item with the pre-tender construction programme and allows the production of value-versus-time and cost-versus-time curves from which 'cash in' and 'cash out' can be calculated. This calculation is too detailed to be repeated for every active project, every quarter or every month and so contractors devise shortcuts when undertaking the company or divisional cash flow forecasts. This chapter describes the principles of the calculations in cash flow forecasting and reference is made to shortcuts that will allow cash flow forecasts to be made more regularly than just the one detailed forecast at the estimating stage.

The requirements of a forecasting system

Cash flow forecasting is strongly advisable. For the forecasts to be meaningful, they must be done regularly and, for forecasting to be done regularly, the method must be simple and easy to do and yet accurate enough for the purpose. It is essential therefore to reduce the data required for cash flow forecasting to the minimum possible compatible with reliable forecasts and to streamline the necessary calculations.

The data needed

In construction companies the most appropriate approach is to calculate cash flows on a project basis and aggregate the cash flows from all projects and head office to form the overall company cash flow. This can be structured into divisions or areas for larger companies. The data required for a project comprises:

(1) A graph of value versus time, value being the monies a contractor will eventually receive for doing the work
(2) The measurement and certification interval
(3) The payment delay between certification and the contractor receiving the cash
(4) The retention conditions and retention repayment arrangements
(5) A graph of cost versus time, the contractor's cost liability arising from labour, plant, materials, subcontractors and other cost headings as necessary
(6) The project costs broken down into the above items
(7) The delay between incurring a cost liability under each cost heading and meeting that liability

The data required for the company's head office comprises:

(8) The head-office outgoings and the time of their occurrence
(9) The head-office incomings and the time of their occurrence

The project-based data items (2), (3) and (4) are readily available. Item (6) can easily be calculated, but after study of several contracts of the same type, appropriate cost breakdowns can be derived. Item (7) should be known by the buying department. Items (1) and (5) are a little more difficult to obtain.

Value versus time

Item (1), the graph of value versus time, is needed in order to derive 'cash in'. This can be obtained by producing project plans, in network or bar-chart form, and calculating the value of each activity and summing the value in each time period of either weeks or months. Figure 12.1 shows a simple example of the value versus time calculated from the project plan. Usually calculations are done on a cumulative basis and so the cumulative value versus time is produced by a running total over each time period.

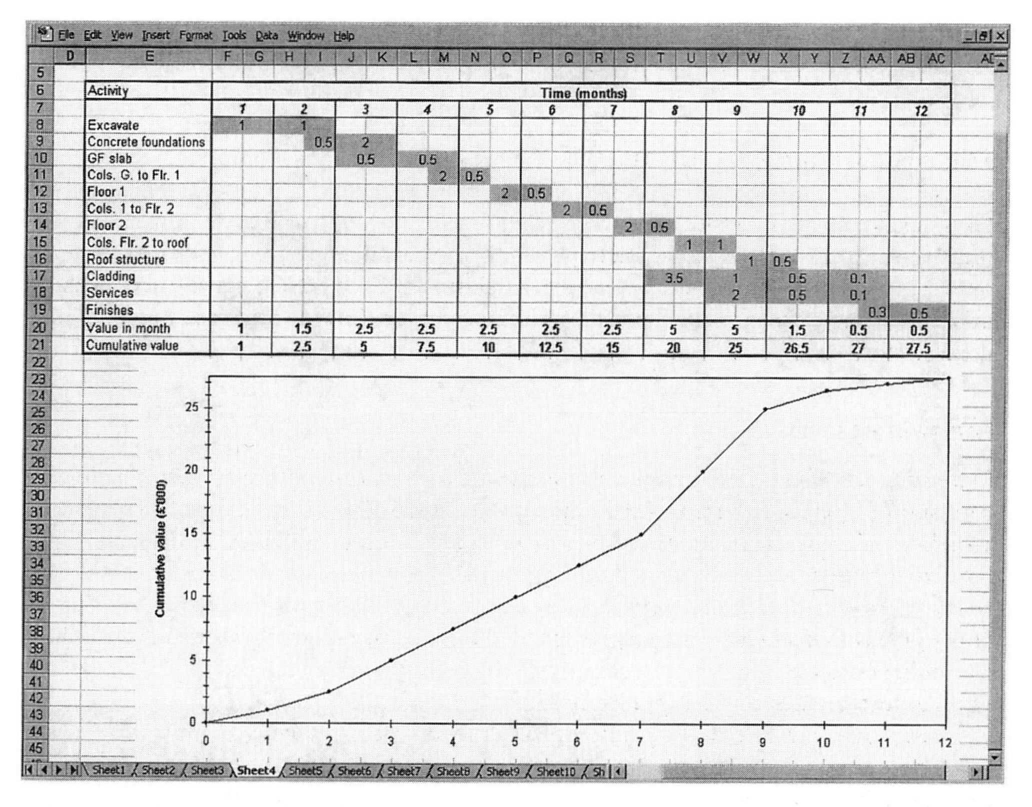

Activity	Time (months) 1	2	3	4	5	6	7	8	9	10	11	12
Excavate	1	1										
Concrete foundations		0.5	2									
GF slab			0.5	0.5								
Cols. G. to Flr. 1				2	0.5							
Floor 1					2	0.5						
Cols. 1 to Flr. 2						2	0.5					
Floor 2							2	0.5				
Cols. Flr. 2 to roof								1	1			
Roof structure									1	0.5		
Cladding								3.5	1	0.5	0.1	
Services									2	0.5	0.1	
Finishes											0.3	0.5
Value in month	1	1.5	2.5	2.5	2.5	2.5	2.5	5	5	1.5	0.5	0.5
Cumulative value	1	2.5	5	7.5	10	12.5	15	20	25	26.5	27	27.5

Fig. 12.1 Value versus time from a bar-chart (values shown as £000).

Cost versus time

Item (5), the graph of cost versus time, is needed to derive 'cash out'. The contribution margin, i.e. the margin for profit and head office overheads added to the estimated direct costs, may be spread uniformly throughout the project. In other words, each bill item carries the same percentage for profit and overheads. In this case the cumulative cost versus time is a simple proportion of the cumulative value-versus-time figures. For example, if 12% was added uniformly to all items for profit and overheads and the cumulative value was calculated from the project plan, the cumulative costs would be 0.89 times cumulative value:

$$\text{Value} = \text{cost} + 12\% = 1.12 \times \text{cost}$$

Therefore

$$\text{Cost} = \text{value} \div 1.12 = 0.89 \times \text{value}$$

Alternatively the contribution margin may not be uniformly distributed because either rate loading had occurred and early activities were carrying greater margin than later activities, or some setting-up costs were adversely reducing the margin on early work. In this case, the cumulative cost could be produced in the same way as the value versus time from the project plan by costing each activity.

Although the cash required by a project is dependent on several factors, one of the crucial factors is the margin achieved by the project. It is important to have a realistic assessment of this margin. Chapter 10 on *Competitive bidding* explains at least one reason why applied mark-up (or margin) and achieved mark-up (or margin) are different, with the latter on average being less. The historical record of the company should be used as guidance. It is also important to separate out claims when assessing the achieved margin. If, for example, a project had an applied mark-up of 10% and achieved this eventually in the form of 5% from the monthly measurements and 5% through claims, the cash flow would be substantially worse than an achievement of 10% through the monthly measurements. In these cases, it is suggested that the margin used in the cash flow calculations is 5% and the claims income be treated as a lump sum coming at a predicted future date, say the end of the contract.

A suggested simplification

The production of cumulative value-versus-time and cost-versus-time figures based on project plans is time-consuming because it requires every project in the forecast, i.e. every active or new project in the company's workload, to be represented by a network or bar-chart. However, it is possible to bypass the 'plan' stage and use 'standard' cumulative value-versus-time curves that adequately represent each project type. These cumulative value-versus-time curves, presented in percentage terms in Fig. 12.2, can be obtained from certification records of past projects. Hardy demonstrated that, within a company, projects of the same type had similar shapes of

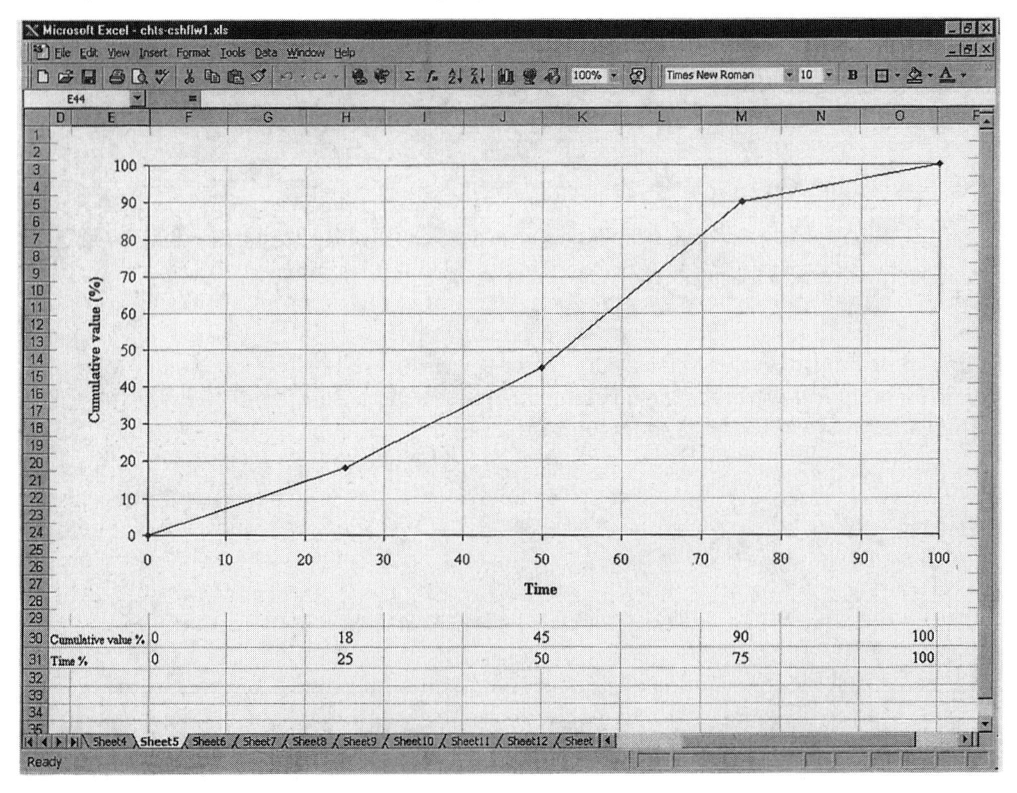

Fig. 12.2 Cumulative value versus time in percentage terms.

cumulative value versus time when expressed in percentage terms. To make a standard curve from a library unique to a specific project the time scale has to be multiplied by the project's estimated duration and the value scale by the project's estimated value.

When the contribution margin is uniformly spread throughout the project, the cumulative cost versus time is easily derived from the standard cumulative value curve. When contribution is not uniformly spread, some calculation is necessary before the cumulative cost-versus-time curve is produced. However, if the information as to how the contribution is distributed throughout the project is obtained from the estimating department or the tendering panel, the cumulative cost versus time figures in percentage terms can be produced without a project plan. McKay demonstrated that this approach of using standard curves did not lead to loss of reliability in the eventual cash flow forecast. See McCaffer (1976) for information on the work of Hardy and McKay.

The calculations

The calculations can be undertaken with a spreadsheet as shown in Fig. 12.3. A time period of one month is used, but weekly time periods could be used if necessary. Figure 12.3 is made up as follows. Row 1 is the cumulative value derived as previously discussed. Row 2 is the cumulative value less the retention (in this example retention is assumed to be 10%). Row 3 is the value less retention (row 2) shifted by an amount equal to the payment delay between valuation and the

	Time (months)													
	1	2	3	4	5	6	7	8	9	10	11	12	13	18
1 Cumulative value	1000	2500	5000	7500	10000	12500	15000	20000	25000	26500	27000	27500	27500	27500
2 Cumulative value less retention	900	2250	4500	6750	9000	11250	13500	18000	22500	23850	24300	24750	24750	24750
3 Cumulative payment received from certification		900	2250	4500	6750	9000	11250	13500	18000	22500	23850	24300	24750	24750
4 Cumulative retention payment													1375	2750
5 Cumulative cost	920	2300	4600	6900	9200	11500	13800	18400	23000	24380	24840	25300	25300	25300
6 Cumulative labour costs (30%)	276	690	1380	2070	2760	3450	4140	5520	6900	7314	7452	7590	7590	7590
7 Cumulative labour payments	276	690	1380	2070	2760	3450	4140	5520	6900	7314	7452	7590	7590	7590
8 Cumulative materials cost (20%)	184	460	920	1380	1840	2300	2760	3680	4600	4876	4968	5060	5060	5060
9 Cumulative materials payments		184	460	920	1380	1840	2300	2760	3680	4600	4876	4968	5060	5060
10 Cumulative plant cost (30%)	276	690	1380	2070	2760	3450	4140	5520	6900	7314	7452	7590	7590	7590
11 Cumulative plant payments		276	690	1380	2070	2760	3450	4140	5520	6900	7314	7452	7590	7590
12 Cumulative subcontractor cost (20%)	184	460	920	1380	1840	2300	2760	3680	4600	4876	4968	5060	5060	5060
13 Cumulative subcontractor payments		184	460	920	1380	1840	2300	2760	3680	4600	4876	4968	5060	5060
14 Cumulative cash out (7 + 9 + 11 + 13)	276	1334	2990	5290	7590	9890	12190	15180	19780	23414	24518	24978	25300	25300
15 Cumulative cash flow (3 + 4 - 14)	**-276**	**-434**	**-740**	**-790**	**-840**	**-890**	**-940**	**-1680**	**-1780**	**-914**	**-668**	**-678**	**825**	**2200**

Fig. 12.3 Cash flow forecasting calculations.

contractor receiving their money. Row 4 is the cumulative retention payments inserted at the time they would be received, in this case half at one month after practical completion and half six months after practical completion.

The cost headings in Fig. 12.3 are treated individually. The cumulative costs for the total project are shown in row 5. The proportion of costs due to labour is calculated and inserted in row 6, for materials in row 8, for plant in row 10 and for subcontractors in row 12. This can be extended to as many cost headings as required. The proportions of the cost breakdown vary throughout the contract period and this variation can be taken into account; however, the introduction of such a refinement increases the calculations required, and the increase in the accuracy of the forecast is not commensurate with the extra effort.

Rows 7, 9, 11 and 13 shift the cost liabilities shown in rows 6, 8, 10 and 12 by an amount equal to the average delay between incurring the cost liability and making the payment. For labour this is taken as zero because labour is usually paid after one week's delay, which is less than one month—the smallest time unit in the example. For plant, materials and subcontractors, the delay is taken as one month. Row 14 is the cumulative cash out and is the sum of rows 7, 9, 11 and 13. Row 15 is the cumulative cash flow from row 3 plus row 4 less row 14.

Another suggested simplification

Much of the calculating effort goes into reckoning the cash out due to each individual cost heading. This can be particularly tedious if the exercise is based on time periods of one week instead of one month. The simplification is to calculate a weighted average payment delay for all cost headings, the weighting being the percentage of cost due to that heading. For example, the figures tabulated below show the contract costs as a percentage of the total cost, broken down into four main groups. The delay experienced between the incurring of a cost liability and the making of a payment is also shown for each of these groups. Using the percentage cost as weighting, the weighted average payment delay is calculated as 5.1 weeks.

This weighted-average delay applied to the total cost gives cash out figures that are accurate over the duration of the contract except during the first few weeks and the last few weeks.

Cost Group	Costs as percentage of total cost	Delay between incurring a cost liability and making payment	Calculation of weighted average payment delay
Labour	30	1 week	30% × 1 = 0.3
Plant	30	8 weeks	30% × 8 = 2.4
Materials	20	6 weeks	20% × 6 = 1.2
Subcontractors	20	6 weeks	20% × 6 = 1.2
			Total, i.e. weighted average payment delay = 5.1

The company cash flow

Each project is treated as above and the cash flows from each project summed together and then added to the company's head-office cash flows to produce the company cash flow. The head-office cash flows are treated separately in the following way:

- The head-office outgoings are entered at the time the outgoings occur. These outgoings are, for example, rent and rates, telephones, office equipment hire charges, payments to shareholders, tax, directors' fees, etc.
- The head-office incomings are treated in the same way. The incomings are derived from selling head-office services; also any claims or retentions unsettled from past projects may have the expected amount entered in head-office incomings.

Capital lock-up

The negative cash flows experienced in the early stages of projects represent locked up capital that is supplied from the company's cash reserves or borrowed. If the company borrows the cash it will have to pay interest charged to the project; if the company uses its own cash reserves it is being deprived of the interest-earning capability of the cash and should therefore charge the project for the interest lost. A measure of the interest payable is obtained by calculating the area between the cash-out and the cash-in curves shown as the '*cash flow (–ve)*' to be financed in Fig. 12.4. The area is in units of £ × weeks (or months) because the vertical scale is in pounds and the horizontal scale in time (weeks or months). This compound measure, say 5000 £ × months, represents the volume of borrowing from the extremes of £5000 for one month to £1 for 5000 months. This measurement has been given the name of *captim*, standing for capital × time. A single interest calculation is all that is required to convert captim into an interest charge, e.g. captim of £25 000 × months, annual interest rate of 15%:

$$\text{Interest payable} = \frac{25000 \times \text{months}}{12\ \text{months/year}} \times 15\%\ \text{p.a.}$$
$$= £312.50$$

This calculation can be done on the negative captim alone, assuming that the cash released by the project does not earn interest, or the interest earned by the positive captim is subtracted from the interest paid on the negative captim. It is likely that the interest paying rate will be different and probably greater than the interest earning rate. The use of this captim measure enables the effect of interest charges to be evaluated.

The factors that affect capital lock-up

The factors that affect the capital lock-up for an individual project or contract are described in the following sections.

Margin

The margin, whether profit margin or contribution (profit plus head-office overheads) margin, is amongst the most important because it determines the excess over costs and it is the excess that controls the capital lock-up. Quite simply, the larger the margin, the less capital that is locked up: conversely the smaller the margin, the more capital that is locked up in the contract.

The margin that should be used in calculating contract cash flows should be the 'effective' margin that is being achieved at the time of executing the contract. This 'effective' margin is

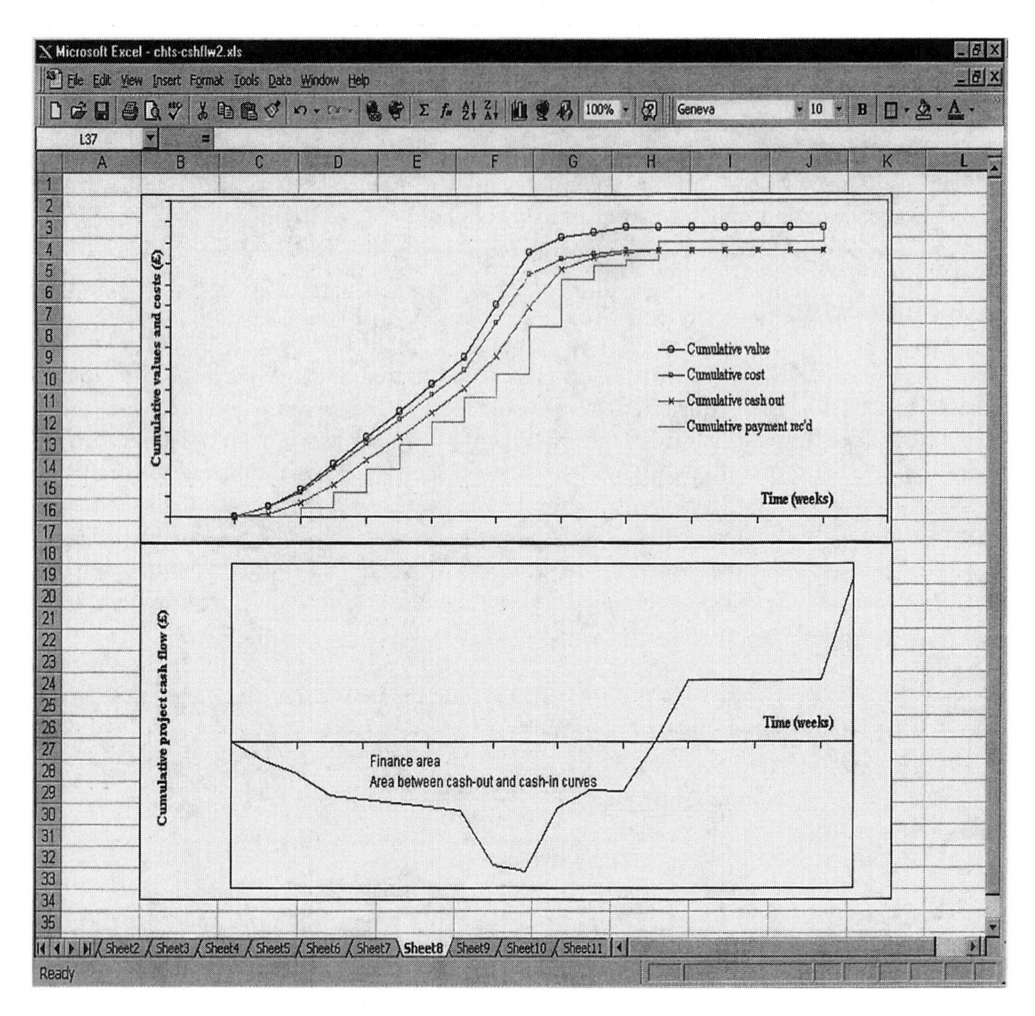

Fig. 12.4 Cash-out and cash-in curves.

neither the margin included at the tender stage nor the margin achieved at the end of the contract when all the claims are settled.

For example, if the tender panel included a margin of 9% and, as a result of variations and other client interference, the costs rose by 4%, this would reduce the 'effective' margin to only 5%. Even if an astute contractor recovered another 5% in claims, making the overall achieved margin at the end of the contract 10%, the 'effective' margin at the time of executing the contract and thus determining the cash flows, would still be only 5%. Table 12.1 presents scenarios of the same project in which the partial effect of variability in various factors are outlined. Table 12.1*a* provides the control project for undertaking the comparisons.

Table 12.1*b* employs a different margin (3%) for the project duration under consideration. The reduction in the margin from 10% (for the control project) to 3% in Table 12.1*b* changes the cash flow of the project from positive to negative. This results in a capital lock-up for the project during the period under consideration, and creates the need of external finance for the project.

Table 12.1 Effects of different factors on cash flow.

a. Control project used for comparing effect of factors on cash flow

Month	1	2	3	4	–	–	15	16
Cost	100	100	100	100	–	–	–	–
Margin (@ 10%)	10	10	10	10	–	–	–	–
Value	110	110	110	110	–	–	–	–
Retention	5%	5%	5%	5%	–	–	–	–
Monies due	104.5	104.5	104.5	104.5	–	–	–	–
Cash flow (monies due less cost)	4.5	4.5	4.5	4.5	–	–	–	–

b. Example of the effect of margin variability on cash flow

Cost	100	100	100	100	–	–	–	–
Margin (@ 3%)	3	3	3	3	–	–	–	–
Value	103	103	103	103	–	–	–	–
Retention	5%	5%	5%	5%	–	–	–	–
Monies due	97.85	97.85	97.85	97.85	–	–	–	–
Cash flow (monies due less cost)	−2.15	−2.15	−2.15	−2.15	–	–	–	–

c. Effect of variability in retention on cash flow

Cost	100	100	100	100	–	–	–	–
Margin	10	10	10	10	–	–	–	–
Value	110	110	110	110	–	–	–	–
Retention	10%	10%	10%	10%	–	–	–	–
Monies due	99	99	99	99	–	–	–	–
Cash flow (monies due less cost)	−1.0	−1.0	−1.0	−1.0	–	–	–	–

d. Effects of claims on cash flow

Cost	100	100	100	100	–	–	–	–
Margin	10	10	10	10	–	–	–	–
Value	110	110	110	110	–	–	–	–
Extra cost from unforeseen work	–	10	–	–	–	–	–	–
Claim	–	–	20	–	–	–	–	–
Claim settled	–	–	–	–	–	–	–	15
Retention	5%	5%	5%	5%		–	–	–
Monies due	104.5	104.5	104.5	104.5		–	–	15
Cash flow (monies due less cost)	4.5	−5.5	4.5	4.5		–	–	15

e. Effects of front loading on cash flow

Cost	100	100	100	100	–	–	100	100
Margin (non-uniform)	15	10	10	5	–	–	0	0
Value	115	110	110	105	–	–	100	100
Retention	5%	5%	5%	5%	–	–	5%	5%
Monies due	109.25	104.5	104.5	99.75	–	–	95	95
Cash flow (monies due less cost)	9.25	4.5	4.5	−0.25	–	–	−5	−5

f. Effects of overmeasurement on cash flow

Cost	100	100	100	100	–	–	100	100
Margin	10	10	10	10	–	–	10	10
Value	110	110	110	110	–	–	110	110
Measured value	120	115	115	105	–	–	104	101
Retention	5%	5%	5%	5%	–	–	5%	5%
Monies due	114	109.25	109.25	99.75	–	–	98.8	95.95
Cash flow (monies due less cost)	14	9.25	9.25	−0.25	–	–	−1.2	−4.05

Note: all cash amounts are in units of (£1000)

Most contractors use the tender margin when calculating contract cash flows and this leads to an optimistic forecast. Although the 'margin' is an important factor in determining a contract's cash flow, the tender margin is usually chosen for market reasons rather than cash-flow reasons.

Retentions

In the UK the system of retentions simply reduces the 'effective' margin during the execution of the contract and the effect of retentions can be included in the calculations as shown in this chapter. In times of very low margins the retention can reduce the 'effective' margin to zero or less. Table 12.1*c* shows how a change in the level of retention can impact on the project's cash flow. Because an additional 5% of payments due is withheld, for the period under consideration the contractor is subjected to a capital lock-up of £1000 a month. Under the New Engineering Contract a contractor's inability to provide on time a first programme of works could result in the Project Manager retaining 25% of the amount due (NEC: ECC Clause 50.3), thus severely affecting the project's cash flow. As retentions are fairly standard in public sector contracts, there is usually little scope for negotiation to reduce retention and improve cash flows.

Claims

As explained above under the section on 'Margin', claims can return a contract to its original intended level of profit. However, as the settlement of claims is normally subject to some delay the actual settlement does not improve a contract's cash flow and the circumstances giving rise to a genuine claim are likely to worsen a contract's cash flow. In Table 12.1*d* the introduction of extra work at a cost of £10 000 to the contractor during the second month, which was unforeseen at the start of the project, results in a claim by the contractor. Although this is settled at a much later time, the impact of the additional work creates a negative cash flow of £5500 that requires external capital to finance.

The settlement of claims is, of course, important to the company's cash flow and therefore it is important that claims are settled as quickly as possible.

Front-end rate loading

Front-end rate loading is the device whereby the earlier items in the 'bill' carry a higher margin than the later items. This has the effect of improving the 'effective' margin in the early stages of the contract while keeping the overall margin at a competitive level. It is in these early stages that capital lock-up is at its worst. Table 12.1*e* shows a more than 100% improvement in cash flow for the first month by front-loading the margin for that month to 15% and reducing the margin for the fourth month to 5%. The additional positive cash flow of £4750 in the first month will be employed to offset any borrowing that the contractor will require at a later stage in order to execute the project. The degree to which front-end rate loading can be done depends on the client's awareness.

Overmeasurement

Overmeasurement is the device whereby the amount of work certified in the early months of a contract is greater than the amount of work done. This is compensated for in later measurements. Thus, overmeasurement has the same effect as front-end rate loading; it improves the 'cash in' in the early stages and reduces the capital lock-up. Table 12.1*f* depicts how overmeasurement can impact on the cash flow of a project. The first two months of the project experience an additional positive cash flow of £9500 and £4750 as a result of an overmeasurement of £10 000 and £5000 respectively. The early additional cash will then allow the contractor to run the project during the third and fourth months when the cash flow of the project is negative. For contracts that are administered under the New Engineering Contract System, the responsibility for measurement lies with the Employer's Project Manager or the Supervisor. There is therefore little scope for the use of overmeasurement by the contractor in improving cash flow on the project.

Back-end rate loading and undermeasurement

Back-end rate loading is the opposite of front-end rate loading. Back-end rate loading is the device whereby the later items in the 'bill' carry a higher margin than the earlier items. Undermeasurement is the opposite of overmeasurement. Undermeasurement is the situation whereby the amount of work certified in the early months of a contract is less than the amount actually done. Both these devices have the effect of increasing the capital lock-up and, in most circumstances, contractors do not seek these situations.

Delay in receiving payment from client

The time between interim measurement, issuing the certificate and receiving payment is an important variable in the calculation of cash flows. Although 'monies out' goes to many destinations, e.g. labour, plant hirers, materials suppliers and subcontractors, the 'monies in' comes from only one source—the client. Thus, any increase in the delay in receiving this money delays all the income for the contract with a resulting increase in the capital lock-up. The time allowed for this payment is specified in the contract and normally interest is charged if payment is late. This may act as an incentive to the client but, if they are slow in paying, it is the contractor who has to find the cash.

Delay in paying labour, plant hirers, materials suppliers and subcontractors

The time interval between receiving goods or services and paying for these is the credit the contractor receives from suppliers. A one-week delay is normal in paying labour and anything between three and six weeks is normal in paying plant hirers and materials suppliers. Any increase in these would reduce the capital required to fund a contract. However, these times evolve as part of the normal commercial trading arrangements and any increase in these times may undermine commercial confidence in the company. Hence, these factors are not usually seen as suitable for controlling capital lock-up in a contract.

Company cash flow

The additional key factor that affects the company cash flow is the timing of contract starts. If several new contracts start within a few weeks of each other then the maximum negative cash flow for each contract may occur at around the same time and the cumulative effect of several contracts demanding cash can cause difficulties.

Interim valuations and cash flow

Interim valuations are a feature of all major forms of contract and the valuations mechanism represents the only source of income to the contractor for that particular project. Thus, the importance of the chore of preparing interim valuations is clear. Differences exist between the different forms of contract as to which of the parties is responsible for the preparation of interim valuations. Under the New Engineering Contract (NEC: ECC 1995) the responsibility for preparing such interim valuation is placed on the employer's appointed project manager (Clause 50.1). The Project Manager's valuation takes account of any application for payment made by the contractor. The valuation also considers the current forecast of actual cost (which represents the direct cost of materials, labour and plant to execute the project) for the works submitted by the contractor. These interim payments are of such critical importance to a contractor's cash flow that no major contractor would be content to leave an interim valuation to the client's representative, even when the contract permits this. Thus, contractors are faced with undertaking interim valuations, usually at monthly intervals, although this can vary.

The importance of these valuations is that they:

- Control the contractor's cash flow
- Provide financial information
- Serve as information on the general progress of the works

The elements in a valuation may include:

- Preliminaries
- Insurances
- Measured works
- Dayworks
- Variations
- Unfixed materials
- Statutory fees
- Nominated subcontractors
- Price adjustments
- Claims
- Retention
- Advanced payments to the contractor
- Ex-gratia awards

The importance of each of the above elements varies with different forms of contract. The New Engineering and Construction Contract, for example, makes provision for method-related and time-related charges, which is helpful in valuing work done under preliminaries (Clause 50.2, 50.3 and 52.1). However, in all forms of contract the largest cash flow is likely to be derived from the *measured works*.

There are three types of site measurement:

- Measurement of work contained in the bill items
- Measurement of activities
- Determination of stages of construction

Measurement of work contained in the bill items

In contracts using bills of quantities the valuation of the work contained in the bill items is obtained by:

(1) Measuring the dimensions
(2) Determining the quantities
(3) Assessing the value earned
(4) Assessing the future expenditure to completion
(5) Assessing the percentages completed
(6) Assessing the completion of milestones

Alternatively, a mixture of all the five methods could be applied. The five methods are described in more detail below.

Measuring the dimensions

The physical dimensions of the works are measured and converted into quantities as used in the bill. Examples would be linear items such as kerbing, superficial area items such as site clearance and cubic items such as volumes of concrete placed.

Determining the quantities

Quantities may be derived from working up dimensions or assessed directly to determine the quantities on which payment is based as follows:

- Volumes (e.g. concrete in pad foundations), areas (e.g. brickwork) or lengths of physical work (e.g. kerbs, road marking)
- Weights (e.g. tonnes of reinforcement)
- Number of itemised units (e.g. compressor unit)
- Time (e.g. weeks of supervision)
- Sums of money (e.g. insurances)

Assessing the value earned

Value earned may be from lump-sum items and assessing the value depends on identifying the completed lump-sum items and totalling the amounts due. All prime cost (PC) items are assessed in this way.

Assessing the future expenditure to completion

One alternative to measuring completed work is to assess the remaining work to be done and deduct this assessment from the amount due. This alternative is particularly suitable where the project does not have a priced bill of quantities.

Assessing the percentages completed

Another alternative to measuring completed work is to visually assess the percentage completion of individual or groups of bill items. This is particularly suitable for enumerated items.

Assessing the completion of milestones

Some projects define appropriate milestones that then form a target for payments. Once a milestone is satisfactorily attained, the related payment is made.

Procedure in a bill of quantities contract

The procedure for preparing and accepting interim valuations under the New Engineering Contract Document: Engineering Construction Contract (NEC: ECC Options B and D) is briefly as follows; references to the pertinent clauses in the NEC: ECC form of contract are given.

Monthly statements (Clauses 11.2; 50.1)

The contractor shall submit to the engineer, at agreed time intervals enshrined within the contract (usually monthly), a statement showing (Clause 52):

(a) The estimated value of permanent works executed up to the end of the month
(b) A list of goods and materials delivered to the site
(c) A list of any goods the property of which has been vested in the employer pursuant to Clause 70
(d) The estimated amounts in respect of temporary works or construction plant and all other matters for which separate amounts are included in the bills of quantities

Monthly payments (Clause 50.2)

The documentation on the progress of works submitted by the contractor should be checked and verified by the employer's project manager or supervisor. This can form the basis of assessment of the works, which the project manager has to undertake at periodic intervals. A monthly assessment interval is the common practice. On some projects, however, the assessment interval is derived from the project's milestones. The Project Manager has to issue a certificate of assessment for work done to date by the contractor within one week of the assessment of works so that the employer can pay the contractor within 21 days after each certification by the project manager. A copy of the certificate of assessment is also sent to the contractor detailing the basis for arriving at the amount that in the opinion of the project manager is due to the contractor.

Retention (Clause 50.3)

Interim payments made to the contractor are subject to the deduction of retention as detailed in Option P of the NEC: ECC Document. The limit of retention is not specified but it is common practice to have 5% of the amount due to the contractor retained until a reserve shall have accumulated in the hands of the employer up to a limit equal to 3% of the tender total.

Examples
(a) On a contract with a tender total of £48 000.00, the retention would be 5% of the total estimated value due (under (a) and (d) above) until the total estimated value under (a) and (d) exceeded £28 800.00; thereafter the retention would remain at £1440.00.
(b) On a contract with a tender total of £400 000.00, the retention would be 5% of the total estimated value due under (a) and (d) until the total estimated value under (a) and (d)

exceeded £240 000.00; thereafter the retention would remain at £12 000.00 (i.e. 3% of £400 000.00).

Other points in connection with interim certificates

The Project Manager is entitled to demand from the contractor reasonable proof that all sums (less retention) included in previous certificates for subcontractors have been paid (NEC: ECC Clause 52). The Project Manager or the Supervisor has power to omit from any certificates the value of any work and materials with which they are dissatisfied (NEC: ECC Clause 50.2). Where there is more than one period of maintenance (such as periods for sectional completion), the expiration of the maintenance period is to be taken as that of the latest period (NEC: ECC Option L). The procedure for preparing and accepting interim valuations under the Joint Contract Tribunal (JCT) contract is briefly as follows; references to the pertinent clauses in this form of contract are given.

Interim payment (Clause 30.1)

The quantity surveyor is given responsibility for preparing valuations subject to the authority of the architect, who can authorise valuations where they consider them necessary.

In practice, however, the frequency of valuations is monthly. The contractor is under no contractual obligation to assist in the preparation of the valuation. The contractor's own self interest is of course a different matter.

The amounts to be included and subject to retention are the values of:

- Main works and variations
- Materials and goods delivered or off-site if authorised by the architect
- The above items with respect to nominated subcontractors
- The contractor's profit, etc. on the nominated subcontracts

Other items included but not subject to retention include statutory fees, testing, royalties, authorised remedial works, general payments to nominated subcontractors, adjustments under fluctuation provisions and similar payments to nominated subcontractors.

The employer is required to pay the contractors within 14 days of the issue of the valuation certificate by the architect. The contractor is obligated to pay the nominated subcontractor within 17 days from the date that the architect issues the certificate.

Retention (Clauses 30.4 and 30.5)

The contract will state the amount, which is normally 5% on contracts up to £500,000 or 3% on larger contracts.

Valuation and certificate forms

Quantity surveyors normally use the forms issued by the RICS for the preparation of valuations. The architect uses a certificate form prepared by the RIBA for the issue of the interim certificate.

Other forms of contract

These include the GC/Works/1 (1998) and the PSA/1 (TBV Consult 1994). The GC/Works/1 family of contract documents is the standard Government forms of contract for major and minor UK building and civil engineering works. The PSA/1 is essentially an amended version of the GC/Works/1 for use by both the private and public sector. Interim valuation procedures for both forms of contract reflect the JCT approach. The main difference is that the project manager replaces the Architect.

Price adjustment

The amount due to the contractor at each valuation is adjusted to reflect variations in the cost of materials and other input resources as a result of changes in the value of money due to inflation. This is done by applying a price adjustment factor to cater for:

- The change in the price for work done to date since the last assessment
- The amount for price adjustment included in previous amount due

Further details and examples on the use of the price adjustment factor can be obtained by referring to the NEC: ECC Option N (N2 and N4), and also to the NEC: ECC Guidance Notes on Option N.

Measurement of work in activities

Some forms of contract in use measure work completed by activities in a schedule and not bill items. These contracts require the creation of a works activity schedule. This is a schedule of activities, which can be derived from the planning schedule of activities. The contractor is paid on the completion of the activities in the schedule.

This type of contract holds the attraction that the contractor's management control documents encompassing the estimate, the plan, the budget and payment documents can relate to the same basic documents—a plan based on the works activity schedule. Contracts of this type are covered in the NEC: ECC by Options A and C.

Computers and cash flow

The basic data required to compute a cash flow for a single project are value versus time and cost versus time. These are represented by rows 1 and 5 in Fig. 12.3. The tedium of cash flow calculations is producing these data. As previously discussed, there are shortcuts to producing these data which depend on the estimation of the value and costs at various time points. If these estimates can be made then cash flow calculations as illustrated in Fig. 12.3, or variations thereof, can easily be set up and executed by any *spreadsheet* package. These will allow the manipulation of the data to provide the cash flow. However, this approach does involve an element of 'estimation' of the value and cost data into the calculations.

Current practice in most contractors' organisation involves generating computerised versions for both the estimate and the plan. This provides the more attractive possibility of combining the estimator's data with the planning data to produce a more accurate cash flow.

In the estimator's data there are for each bill item or group of bill items the following data:

- Total and unit costs
- Allowances for profits, overheads and other additions
- Labour, plant, materials and subcontractors' costs and quantities

The estimator's data may also contain the duration of the operation described in the bill item but it will be unlikely that the dates within which the operation will take place will be present. What is missing as far as cash flow calculations are concerned is the timing of all the operations.

Contained in the planning data are data relating to each activity which will include:

- The description
- The duration
- The relationship with other activities
- The resources allocated to each activity

Nowadays, it is normal to find computer-based planning systems in which provision is made to cost the resources and to add overheads and profits. In these integrated planning systems estimates can be built up activity by activity and hence the cash flow is produced with no greater effort. However, this approach has not proved sufficient for bill-of-quantities type contracts. Thus, the existence of separate estimating and planning packages is still usual. This requires the allocation of bill items to activities in the plan.

The process of allocating bill items to activities is required to determine the timing of each item. Thus, all the costs and revenues associated with each bill item are now fixed into a time slot within the plan. Figure 12.5 illustrates the relationship between the data. The contract details come from the estimating system.

The process of allocating bill items to activities means that the activities now have collections of bill items attached. To produce cash flow forecasts, additional data need to be included in the payment conditions. These are the contract payment conditions, credit delays or trading conditions and bank interest rates. Examples of these are given below:

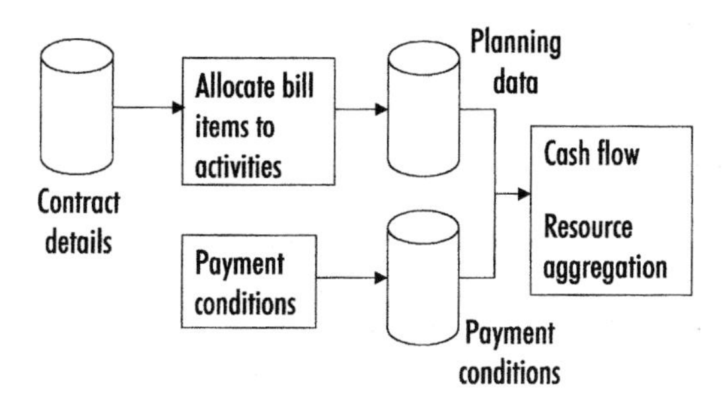

Fig. 12.5 Relationship between data on computer systems for cash flow calculations.

Contract conditions		
		Weeks
	Week of first certificate	4
	Interval between certificates	4.3
	Week of practical completion	20
	Week of end of maintenance period	30
	Delay between certificate and payment	2
	Retention	5.0%
	Maximum retention	5.0%
	Retention during maintenance period	2.5%.
	Trading conditions	
	Current delays	Weeks
	Labour	1
	Plant	4
	Auxiliary plant	4
	Material	5
	Subcontract	4
	Additional	0
	PC/Provisional sums	0

Banking conditions		
	Earning rate of interest (yearly)	4.50%
	Earning rate of interest (weekly)	0.08%
	Paying rate of interest (yearly)	11.0%
	Paying rate of interest (weekly)	0.20%

With the data contained within the estimate and the planning data now combined and augmented by the payment conditions, there is sufficient information to calculate:

- Cost breakdown
- Payments, receipts and values
- Cash flow

Cost breakdowns for labour, plant, materials and subcontracts can now be calculated week by week. Previously within the estimate these would be available, item by item, section by section, trade by trade but **not** on a time basis.

The *payments* due each month could be forecast together with the associated receipts and values. This is really a preliminary report prior to calculating cash flow.

Cash flow report giving the cash week-by-week and the receipts month-by-month, the net cash flow and the cumulative cash flow with or without the effects of interest charges. The importance of a cash flow calculated in this manner is that it is based on the inputs of the estimator and planner. It excludes an assumption of what the value- and cost-versus-time curves might be. If the estimator made certain assumptions, they would be reflected in this cash flow. If the estimator changes the assumptions, the cash flow can be recalculated.

There have been many attempts to produce quickly calculated cash flows. Little evidence exists that for accuracy a cash flow forecast based on a planning schedule and accurately calculated cost and values can be improved upon. Given computing power, this approach can compete with the approximate methods for speed.

Cash flow forecast by standardised models

The availability of databases of project histories has enabled the development and rapid growth in use of cash flow forecasting models. These are designed to use the expenditure and income from past projects to establish the cash flow profile for project types or a company. The developed profile then serves as the framework for making a forecast for any new proposed project over its planned life. Typically, the models present a user interface that enables a user to enter estimated project data, which is then computed by the software to generate a forecast profile of cash inflows and outflows for the proposed project. The underpinning assumption of most of the models is that the cumulative expenditure and income profile for any project will follow the standard S-curve. However, a contractor's history of past projects could be employed to develop more representative algorithms of a generic profile of project expenditure and income that reflects its industry conditions. Once that initial forecast is made, the conditions and assumptions can be varied to generate a new forecast that best reflects the circumstances and real-life conditions of the proposed project.

A good project forecasting model will also produce at least one cash flow for the forecast, along with value drivers and a list of critical variables that drive the forecast. It should be understood that the default forecast produced by any model is usually not as reliable a forecast for the actual project being planned. The advantage presented by the model stems from its capability to provide sensitivity analysis, where the user, after making the default forecast, can alter the inputs and conditions for the proposed project in a 'what-if' analysis to support and identify potential financial risks associated with the project. Examples of such models include the work of Kaka and Lewis (2003) and Khosrowshahi (2000). It is quite common for most project management software to incorporate a certain level of cash flow forecasting capability.

Concluding remarks

Cash flow forecasting provides a valuable early warning system to predict possible insolvency. This enables preventative measures to be considered and taken in good time. Examples of the actions available are:

- Not taking on a new contract if, when the contract is included in the cash flow forecast, the company's projected cash requirements are much more than the overdraft limit
- Renegotiation of overdraft limits supported by reliable forecasts
- The adjustment of the work schedules of existing contracts
- The negotiation of extended credit with some suppliers
- Accepting suppliers' full credit facilities even if it means temporarily losing some discounts

The calculations described are undertaken either in spreadsheet application software or with bespoke computer systems to provide greater calculating capability and make it simpler to include all the company's projects in a regular forecast.

Experience has shown that the discipline of reviewing all projects for cash flow forecasting has a rewarding spin-off in that the attention of senior management is attracted to projects requiring their attention. Another by-product includes total current workload figures, which are useful as a guide to the intensity of tendering efforts required. Nowadays, most commercial computer packages for project management and estimating include facilities for generating and updating cash flow forecasts.

In a cash flow forecasting system it may be useful to set up a funding account for the plant department and pay the depreciation recovered from internal hire into the funding account for replacement. This will alert the company to the danger of replacement costs exceeding the sum of depreciation.

References

Kaka, A. & Lewis, J. (2003) Development of a company-level dynamic cash flow forecasting model (DYCAFF). *Construction Management and Economics*, **21**(7), 693–705.

Khosrowshahi, F. (2000) *Proactive Project Financial Management*. Chartered Institute of Building, Ascot.

TBV Consult (1994) *General Conditions of Contract for Building and Civil Works PSA/1 (with Quantities)*. HMSO, London.

McCaffer, R. (1976) *Contractors' Bidding Behaviour and Tender Price Prediction*. PhD thesis. Department of Civil Engineering, Loughborough University of Technology.

13
Economic assessments

Summary

Interest relationships, present worth (PW), equivalent annual costs, discounted cash flow (DCF) yield, inflation, accuracy of future estimates, sensitivity and risk analysis, tax allowances, worked examples, life-cycle costing, cost–benefit analysis and financial modelling.

Introduction

Economic assessments, or investment appraisal techniques, form the basis for answering two main questions:

- Which proposed scheme is the most economic?
- Are any of the proposed schemes profitable enough?

This chapter briefly describes investment appraisal techniques and gives examples appropriate to the construction industry. The essential difference between traditional economic calculations and those of modern investment appraisal is the latter's recognition of the effect of the time value of money: that is, the effect of interest rates on the method of assessment. In more recent times, the use of other non-financial criteria has become commonplace in the appraisal of projects, particularly the ones funded by the public. Current use of these techniques has been predominantly by the industry's clients or their professional advisors for appraising the feasibility of projects. However, the use of these techniques is growing among contractors because of the changing nature of the contracting business within the construction industry. For example, the involvement of contractors in the Private Finance Initiative and the increasing reliance on design-and-build contracts by the industry's clients necessitates the use of such appraisal techniques on the part of the contractor. Equally, the client's search for greater economy often results in contractors providing alternative bids that offer greater economic viability. In addition, there are a number of specific situations for which the application of economic appraisal techniques can be useful to the construction contractor. These are discussed below:

(1) *The economic comparison between different items of plant or methods.* Many of these comparisons relate to the short term and therefore require no more than a straight comparison of cost. At site level the engineer will be constantly appraising the different methods possible without involving interest rates in such comparisons. This is entirely valid since most site operations are of short duration and the effect of interest calculations is minimal. On

choices of method relating to activities of long duration, that is several years, or in selecting between items of plant that will operate for several years, then the principles of modern investment appraisal should be used.

(2) *Recovery of capital invested in plant through either external or internal hiring.* It is sometimes alleged that the contractor's projects make capital, which is subsequently reinvested in plant, which is then run on a breakeven basis. It is essential that all capital, including the capital invested in plant, earn an adequate return.

(3) *Assessing the replacement age of plant items.* Although in some cases the mechanical performance of plant determines when it should be replaced, in other cases the economic factors prevail. The techniques of modern investment appraisal can be used to identify the age range at which the plant should be replaced.

(4) *Decisions relating to utilisation.* The principles of sensitivity analysis described in this chapter can be used to determine the effect of utilisation on the economic viability of plant items. Sensitivity analysis is also useful for determining the utilisation rate below which it is uneconomic to keep an item of plant.

(5) *The effect of corporation tax and development grants on the economics of plant ownership.* A plant item can be seen as an investment leading to a series of returns; these returns will be subject to corporation tax and the effect of this will reduce the plant items' return on capital. In some regions development grants may operate and these will improve the plant items' return on capital.

The material presented in this chapter will provide the theoretical basis for all the above calculations. The chapter will be of interest to engineers who spend some time with consulting engineers (usually preparing for their professional interview). However, the specific items that will help contractors' staff to answer the problems set out above are as follows:

(a) For economic comparison of long-duration construction methods and choice between plant items—present worth and equivalent annual cost techniques. These techniques are also useful in determining the form of ownership, e.g. hire or buy or lease.

(b) For capital recovery of capital invested in plant—capital recovery factors are tabulated in the appendix to the chapter and explained within the chapter. Also, the setting of plant hire rates requires knowledge of interest calculations. The rate of return on any investment is now commonly measured by discounted cash flow (DCF) yield; the calculation of DCF yield and the meaning of this measure of profitability are fully explained.

(c) For determining the economic replacement age of plant items—equivalent annual costs are used to find the balance between increasing repair and maintenance costs and decreasing resale value. An example is given in this chapter.

(d) For frequency of maintenance—sensitivity analysis producing sensitivity graphs. Maintenance is just one factor in determining the costs of running an item of plant. The effect of this one factor on the rate of return can be determined by producing a sensitivity graph from calculations based on a range of different maintenance costs. An example of the sensitivity of rate of return to the utilisation factor is given in Fig. 13.6(b). The utilisation factor is of crucial importance to the profitability of most hired plant.

The sensitivity analysis can be extended from the effects of variability in one factor into a risk analysis where the variability of all the factors that affect plant profitability can be studied. These factors would include maintenance cost, utilisation rate, hire rate, capital cost, resale value, fuel and running costs, etc. This chapter describes the principles of both sensitivity and risk analyses.

The contents as set out explain the basis of these calculations and where appropriate give suitable examples.

Interest

The major difference between modern investment appraisal and historical rule-of-thumb methods of appraising projects is the recognition that interest rates are substantial enough to warrant being included in the appraisal. The inclusion of interest calculations results in the need to know when cash transactions take place. A cash flow is the transfer of money out of (negative cash flow) or into (positive cash flow) the project in question at a known or forecast point in time. Knowledge of the amount and time of cash flows representing a project permits the use of modern investment appraisal techniques. Previously, if interest calculations were ignored, a choice between two pumping schemes might be made on, say, the initial investment, which could lead to installing the smallest pump possible. Alternatively, the choice could be made on the total projected expenditure, made up of the initial investment plus the projected annual running costs multiplied by the number of years the scheme would operate. This could lead to installing the largest pump possible. Neither method of appraisal is satisfactory since the first ignores running costs and the second exaggerates their effect. To balance the effect of initial sums and the future running costs, the time value of the money involved must be taken into account through interest rates representing the cost of capital. The cost of capital is the interest rate paid for borrowing capital; to a company it is a weighted average of the dividends paid to shareholders for equity capital and interest paid on loans.

The manipulations of cash sums and interest rates are achieved by the use of the tabulated factors given in Appendix A13.1 at the end of this chapter. This appendix gives only the factors for 10% and 15%, but the factors for other interest rates can be calculated and the expressions needed are also given.

The use of the tables can be described by six short examples, which are given in the same order as the factors are tabulated.

1. Compound amounts

Figure 13.1 presents a graphic illustration of compound amount. If a sum of money, £1000, is invested in an account for some years, say five, at an interest rate of 10%, the amount that can be withdrawn at the end is:

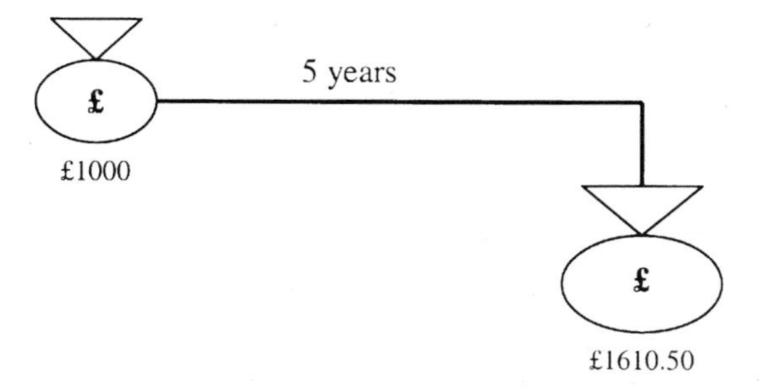

Fig. 13.1 Relationship between present and future value of an amount of money.

$$£1000 \times 1.6105 = £1610.50$$

The factor 1.6105 is obtained from the tables.

2. Present worth

This is the inverse of the compound amount. If a sum of money, £1610.50 is required in five years time, what amount has to be invested today to generate this amount if interest is 10%?

$$£1610.50 \times 0.62092 = £1000$$

The factor 0.62092 is obtained from the tables.

Figure 13.1 also illustrates the relationship between the compound amount and the present worth. The £1000 is called the present worth (PW) of £1610.50 in Year 5, as it is today's equivalent of £1610.50 in year 5. Having £1000 today is the same as having £1610.50 in five years because the £1000 could earn £610.50 of interest in five years. This provides a convenient way of comparing two different sums of money occurring at different times.

3. Compound amount of a regular or uniform series

Just as example 1 gave the amount generated by investing one sum for a specified period at a given interest rate, this factor gives the amount generated by investing a regular sum each period for a specified number of periods at a given interest rate, as illustrated in Fig 13.2. Each sum earns interest for a different time period. It would be possible to use factor 1 to calculate the amount generated by each individual sum and finally add them all together. The compound amount factor for a regular series does the calculation in one step. Thus investing £100 each year of six years at an annual rate of 15% will generate a total amount of £875.30 i.e.:

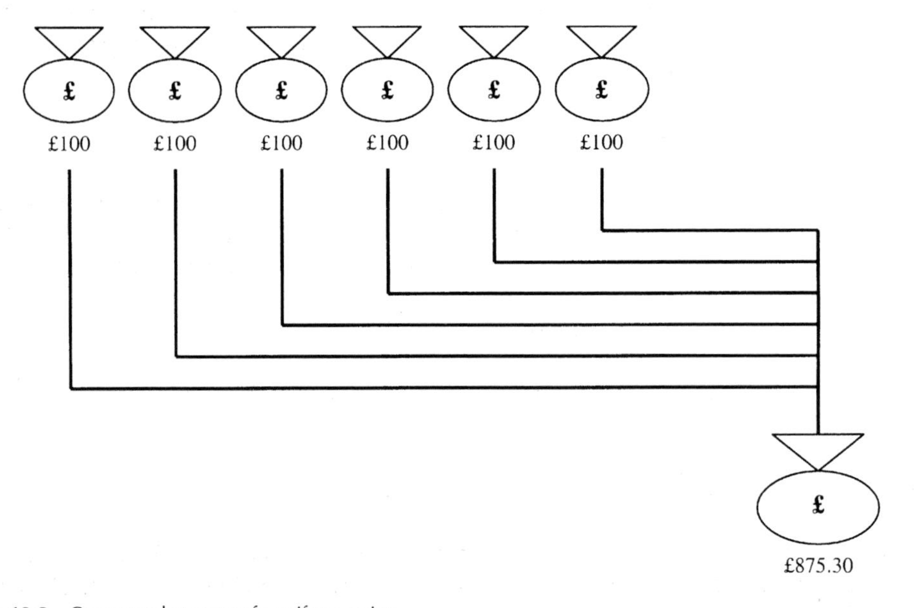

Fig. 13.2 Compound amount of a uniform series.

$$£100 \times 8.753 = £875.30$$

where 8.753 is obtained from the tables.

4. Sinking fund deposit

This factor allows the calculation of the amount needed to be invested each period in order to generate a specified amount at some future date. It is the amount required to be regularly deposited in a sinking fund to pay off a debt. For example, if £500 was owed in five years' time, the amount that would have to be deposited each year is £81.90, provided 10% can be obtained.

$$£500 \times 0.16379 = £81.90$$

where 0.16379 is tabulated. Note that £81.90 is less than the £100 per year needed if no interest was earned.

5. Present worth of a regular or uniform series

Just as factor 2 calculated the PW worth of an individual or lump sum, this factor calculates the present worth of a regular series. Thus the PW of £100 each year for five years, given an interest rate of 15%, is £335.21.

$$£100 \times 3.3521 = £335.21$$

where 3.3521 is tabulated. That is, £335.21 is the equivalent today of £100 each year for the next five years; or £335.21 is the amount needed to be invested today to pay £100 each year over the next five years, provided an interest rate of 15% can be obtained.

This factor provides a method of comparing different regular series in combination with different lump sums. Together with factor 2, this factor provides the basis for the comparison of different projects on a PW basis.

6. Capital recovery

This factor calculates the amount to be recovered each period from an investment in order that the invested capital and a specified return can be recovered. If £500 is invested, the amount that must be recovered each year for the next five years to ensure a return of 10% is £131.90.

$$£500 \times 0.26379 = £131.90$$

where 0.26379 is tabulated. This factor is useful in calculating plant hire capital.

Economic comparisons

The object of economic comparison is to compare the cash flow that will result from one course of action with the cash flows that will result from another course of action. An example is in comparing the cost of buying an item of plant with leasing it. On comparing two or more different plant arrangements for, say, a pumping scheme where the initial capital expenditure and running costs are different. The only difficulty in comparing such cash flows is that they occur at different times. It is necessary, therefore, to manipulate the cash flows into a form that can be compared. The most popular is the PW method of comparison and an alternative method is that of comparing expenditures on a basis of *equivalent annual costs*. The PW method simply converts all cash flows to time now. Equivalent annual cost allocates the initial capital investment to each year of operation.

Present worth

The typical use of PW is to compare two (or more) schemes, each with a different initial investment and different running costs. For example, consider Schemes 1 and 2 and the cash flows estimated at today's prices pertaining to buying and operating an item of equipment.

	Scheme 1	Scheme 2
Initial plant cost	£5000	£4000
Running cost	£500 p.a.	£800 p.a.
Life	6 years	6 years

If interest rates are 10%, the present worth of these schemes are

Scheme 1 PW = £5000 + £500 × 4.3552 = £7177.60

Scheme 2 PW = £4000 + £800 × 4.3552 = £7484.16

The factors were obtained from the tables.

Thus, since Scheme 1 has the smaller present worth, it is said that Scheme 1 is the more economic. What is being compared is the £1000 extra initial investment of Scheme 1 with the £300 extra running costs in Scheme 2. In other words, is £1000 now more or less than £300 each year for the next six years with obtainable interest rates being 10%? From previous calculations £1000 now is £306.56 less than £300 each year for six years.

The above comparison is valid because both schemes have the same life. However, if two schemes with different lives are being compared, it is necessary to take steps to have an analogy that presents two schemes of the same life. For example, compare Scheme 1 from above and Scheme 3 below.

	Scheme 3
Initial plant cost	£3000
Running cost	£500 p.a.
Life	3 years

The PW of Scheme 1 is £7177.60; the PW of Scheme 3 is £3000 + £500 × 2.4868 = £4243.40. However, this £4243.40 represents only three years of service whereas £7177.60 represents six years of service. It is necessary to consider a replacement for Scheme 3, and since all these

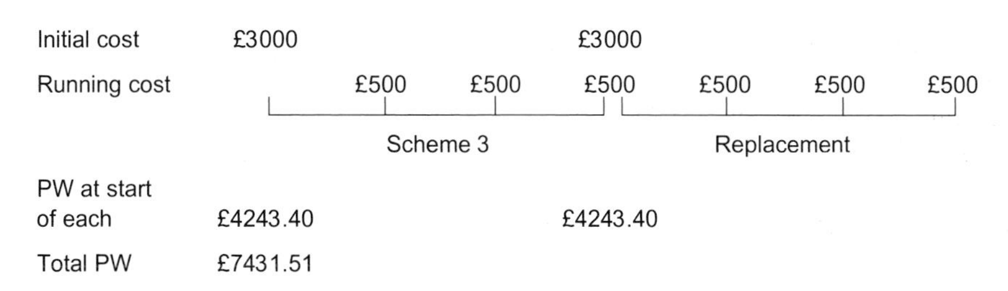

Fig. 13.3 Relationship between total present worth and the present worth of the scheme and its replacement.

estimates have been made at today's prices the replacement costs the same as the first plant item. The PW of Scheme 3 and its replacement becomes

$$\text{PW Scheme 3 and replacement} = £4243.40 + £4243.40 \times 0.75131 = £7431.51$$

where 0.75131 is the PW factor at Year 3. Figure 13.3 sets out the reasoning for this calculation. Clearly six years and three years have been chosen to illustrate the above. If six years and four years were being considered, then one replacement of the six-year scheme and two replacements of the four-year scheme would produce two schemes of the same life.

Present worth example

As a further example, the following cash flows pertain to buying and running an item of plant and the alternative of hiring it. The PW calculations are based on an interest rate of 10%.

		Buying		Hiring
Initial cost		£8000		
Running cost		£800 p.a.		£2 100 p.a.
Life		7 years		
PW of buying	=	£8000 + £800 × 4.8684	=	£11 894.72
PW of hiring	=	£2100 × 4.8684	=	£10 233.64

Therefore hiring is more economical.

Equivalent annual costs

These comparisons are based on calculating the equivalent capital cost each year of the project. Using Scheme 1 the equivalent annual cost is

$$£500 + £5000 \times 0.2296 = £500 + £1148 = £1648$$

The £500 is already an annual running cost; the £1148 is the annual equivalent over six years of £5000 at 10%. The factor used is the capital recovery factor for 10%. This, compared to the equivalent annual cost of £800 + £4000 (0.2296) = £1718.40 for Scheme 2, shows that as a means of differentiating between two projects the result is the same as a present worth comparison.

The equivalent annual cost for Scheme 3 is £500 + (£3000 × 0.40211) = £1706.33. Even if a replacement is considered, the annual cost will still be £1706.33.

Equivalent annual cost example

The example used is the same as that for the PW example above.

Equivalent annual cost of buying = £800 + £8000 × 0.2054 = £2443.20

Equivalent annual cost of hiring = £2100

Thus hiring is more economical, as the PW comparison also showed.

Profitability measures

The purpose of economic comparison is to differentiate between two or more projects in order to choose the least expensive. Profitability measurement applies to projects where an investment leads to a return; it is important to determine if this return is adequate. A trite example is an investment of £100 and a return of £110 compared with an investment of £1000 and a return of £1011. While the net profit is larger in the second case, the return on capital will not be satisfactory. Since capital costs money in the form of interest charges or dividends to shareholders, it is necessary to measure the return on capital and compare it to the cost of capital, thereby determining whether the return is adequate. The generally accepted way of measuring return on capital is the DCF yield, sometimes called the internal rate of return.

DCF yield

One definition of DCF yield is that it is the 'maximum' interest rate that could be paid on borrowed capital assuming that all the capital needed to fund the project is acquired as an overdraft. An example of this definition is given below:

	Year	Project's cash flow (£)	Interest paid on borrowed capital at 8% (trial estimate) (£)	Borrowing account (£)
Investment	0	–1000		–1000.00
Return	1	+400	–80.00	–680.00
	2	+400	–54.40	–334.40
	3	+400	–26.75	+38.85

Explanation At the outset £1000 was borrowed; in Year 1 this costs £80; at the end of Year 1 the account is –£1000 – £80 + £400 = –£680.

This £680 costs £54.40 in Year 2, and at the end of Year 2 the account is –£334.40. This –£334.40 costs £26.75 and at the end there is £38.85 remaining. The definition requires the maximum interest rate that could be paid on borrowed capital. Since charging at 8% leaves £38.85, then 8% is not the maximum interest that could be paid for borrowed capital by this project. A second trial would have to use a higher interest rate, as below:

Year	Project's cash flow (£)	Interest paid on borrowed capital at 10% (£)	Borrowing account (£)
0	−1000	–	−1000
1	+400	−100	−700
2	+400	−70	−370
3	+400	−37	−7

Since the amount remaining is very close to zero at −£7, it can be assumed that the maximum interest rate that can be paid for capital is almost 10%. The exact rate is 9.69%.

This figure of 9.69% represents the DCF yield; it represents the maximum that could be paid for borrowed capital. If the cost of capital is greater than 9.69% then this project does not yield enough to be profitable. Measuring the maximum that could be paid for borrowing capital is a way of measuring the amount being generated by the project. The DCF yield is therefore a profitability measure.

The more usual way of calculating the yield is to use the present worth factors as tabulated in the interest tables (Appendix A13.1) and to *redefine DCF as the interest rate which, if used to discount all cash flows, would produce a net present worth of zero.* The above project's DCF yield would be calculated as follows:

Year	Project's cash flow (£)	First trial		Second trial	
		PW factors at 9% (£)	PW (£)	PW factors at 10% (£)	PW (£)
0	−1000	1.000	−1000.00	1.000	−1000.00
1	+400	0.917	366.80	0.909	363.60
2	+400	0.842	336.80	0.826	330.40
3	+400	0.772	308.80	0.751	300.40
	Net present worth		+12.40		−5.60

Interpolating between 9% and 10% gives the interest rate that results in a net present worth of zero.

$$\text{DCF yield} = 9 + (10 - 9) \times \frac{12.40}{12.40 - (-5.60)}$$

$$= 9.69\%$$

The graph in Fig. 13.4 shows that the DCF yield is the interest rate which gives a net present worth of zero.

Net present worth

Net PW itself can be used as a means of determining whether a project is profitable enough to be considered a worthwhile investment. The interest rate used in the calculation of the net PW is usually called the criterion rate of return. The criterion rate is the assessed cost of capital to the company plus a margin in excess of cost to allow for risk. If the project's DCF yield is greater than the criterion rate, the project's net PW at the criterion rate will be positive. If the project's DCF yield is less than the criterion, then the project's net PW at the criterion rate will be negative. In

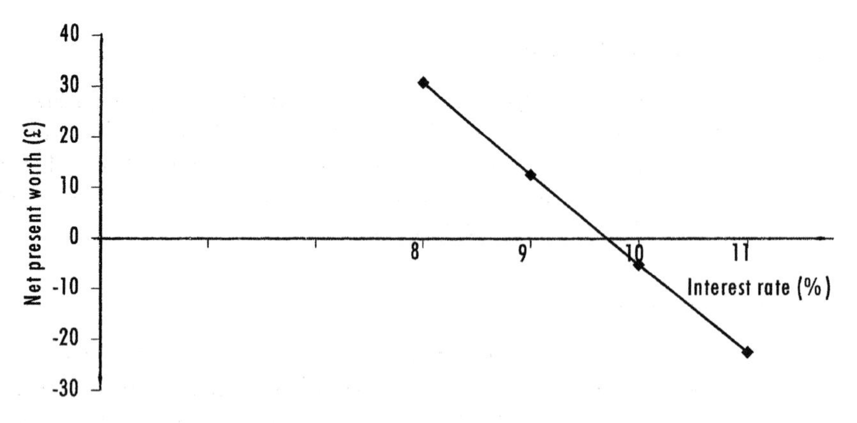

Fig. 13.4 Graph of net present worth vs. interest rate.

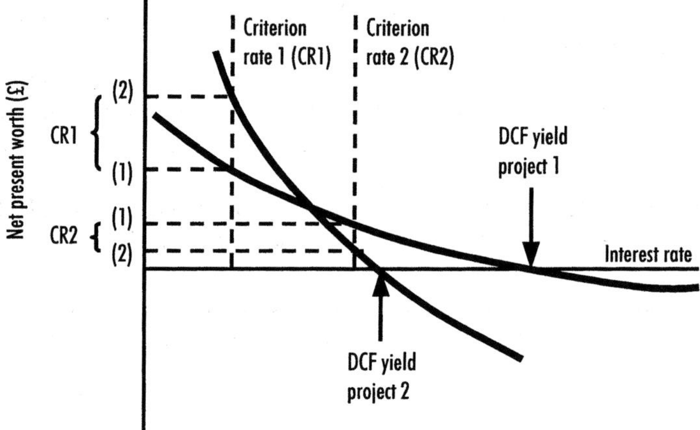

Fig. 13.5 Net PW vs. interest rates showing the differing order of net PW that can be obtained from different interest rates.

the example shown below, Project 1 has a positive net PW at the criterion rate of 15%. Project 2 does not. Therefore, in selecting projects this method is useful in rejecting the unprofitable ones.

Example—Criterion rate 15%

Project 1		Project 2	
Year	Cash flow (£)	Year	Cash flow (£)
0	−4000	0	−4000
1	+2000	1	+1700
2	+2000	2	+1700
3	+2000	3	+1700
Net present worth	+566.40		−118.56

Limitation of net present worth

It is inadvisable to depend exclusively on net PW to select between different projects. Figure 13.5 shows two projects. If the criterion rate was CR1, Project 2 would, at that rate, have the largest net PW. If the criterion rate was CR2, Project 1 would, at that rate, have the largest net PW. Since Project 1 has the larger yield, it would be inappropriate to use the single calculation of net PW at rate CR1 and imply that Project 2 has the larger yield.

Limitation of DCF yield

DCF yield calculations under some cash flow patterns can have two or more interest rates that both produce a zero net PW. These patterns are the ones that have large negative flows in the cash flow stream at times other than at the beginning. Examples are opencast coal mining and quarrying work where expensive land reinstatement is required at the end of the project and also when, as a result of tax time lags, a negative tax bill occurs at the end of a project. The solution is a dual rate calculation, and most computer packages which are available to undertake investment appraisal calculations offer the dual rate calculations as a facility.

The dual rate method introduces a second interest rate, an earning rate, assumed to be the rate at which interest can be earned on capital in a safe investment. The calculations proceed to find the maximum paying rate, as before for the fixed earning rate. The dual rate approach produces a unique yield for a given earning rate.

The example below illustrates the calculations. The paying rate calculated by the dual rate method must always be quoted with the earning rate used in the calculation.

Year	Cash flow (£)	Paying rate 10% (found by trial and error) (£)	Earning rate 5% (£)	Borrowing account (£)
0	−2000			−2000
1	+1000	−200	+32.40	−1200
2	+1000	−120	+84.02	−320
3	+1000	−32		+648
4	+1000			+1680.40
5	−1765			−0.58

(Note 0.58 taken as nearly zero)

An alternative approach for when the negative amount occurring at the end of cash flow stream is not large is to discount the negative amount by one year and add it to the previous year's cash flows, and to repeat until a positive amount is obtained. This allows the single-rate calculations to be undertaken and is the method used in the example involving corporation tax later in this chapter.

Other methods of project evaluation

The methods of DCF yield or net PW have largely replaced other methods of project evaluation or project selection. Among the other methods that exist are payback and average annual rate of return.

Payback period

This involves calculating how long it takes for a project to repay its original invested capital. The shorter the payback period, the greater is the likelihood that the project will be profitable. Projects 1 and 2 below are examples of projects with payback periods of 4 years and 2.67 years respectively; therefore, Project 2 is to be preferred to Project 1. The weakness in the method is that it does not take account of the cash flows that occur after the payback period.

Example—Criterion rate 15%

Project 1		Project 2	
Year	Cash flow (£)	Year	Cash flow (£)
0	–8000	0	–8000
1	+1000	1	+3000
2	+1000	2	+3000
3	+3000	3	+3000
4	+3000	4	+1000
5	+3000	5	+1000
Payback period:	4 years	Payback period:	2.67 years

Average annual rate of return

This method involves calculating the average return and expressing it as a percentage of the invested capital. For Project 1 above, the average annual return is:

$$\frac{1000+1000+3000+3000+3000}{5} = £2200$$

which is 27.5% of £8000. Therefore, the average annual rate of return is 27.5%. The weakness of this method is that it does not discriminate between the timings of the cash flows. This is illustrated by the fact that Project 2 above also has an average annual rate of return of 27.5%; although it is already known and simple inspection will show that Project 2 is clearly the more profitable project, because of the timings of the cash flows. The average annual rate of return should only be used in conjunction with other more discriminating methods of evaluation.

Inflation

So far it has been assumed that the forecasts of future costs and revenues have been at today's prices. Clearly, it is easier to forecast at today's prices since this removes at least one more variable from the forecast, although it also brings into question whether this is a valid approach. However, the effect of inflation can be taken into account when undertaking PW and yield calculations.

Present worth

In calculating PWs for the purposes of comparing the costs of alternative projects the effects of inflation can be allowed for in three ways:

- By ignoring inflation
- By adjusting the estimated cash flow
- By adjusting the interest rate used to discount the future cash flows

Ignoring inflation

If the purpose of calculating PWs is simply to compare proposals and to select the most economic, and if the inflation rate assumed is small, then this comparison will not be affected dramatically by the inclusion of inflation. If small allowances for inflation are added into each proposal under consideration, the effect will be to increase the PW of each proposal and the ranking of the proposals is likely to remain unaltered.

Adjusting cash flows

The simplest way to allow for inflation in PW calculations is to take the cash flows estimated at today's (i.e. year zero) prices and adjust them for an assumed inflation rate.

As an illustration, consider Scheme 1 in the original example of economic comparisons. The cash flows estimated at year zero prices and those adjusted for an assumed rate of 8% inflation each year are:

Scheme 1 Cash flows at year zero		Scheme 2 Cash flows adjusted for an annual rate of 8%	
Year	Cash flow (£)	Year	Cash flow (£)
0	5000.00	0	5000.00
1	500.00	1	540.00
2	500.00	2	583.20
3	500.00	3	629.86
4	500.00	4	680.24
5	500.00	5	734.66
6	500.00	6	793.44

The PW of the original Scheme 1 was £7177.60 calculated at an interest rate of 10%. The PW of Scheme 1 adjusted for an annual inflation rate of 8% is £7814.76. The PW is larger because it represents the capital required to be invested now, at 10%, to pay for the initial investment of £5000.00, the running costs of £500.00 per year and also the extra costs due to inflation as represented in the adjusted cash flows. If the cash flows for Scheme 2 were adjusted similarly for inflation and the PW calculated then comparison of the PWs for the two inflation-adjusted proposals would be possible.

Adjusting the interest rate

In the example above, the inflation adjustment simply increased the cash flows by 8% each year. The adjustment was made as follows, d representing the inflation rate ($d = 0.08$ for 8%).

Scheme 1		Scheme1	Scheme 1	
Year	Original cash flows (£)	Adjusted cash flows (£)	Original cash flows: inflation adjustment (£)	
0	5000.00	5000.00	5000.00	
1	500.00	540.00	500.00	$\times (1 + d)^1$
2	500.00	583.20	500.00	$\times (1 + d)^2$
3	500.00	629.86	500.00	$\times (1 + d)^3$
4	500.00	680.24	500.00	$\times (1 + d)^4$
5	500.00	734.66	500.00	$\times (1 + d)^5$
6	500.00	793.44	500.00	$\times (1 + d)^6$

The PW for each adjusted cash flow is calculated by multiplying the adjusted cash flow by the PW factor as, for example, for Year 4:

Year	Cash flow		PW factor
4	$£500 \times (1 + d)^4$	$\times$	$\frac{1}{(1+i)^4}$

where i is the interest rate.

This can be generalised for any year as:

Year	Cash flow		PW factor
n	$£500 \times (1 + d)^n$	$\times$	$\frac{1}{(1+i)^n}$

This calculation can be simplified by substituting $(1 + d)^n (1 + e)^n$ for $(1 + i)^n$, where d is the inflation rate and e is calculated so that:

$$(1 + i)^n = (1 + d)^n (1 + e)^n$$

giving

$$(1+e) = \frac{(1+i)}{(1+d)}$$

and

$$e = \frac{(1+i)}{(1+d)} - 1$$

Using this substitution, the PW calculation becomes:

Year	Cash flow		PW factor
n	£500 × $(1 + d)^n$	×	$\frac{1}{(1+d)^n(1+e)^n}$

The elements $(1 + d)^n$ cancel and the calculation is reduced to:

$$£500 \times \frac{1}{(1+e)^n}$$

This is the same as the calculation of PW on the original cash flows except that the interest rate used is *e* instead of *i*. The effect of inflation has, therefore, been accommodated by adjusting the interest rate. If *i* were 10% and *d* were 8% then:

$$e = \frac{(1+0.1)}{(1+0.08)} - 1 = 0.01852 = 1.85\%$$

If the PW of the original cash flows were calculated using an interest rate of 1.85% the PW would be £7814.76, which is the same as that achieved by adjusting the cash flows themselves. Adjusting the interest rate is a quicker method of achieving the same effect. The 1.85% calculated is measuring the interest earned in excess of the inflation rate.

If *i* were 10% and *d* were 10% then *e* would be zero and the PW including inflation would be simply the sum of the original cash flows. If *i* were 10% and *d* were 15% then *e* would be –4.34% and the PW factors would be greater than unity.

This method of adjusting the interest rates is the most common method employed in allowing for inflation in economic comparisons.

Yield calculations

If the cash flows used to calculate the DCF yield were estimated at present day prices then this yield would not reflect the effect of inflation. If the cash flows were adjusted for inflation then the yield calculated would be larger than the yield calculated on the original cash flows by an amount equivalent to the inflation rate. The example presented here shows the original example used with a yield of 9.69% on the original cash flows and the original cash flows adjusted for inflation at the rate of 8% per annum which gives a yield of 18.45%:

	Original cash flows		Cash flows adjusted for inflation at 8% per year
Year	(£)	Year	(£)
0	–1000.00	0	–1000.00
1	+400.00	1	+432.00
2	+400.00	2	+466.56
3	+400.00	3	+503.88
Yield =	9.69%	Yield =	18.45%

The yield calculated on the *uninflated cash flows* is called the real yield or real rate of return because it contains no effects from inflation and the yield calculated on the *inflated cash flows* is called the apparent yield or apparent rate of return because it appears larger than it actually is, due to inflation. The relationship between the real rate (i_r), the apparent rate (i_a) and the inflation rate (i_d) is similar to the relationship used in the PW calculations and is:

$$(1 + i_a) = (1 + i_r)(1 + i_d)$$

Most proposed investments are appraised on estimates based on today's or year zero prices and the rate of return calculated and used to judge a proposal's viability is therefore normally the real rate of return. It is assumed that, as the project is executed, inflation will push up both the costs and the revenues so that the achieved apparent rate will become larger than the originally calculated real rate; also, when the effects of inflation are removed the achieved real rate will be at or near the original value. However, if inflation pushes up costs more than revenues, the apparent rate may not be sufficiently large to give the original real rate. The cash flows recorded as projects are taking place are normally based on the transactions that occur at current prices and thus the recorded cash flows have inflation included as a matter of course. The yield calculated on these recorded cash flows, therefore, is the apparent rate of return. It is therefore important to distinguish whether the cash flows are based on constant, year zero, prices or current prices when interpreting the calculated yield.

If, for example, the apparent yield calculated on recorded cash flows were 18.8% and the rate of inflation were 10% then the real rate of return would be 8%. Or, if the apparent rate calculated on recorded cash flows were 14.5% and the inflation rate were 10% the real rate would be only 4.09%. The most frightening example is if the apparent rate, calculated on recorded cash flows, were only 8.0% and the inflation rate were 10% the real rate would be –1.8%, that is, a *negative* return.

Accuracy of future estimates

The predicted cash flows used in economic appraisals of projects will be subject to error, as is any estimate. Accepting that these errors exist, appraisal techniques can be extended to evaluate the effect of these errors and produce forecasts of yields in probabilistic terms. The two main techniques are known as sensitivity analysis and risk analysis.

Table 13.1 Thirty-six combinations of construction time and construction cost.

Construction time (years)	Construction cost (£)					
1.5	450 000	500 000	550 000	600 000	650 000	700 000
2.0	450 000	500 000	550 000	600 000	650 000	700 000
2.5	450 000	500 000	550 000	600 000	650 000	700 000
3.0	450 000	500 000	550 000	600 000	650 000	700 000
3.5	450 000	500 000	550 000	600 000	650 000	700 000
4.0	450 000	500 000	550 000	600 000	650 000	700 000

Sensitivity analysis

This technique requires the DCF yield (or PW) to be calculated for different values of the cost and incomes estimates. If a hotel was being constructed and the effect on the DCF yield of two variables, 'construction cost' and 'construction time', was being examined, the DCF yield could be calculated for different estimates of the construction cost and time. Table 13.1 shows 36 different combinations of construction cost and time, and Fig. 13.6(a) shows the 36 different yields resulting from using these 36 different estimates of cost and time. The graph is known as a sensitivity chart and demonstrates how sensitive DCF yield is to changes in the two variables examined.

For example, if the duration was increasing as a result of a lack of proper planning and site control, the reduction in DCF yield could be predicted. If the construction cost was increased in order to finish more quickly, the loss in DCF yield due to increased cost could be predicted and the improvement in DCF yield due to reduced construction time could also be predicted. Whether the increase in cost and reduction in duration leads to an improved DCF yield can be seen from these graphs. It is possible to draw on the graph a line representing the criterion rate of return and thus identify the cost and time combinations that will produce an acceptable return.

Figure 13.6(b) shows a sensitivity graph of rate of return vs. utilisation rate for an excavator. (This graph is the result of an analysis on a specific item of plant; it is not generally applicable to all plant items.) The limitations of this technique are that it can only cope with a limited number of variables on one graph and that it does not identify the probability of the various combinations occurring.

It is unlikely that the smallest construction time would occur with the smallest construction cost; sensitivity graphs do not identify this. The process of risk analysis extends the calculations to include more variables and to take in the above probabilities.

Risk analysis

Risk analysis requires a 'model' of the cash flows that are used in the DCF calculation. This model consists of breaking down the cash flows until a level of detail is reached whereby estimates for the element can be produced. Figure 13.7 shows such a cash flow model for a proposed small factory. The project being modelled is to buy some land, build a factory, buy equipment, produce electronic goods and sell them. The model has been kept simple for illustration purposes. The estimates produced for each element are not just single sums, unless the cash flow is known and will not change; the variability of the cash flow for that element is also estimated. Figure 13.8 shows such an estimate for the construction cost. It is in the form of a histogram because risk analysis requires that weights or relative likelihood of the cash flow occurring be attached to the

a)

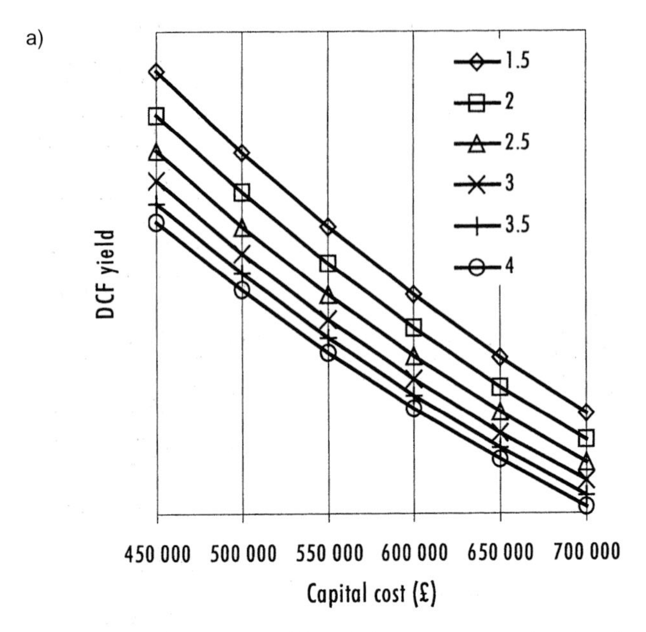

Fig. 13.6(a) Sensitivity chart of DCF yield, capital cost and construction times from Table 13.1.

b)

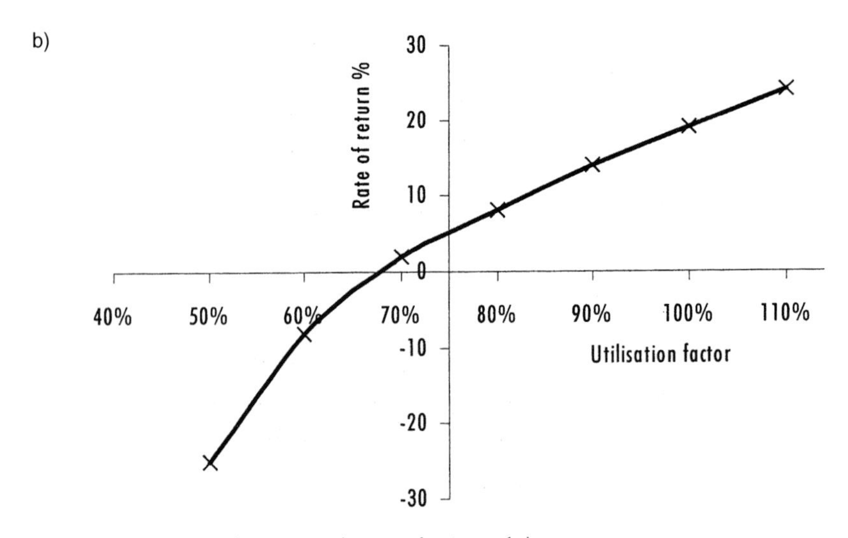

Fig. 13.6(b) Sensitivity of rate of return to utilisation of an item of plant.

estimates. What the estimates in Fig. 13.8 are reflecting is an expectation that the construction cost will most likely be £500 000; the estimator has given this a 50% chance. The estimator allows for the possibility that the construction cost will be as low as £450 000, but only gives this a chance of 25%, and similarly acknowledges that the cost could be as high as £600 000.

Those who are new to this type of cash flow modelling with probabilities attached to estimates typically have two reactions; one is to declare such estimating as impossible and the other

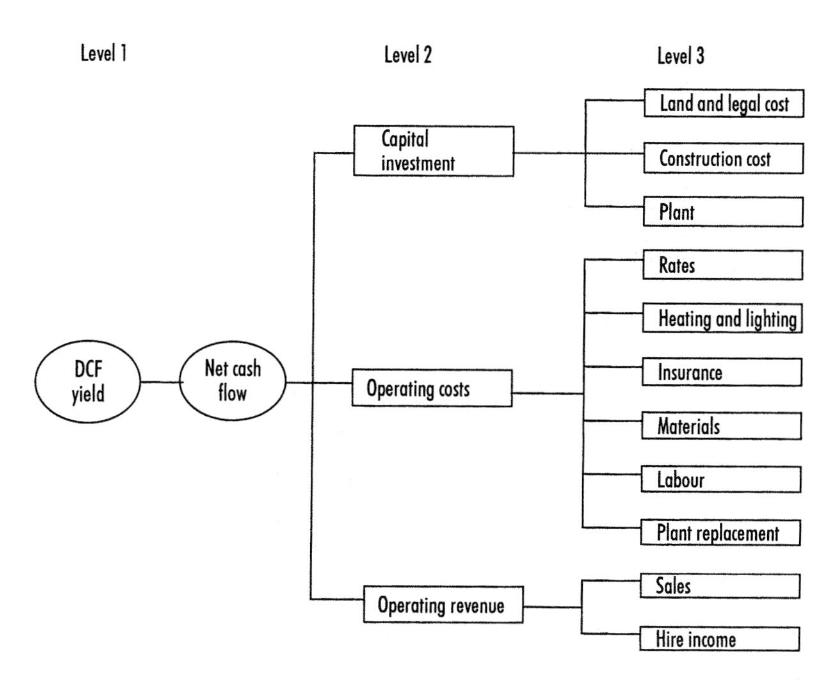

Fig. 13.7 Cash flow model for proposed factory.

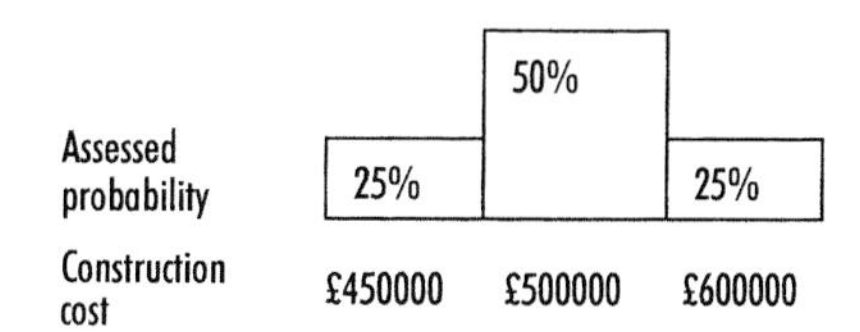

Fig. 13.8 Estimates of construction cost and the variability of the cost.

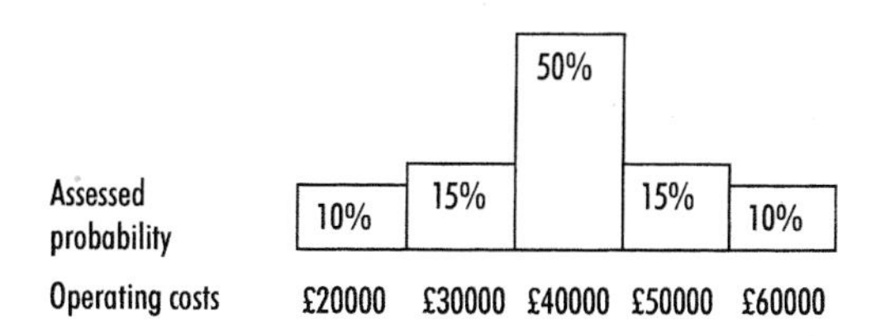

Fig. 13.9 More detailed assessment of a cost's variability.

is to say it is not detailed enough to reflect the real world. In answer to the first point, the argument usually takes the line that estimates have to be made, that estimates almost by definition will be in error and that if the normal estimating process produces £500 000, then it is no great step to examine estimates above and below this. If the element being estimated is too general for this treatment, it can always be broken down into smaller elements. The answer to the second criticism is that if the estimates are made more detailed, as in Fig. 13.9, the effort required to produce such detailed estimates may outweigh the information to be gained from the analysis. A balance between too much detail and too broad an approach must be found and this will depend

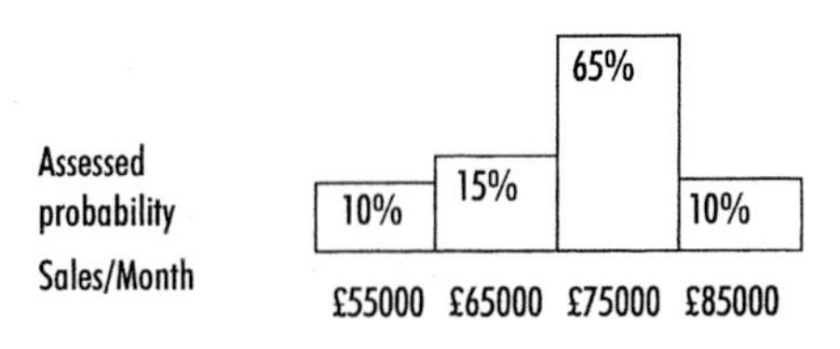

Fig. 13.10 Non-symmetrical estimate of variability.

on the specific case being analysed. Figures 13.8 and 13.9 may imply that estimates of variability are always symmetrical about a mean, but this is not so and Fig. 13.10 shows an estimate of variability that is skewed.

Risk analysis requires that estimates for each element be made in the form above. Estimates are only made for the smallest, most detailed elements in the model since the other elements can be calculated. The elements to be estimated in Fig. 13.7 are those shown in Level 3.

There are 11 elements in Level 3 and, if five estimates were made for each element, then a total of 5^{11}, or 48 828 125, different cash flows and hence yields could be calculated for this small model. Whereas such computations would have been unthinkable a few years ago, the development of more powerful and cheaper computers has put the capability of such analysis within the means of every construction organisation. Most software programmes for undertaking the analysis by generating different combinations of possible outcomes are based on a random selection of the estimates for each element. Each random selection will be weighted according to the probabilities attached to the estimates. To explain the random process further, examine the very simple model shown in Fig. 13.11(a). The elements in Level 3 have the estimates as shown in Figs 13.8, 13.9 and 13.10. The total number of possible cash flows are:

$$3\ (\text{from Fig. 13.6}) \times 5\ (\text{from Fig. 13.7}) \times 4\ (\text{from Fig. 13.8}) = 60$$

The first selection takes a construction cost at random from Fig. 13.6. In every 100 selections 25 of these will be £450 000, 50 will be £500 000 and 25 will be £600 000. This single selection made at random will be used as the capital cost element when calculating the cash flows. Similarly with the operating cost a random selection will be made from Fig. 13.9. This selection will not have any relationship with the selection for construction cost or sales. The sales figure will be selected from Fig. 13.10.

These three selections will be used to calculate the net cash flow and hence the DCF yield. This yield is just one of the 60 possible yields. If the process was repeated 10 times, a representation of all the possible yields and the variability of these yields would be obtained. Similarly with the larger example in Fig. 13.7, if the process of random selection and calculation of yield was done, say 200 or even 500 times, a representation of the possible yields and its variability would be obtained. The results of such an analysis are usually presented in a histogram form as shown in Fig. 13.12. The cumulative curve allows the calculation of the probability of achieving a certain yield. This requires a more sophisticated definition of criterion rate of return. Previously it was a simple rate, say 15%; now it has to have a probability attached—say the criterion becomes 15% provided there is an 85% chance of achieving 15% or better. Figure 13.11(b) shows a similar simple cash flow model for plant hiring, which illustrates that the principles of risk analysis described can equally be applied in this case.

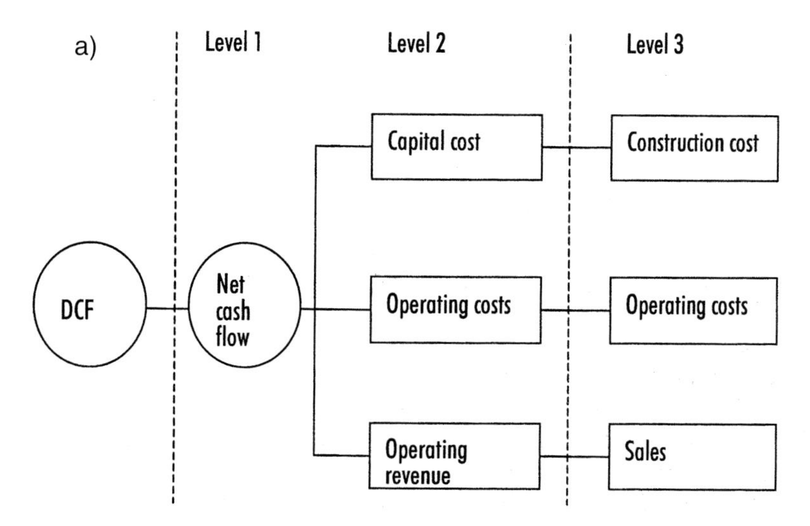

Fig. 13.11(a) Simple cash flow model.

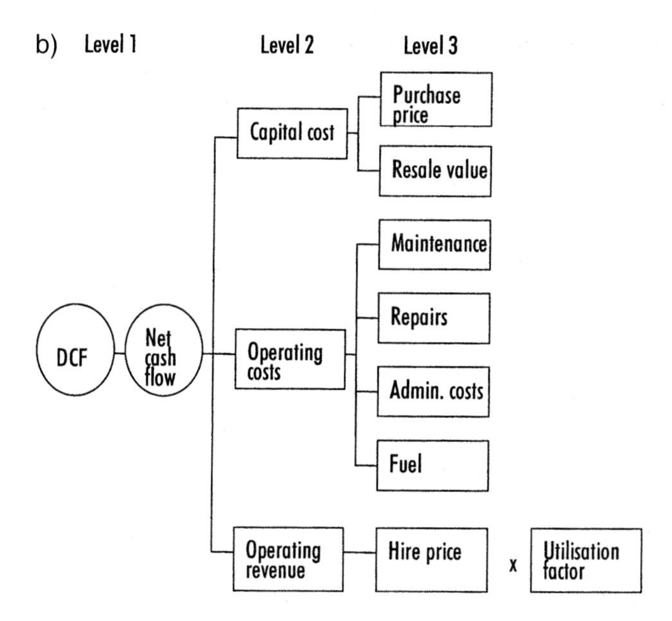

Fig. 13.11(b) Simple cash flow model for a plant hire operation.

Financial modelling

The cash flow models in Figs 13.11(a) and (b) and the concept of each element in the cash model having a range of estimates introduces the topic more generally known as financial modelling. Financial modelling has grown with computer use to the point where it is now a powerful tool in the hands of many business managers. There is a complete range of financial models, from the use of a general-purpose spreadsheet to the purpose-written software package. The following is a brief summary of the capability of the range of financial modelling packages.

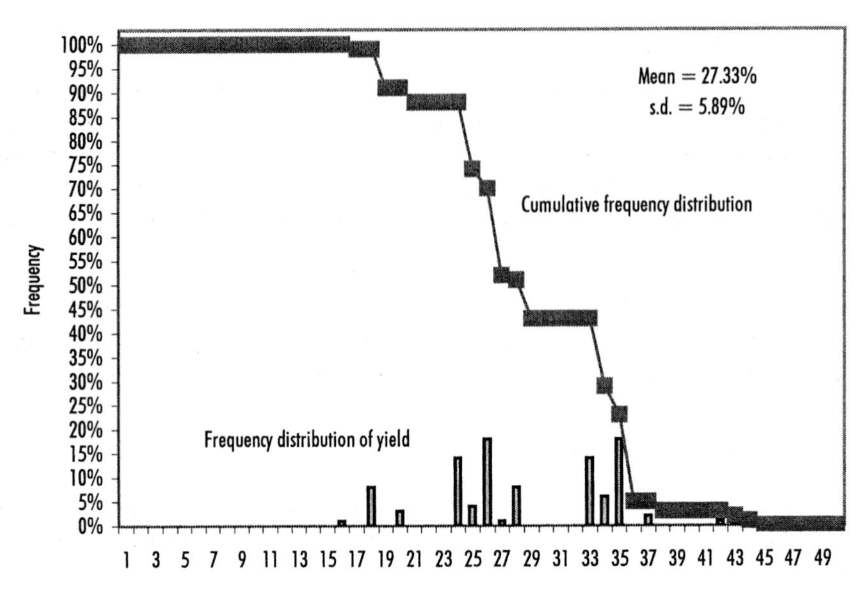

Fig. 13.12 Simulation output.

Financial modelling packages

Report generator

This is a computer program operating on historical/actual information usually stored in a computer file. It executes calculations such as additions and subtractions of rows and columns but no attempt is made to produce forecasts using assumptions or relationships pertaining to the firm.

Deterministic simulation

A deterministic model has the property that all the variables included are precisely specified by a single value. A simulation model is one that imitates a more complex reality. Therefore, a deterministic simulation model is a representation of all or part of a project.

Probabilistic simulation

This is a model, as explained previously, for coping with uncertainties or risk by using multiple estimates as input data with a relative frequency or assessed probability of their likelihood of occurrence. The use of probabilistic estimates of this kind is the subject of risk analysis, which was described earlier.

'What-if' model

Deterministic models operate on a 'case study' basis and give the answers to 'what-if' questions. The approach in the 'what-if' models is determined by the questions to be answered. They are used primarily to explore the specific effects of individual variables. The construction of the sensitivity graphs in Figs. 13.6(a) and 13.6(b) are examples. The question explored can of course be much more specific.

Optimising model

The model in this instance derives an 'optimum' solution of some specified objective using mathematical procedures. By adjusting the controllable variables, a model builder may be able to search for such an optimal solution. However, all models, to a certain extent, allow a model builder to search for better alternative decisions. The non-optimising model builder wishes to know 'what is the likely outcome in a range of alternatives for a given set of different variables'.

Statistical forecasting model

Statistical techniques can be used to provide forecasts by operating on historical data. This type of model is often used in conjunction with a deterministic model.

Analytical model

Analytical models are models in which some type of algorithm is used to produce a solution or output. Most common types of analytical models are linear programming and discounted cash flow models.

Modelling and risk analysis

Risk analysis is coped with by the models that allow probabilistic simulation. The description earlier in this chapter of how to cope with inaccuracy in estimates is one example. This example gives only one method of presenting the variability in the input data, by the use of multiple estimates, each with an assessed probability. There are, however, alternative approaches to inserting the variability. The main alternative is to assume a frequency distribution to describe the variability of each element in the model. Figure 13.13 illustrates this. This in turn opens the question of the description of the distributions that represents each element. The distributions can include normal distributions, a skewed distribution, a triangular distribution or, in the description given earlier, a user-defined distribution. Figure 13.14 gives examples of these. One interesting feature of experimenting with the different means of describing the variability of each element is that it will produce some differences in the simulated outcome. While most simple spreadsheets have a limited capability for undertaking the analysis for such variability, it is not uncommon to find finance directors in construction companies who utilise off-the-shelf software to support the computations.

Life-cycle costing

The use of financial modelling has allowed the concept of life-cycle costing to evolve. Life-cycle costing is the use of the techniques of discounted cash flow described earlier to allow, in the case of buildings, alternative designs to be evaluated at an earlier stage, taking into account not only the initial capital cost, but also all the future running costs and income. A previous term used in this area has been 'cost in use', again implying that the cost of operating the building is to be included in the appraisal.

The argument for using life-cycle-costing techniques rests on the grounds that at the early design stage the client or his/her advisors exercise total control over the building design, no commitment has been made and it is at this stage that a full appraisal should be conducted. If left to

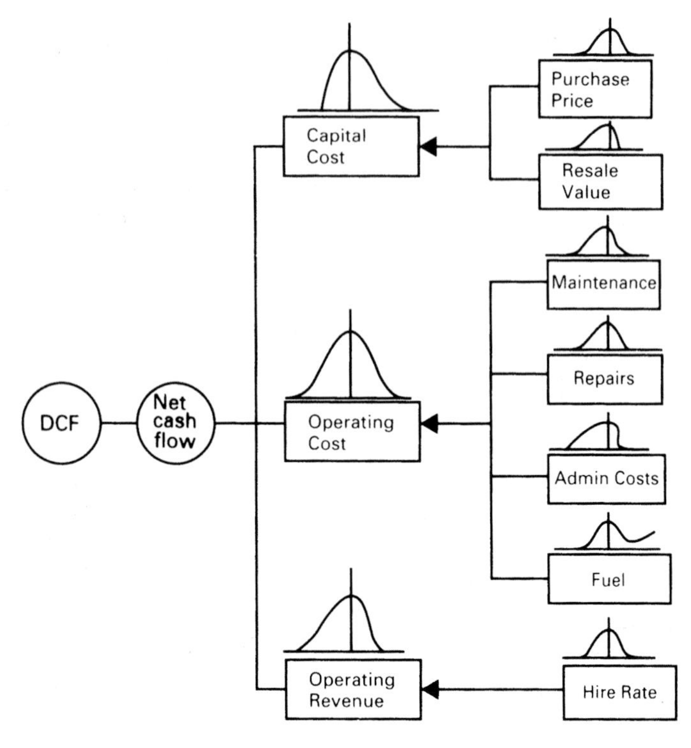

Fig. 13.13 Cash flow model illustrating the use of frequency distributions in estimates.

later in the design process, the cost of changing course is at least the cost of abortive, but expensive, design work. The argument for including running cost is obvious; an appraisal based on capital costs alone is unsatisfactory.

Thus financial modelling techniques are employed to evaluate a building project.

The steps in a life-cycle costing approach are as follows:

- Define objectives — e.g. building 3000 m^2, define quality
- Approach — new building, conversion, leasing, etc.
- Identify — costs, revenues, influencing factors (e.g. fuel, inflation, tax, required rate of return, etc.)
- Collect data — assemble the data required on the individual variables and factors
- Define relationship between variables
- Construct model
- Run model — experiment with initial data, produce reports

References in the bibliography at the end of the book recommend further reading in this area. The value of life-cycle costing is in:

- Assessing the total cost and resources required in the project
- Identifying the required funding in relationship to any limitation
- Conducting trade-off studies between alternatives
- Estimating the revenue levels required to produce a required rate of return

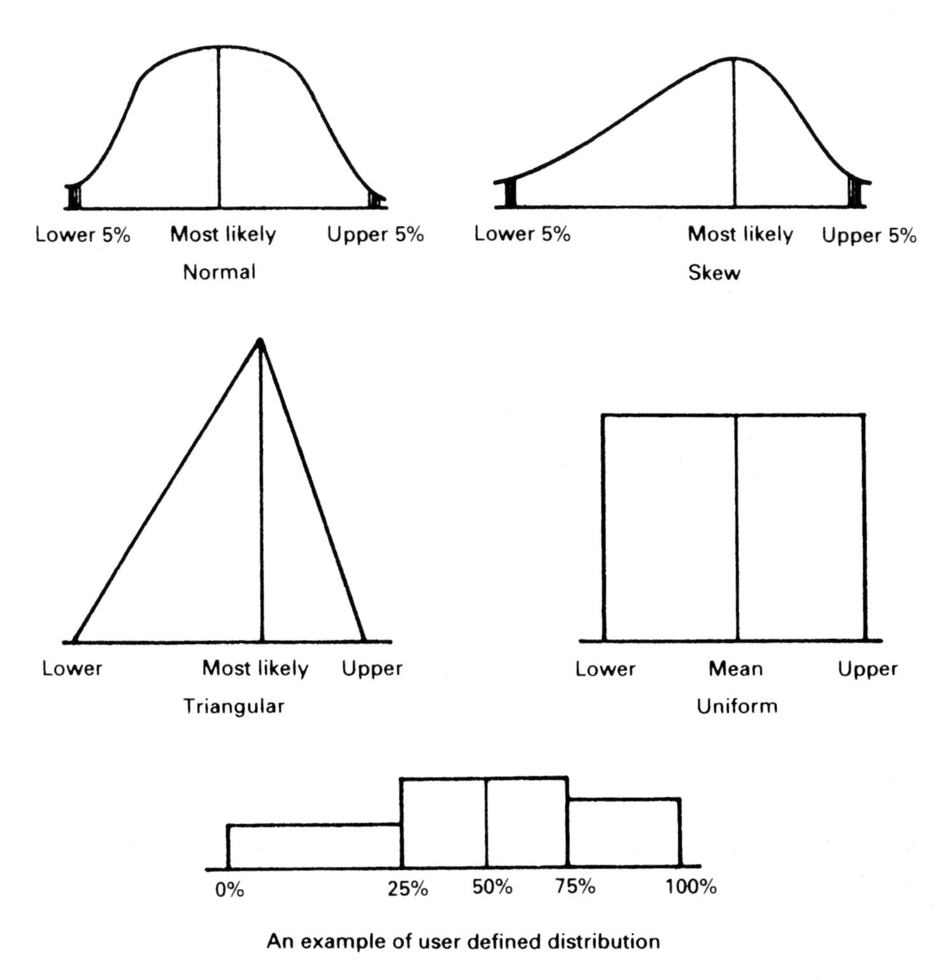

Fig. 13.14 Types of distribution describing variability in estimates.

Financial modelling, risk analysis and life-cycle costing are finding increased applications in civil engineering and in electro-mechanical, process plant, power and building projects. This is attributable to developers and owner organisations placing a greater emphasis on value for money.

Cost–benefit analysis

Cost–benefit analysis is another technique of project appraisal using discounting techniques to aid the evaluation and selection of projects. The appraisal of projects using cost–benefit analysis techniques is based on a time profile of costs and benefits for the parties affected by the project. It has been developed to deal with public-sector type projects where there is no commercial return. The construction of a road, for example, brings many benefits to the community. These benefits may include time savings, operating cost savings and accident savings. Whilst all these benefits undoubtedly exist and a road project under consideration may be very worthwhile, the benefits to the community do not show themselves as revenues or positive cash flows. Thus, the techniques of calculating net present value to determine if the future income justifies the investment at a given rate of return is not readily applicable, unless the benefits are translated into monetary terms. This is achieved by one of two ways:

- Using *direct money values* that reflect the goods and services involved in providing the investment project
- Relying on *quasi-market values* of the benefit elements using replacement-cost methods and revealed market preference

In application areas where cost–benefit techniques have become well established, e.g. in highway project evaluation, the mechanisms for evaluating the costs and benefits have been developed. An example is provided by the Highways Agency's *Design Manual for Roads and Bridges*, Vol.13, *Economic Assessment of Road Schemes* (1996). In this method of appraisal the *user benefits* are defined as the difference between the *user cost* on the existing road system discounted over 30 years and the *user cost* on the improved network discounted over 30 years. This calculation gives the net user benefit expressed as a reduction in user cost with the improved scheme. The *construction cost* of the improved scheme is compared with the *user benefit* and the net present value calculated. It is this net present value that is used in the appraisal of the proposed improvement.

The *manual* evaluates the user costs and benefits under the headings of:

- Time saving
- Vehicle operating costs
- Accidents
- Accidents on links
- Accidents at junctions

To assist in the evaluation the manual gives advice on the values to be applied to these factors. Table 13.2 shows an example summary of a cost–benefit study for a roads project. A roads project is only one example and the techniques of cost–benefit studies can be found in other areas of public buildings and works and strategic investment decisions of construction companies. Examples of areas where construction companies can apply the technique of cost–benefit analysis include:

- Justifying major equipment investments—such as leasing versus buying
- Examining how the use of IT could improve an existing business and operational processes
- Measuring life cycle costs in real financial terms
- Determining the impact and most cost-effective way to meet regulatory mandates (e.g. health and safety, ISO 9000)

The key point is that cost–benefit analysis works well in areas where the values used to quantify the benefits in 'cost' terms have been well developed.

Some worked examples

The following worked examples are presented as a guide to potential users of these techniques. The examples are:

Table 13.2 Summary of cost–benefit analysis for proposed road scheme.

	Proposed scheme £M at present value	
Construction costs	12.30	
Land and compensation costs	2.80	
Maintenance costs	0.45	
Total costs	15.55	
	Do minimum costs £M	
Total costs	0	
	Benefits	
	High traffic forecast £M at present value	Low traffic forecast £M at present value
Total quantified monetary benefits	24.80	17.30
Therefore net present value of proposal	9.25	1.75

1. Benefits include time and vehicle operating costs.
2. All benefits and costs discounted to same base year.

- A DCF yield calculation allowing for corporation tax and development grants
- A replacement calculation based on minimising the equivalent annual costs
- A PW calculation showing the effect of changes in interest rate

An example demonstrating the inclusion of corporation tax

A company with annual profits of over £1 million purchases and installs a new item of plant. The plant is classed as industrial plant and attracts tax allowances of 25% written down. The cost of equipment, which has no resale value, and the net revenue are shown in Columns 1 and 2 of Table 13.3(a). If corporation tax is 30% what is the net of tax DCF yield? A tax time-lag of one year exists. The minimum rate of return the company expects to earn on its capital is 5%.

Cash flows from Table 13.3 have a negative sum, –£1800 in Year 9. Before discounting by the trial interest rates it is necessary to discount £1800 by one year at the minimum return and add it to the +£3901.02 of Year 8 (Table 13.3(b)). This produces a positive cash flow and permits a single rate calculation.

Interpolating to find the interest rate that gives a net PW of zero:

$$\text{DCF yield} = 8 + 1 \times \frac{1955.46}{1955.46 - (-979.86)} = 8.67\%$$

Including an investment grant

If the plant item in the previous example was installed in a region which attracted an investment grant of say 20%, the cash flows would become as shown in Table 13.4(a). An investment grant lag of one year exists. With the investment reduced by the amount of the grant and the returns the same, the DCF yield will be higher.

The same assumption regarding the take-up of tax allowance applies in this case. Following the same procedure as in the previous case, the negative sum from Year 9 is discounted to Year 8 at the minimum return of 8% to obtain a net cash flow of £2362.56. 8% is assumed as the earning rate for the company's capital invested in a safe venture (Table 13.4(b)).

Table 13.3(a) DCF example including corporation tax.

Year	Investment in Plant (£)	Net revenue (£)	Corporation tax at 30% payable on previous year's profits (£)	Plant tax allowances (£)	Tax saved by allowances (£)	Tax paid (£)	Net cash flow (£)
0	120000.00			30000.00	9000.00	–9000.00	–111000.00
1		30000.00	0.00	22500.00	6750.00	–6750.00	36750.00
2		30000.00	9000.00	16875.00	5062.50	3937.50	26062.50
3		30000.00	9000.00	12656.25	3796.88	5203.12	24796.88
4		30000.00	9000.00	9492.19	2847.66	6152.34	23847.66
5		20000.00	9000.00	7119.14	2135.74	6864.26	13135.74
6		12000.00	6000.00	5339.36	1601.81	4398.19	7601.81
7		10000.00	3600.00	4004.52	1201.36	2398.64	7601.36
8		6000.00	3000.00	3003.39	901.02	2098.98	3901.02
9			1800.00			1800.00	–1800.00

Note: It is assumed that the tax saving of £9000 can be taken in Year 0 because the company offset this tax saving against its other profits. Capital purchases are added to a company 'pool' and resale values deducted in order to calculate capital allowances. As this case has no resale value there are still effectively capital allowances after disposal, which have been ignored.

Interpolating to find the interest rate that gives a net PW of zero:

$$\text{DCF yield} = 17 + 1 \times \frac{323.75}{323.75 - (-1891.06)} = 17.15\%$$

Example of optimal replacement age based on minimum equivalent annual costs

The purchase price of a small electricity plant is £20 000. The operating costs based on the annual average estimated hours of operation are £800 in the first year, when manufacturers' warranties operate, and £1200 in the second year, rising by £300 each year thereafter. The resale value of the plant can be assumed to be as predicted in Table 13.5. The cost of capital is 15%.

The calculations involve determining the equivalent annual cost of keeping the generator for one year and for two years and for three years, etc. (see Table 13.6).

The optimal replacement age is three years because this produces the minimum equivalent annual costs.

The calculations in Table 13.6 are seeking the balance between rising running costs and declining resale values. The method adopted is self explanatory from the headings in the table.

Example showing the effect of interest rates on present worth calculations

A pumping scheme being developed has three different possible systems of pumps and pipework. If the life of the scheme is 20 years, which scheme should be recommended as the most economic?

Table 13.3(b) The effect of different trial interest rates.

Year	Net cash flow (£)	Trial Interest Rate 8%			Trial Interest Rate 9%		
		Net cash flow with discounting of –£1800 (£)	PW factors at 8%	PW (£)	Net cash flow with discounting of –£1800 (£)	PW factors at 9%	PW (£)
0	–111 000.00	–111 000.00	1.000	–111 000.00	–111 000.00	1.000	–111 000.00
1	36 750.00	36 750.00	0.926	34 030.50	36 750.00	0.917	33 699.75
2	26 062.50	26 062.50	0.857	22 335.56	26 062.50	0.842	21 944.63
3	24 796.88	24 796.88	0.794	19 688.72	24 796.88	0.772	19 143.19
4	23 847.66	23 847.66	0.735	17 528.03	23 847.66	0.708	16 884.14
5	13 135.74	13 135.74	0.681	8 945.44	13 135.74	0.650	8 538.23
6	7 601.81	7 601.81	0.630	4 789.14	7 601.81	0.596	4 530.68
7	7 601.36	7 601.36	0.583	4 431.59	7 601.36	0.547	4 157.94
8	3 901.02	2 234.22	0.540	1 206.48	2 234.22	0.502	1 121.58
9	–1 800.00*						
		Net present value		+1 955.46	Net present value		–979.86

* By discounting –£1800 from Year 9 at 8% and adding to the cash flow for Year 8, 8% is assumed as the earning rate that the company can obtain from a safe investment.

Table 13.4(a) DCF example including development grant.

Year	Investment in plant (£)	Net revenue (£)	20% Development grant	Corporation tax at 30% payable on previous year's profits (£)	Plant tax allowances (£)	Tax saved by allowances (£)	Tax paid (£)	Net cash flow (£)
0	120 000.00				30 000.00	9 000.00	−9 000.00	−111 000.00
1		30 000.00	24 000.00	0.00	22 500.00	6 750.00	−6 750.00	60 750.00
2		30 000.00		9 000.00	16 875.00	5 062.50	3 937.50	26 062.50
3		30 000.00		9 000.00	12 656.25	3 796.88	5 203.12	24 796.88
4		30 000.00		9 000.00	9 492.19	2 847.66	6 152.34	23 847.66
5		20 000.00		9 000.00	7 119.14	2 135.74	6 864.26	13 135.74
6		12 000.00		6 000.00	5 339.36	1 601.81	4 398.19	7 601.81
7		10 000.00		3 600.00	4 004.52	1 201.36	2 398.64	7 601.36
8		6 000.00		3 000.00	3 003.39	901.02	2 098.98	3 901.02
9				1 800.00			1 800.00	−1 800.00

Table 13.4(b) The effect of different PW factors.

	Net cash flow (£)	Net cash flow with discounting of −£1800 (£)	PW factors at 17%	PW (£)	Net cash flow with discounting of −£1800 (£)	PW factors at 18%	PW (£)
0	−111 000.00	−111 000.00	1.000	−111 000.00	−111 000.00	1.000	−111 000.00
1	60 750.00	60 750.00	0.855	51 941.25	60 750.00	0.847	51 455.25
2	26 062.50	26 062.50	0.731	19 051.69	26 062.50	0.718	18 712.88
3	24 796.88	24 796.88	0.624	15 473.25	24 796.88	0.609	15 101.30
4	23 847.66	23 847.66	0.534	12 734.65	23 847.66	0.516	12 305.39
5	13 135.74	13 135.74	0.456	5 989.90	13 135.74	0.437	5 740.32
6	7 601.81	7 601.81	0.390	2 964.71	7 601.81	0.370	2 812.67
7	7 601.36	7 601.36	0.333	2 531.25	7 601.36	0.314	2 386.83
8	3 901.02	2 234.22	0.285	636.75	2 234.22	0.266	594.30
9	−1 800.00						
		Net present value		+323.45	Net present value		−1 891.06

Table 13.5 Predicted resale values.

Year	Predicted resale values (£)
1	18 000
2	16 000
3	15 000
4	12 000
5	8 000
6	5 000
7	2 000

Scheme	Pipe diameter (mm)	Installation costs (£)	Annual running costs (£)
A	500	18 250	7250
B	600	20 200	4600
C	750	24 000	4000

Using 15% to represent the cost of capital:

(1) Scheme A

PW of installation cost = £18 250

PW of £7250 each year for 20 years = 7250 × 6.2593 = £45 379.93

Total PW = £63 629.93

(2) Scheme B

PW of installation cost = £20 200

PW of £4600 each year for 20 years = 4600 × 6.2593 = £28 792.78

Total PW = £48 992.78

(3) Scheme C

PW of installation cost = £24 000

PW of £4000 each year for 20 years = 4000 × 6.2593 = £25 037.20

Total PW = £49 037.20

Therefore at 15% Scheme B is the most economical. If the calculations are repeated at 10%:

(1) Scheme A

PW of installation cost = £18 250

PW of £7250 each year for 20 years = 7250 × 8.5135 = £61 772.88

Total PW = £79 972.88

Table 13.6 Calculating the optimal replacement age.

Year	A	B	C	D	E	F	G	H	I	K	L	M	N
	Purchase price	Capital recovery factors at 15%	Equivalent annual costs of purchase price	Running costs	PW factors at 15%	PW of running costs	Sum of PW of running costs	Equivalent annual cost of PW of running costs	Equivalent annual costs of purchase price and running costs	Resale value	PW of resale value	Equivalent annual costs of PW of resale	Equivalent annual costs of purchase running and resale
			(A × B)			(D × E)		(B × G)	(C + H)		(K × E)	(L × B)	(I — M)
	(£)	(£)	(£)	(£)	(£)	(£)	(£)	(£)	(£)	(£)	(£)	(£)	(£)
0	20 000				1.000								
1		1.150	23 000.00	800.00	0.869	695.00	695.00	799.25	23 799.25	18 000	15 642.00	17 988.30	5 810.95
2		0.615	12 300.00	1 200.00	0.756	907.00	1 602.00	985.23	13 285.23	16 000	12 096.00	7 439.04	5 846.19
3		0.437	8 740.00	1 500.00	0.657	985.00	2 587.00	1 130.51	9 870.51	15 000	9 855.00	4 306.64	5 563.88
4		0.350	7 000.00	1 800.00	0.571	1 027.00	3 614.00	1 264.90	8 264.90	12 000	6 852.00	2 398.20	5 866.70
5		0.298	5 960.00	2 100.00	0.497	1 043.00	4 657.00	1 387.78	7 347.78	8 000	3 976.00	1 184.85	6 162.93
6		0.264	5 280.00	2 400.00	0.432	1 036.00	5 693.00	1 502.95	6 782.95	5 000	2 160.00	570.24	6 212.71

(2) Scheme B

PW of installation cost = £20 200

PW of £4600 each year for 20 years = 4600 × 8.5135 = £39 162.10

Total PW = £59 362.10

(3) Scheme C

PW of installation cost = £24 000

PW of £4000 each year for 20 years = 4000 × 8.5135 = £34 054.00

Total PW = £58 054.00

At 10%, Scheme C is the most economical.

The difference between the two calculations is the interest rate. Scheme C requires an extra investment of £3800 in comparison to B. The saving in running costs is £600 per year. When the interest rate is at 15% the savings in running costs are not sufficient to justify the extra expenditure, i.e. the extra expenditure can earn more than £600 per year. When the interest rate is only 10% the extra expenditure cannot earn more than £600 per year and would therefore be better employed earning this saving.

Reference

Highways Agency (1996) *Design Manual for Roads and Bridges*. Vol.13. *Economic Assessment of Road Schemes*. HMSO, London.

Appendix A13.1 Tabulations of interest and time relationships

Terms	i	Rate of interest per period (usually years)	n	Number of time periods

Interest tables for 10%

	$(1+i)^n$	$\frac{1}{(1+i)^n}$	$\frac{(1+i)^n-1}{i}$	$\frac{i}{(1+i)^n-1}$	$\frac{(1+i)^n-1}{i(1+i)^n}$	$\frac{i(1+i)^n}{i(1+i)^n-1}$
Year or period (n)	Compound amount of a single sum	Present value of a single sum	Compound amount of a uniform series	Sinking fund deposit	Present worth of a uniform series	Capital recovery
1	1.1000	0.90909	1.000	1.00000	0.9090	1.10000
2	1.2099	0.82644	2.099	0.47619	1.7355	0.57619
3	1.3309	0.75131	3.309	0.30211	2.4868	0.40211
4	1.4640	0.68301	4.640	0.21547	3.1698	0.31547
5	1.6105	0.62092	6.105	0.16379	3.7907	0.26379
6	1.7715	0.56447	7.715	0.12960	4.3552	0.22960
7	1.9487	0.51315	9.487	0.10540	4.8684	0.20540
8	2.1435	0.46650	11.435	0.08744	5.3349	0.18744
9	2.3579	0.42409	13.579	0.07364	5.7590	0.17364
10	2.5937	0.38554	15.937	0.06274	6.1445	0.16274
11	2.8531	0.35049	18.531	0.05396	6.4950	0.15396
12	3.1384	0.31863	21.384	0.04676	6.8136	0.14676
13	3.4522	0.28966	24.522	0.04077	7.1033	0.14077
14	3.7974	0.26333	27.974	0.03574	7.3666	0.13574
15	4.1772	0.23939	31.722	0.03147	7.6060	0.13147
16	4.5949	0.21762	35.949	0.02781	7.9237	0.12781
17	5.0544	0.19784	40.544	0.02466	8.0215	0.12466
18	5.5599	0.17985	45.599	0.02193	8.2014	0.12193
19	6.1159	0.61350	51159	0.01954	8.3649	0.11954
20	6.7274	0.14864	57.274	0.01745	8.5135	0.11745
21	7.4002	0.13513	64.002	0.01562	8.6486	0.11562
22	8.1402	0.12284	71.402	0.01400	8.7715	0.11400
23	8.9543	0.11167	79.543	0.01257	8.8832	0.11257
24	9.8497	0.10152	88.497	0.01129	8.9847	0.11129
25	10.8347	0.09229	98.347	0.01016	9.0770	0.11016
26	11.9181	0.08390	109.181	0.00915	9.1609	0.10915
27	13.1099	0.07627	121.099	0.00825	9.2372	0.10825
28	14.4209	0.06934	134.209	0.00745	9.3065	0.10745
29	15.8630	0.06303	148.630	0.00672	9.3696	0.10672
30	17.4994	0.05730	164.494	0.00607	9.4269	0.10607
35	28.1024	0.03558	271.024	0.00368	9.6441	0.10368
40	45.2592	0.02209	442.592	0.00225	9.7790	0.10225
45	72.8904	0.01371	718.904	0.00139	9.8628	0.10139
50	117.3908	0.00851	1163.908	0.00085	9.9148	0.10085

Appendix A13.1 *continued*

Tables are reproduced from the examination tables of Loughborough University, Department of Civil Engineering

Interest tables for 15%

	$(1+i)^n$	$\frac{1}{(1+i)^n}$	$\frac{(1+i)^n-1}{i}$	$\frac{i}{(1+i)^n-1}$	$\frac{(1+i)^n-1}{i(1+i)^n}$	$\frac{i(1+i)^n}{i(1+i)^n-1}$
Year or period (*n*)	Compound amount of a single sum	Present value of a single sum	Compound amount of a uniform series	Sinking fund deposit	Present worth of a uniform series	Capital recovery
1	1.1500	0.86956	1.000	1.00000	0.8695	1.15000
2	1.3324	0.75614	2.149	0.46511	1.6257	0.61511
3	1.5208	0.65751	3.472	0.28797	2.2832	0.43797
4	1.7490	0.57175	4.993	0.20026	2.8549	0.35026
5	2.0113	0.49717	6.742	0.14831	3.3521	0.29831
6	2.3130	0.43232	8.753	0.11423	3.7844	0.26423
7	2.6600	0.37593	11.066	0.09036	4.1604	0.24036
8	3.0590	0.32690	13.725	0.07285	4.4873	0.22285
9	3.5178	0.28426	16.785	0.05957	4.7715	0.20957
10	4.0455	0.24718	20.303	0.04925	5.0187	0.19925
11	4.6523	0.21494	24.349	0.04106	5.2337	0.19106
12	5.3502	0.18690	29.001	0.03448	5.4206	0.18448
13	6.1527	0.16252	34.351	0.02911	5.5831	0.17911
14	7.0757	0.14132	40.504	0.02468	5.7244	0.17468
15	8.1370	0.12289	47.580	0.02101	5.8473	0.17101
16	9.3576	0.10686	55.717	0.01794	5.9542	0.16794
17	10.7612	0.09292	65.075	0.01536	6.0471	0.16536
18	12.3754	0.08080	75.836	0.01318	6.1279	0.16318
19	14.2317	0.07026	88.211	0.01133	6.1982	0.16133
20	16.3665	0.06110	102.443	0.00976	6.2593	0.15976
21	18.8215	0.05313	118.810	0.00841	6.3124	0.15841
22	21.6447	0.04620	137.631	0.00726	6.3586	0.15726
23	24.8914	0.04017	159.276	0.00627	6.3988	0.15627
24	28.6251	0.03493	184.167	0.00542	6.4337	0.15542
25	32.9189	0.03037	212.793	0.00469	6.4641	0.15469
26	37.8567	0.02641	245.711	0.00406	6.4905	0.15406
27	43.5353	0.02296	283.568	0.00352	6.5135	0.15352
28	50.0656	0.01997	327.104	0.00305	6.5335	0.15305
29	57.5754	0.01736	377.169	0.00265	6.5508	0.15265
30	66.2117	0.01510	434.745	0.00230	6.5659	0.15230
35	133.1755	0.00750	881.170	0.00113	6.6166	0.15113
40	267.8635	0.00373	1779.090	0.00056	6.6417	0.15056
45	583.7692	0.00185	3585.128	0.00027	6.6542	0.15027
50	1083.6593	0.00092	7217.715	0.00013	6.6605	0.15013

Section three

Administration and company management

14
Company organisation

Summary

The role and responsibilities of the manager, the lines of communication in an enterprise and the structural arrangements to be found in the managerial organisation of some contracting companies are discussed and described.

Introduction

Much has been written theorising on the recommended methods to be adopted by management for successfully leading and controlling companies for most types of industry, but it is emphasised that many firms have put much reliance on the dynamism and entrepreneurial skill of an ambitious individual for business success. As yet, there is no set of rules and regulations to guarantee commercial security.

The purpose of this chapter is to highlight the managerial arrangements that prevail in most construction companies and have proved necessary to ensure their continued viability.

The function of a manager

A business is merely a fairly efficient way of combining the skills and talents of people into an organisation that can produce goods and services in sufficient quantity to satisfy the material desires of the community in which it exists, and importantly provide sufficient return on the capital invested. Arguing further, Peter Drucker (1999) explains the primary responsibility of management is to ensure that the resources available to the firm for producing economic wealth are used in the best possible way within the social and legal frameworks that shape its business context. The practical implications, however, depend on each individual enterprise, for example environmental, social and ethical issues arising from business activity are currently of growing concern to consumers. Clearly, any impact on customer purchasing mores would require attention, probably exercised through more serious corporate social responsibility (CSR) policies than hitherto. Under such potentially changed society values and market conditions, continued demand for profit-generating products would at least provide a guiding signal that the devised business model was properly using resources for the commercial purpose.

To these ends, certain broad organisational arrangements are desirable if management is to exercise its essential duty of welding the various parts of the enterprise together, so that all elements can operate in unison.

Company objectives

First, top management needs to define and interpret the mission/purpose of the enterprise in terms of tangible objectives, i.e. the points to which policies and strategies for action are directed — see Chapter 15 for an example of the process — and are typically those described below.

Best kind of business to operate in

The entrepreneur often bases this decision on instinct and personal experience, but more careful analysis is needed for the established and growing company. Indeed reliance on the status quo product or a core competence does not ensure survival as a cursory look over any 20-year period at company names listed on the Stock Exchange will reveal.

Kind of goods and services to be offered

As far as the construction market is concerned this may mean taking decisions about whether to concentrate on building schools or houses, speculative or tendered contracts, negotiated work such as design and build, PFI/PPP, management-type contracting, or specialist subcontracting etc., as well as laying emphasis on innovation, quality, delivery on time and developing good and continuing relations with the client.

Desired share of the market

The likely markets need to be based on careful evaluation of economic trends and opportunities, all married to a detailed assessment of the firm's growth prospects and potential capability to assemble the appropriate commercial, managerial and technical skills, resources, etc.

Development of know-how

Creating sales increasingly relies on innovation in products, services, construction processes and management, where continuous attention to improving *skills and expertise* of the workforce and suppliers are essential in adding value.

Possible changes and fluctuations of the market in future years

Political, economic and social shifts in consumer outlook, particularly concerning the environment and quality, all vary between different markets and over time.

Company policies

Policies are general statements to guide decision-making and strategic planning towards assuring consistency with the organisation's objectives, often written up in manual form, and these may be broad in scope or very specific, for example:

- Generate profit responsibly, e.g. it is unlikely that the safety or welfare officer can pay the same single-minded attitude to this objective as the project manager in charge of a construction site.

- Continuously improve the quality and value added.
- Train and develop the workforce to a managed plan.
- Look for profitable growth through negotiated contracts.
- Operate in a particular geographical region.
- Purchase IT resources centrally.
- Conduct in-house R&D.
- Source timber supplies from sustainable managed woodlands.
- Recycle waste in line with recognised environmental targets.
- Deal only with environmentally sustainable building/construction development.
- Meet the standards and obligations desired by the society in which the company operates.
- Comply with modern corporate social, environment and sustainability responsibilities.

Company strategy

Achieving the desired objective(s) and policies demands clear and focused strategies (i.e. employment of plans). In the case of a market share objective concerning, say, speculative housing development, alternative strategies for securing superior performance deemed vital to gain and hold onto the perceived competitive advantage might be conceived as factory-sourced modular elements rather than conventional construction, as customer-bespoke system/modular/prefabricated housing construction or as selling speculative developments through e-commerce, etc.

The procedure for developing a detailed input/output *business model* for comparing the alternative strategies for profitable implementation is described in Chapter 15, *Market planning and business development.*

Importantly, strategy subsequently determines the scope (i.e. produce in-house or outsource), scale and structure of the organisation, which in turn impacts on the conception of fresh objectives and new policies. Best results are more likely in an enterprise with a culture of aiming high and possessing flawless execution of technology and management systems, as well as a willingness to respond flexibly to unforeseen changes in the market.

Company organisational structures

The choices of managerial organisation most suited for the company are many and depend upon many factors such as size, geographical location, type of work done, managerial/technical skills available and extent of subcontracted work/outsourced supplies. Noticeably in contrast to selling enterprises, with the attendant need for a complex management structure to penetrate markets wholesale, the construction-only function dictates a relatively flat arrangement with few levels of management. And, unlike manufacturers with sole responsibility for a product, contractors rarely need comprehensive in-house design capacity, so in effect rendering even the international contractor's structure, at least at head office, relatively uncomplicated.

However, on a very large scheme the complex tiers of management commonplace in the factory or production plant may be partly evident at site level (see Fig. 14.7). Furthermore, as more design-and-build work develops, firms are upgrading business development departments in particular and strengthening organisational mechanisms for co-ordinating design partners similar to the project management arrangements described in Fig.8.9, where, for example, teams are established to oversee projects from inception to completion. Structures are also evolving to manage more intense quality assurance procedures and better control of subcontractors and external suppliers through strengthened management of the purchasing function (see 'Supply chains and networks' in Chapter 3), all linked to a greater emphasis on health, safety and environmental issues in the more stringent climate governed by the Construction Design and Management (CDM) regulations and corporate social responsibility issues.

Small company

The simple form of organisational structure, shown in Fig. 14.1 divided by functions, is common to the smaller company with construction interests limited to a few specialties only, for example:

- Concrete structures
- Steel erection
- House building
- Earthmoving
- Pipelaying

It is also well suited to the small and medium-sized local construction company, where technical specialisms are few, departments controlled by a single manager and all the contracts handled from head office, with perhaps only a general foreman/supervisor on site.

Such companies are usually headed by the major and controlling shareholder appointed as Managing Director (broadly, President in US terminology) whose duties typically embrace the following responsibilities:

- Make policy and provide strategic guidance.
- Manage the company in compliance with the stated objectives and articles.
- Co-ordinate the activities of the various business departments.
- Establish staff functions procedures.
- Define and implement the health, safety and social responsibilities policy of the company.
- Develop and standardise cost-effective procedures for departments.
- Develop the company's business, revenue base and profitability.
- Foster a customer-oriented culture.
- Establish sound working relations with clients and shareholders.
- Promote the company's identity.
- Direct and improve operations and management systems.

Medium-sized company

As a company increases the volume of its turnover, the contracts undertaken tend to become more complex and extra skills need to be recruited. The head-office structure, therefore, not only expands but also subdivides into separately managed elements. The structure begins to take on the form shown in Figs. 14.2, 14.3 and 14.4, with the major functions typically led by salaried senior managers, sometimes having director status. A director is an officer of the company entitled to a modest fee for services rendered and as a member of the Board of Directors is collectively responsible for running the company as set out in the Articles of Association. Posts such as the managing director or finance director are usually contracted to carry out specified duties in a salaried executive capacity. Some directors could be major shareholders.

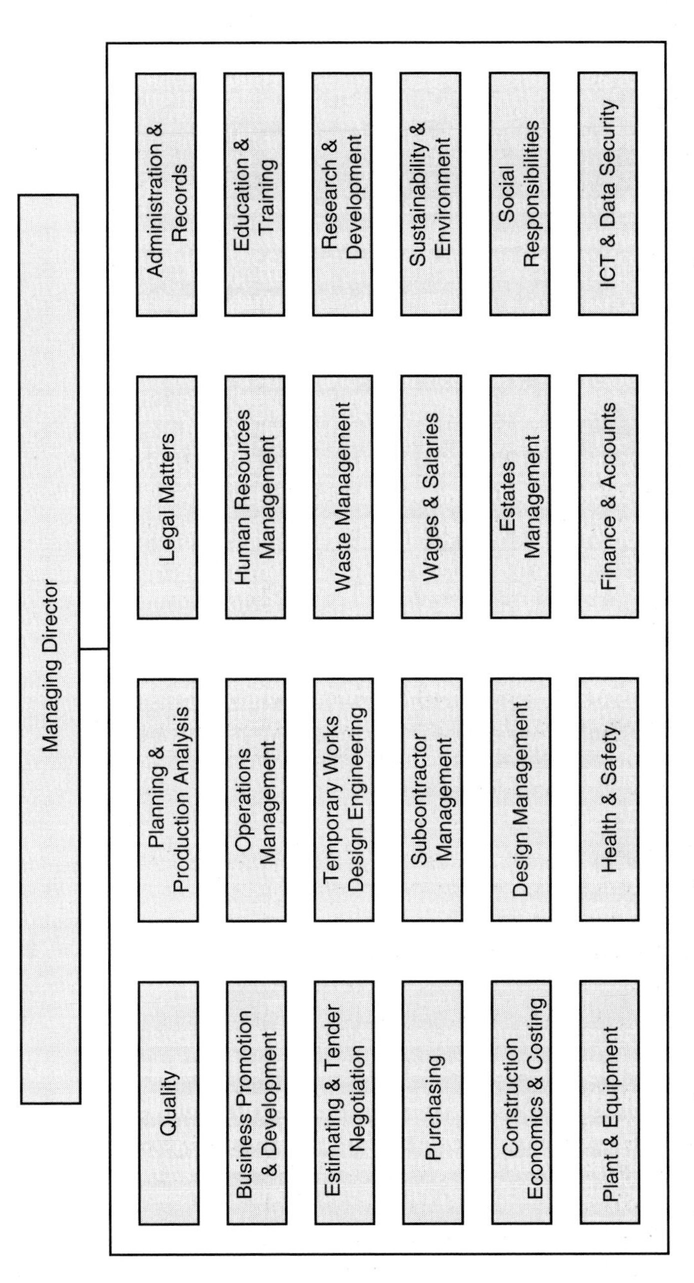

Fig. 14.1 Typical construction company functions.

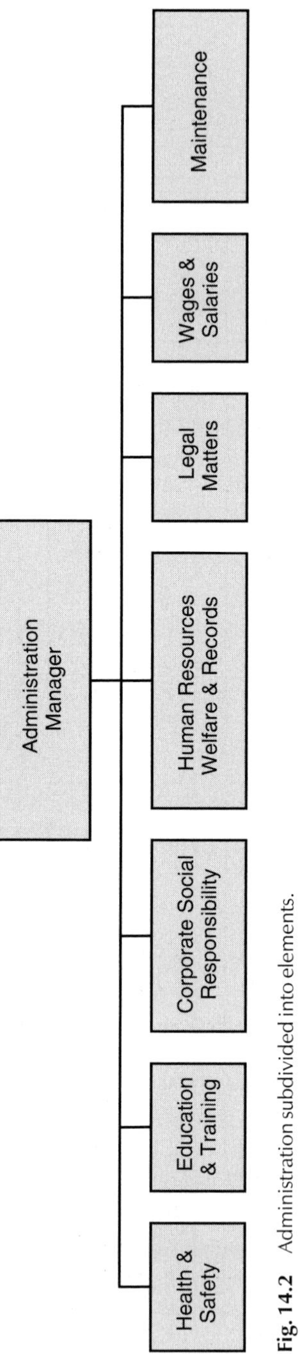

Fig. 14.2 Administration subdivided into elements.

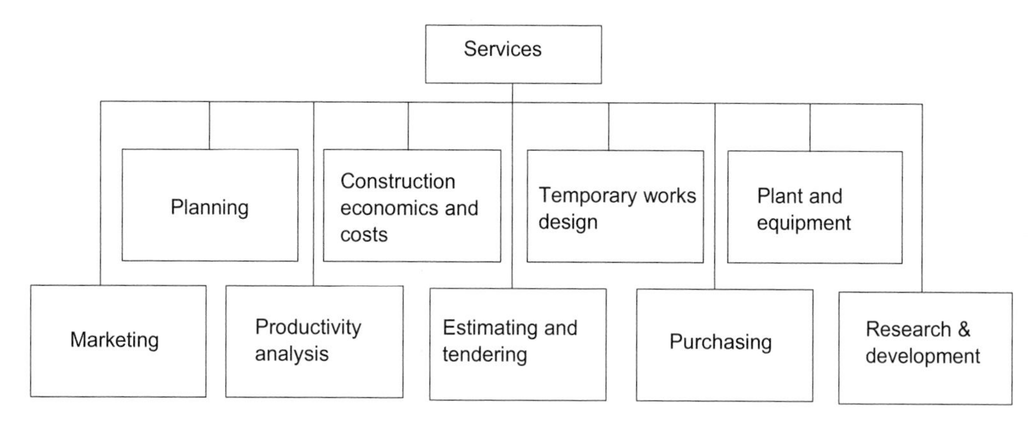

Fig. 14.3 Services subdivided into elements.

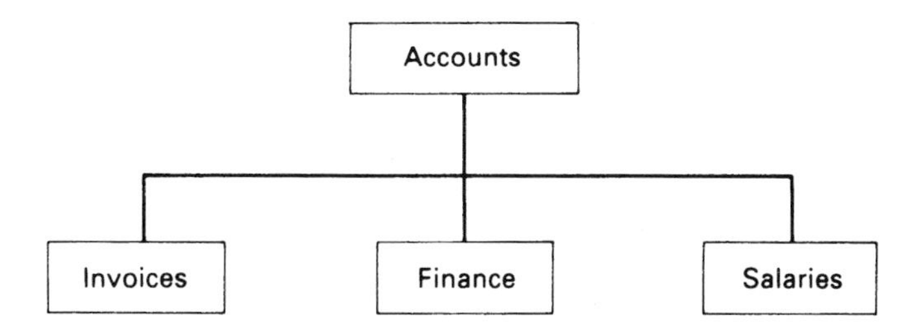

Fig. 14.4 Accounts subdivided into elements.

Large private company

The underlying principle of the simple form of structure is retained but commonly decentralised into specialist divisions, units and departments, for example civil engineering, building, housing, groundworks, international construction, geographical regions, general contracting, management/design and build, special products, etc., executed through the board of directors (vice-presidents) led by the managing director (see Fig. 14.5). However, the appropriate degree of autonomy of functions or units, particularly services and administration departments, depends upon the market, geographical spread, supply chain, communication demands, etc. Indeed, for diversified group holdings, variants of the matrix-type organisation may be more appropriate for achieving a lean cost base.

The large company generally undertakes contracts of many types and sizes, which demands considerable resources and a variety of skills, especially competent directors, managers, staff, workforce and suppliers. The mechanisms needed to bring about expansion vary in relation to market opportunities coupled with the ability of the organisation to secure work at competitive prices delivered to budget, on time and to the client's satisfaction. Since inherent growth is restrictive in an industry overburdened with competitors, raising market share is more commonly sought through takeovers and diversification into different commercial interests. But for the private firm opportunities are limited as an accurate valuation of the company to be acquired needs to be assessed, owners persuaded to sell and sufficient funds obtained to secure the controlling proportion of shares in the new subsidiary.

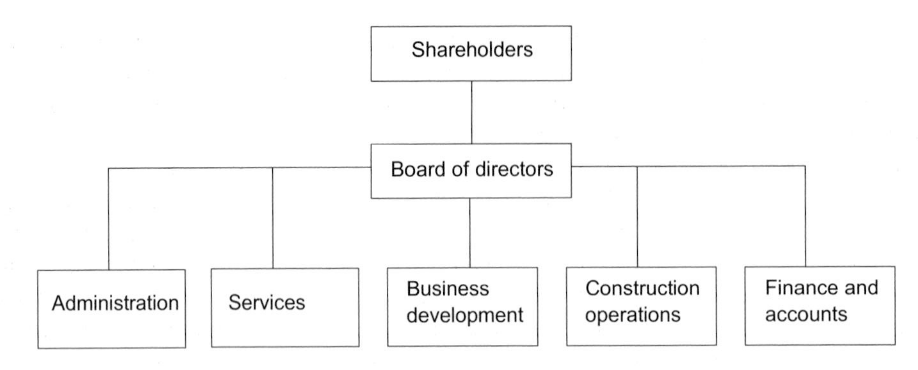

Fig. 14.5 Large company subdivided by function.

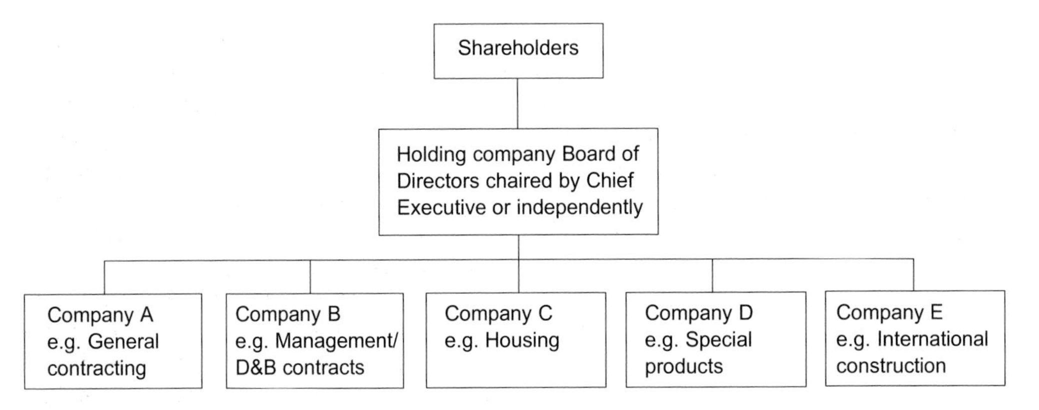

Fig. 14.6 Holdings of a group of companies.

Investigations suggest that such strategies have in general failed to produce better balance sheet results than firms not pursuing growth by acquisition. Growth nevertheless helps build up assets paramount as collateral in gaining access to major construction work such as Design Build Finance Operate (DBFO) and Build Own Operate Transfer (BOOT) projects where financial support from the banks and financial houses is essential in raising capital to finance the significant long-term investments involved.

Public liability company or corporation

Notably the top 20% of companies across all industries measured by market capitalisation account for about 80% of added-value, which in many cases has been achieved by a progressive strategy of creative destruction through mergers and acquisitions until dominant in their market.

In similar manner to the large private enterprise, an ambitious company can through share-acquisition activity typically develop into an international corporate group, structured, for example, as illustrated in Fig. 14.6. The PLC, normally listed on the home stock exchange (see Chapter 18), has publicly tradable shares and is commonly managed as a holding company by the Board of Directors responsible to the shareholders for strategy, operations and governance. Usually a salaried Chief Executive Officer is contracted to steer the enterprise and delegated with power of management and authority. However, in order to contain excessive independent action of the CEO, a very senior director is sometimes nominated as Chairperson to preside over business affairs, represent the company externally and internationally, act as its senior ambassador and deal with stakeholders and outside bodies.

Directors are approved and appointed or removed, subject to breach of contract, by the shareholder(s) at a General Meeting, but the Board may independently fill vacancies and engage additional directors. Consequently, in practice shareholders have limited influence, as resolutions and board membership nominations for the annual general meeting (AGM) are largely the board's prerogative. Likewise, proxy votes delegated to the chair provide the board further poll advantage, although a majority of shareholder(s) can call an extraordinary general meeting (EGM) to vote in a completely new board of liking if necessary. Indeed, the voting rights of shareholders vary markedly according to the stated articles of association and, indeed, with variants in some EU two-tier board arrangements, which give overriding authority to the Supervisory Board for certain decisions.

The Board of Directors, who may also be shareholders, varies according to need, for example salaried or contracted officers comprising the CEO, Chief Operating Officer (COO), senior executives responsible for oversight of major strategic segments, Chief Financial Officer (CFO), and maybe other key functional officers, plus invited non-executive directors (ostensibly independent and the majority in US corporations) paid a fee to serve on the main board. The main role of the latter is to offer specific expertise emanating from influence in banking, politics, the legal profession, business development, shareholder interests, etc. Such knowledge and contacts can be invaluable when defending against hostile takeover bids, particularly with mechanisms such as 'poison pills', 'white knights', staggered boards and other legal impediments. The independents also usually chair the senior appointments, remuneration and audit committees, to ensure that good governance structures are in place to maximise shareholder value. Indeed, good governance and corporate social responsibilities are becoming more onerous tasks for the public company, notably monitoring of: ethical standards; formalised executive responsibilities for the published accounts and non-executive directors' responsibilities for the audit, remuneration, compensation, nominations and data security committees; appraisals of board members; the appointment of consultants and advisers; shareholder and employee democracy; the role of trade unions on the company board; and environmental policy.

Divisional MDs (Vice-Presidents) usually sit on an operational board at secondary level, typically headed by the COO and responsible to the main board.

In some European Union states the main board is split into a Supervisory Board of independent/non-executive directors (including worker or union and shareholder representatives) tasked with strategic oversight and governance and a Management Board of executive directors (appointed by and responsible to the former) separately concentrating on company operational management.

Acquisition integration options

Acquisitions by various means are normally undertaken to open up a fresh route to the market, a new product or an innovative technology. For example:

(1) The vertically (or backwards) integrated company acquires firms in the same category of business but at a different production process stage, for example the contractor with a supply chain comprising mostly subcontractors under ownership.

(2) Horizontal integration represents mergers or takeovers of companies in both the same line of business and production process stage, for example a house builder buying a competitor to increase market share.
(3) A conglomerate is merely a diverse holding of unrelated businesses, usually acquired to diversify income and profit sources.

The span of control

A company needs to grow; otherwise the resulting stagnation may eventually lead to absolute decline. With expansion, organisational structures tend to develop as shown above, necessitating differing levels of management. The task of co-ordinating the various activities then soon exceeds the capacity of the individual, requiring management to delegate responsibility, coupled with some authority, to subordinates. Practice suggests that a manager can cope with between five and eight subordinates; indeed, construction companies typically allocate on average one contracts manager to about five site managers. Clearly, therefore, the extent of a flat or tiered management structure appropriate for the business requires careful consideration.

Pattern of communications

The pattern of communication and control in the industry is much the same as that used by the military, with *direct lines* of command between manager and assistants in each department. In most companies, however, service functions are necessary. The usual procedure is for the service manager to communicate through the head of the department that is receiving the service, e.g. formwork designs may be undertaken at head office for implementation on site. Officially head office only provides a service and will have no authority on site, but in practice considerable authority is vested by virtue of specialist skills. Such relationships are called *functional* ones, and their success depends very much upon co-operation between individuals. Managers at similar levels of responsibility in the organisation, who report to the same superior, or to a different superior at a common level, need to co-operate with one another on a *lateral* basis. Since much communication is at this level, workable relationships are thus of the utmost importance if the company is to operate successfully. Typically the important lateral relationships below directorate level are business development, construction operations, engineering services, administration and finance.

While many companies lay down a formal management structure in the form of a 'family tree', in practice many informal relationships are demanded by the organisation; hence the newly appointed manager is well advised to settle in slowly and understand these lines of communication before making many decisions.

Matrix organisation

In a business engaged in total-package contracts and/or operating globally, the matrix management structure offers resources allocation advantages, by combining both functional and project management structures. For example, in the engineering-type organisation shown in Fig. 14.7, specialist/expert functional managers heading mechanical, electrical engineering departments, etc., *co-ordinate* with individual project managers responsible for delivering the products on site, by providing the required information and specialist resources support. Importantly, centralised communication is essential to effectively co-ordinate, monitor and control the

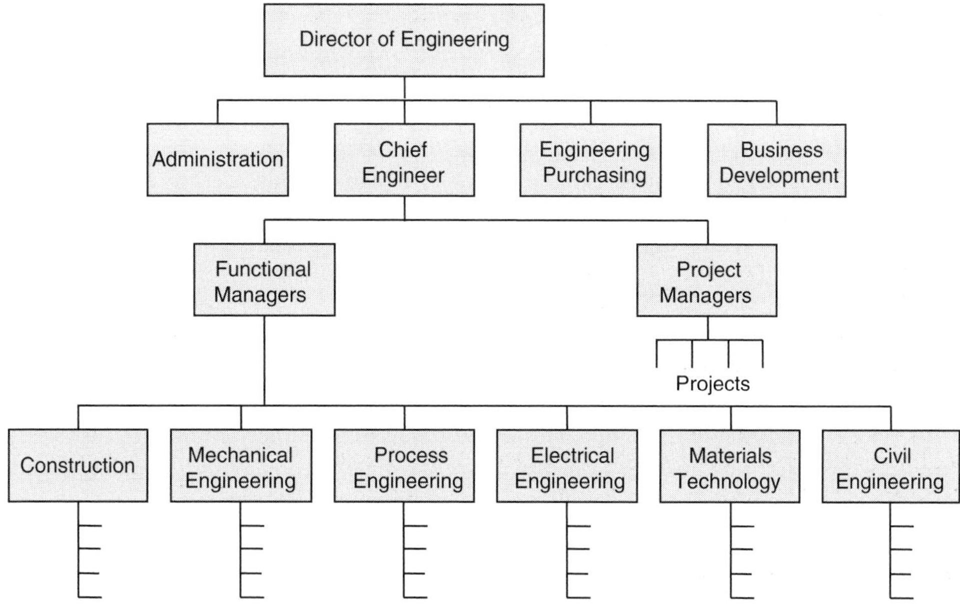

Fig. 14.7 Matrix organisation.

disparate elements involved. Hence, for construction work in particular, the paramount pressure to ensure tight end results for an individual project, favours *lead* responsibility being placed with the project manager, who in effect 'buys in' the required services.

Clearly, the balance of responsibilities between functional heads and project managers will vary depending on the nature of the end product. For example, in business development the marketing department might more effectively take the lead role. (See Chapter 8 for a more detailed description of project procurement and project management.)

Departments/functions

Similar functions are normally grouped together for managerial effectiveness, which is likely to be a compromise as many will conflict. Each department should be given clear and, if possible, measurable objectives, which can be either short or long term, so that it can work within the company's overall scheme. To this end, construction companies typically adopt the strategy of subdividing into the following broad functions:

- *Construction operations*—production management and control of contracts, including design management in design-and-build work.
- *Business development/marketing*—covers the interdependent mutually supportive functions of selling, estimating, tendering, pre-contract planning and the increasingly important area of identifying, negotiating new work and leading projects through to implementation.
- *Services*—construction planning, temporary works design, R&D, construction economics, production analysis, purchasing, plant.
- *Financial control*—capital procurement, cash flow, book-keeping, accounts, statutory returns.

- *Administration*—health and welfare, wages and salaries, information technology, quality management, training and education, corporate social responsibility, public relations, records and data protection and security, maintenance and legal matters.
- *Outsourcing*—some functional segments may be effectively outsourced to (offshore) contractors when costs are lower and service quality deemed satisfactory. Services such as IT-related, finance and accounting, human resources, design and engineering, etc. are typical candidates, but caution needs to be exercised in not transferring to third parties strategic functions or those requiring careful understanding and management.

Market planning and development

The business development function is primarily concerned with marketing, particularly developing growth by identifying and seeking out new opportunities, promoting the firm's products and services, including responsibility for technical appraisals and supply of advice to customers on all aspects of the firm's business from procurement to final delivery. With demanding clients increasingly emphasising improved value for money and better quality in the provision of their construction requirements, the function now extends beyond mere selling or simply tendering for workload to embracing the key elements needed to operate successfully in the expanding markets for design and build work, partnering, management contracting, etc. Consequently, for the larger concern, all the major technical and managerial elements required to draw clients into meaningful contracts are essential. Indeed, to some degree a well-founded department would contain a wide variety of the company's expertise, acting independently in a business development role and embracing capital sourcing expertise, scope/feasibility design services, cost planning, estimating, construction planning, buying, legal and contractual services, etc.

As the function develops in importance the mechanisms for apportioning its costs in the company will need to be carefully evaluated so as not to over-burden the allocations made to conventionally generated turnover.

Construction operations

Production

Once a contract is secured, the site manager responsible for the efficient execution of construction is fairly independent of the rest of the company, relying on service departments only when necessary. Overall authority is exercised through a contracts manager based at head office, which can generally make specific demands on service managers, although the official relationship is only lateral. Indeed, with the trend towards more subcontracting and associated supply chain management, co-ordination of the latter functions, particularly planning, purchasing, cost control and IT, is becoming a stronger feature of the production responsibility needing director-level leadership.

Subcontractors and suppliers

Few contractors are sufficiently vertically integrated with subsidiary companies able to supply all the materials, equipment, components and technical installation expertise needed for a given project and consequently rely on specialist subcontractors. Unfortunately, the ready availability and willingness of many to undertake work, particularly labour-only subcontractors, has hindered progress in providing clients with guaranteed quality. The introduction of Quality Standards such as ISO 9000 is an attempt to improve performance in this respect with the more enlightened clients and contractors increasingly requiring all suppliers and subcontractors to have secured approved quality assurance procedures.

Moreover, like the manufacturing sector, particularly the automobile producers, supply chains or networks of subcontractors committed to achieving quality, and full co-operation by sharing information while also lowering costs, enable best practices to be identified and spread throughout. Only those subcontractors and suppliers capable of operating in this way and willing to do so are then invited to join in future contracts.

Major clients, developers and contractors are increasingly applying the concept through closer involvement in, for example, the modernistic forms of contract such as design and build, management-type contracts and partnering, particularly where long-term agreements are involved. Notably the increasing share of design-and-build contracts is helping alter the culture of the construction industry, where contractors are embracing design firms in soundly integrated managerial and technological relationships—essential for developing high-quality products and services. Nevertheless, evidence of retained in-house construction competencies and resources continues to be sought by many clients seeking managing-type contractors.

Construction site management

A family tree illustrating the formal lines of communication of a typical construction project is given in Fig. 14.8. However, the precise project arrangements will vary depending upon the degree of the contractors' involvement in actual construction, for example in the case of general contractors, subcontractors and similar, the functions shown are all likely to be present. While in other types of contracting such as management contracts the site labour force responsibilities are largely eliminated, conversely there will be a need for strengthened design, planning and cost management teams to work closely with the client's advisors because design tends to continue alongside construction and works/subcontractors have to be carefully co-ordinated.

Notwithstanding these differences, the contracts manager, usually head-office based on all but the very large project, is responsible to the board for the overall successful management of the project(s), particularly the financial and legal aspects, together with dealings with client or the representative. The site manager oversees the day-to-day control of the processes conducted on site including liaison with the architect/civil engineer regarding instructions, payments, progress meetings, and commercial dealings with subcontractors, etc. The construction engineer can have manifold duties ranging from construction planning, materials control through to setting out. The site supervisor carries the onerous task of coordinating the labour force through the crew leaders, whether directly or self-employed. Both would need to have close functional ties with respect to supervising subcontractors and suppliers. The project quantity surveyor normally deals with construction economics, embracing valuation and agreement of completed work for payment, costing of variations, cost control itself and commonly bonus payments. The other functions are reasonably self-explanatory.

Services

These functions have traditionally provided support to construction operations, but in recent

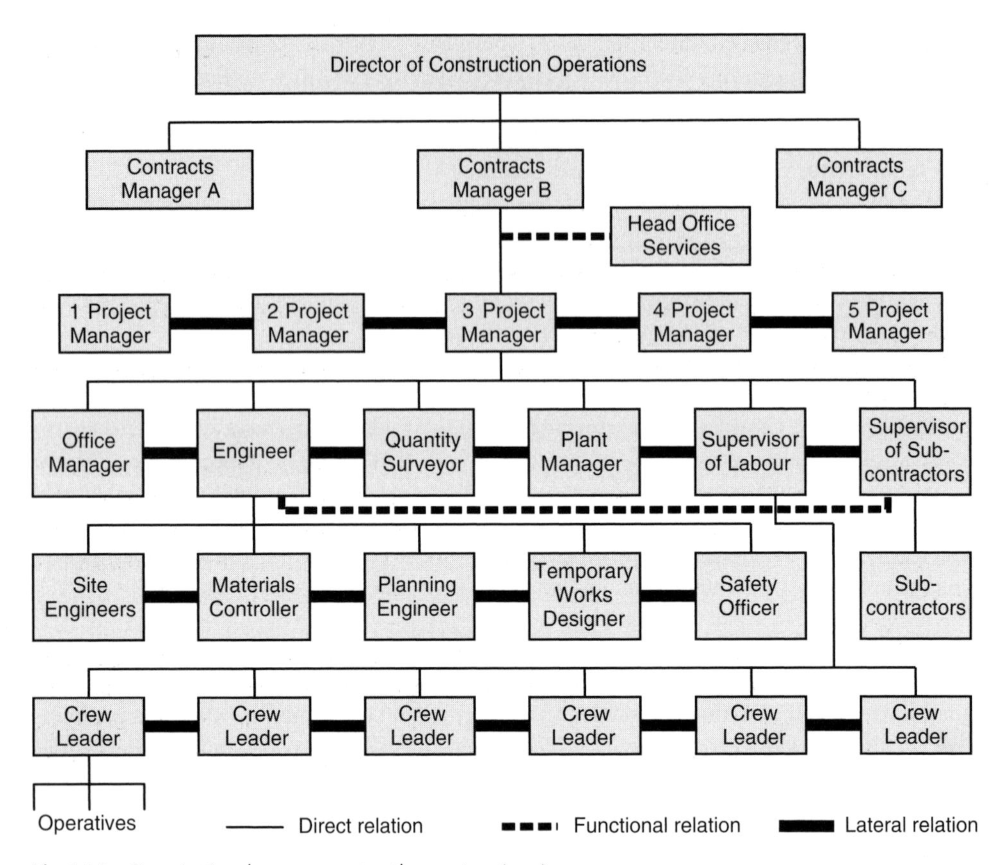

Fig. 14.8 Organisational arrangement on the construction site.

years with the trend towards more design and build, etc. a much fuller lead role at board level is increasingly necessary to cover the range of expertise needed in developing projects from enquiry through design to construction including possibly maintenance and even operation of the facility. Indeed, complete project management teams may be necessary for major projects.

Estimating and tendering

Commonly located under services for traditional contracting but more appropriately allied to the business development function where negotiated projects dominate, the usual procedure is for an estimator to be responsible for pricing work, whether negotiated enquiries or tender estimates involving a bill of quantities to which a mark-up is added for the final bid, this latter amount being decided at directorate level. As a rule of thumb, about one in ten bids result in actual contracts, leavened by a good track record in obtaining negotiated projects from favoured clients.

The usual practice is for the estimator to be given one or two months, depending upon the size of the project, to prepare the estimate. Assistance is required from the planning and temporary works departments in preparing the tender programme and from the buying department for obtaining quotations for materials supplies and subcontractors' work unless the business development function is sufficiently comprehensive to also embrace these elements at the pre-project implementation level.

Head-office planning

The functions of central planning are usually twofold:

- To support the estimating department, which in traditional contracting is often located under services in providing tender programmes, method statements and generally co-ordinating the whole tendering exercise
- To provide planning and co-ordinating services to site for those tenders that turn into contracts

Such arrangements have created problems for virtually all planners at head office, irrespective of the company. The problems essentially arise as a result of lack of authority. In situation (a) the estimator alone feels responsible for the final tender, and in case (b) site managers often prefer to plan the work their way, which is compounded by the difficulties of distance and communications with head-office planning. Indeed, site personnel commonly mistrust head-office interference.

Furthermore, the tendency is for management to locate spare engineers in a planning department for a few months until new contracts materialise. Such a policy has led to poor planning and a lack of confidence in the planning function by those who carry greater responsibilities, such as the estimator and site managers. The result is that few personnel wish to make head-office planning a career.

Purchasing

The buying department is responsible for obtaining all materials quotations and subcontractor procurement for both the tender and contract stages. At the tender stage the usual procedure is for a buyer to be responsible, so that the necessary quotations are at hand when required. Once a tender turns into a contract, the purchasing responsibility is then handed over to a more experienced buyer, who can hopefully negotiate the most favourable terms for the contract before placing the final orders. Increasingly, the function is playing a more central control and audit role in the supply chain. For a detailed analysis of the procedures for initiating an order, see Fig. 6.4 in Chapter 6.

The advantages of centralising purchases are:

- Provides a central point of responsibility for several contracts.
- Standard procedures can be adopted.
- Bulk purchasing is possible.
- Experience is developed.
- Head-office administration facilities are readily available.
- Monitoring and audit procedures can be readily applied.
- Supply-chain participants' performance and costs may be more effectively compared against target indicators.

Construction economics

This function is customarily carried out by the quantity surveyor (QS) and embraces responsibilities for the measurement of and the subsequent correct payment for work done on site, including negotiation of new work, variation orders, subcontractor payments, claim situations for each contract and commonly the budgetary and cost control elements. Usually the quantity surveyor is site based, with the QS department at head office executing unfinished business on completed contracts. Furthermore, co-operation and central co-ordination with purchasing is becoming essential for efficient supply chain management and auditing.

Engineering support

Unless the project is very large, functions such as formwork design, temporary works design and work study are located at head office.

Research and development (R&D)

R&D activities tend to feature strongly with the suppliers of manufactured goods to the industry rather than construction companies themselves. Indeed, the sector generates relatively little patenting of technology or intellectual property for commercial exploitation. However, the trend towards more design and build should provide better incentives for construction firms to invest in near-market research centred around improving commercial and management expertise and product performance, particularly development of manufactured standard components and units to maximise assembly and reduce *in-situ* operations on site. To this end, collaboration with the universities through staff development projects such as Knowledge Transfer Partnerships (KTP) involving research undertakings leading to advanced qualifications provide valuable opportunities for building a research culture where little existed previously.

Plant

The location of equipment control in the company organisational structure depends upon many factors, some companies preferring the plant department to operate as a profit-making division, hiring out both internally and in the open marketplace. Other companies use the plant department as a service, providing only those common items of equipment necessary on most contracts. In this case, the plant responsibility is often attributable to the services director. Whatever the choice of structure, plant ownership represents relatively large capital investments, which pose different types of problems compared to the management of construction work and therefore should be an option that is taken up only after considering all the consequences for the enterprise, as described in Chapter 7. The alternative arrangements for operating plant and some of their management features are shown in Fig. 14.9.

Plant organisation options

Independent plant hire

The firm operates in effect as a plant hire concern, in which plant is owned by either an independently formed company or a separate division of a construction company, or similar, and

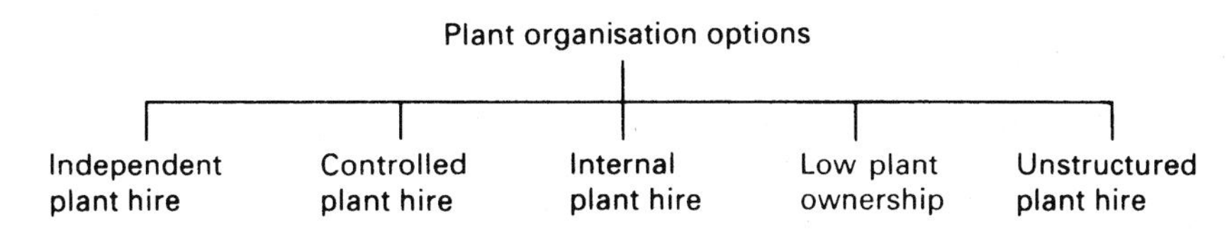

Fig. 14.9 Plant organisation options.

hired out to clients with the main objective of making a profit. Should the needs of the parent construction company arise, then this would be met on a similar basis to any other client.

Such plant hire companies would usually hold a range of the popular items in demand, although there is an increasing tendency towards specialised plant. The independent plant hire sector accounts for about 50% of the turnover value of plant used by the construction industry.

Controlled plant hire

Few construction companies can build up sufficient and diverse equipment holdings to provide an independent plant hire service. However, some of the large concerns possess considerable plant assets and, to encourage profitability, external hire is sometimes undertaken to raise utilisation levels. But, because outside hire must meet the demands of the market, there exists the danger that such matters as servicing and maintenance of the externally hired plant may receive priority over the internal needs of its own construction sites. As a 'rule of thumb' the ratio of 'hired internally' to 'hired externally' should not fall below two to one, to reduce this tendency.

Internal plant hire

Some contracting companies prefer to restrict their plant holdings to a purely service function by supplying only the internal needs of the company. A suitable rate of return on capital is required and the plant department is operated as a cost centre in order to encourage profitability. But, since there is little external competition, frequently items of plant are acquired to suit the needs of the construction contracts and thus economic utilisation levels cannot always be achieved and consequently hire rates get distorted. Unfortunately for many companies where equipment plays a relatively small role in relation to the overall operations of the company, this option is often the only viable arrangement.

Low plant ownership

Plant ownership requires workshop facilities, experienced and mobile maintenance crews and administrative facilities, etc. Many companies prefer to avoid this problem by operating only a very few equipment items, such as concrete mixers, small excavators, etc. Any major requirements are simply rented from the vast choice of hire firms available in the independent sector. Where there is a steady demand for hired-in plant, the company may sometimes institute a policy of hiring through a central administration to take advantage of competitive hire rates and even facilitate the transfer of plant between sites. The major disadvantage of this system arises when market rates change markedly between the time of tender for work and the actual hiring of the item on the construction project.

No plant structure

The final option is to have an unstructured organisation, whereby individual contracts purchase their plant requirements and are credited with resale values when the plant leaves the site. In this case care has to be exercised in assessing equitable sums when purchases and re-sales are internal transactions. This alternative method is usually confined to special items, e.g. grouting pumps, which are unlikely to be used subsequently by other general contracts.

Administration

The administration function is often too large to control as a single entity and is to be found in separate elements in most companies. The list includes personnel, health and safety, welfare, training, social facilities, postal services, legal and insurance facilities, wages and salaries, records, data protection and security, maintenance and corporate social responsibilities. The function is headed by an administration manager or director with subordinate managers responsible for groups of the above elements, reporting directly. Few of the duties are site based, possibly excepting safety, wages and canteen provision.

Information technology

Construction presently compares unfavourably with other sectors in the application of information technology, occupying a position towards the bottom of the industrial league table of IT spenders per employee. Notwithstanding, the growth of IT in recent years demands that construction companies largely manage these particular resources centrally in order to provide a degree of standardisation and avoid incompatibilities within the organisation, which can become an especially acute problem if departmental managers are allowed to acquire computers and software products without reference to an agreed company strategy and administration procedure. It also becomes imperative to provide for reliable data protection for all aspects of the firm's business.

Health and safety

The task of setting up health and safety (H&S) assurance procedures for the company and subsequent assessing provision in the workplace is best separated under a different chief executive officer/director from the departmental and project managers actually responsible for operations on site under, for example, the CDM and COSHH (Control of Substances Hazardous to Health) regulations. In this way, any conflict of interest between the needs of production and safety are minimised, with systems in place and problems avoided if and when visited by the Health and Safety inspectorate or in meeting requirements of the CDM H&S co-ordinator.

Quality management

Similarly, with the introduction of ISO 9000 quality standards into a company, the role of the quality assessment unit should be to advise departmental managers in preparing quality assurance procedures and work instructions and thereafter to follow up with inspections and assessments to sustain the expected performance levels. Indeed, where the firm has gained approval by external bodies such as the BSI, Investors in People Kite marks, etc., regular internal audits of functions and departments are essential as preparation for external assessment visits. Importantly, quality enhancement is increasingly requiring firms to give full commitment to education and training, necessitating clear objectives and strategies, approved by the Board of Directors, implemented through an annual appraisal scheme for all key employees, underpinned by a training plan of action for each employee and intended new recruits.

Training and qualifications

National and Scottish Vocational Qualifications (S/NVQs)

While the well-established programmes of learning in the further/higher education (FE/HE) sector provide specific knowledge, testing of understanding and qualifications for starting out on trade, technical, specialist and professional careers after schooling, effective job execution also demands particular competencies delivered to recognised standards. In this respect the NVQ scheme and SVQs in Scotland, currently being progressively developed by employers for their particular service, business sector or industry, help employees to reach the appropriate qualifying competency standard for the variety of occupations, skills and responsibility categories needed in the work place. Similarly, the intended enhancement of vocational study options to the core element in the school curriculum to include work-based learning pathways developed in conjunction with awarding bodies, employers and Sector Skills Councils, has the potential to provide much improved foundation level knowledge and qualifications, in preparing the 14–19-year-old cohort for further S/NVQ career progression.

The S/NVQ award scheme is organised into five competency levels, which map onto the National Qualifications Framework (October 2004) as follows:

S/NVQ level	NQF level equivalent	National Qualifications Framework level indicators with respect to S/NVQ requirements	Associated qualifications framework in education (EQF)*
Entry	Entry	Entry level qualifications recognise basic knowledge and skills and the ability to apply learning in everyday situations under direct guidance or supervision. Learning at this level involves building basic knowledge and skills and is not geared towards specific occupations.	Entry certificate in literacy
Level 1	Level 1	Level 1 qualifications recognise basic knowledge and skills and the ability to apply learning with guidance or supervision. Learning at this level is about activities which mostly relate to routine, predictable work, everyday situations and may be linked to job competence.	GCSE grades D–G, equivalent diplomas in education
Level 2	Level 2	Level 2 qualifications recognise the ability to gain a good knowledge and understanding of a subject area of work or study, and to perform varied tasks with some guidance or supervision. Learning at this level involves building knowledge and/or skills in relation to an area of work or a subject area and is appropriate for many job roles involving variable and complex work, some of which is non-routine and may require collaboration in a group or team.	GCSE grades A–C, equivalent diplomas in education

S/NVQ level	NQF level equivalent	National Qualifications Framework level indicators with respect to S/NVQ requirements	Associated qualifications framework in education (EQF)*
Level 3	Level 3	Level 3 qualifications recognise the ability to gain and, where relevant, apply a range of knowledge, skills and understanding. Learning at this level involves obtaining detailed knowledge and skills needed in taking autonomous responsibility for variable and complex work, most of which is complex and non-routine, including guiding, training, supervising and control of other workers.	National diplomas and certificates in further education, Scottish Higher, GCE A levels
Level 4	Level 4	Level 4 qualifications recognise specialist learning and involve detailed analysis of a high level of information and knowledge in an area of work or study. Learning at this level is appropriate for people working in technical and professional jobs, performed in a variety of contexts, including allocation of resources and/or managing and developing others, and control of other workers.	Certificates of higher education
Level 4	Level 5	Level 5 qualifications recognise the ability to increase the depth of knowledge and understanding of an area of work or study to enable the formulation of solutions and responses to complex problems and situations. Learning at this level involves the demonstration of high levels of knowledge, a high level of work expertise in job roles and competence in managing and training others. Qualifications at this level are appropriate for people working as higher grade technicians, professionals or managers. Typically provide access to postgraduate programmes.	Diplomas of higher education and further education, foundation degrees, higher national diplomas
Level 4	Level 6	Level 6 qualifications recognise a specialist high-level knowledge of an area of work or study to enable the application of an individual's own ideas and research in response to complex problems and situations. Learning at this level involves the achievement of a high level of professional knowledge and is appropriate for people working as knowledge-based professionals or in professional management positions.	Bachelors' degrees, graduate certificates and diplomas
Level 5	Level 7	Level 7 qualifications recognise highly developed and complex levels of knowledge that enable the development of in-depth and original responses to complicated and unpredictable problems and situations. Learning at this level involves the demonstration of high-level specialist professional knowledge and is appropriate for senior professionals and managers.	Masters' degrees, postgraduate certificates and diplomas
Level 5	Level 8	Level 8 qualifications recognise leading experts or practitioners in a particular field. Learning at this level involves the development of new and creative approaches that extend or redefine existing knowledge or professional practice.	Doctorates, specialist awards

* See SCQF for Scotland

Examples of construction occupation and S/NVQ level

Level 1—Construction site operative
Level 2—Experienced worker or completed apprenticeship
Level 3—Skilled worker or Supervisor (Technician equivalent)
Level 4—Manager (Chartered Profession equivalent)
Level 5—Contracts manager or similar senior positions

S/NVQ quality monitoring

In England (Northern Ireland, Scotland and Wales have separate organisations) the Qualification and Curriculum Authority (QCA) regulates NVQ qualifications to the criteria set out in the *National Qualification and Occupational Standards* framework and ensures that qualifications are broadly comparable across different sectors, including responsibility for quality audit, assurance and management. QCA also formally recognises (accredits) specific proposals for NVQ awards developed by Awarding Bodies working in conjunction with Sector Skills Councils. SSCs are responsible for identifying, defining and updating employment-based standards of competence for the specific sector occupations.

Sector Skills Councils (SSCs)

SSCs are led by employers and bring together industry leaders, trade unions and government to address skills issues directed towards:

- Improving productivity, business and public service performance
- Reducing skills gaps and shortages
- Increasing opportunities for employees to improve skills
- Raising the availability of funded apprenticeships and training provision

Much of the UK workforce is represented within the Skills for Business network of about 24 SSCs, the construction-relevant councils being as follows:

- ConstructionSkills (construction)
- Lantra (environmental and land-based industries)
- SummitSkills (electro-technical, heating, ventilating, air conditioning, refrigeration and plumbing industries)
- ProSkills (processing and manufacturing of glass, building products, etc.)
- SEMTA (science, engineering and manufacturing technologies)
- Cogent (chemicals, nuclear, oil and gas, petroleum and polymers)
- AssetSkills (property, housing, cleaning and facilities management)

Procedure for approving and operating an S/NVQ award

(1) Experienced Awarding Body (e.g. City & Guilds, EdExcel/BTEC, OCR, CITB-Construction Skills, ECITB, professional institute(s), trade association, etc., or a combination of such bodies) designs the S/NVQ content, assessment and quality assurance procedures.

(2) Awarding Body obtains the specific SSC endorsement prior to submission to the relevant QCA for formal accreditation of the qualification (e.g. ConstructionSkills; also relevant to construction are Lantra, Cogent, Energy & Utility Skills, SEMTA, SummitSkills, AssetSkills and ProSkills).
(3) QCA accredits the proposals for the qualification submitted by the Awarding Body.
(4) Awarding Body approves the assessment centre (e.g. college, training organisation, etc.) to run the S/NVQ, and implements quality assurance and management responsibilities.
(5) Assessment centre assesses the NVQ to the criteria set out by the awarding body.
(6) QCA quality monitors the awarding body's performance.

Registration of the candidate

The S/NVQ registration process is flexible and open to all grades of full-time employee, part-time worker and school or college student with a work placement. Qualification is normally achieved through on-the-job assessment and training, usually conducted as follows:

(1) Candidate registers with the approved assessment centre of the awarding body offering the accredited qualification.
(2) Assessment centre Advisor and internal Verifier are allocated for the candidate.
(3) Job activities are jointly evaluated and the appropriate S/NVQ level to be achieved ascertained.
(4) Candidate's current expertise in relation to the job is identified.
(5) Means for meeting missing skills are determined and specific training, development programmes recommended, e.g. school/FE/HE courses, on-the-job coaching, job-shadowing, job-swapping, etc. (see EQF above for programme levels).
(6) Plan of required tasks/work is established and target undertakings/submission/completion dates programmed.
(7) Candidate's portfolio of captured evidence demonstrating ongoing achievements is progressively presented for assessment by the advisor.
(8) Candidate's knowledge, understanding and work-based performance is assessed against the competency criteria.
(9) Programme is amended, adjusted and updated as necessary as the candidate develops in competency.
(10) S/NVQ awarded when all units have been successfully achieved, internally assessed/verified and externally verified by the AB.

S/NVQ units are presently undergoing development towards establishing standardised credit values and assessment regulations, aimed at facilitating a universal credit accumulation and transfer scheme for adult learners. The available construction S/NVQ awards and syllabi are listed on the Construction Skills website (www.constructionskills.net/).

Training programmes and continuous professional development (CPD)

(1) The construction and engineering construction training boards, and professional/trade/employer associations, are active in S/NVQ award development; particularly the operative and trade occupation levels 1 & 2/3 for the 14–19 year age group. Much of the training itself is directed through national/local government-funded initiatives, such as

'on-the-job training' programmes, etc. Commonly approved industry training centres/providers and colleges work in concert with co-operative employers to deliver the accredited S/NVQ award. In the fully developed form, modern apprenticeships comprising several weeks of intensive training followed by guided practical experience may be part funded by the Learning and Skills Council to assist employers to defray some of the costs (see www.apprenticeships.org.uk). Key Skills and City & Guilds certificates, BTEC diplomas, graduate awards, etc. may be incorporated into the S/NVQ as appropriate for the specific level.

(2) Firms paying the CITB (Construction Industry Training Board) levy are able to secure grants and assistance for a variety of training and CPD activity, which may sometimes be suitable for incorporation into the S/NVQ scheme, when carefully planned and organised.

(3) The Armed Services, whose personnel often enter the construction industry after demobilisation, are important S/NVQ training providers.

(4) Qualifying standards for the chartered professional and technician occupations are in the main separately defined by each professional institute. Consultant firms, services organisations and contractors may offer HE graduates personalised training agreements (TA) or participation in a company-approved training scheme tailored to the particular institute route to membership, where an appropriate S/NVQ may be accepted as part of the process.

(5) Training programmes are sponsored by government, designed to bring/induce the unemployed back into work and also provide useful competency qualifications, e.g. OCR-CLAIT computer literacy and information technology skills.

Construction Skills Certification Scheme (CSCS card)

Significantly, in demonstrating support for properly assessed competencies for construction personnel, companies participating in the CSCS scheme require possession of the appropriate CSCS card to work on the construction site. All personnel, including the design and other professional consultants are embraced, with even delivery personnel, etc. required to apply for the Regular Visitor card. Passing the CSCS Operative Health and Safety Test is also a mandatory requirement, unless exempted by evidence of other approved health and safety awareness training and testing.

The scheme is owned by CSCS Limited and controlled by a management board consisting of members drawn from the Construction Confederation, Federation of Master Builders, GMB Trade Union, National Specialist Contractors Council, Transport and General Workers Union and Union of Construction Allied Trades and Technicians. Observer status extends to the Department for Education and Employment, the Department for the Environment, Transport and the Regions, the Health and Safety Executive and the Confederation of Construction Clients.

The CITB-Construction Skills administers the scheme and verifies documented formal evidence of competency. The CSCS website (www.constructionskills.net) lists the currently recognised construction occupations and provides an appropriate job description for each.

Application for registration for a CSCS-listed occupation may be made individually after having achieved the appropriate S/NVQ qualification or equivalent, whereupon the approved colour-card (similar to a credit card), valid for three or five years, is issued following payment of the stipulated (small) fee.

A Certificate of Achievement is also available for individual companies depending on the proportion of their on-site registered card-holder personnel.

About 10% of the core construction trades workers are currently CSCS registered. Ultimately all construction personnel are intended to be included, which will clearly necessitate continued government financial support for training initiatives and enlightened client and developer procurement policies relating to qualified workforces in contracts, but most essentially the vigorous commitment to training and registration enforcement by the major contractors and their supply networks/subcontractor chains.

Finally, as a point of caution, the all too prevalent poaching of trained staff by firms unwilling to invest in training is likely to continue to plague parts of the industry.

Corporate social responsibility (CSR)

In general, government through elected politicians is accountable to citizens for determining the goals of regulators, dealing with externalities, mediating between different interests, attending to the demands of social justice, providing public goods, collecting taxes, prioritising and organising the necessary resources, etc. Nevertheless, under intensifying globalised trading conditions, enlightened company boards, particularly of multinational corporations, are gradually paying more attention to social responsibility policy beyond the minimum requirements stated in their articles of association, as a means of addressing pressing international and public concerns specifically related to business ethics, while also meeting normal obligations to shareholders and other stakeholders for profit generation. A CSR initiative, consultant, officer or department, is typically charged with the task of advising an appropriate senior executive on matters such as business ethics, relevant corporate governance and integrity, accommodation of sustainability and environmental issues, ethical management practice—avoiding dealings and investments in socially unacceptable sectors of commerce or industry, denying contracts with companies aggressively exploiting its workers and similar social injustices, etc.

Finance and accounts

The financial management function is responsible for book-keeping, payment of invoices, provision of Companies Acts information and preparation of the trading, profit/loss accounts and balance sheet results. Most importantly the financial officer or director needs to work closely with the MD or CEO in controlling the overall financial affairs of the company. Furthermore, although each individual contracts manager is informed about the financial position of their project and must endeavour to keep within tender figures, it is the finance manager's concern to monitor the overall financial position of the company.

Management attitude

Opportunity

Most employees favour the firm that provides good opportunities for promotion. Indeed, in most organisations virtually 80% of value added is achieved by only 20% of the workforce; hence a good company should have a sound policy of manager development to assist promotion of this more dynamic element from within. Nevertheless care must be exercised to ensure that sufficient 'outsiders' are also introduced to resist the tendency to loss of drive when relationships become stagnant.

While people from other companies and industries often bring fresh ideas and attitudes, manager selection can become problematical if the company is in a rapid growth situation. Extensive outside recruitment often then becomes unavoidable, potentially leading to a shambles of conflicting personalities, values and attitudes, resulting in a poor public image of a 'flying by the seat of the pants' type of enterprise.

Personnel development

Manager development first implies recruitment of well-educated people having common purpose with the enterprise, followed by attention to career development and giving opportunity to gain experience in as many parts of the organisation as possible. The young appointee permanently on site will gain little knowledge of the working of the company as a whole. To become a well-rounded manager, it is essential for them to spend one or two years in a head-office-based department, well supplemented with short courses providing information on modern management techniques. A continual process of CPD and training, ideally leading to a relevant NVQ, enables potential candidates to be tested for top jobs, or alternatively located in specialist functions where technical skills and up-to-date knowledge can be of value or ultimately encouraged to move on as appropriate.

Motivation

A core duty of a manager is to coach assistants, who will then go a long way in encouraging motivation, essential if the members of the team are to pull their weight effectively and wholeheartedly. Morale must be high; nothing is more inspiring than a feeling of confidence. Many managers unwittingly destroy this confidence through excessive egotism, arrogance, being over-secretive, aloof and not communicating small but significant information about the department to their assistants.

Thus, it is of vital importance that managers know where they stand in relation to assistants and gain their fullest respect. In return the assistants will expect some interest to be shown in their careers and promotion prospects; otherwise able people will be quickly disillusioned and look elsewhere for employment. A negative attitude in this respect can lead to an unhealthy cluster of people at the top, leaving few with experience to take over when the time arrives.

Leadership

Leading a team demands the ability to combine human resources and obtain the best performance possible. To do this, requests and orders are necessary and these can only be really effective if assistants are willing to carry them out efficiently, although they may not fully respond without respect for their superior. Leaders, by implication, should therefore be inspiring people of experience, understanding and vision, and have enough confidence to delegate responsibility and stand by decisions, while also instilling discipline. *Good management rests on building a confident team, demonstrated through mutual respect, commitment and discipline.*

Reference

Drucker, P.F. (1999) *Practice of Management*. Butterworth-Heinemann, Oxford.

15

Market planning and business development

Summary

All companies need to seek new markets in order to survive in an increasingly competitive situation. The principles involved in formulating a business development strategy are described, providing the framework upon which a detailed trading analysis and corporate plan can be constructed.

Introduction

The purpose of every construction company is to be able to obtain sufficient business orders in the form of projects, and to execute these orders efficiently. Obtaining the business orders involves market planning and business development and promotion.

Market planning

Marketing is the management function that organises and directs all those business activities involved in assessing and converting customer purchasing power into effective demand for a specific product or service, and in moving the project or service to the final customer or user so as to achieve the profit target or other objectives set by the company.

Accordingly, business is developed by identifying and seeking out new opportunities, promoting the firm's products and services, including responsibility for technical appraisals and supply of advice to customers on all aspects of the firm's business from procurement to final delivery.

With demanding clients increasingly emphasising improved value for money and better quality in the provision of their construction requirements, the function now extends beyond mere selling or simply tendering for workload to embrace the key elements needed to operate successfully in the expanding markets for design-and-build work, partnering, management contracting, etc. In particular, companies involved with modernistic types of contract should benefit considerably by carrying out marketing studies since much turnover is secured from negotiated contracts such as BOOT, DBFO and PFI/PPP projects. It is in this field that contractors can offer attractive packages to the client and thereby have more control over revenue. Consequently, for the larger concern all the major technical and managerial elements required to draw clients into meaningful contracts are essential. Indeed, to some degree a well-founded marketing department would contain a wide variety of the company's expertise acting independently in a business development role.

While it can be inferred that, unless the company is large and able to offer a wide range of services, there is little place for marketing, nonetheless virtually all firms are subject to unpredicted changes in business conditions and therefore in practice need to carry out market analysis in some form. The details of the approach will vary from company to company, but in essence three stages are clearly definable: *identifying*, *promoting* and *satisfying* the customer's needs at a profit. In simple terms, development of the business involves seeking out the customer's material wants and desires, so that the company can organise itself in the best possible way to satisfy those tastes and requirements.

Identifying customers' needs

Drucker (1999) writes that the primary concern of a company is to survive, but, in practice, many other objectives are necessary to satisfy the desires of the shareholders, management, workforce and society. Quite naturally, in a capitalist market, the need to return a profit on the financial investment of shareholders and lenders is very important, but other objectives may include the desire of top management to see the company grow and continually increase turnover, become a well-known and respected company in its field, operate in more stable markets, keep up with competitors and demonstrate impressive corporate responsibility; these are but a few aims. However, the ability of the company to achieve its goals is continually being influenced by business changes as indicated by pressure on profits, supply outpacing demand or vice versa and competitors being more successful, as well as through alterations in customer tastes and requirements. Business development should therefore be the management function that enables the company to keep in touch with customer needs and thus plays an important part in ensuring that the enterprise remains profitable.

The business development process

Most construction projects take one or more years to complete; they are often unique and require specialist equipment. The alternatives open to a construction company, once it has embarked upon a particular type of work, are likely to be those which can only be introduced slowly and in the long term, e.g. expanding into civil engineering works from a building background. To be fully confident in setting the new objectives, the company must be prepared to undertake the most thorough investigation of itself and business opportunities. Broadly, the investigation should involve two separate stages:

(1) Formulating a business forecast
(2) Assessing the strengths and weaknesses of the company

The matching of these two exercises enables senior management to set the new policies (see Fig. 15.1).

The business forecast

All companies buy or hire factors of production, in the form of labour, materials and equipment. Initially, capital is borrowed or saved to purchase the goods, which are then turned into other products, which are then sold at a profit, the lender of the capital thus being rewarded for putting

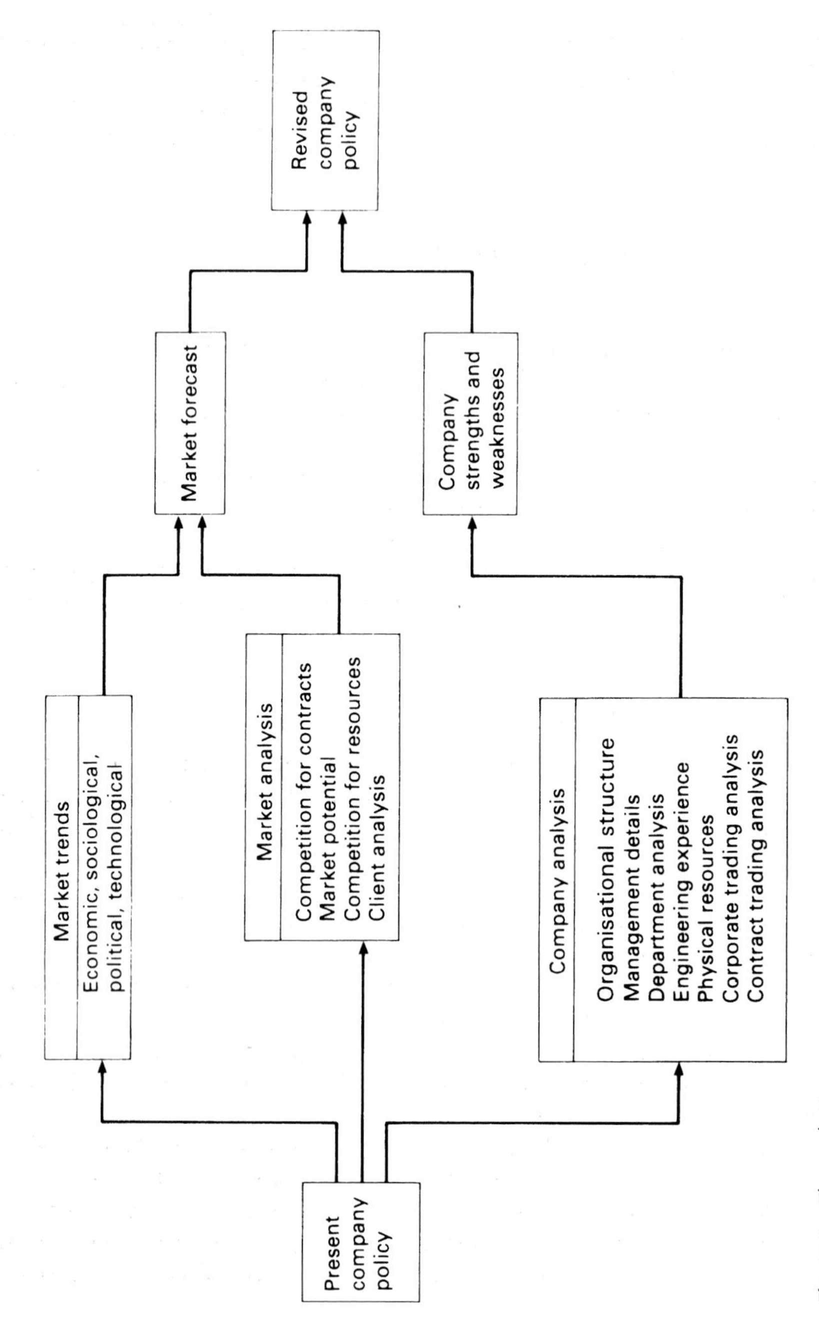

Fig. 15.1 The marketing process.

money at risk. The company that knows its market well is thus in a better position to improve its competitiveness and profitability.

The business and market forecast is made possible only after considering many separate factors, the main areas for a typical construction company being identified in the following sections.

Analysis of the business opportunities

Analysing competition for contracts

A survey of the past performance of other companies may bring to light those areas of contracting that have been successful and are therefore favourable for development, or conversely areas to avoid because of fierce competition. The main points to determine are:

- The present share of the company for each of the different classes of work carried out
- The share and turnover for each of the major competitors
- The recent growth record and profitability of the major competitors
- The record of obtaining work of each type in competition, related to the past mark-up levels; this information may indicate the improvement necessary in order to become more competitive

Analysing potential for new business

Although a thorough knowledge of the competitor's position is essential before entering new markets or expanding old ones, specific opportunities must be thoroughly researched in the hope of finding out patterns and trends for the future. The information required should include:

- The type and volume of work available for tender and negotiation during the past years.
- The number of competitors tendering for different contracts put on the market (this information is often difficult to obtain), the location of the work and clients involved.
- A forecast of the likely growth of opportunities sector by sector, particularly those introducing innovative technologies and new business models.
- Government-influenced development areas potentially generating work both directly in the form of cash injections and indirectly through the additional services demanded by a modern growing community once established
- Government forecasts of expenditure, the effects of regional trading and/or political groupings, e.g. European Union membership, the size of the budget deficit, the rate of growth of the gross domestic product and the trade balance, which all need to be taken into account. These latter points are particularly important for assessing the future requirements in the private sector, as the rate of investment is often governed by the confidence that industry has in the prospects for the economy as a whole.

The above findings will help to establish which are the buoyant and potential sectors and the types of company operating in them. Furthermore, the information will indicate which sectors are investing and therefore likely to require new facilities.

Analysing competition for resources

There is very little point in making elaborate plans for a new venture if the resources in the form of capital, materials and labour are not there to supply it.

Capital

Most construction enterprises, because of their small asset base, enjoy very limited access to investment capital since little collateral security can be offered to offset the lending risk; hence private sources or bank overdrafts are likely to be the principal providers. The larger concern, notably the company listed on the Stock Exchange, has more opportunities to raise capital through new share offerings, rights issues, etc., but in comparison to manufacturing and other commercial enterprises with the capability to raise funds by issuing their own bonds, even in international currencies, the small asset base is again likely to hinder the raising of significant finance to fund expansion opportunities. Consequently strategically well-placed construction companies wishing to open up opportunities in bigger markets such as DBFO tend to face takeover pressures from more powerful international concerns with access to long-term capital and business development skills.

Materials, labour resources and subcontractors

Access to reliable sources of labour, materials/suppliers and quality subcontractors carries risks of availability and indeed security. In the advanced economies labour supply, especially experienced managers, engineers and skilled workers is very competitive, requiring major commitments from the construction sector acting in conjunction with government training and education schemes to try to ensure adequate recruitment. These are not short-term problems and any company, including its chain or network of suppliers and subcontractors, not willing or unable to devote the necessary resources to training is best advised to leave alone the sectors where skill and/or technical competence predominate. In the international arena these difficulties can sometimes be alleviated by virtue of globalisation, particularly countries grouped together for regional trade, so alleviating labour shortages by engaging foreign nationals and staffing projects with them. Moreover, for other resources trading arrangements are gradually being influenced by the World Trade Organisation control of tariffs through rules regulating imports and exports of goods and services, making it possible for projects to be supplied from the most convenient and competitive source rather than just the local market.

Analysing the client

Any records kept on clients should be very revealing and may have considerable bearing on the final decision to enter a market. For example, profitability may have varied with different clients, depending upon the attitude to claim situations or interpretation of the contract documents. Some clients prefer to select on the basis of tenders, others by negotiation, partnering etc.; if the company has been dropped from a client's tendering list, the reasons why may be important. Also, until recently it was the contractor's financial position that was questioned; today, even the client has money problems and there is little point in obtaining contracts if payment by the client is open to doubt.

Company strengths and weaknesses

A change in company policy regarding new directions will be unsuccessful if the new objectives are frustrated by fundamental weaknesses in the company structure. Therefore, any search for new business should be coupled with an equally thorough examination of the company itself.

The company analysis

Organisation structure

Most companies have a 'family tree', which represents the official structure of the management organisation. In practice, the actual lines of command and communication are likely to be more subtle than those formally recognised. However, this 'family tree' is a good starting point in highlighting potentially weak structural arrangements.

Management details

The quality of the present managers will be tested upon entering new markets. The personnel department often holds much information on such matters as salary, qualifications, education, training and experience. This data helps to identify potentially strong management areas and those that have failed to develop a healthy ladder of achievement for the younger staff to gain experience. If the process is repeated for each department, gaps and stagnant areas become apparent.

Financial and operational control departments

This review should be extensive and probably not undertaken until policies have been tentatively made. The likely candidates for investigation are accounts, buying, estimating and IT departments as these contribute significantly to overheads, which may rapidly increase when the company expands into new markets. Overheads should also be borne in mind when moving from a fairly straightforward market, such as housing, to, say, D&B/PFI requiring high technical competence and support from service departments such as design, temporary works, centralised purchasing, cost management, head-office planning, financial control and IT.

Supply chain or network

The company extensively engaged in design and build or similar projects requires the development of a supply chain or network of subcontractors committed to achieving best quality practices. These need to be identified with only those subcontractors and suppliers capable and willing to fully co-operate by sharing information while also lowering costs, invited to join in future contracts.

Engineering experience

The management and operational control surveys may field much information about the nature of the company and its employees. However, the construction business requires good managers to be good engineers also. Any change in policy should spring from a sound base of experience; it is far too risky to rely entirely on imported skills when undergoing change. Hence, a careful analysis is required of the existing skills within the company to see if they will provide an adequate basis. Indeed, a solid base of architectural, design, engineering, IT and commercial expertise will be essential for companies involved in design and build and similar work, which demands considerable knowledge and competency in negotiation with clients, suppliers and subcontractors.

Physical resources

To put new objectives into practice may necessitate new office space and stores facilities. This is a very important consideration if bottlenecks are to be prevented. However, the acquisition of land and the construction of new facilities take time, are expensive and require careful planning of the location.

Corporate trading analysis

To assess the financial strength of the company and to compare its performance with that of major competitors, the following financial ratios yield important information:

- Return on capital employed
- Profit on turnover
- Turnover of capital
- Growth in capital employed and in net profits
- Current assets to current liabilities
- Stock values to sales
- Debtors to sales, converted to time periods
- Profit per employee

By comparing figures over the past five years with other companies in similar fields, some judgement is possible on the viability of the firm and its ability to take on new ventures successfully.

Contract trading analysis

The contract trading analysis means looking at individual contracts in a fair degree of detail. The sorts of questions to be asked are:

- What was the effect on profits of contract type, size, location, duration and client?
- What trends in profitability, say during the past five years, can be seen in different construction sectors?
- How did actual profit compare with the bid mark-up for individual contracts? (Once again, trends might be spotted, indicating perhaps partnering to be more reliable than open tendering.)
- What effect would the changing of mark-up on each tender have had on the turnover and profit expectations for the various types of work?

The results, however, need to be considered with care, since those contracts where profit is high tend to be those that have high risk or require special technical skills.

Marketing policy formulation

After the forecast patterns of demand and competition are assessed and the analysis made of opportunities and risk, together with evaluation of the company strengths and weaknesses, the revised corporate policy can be established.

We have now arrived at the last box of Fig. 15.1 and, once the relevant facts are known and synthesised, senior managers will gather together the appropriate expertise to help ascertain what changes need to be made. To this effect, different scenarios are generated via brainstorming sessions, workshops, etc., taking into account the possible future impacts of potential movements of social, cultural, political, economic and technology trends over the medium term. The important point is that policies and strategies be specifically formulated and recorded. Here, the systematic decision-making technique developed by Kepner and Tregoe (1997) described in Chapter 7 can usefully assist the process. The identified factors deemed important in meeting the marketing objectives are subjectively ranked, the perceived performance of each strategy/policy option then rated and its weighting calculated. The grand total for each thus provides individual numerical values for comparison. (Also see the section on 'Business objectives' in Chapter 14 *Company organisation*.)

Business model

The potential commercial viability of each strategy or policy may be further evaluated by constructing an appropriate model comprising the conceived business plan manipulated for example on a spreadsheet with product or service, partner, supplier and client or customer data. Simulated results are subsequently produced for sensitivity analysis to compare the alternative options and ultimate field-testing of the preferred option, typically formulated as follows:

(1) Set the market policy or strategy option for the product or service.
(2) Ascertain the marketing data and procurement information.
(3) Assess the means of dealing with partners, suppliers and client or customers.
(4) Enumerate the requirements for achieving the policy or strategy.
(5) Devise outline production, management and business processes and methods.
(6) Develop and manipulate the relationships between appropriate economic variables.
(7) Determine/simulate the resources schedule and time plan.
(8) Establish the sales plan.
(9) Derive the anticipated business and financial results.
(10) Repeat the procedure on alternative models and conduct sensitivity analysis.
(11) Select the preferred option.
(12) Monitor and measure out-turn results on the selected option.
(13) Adjust, modify or replace the policy/strategy or model as necessary, and re-examine.

Trends that may upset forecasts

No matter how good the information on which the new policies, strategies and business model or plan are formulated, new facts soon emerge and errors appear in the estimates. Some could be due to normal fluctuations but other influences may be disrupting past trends and demand re-engineering and strategic redirection of the firm, for example:

- *Political changes*, e.g. stricter legislation affecting company training policy, more rigorous safety requirements, limits on urban sprawl, more use of 'brown field' sites, tighter legislation on the generation of waste and mandatory recycling
- *Sociological and cultural changes* such as shifts in consumer outlook on fundamental issues concerning quality, pollution, corporate governance, corporate social responsibility, environment and design for sustainability
- *Economic influences* caused by movements in the economies of other countries and powerful regional developments like the European Union and increasingly influential East Asian economic co-operation arrangements
- *Technological changes*, for example towards concurrent/integration of design and construction, application of CAD/CAM technology, customer bespoke system/modular/prefabricated construction, business process outsourcing (BPO) of back-office functions, e-commerce advertising, transactions, payments, online tracking and delivery

While most of these factors are difficult to quantify, they should be kept under continuous review and the policies adjusted when necessary, with care taken not to over-react to new evidence as this may cause loss of confidence at middle management level.

Promotion of product or service

Promotion is the activity that takes opportunity identification to the moment of sale and covers all the aspects of persuading the customer to buy, including the simple, inexpensive methods of communication discussed in the following.

Advertising

Advertising can take many forms, such as:

- Eye-catching yet pleasing advertisements in the press for new management appointments and other employees
- A standard and attractive colour scheme for all company plant and goods vehicles embellished with clearly identifiable trade markings
- Good quality notepaper and attractive letter headings
- Advertising in the trade journals, on TV and the internet, etc., even extending to corporate web 'blogging' PR
- Sales brochures outlining the activities of the company, which can be sent to prospective clients and other interested parties
- Hoardings installed at football grounds, etc. in the hope of catching the television cameras
- Topping-out ceremonies photographed in the local press

Public relations

Good public relations are fostered by:

- Exhibitions, particularly encouraging students at the institutions, universities and colleges to visit the company's construction works.

- Lecturing to interested parties on subjects and topics which the company is qualified to do. Such activities help to enhance the professional stature of the company in the industry generally, especially when presented by competent staff well trained in presentation skills. Notably, when distributed via the media, e.g. tv, radio, film, road show etc. 'received prononciation' is generally preferred, but occasionally an appropriate regional accent may be more effective, depending on the context.
- Writing articles for learned journals and magazines by company specialists.
- Spreading the news through all available methods about the company's successful contracts, training schemes, health and welfare facilities, safety record, etc.
- Producing films illustrating the best features about the company and its technical abilities. By keeping the advertising low key, the films, if well made, will be suitable material for teaching and lecturing purposes at universities and colleges.
- Developing a reputation for good quality, early completions, few labour problems and low emphasis on contract claims. Correspondingly, many prospective clients are impressed by support facilities from head office, e.g. planning, temporary works design, cost control, etc.
- Keeping neat well-established sites with clear directional and other signs. The Japanese in particular impress clients by allowing the labour force about 15 minutes before finishing time to clear away rubbish, waste, off-cut materials, etc., insisting that the task is carried out. The results are indeed quite impressive, with neat, orderly sites the outcome.
- Projecting and branding the image of the company in terms such as: commercially and technically advanced in its market segment, IT competent and well resourced, engages well-qualified experienced managers, employs a trained workforce, has a well-developed quality chain of suppliers, operates a respectable corporate social responsibility policy, has excellent health and safety and environmental management achievements, provides good employee development opportunities, etc.

Selling

In general, companies particularly involved in speculative housing, negotiated contracts, package deals and similar have real opportunities to engage in the 'high-powered' selling techniques typical of the consumer industries.

Customer relations

Above all others, a well-satisfied customer provides an excellent ambassador for future sales.

Implementing the market plan

The *business development function* is primarily concerned with implementing the market plan, particularly by seeking out the newly identified opportunities, promoting the firm's products and services, including responsibility for technical appraisals and supply of advice to customers on all aspects of the firm's business from procurement to final delivery. With demanding clients increasingly emphasising improved value for money and better quality in the provision of their construction requirements, the function now extends beyond mere selling or simply tendering for workload to embrace the key elements needed to operate successfully in the expanding markets for design and build work, partnering, management contracting, etc. Consequently, all the major technical and managerial elements required to draw clients into meaningful contracts are essential and these include capital sourcing, scope/feasibility design services, cost planning, estimating, construction planning, buying, legal and contractual matters, etc. Indeed to some degree a well-founded department would contain a wide variety of the company's expertise acting independently in a business development role.

Satisfying the customer's needs at a profit

Some of the obvious opportunities for consideration are:

(1) Advising clients of their facility needs and subsequently bringing projects successfully to implementation will help secure follow up orders. Added values such as back-up services of experienced and well-qualified staff experienced in areas such as design and building surveying, tendering and estimating, construction planning and programming, temporary works design, law and insurance, industrial relations, research and development and, not least, sound finances are also likely to foster business development. Moreover, the internet, web sites, intranet/extranet electronic data interchange services and e-commerce transactions are essential routine tools expected in modern business communications.

(2) Where materials, finishes and, particularly, precise client requirements are tightly specified, subsequent careful investigation by the contractor(s) of market alternatives may yield better solutions, goods at more competitive prices, guaranteed supplies and services on time, discounted prices for bulk purchases, etc. Adhering to the specified construction programme, so that all the parties involved can align activities in advance is also likely to be well appreciated by the client with deadlines to meet. For example, the supermarket company wishing to take advantage of a particular part of the selling season may pay more for early completion. In contrast, the public client will be perhaps more concerned with cost overrun (the implications being the use of taxpayers' money) and may remember excessive profit maximisation by the contractor. Several large companies in the past have lost their place on tender lists by a too-aggressive stance to claim situations.

(3) In a market of continually evolving consumption trends, reflected in modernistic forms of project procurement, better control of workload becomes possible through the greater empowering of influence for the bidder in decision-making with developers, funders, public sector bodies and, not least, the general public. Hence, the marketing and business development processes need to focus sharply on such aspects, particularly when a financially funded element can be incorporated into the bid proposal.

Evidence of the ability to bring together and manage a wide variety of project elements, together with clear articulation and (simulated) demonstration of the provided benefits of intangibles such as design skills, time, cost, quality, safety and fitness-for-purpose results, are further essential elements of persuasion in the promotion exercise. Also good after-sales service during and after the maintenance period, including co-operation with the client in improving designs and contract procedures in the light of the experience gained, represent quality indicators. Major clients may also look for a credible corporate social responsibility policy, successfully implemented throughout the contractor's network.

Necessarily, therefore, the development and sustaining of specific market relationships (*relationship marketing*) are paramount features in modern supply-chain or network arrangements, where individual programmes need to be specifically tailored for the prospective, active and former customer, representative or partner and/or stage of the process, etc. as appropriate. The seven-point procedure developed by the Skanska company demonstrates a practicable approach to RM as follows:

(1) Partner selection—carefully ascertain with whom to work.
(2) Nature of the contract—assess potential relationship impacts.
(3) Understanding each other—anticipate and comprehend needs, expectations and perceptions.
(4) Interpersonal relationships—develop these both at work and socially.
(5) Ways of working—understand and develop relationships at the organisation level.
(6) Performance management—set key measures to drive action and improvement.
(7) Dealing with problems—learn from experiences and implement to improve the business.

(4) Climate change—The United Nations 'Kyoto' protocol, directed towards better conservation of the environment and being implemented by developed countries, requires each participating government to reduce national atmospheric pollutants, notably carbon, during 2008–2012 to an average 5.2% below the 1990 level. Notably, each member state of the EU (which annually consumes approximately 15% of world primary energy supplies; see Chapter 3 for international regional statistics) has accepted an emissions target and agreed to issue emissions allocations to every large industrial facility in relation to current amounts produced. Clearly, power generation, refining, extraction, petrochemicals, agribusiness, ceramics, metals, synthetics, engineering and transport industries are likely to be significantly affected.

Thus, these and other clients heavily involved with commissioning facilities consuming large quantities of energy for their operation, or built from highly processed materials, e.g. concrete/steel/brickwork/ceramics/synthetics and similar, may offer challenging opportunities to the construction sector, particularly marketing and development of technically innovative design, construction, maintenance and sustainability packages emphasising energy saving.

References

Drucker, P.F. (1999) *Managing for Results*. Butterworth-Heinemann, Oxford.

Kepner, C.H & Tregoe, B.B. (1997) *The New Rational Manager—An Updated Edition for a New World*. Kepner–Tregoe, Princeton, NJ.

16
International construction logistics and risks

Summary

This chapter provides an overview of the logistical-type problems likely to be encountered by promoters, designers and contractors when operating internationally. Raising finance, dealing with unfamiliar conditions of contract and legal systems, transport and shipment of goods, payment procedures and managing local labour are highlighted.

Introduction

Some firms in sectors such as automobiles, construction equipment, consumer electronics and pharmaceuticals operate on a truly global scale, with just a few companies dominating trade in a significant proportion of world markets, able to access international finance and move products and components around in order to optimise manufacturing costs. Plants are located purely for economic reasons and supplies procured from bespoke subcontractors in a variety of countries able to produce to exacting standards. In construction only a very few large group holding companies operate on a similar global basis, but largely through acquisition of local firms. More realistically, companies either enter the international arena by limiting themselves to a regional strategy or more typically range further afield as specialists, for example in engineering construction, heavy civil engineering and complex building, where the scale of operations, financing, etc. or the level of technology is beyond the capability of local firms.

Increasingly both designers and construction contractors are seeking more turnover internationally in this manner, with all types of project being undertaken from housing through to major infrastructure and industrial engineering work. Much is fairly conventional, while some involves financing the complete scheme and increasingly even operation of the facilities for a period after the construction phase. Most require adjustments in management procedures, construction methods and especially procurement practices.

The international environment

International construction tends to confront the parties involved with problems ranging from the practical, such as extended transport arrangements and communications systems, to more subtle aspects relating to different economic, political, social and cultural conditions. Historical background, language and even the written form are likely to be unfamiliar, where all have to be effectively overcome including changed geography, climate, legal and social customs. The normal expectations of receiving immediate supplies, readily available spare parts, experienced advice and abundant well-qualified staff are unlikely to be routinely met, and all generally require more management effort for effective resolution. Hence, expatriates occupying key positions and functions need to be carefully selected and prepared for international work, especially in local customs and practices where corruption and normal business procedures may be difficult to disentangle. Indeed, without experienced staff much construction work becomes very difficult to manage, resulting in errors, poor quality and low productivity. Nevertheless, wrong decisions and poor choices cannot always be avoided; indeed, scheduling of orders, organising goods despatch and dealing with delivery errors and delays are likely to compound an already overstretched staff and physical resources, causing even the simplest management systems to falter.

Customs procedures and import controls also commonly vary markedly even in neighbouring countries, many being very restrictive, particularly for manufactured items with locally available alternatives. Furthermore, currency transactions may be limited or even impracticable if the country has a weak trading base without foreign exchange facilities.

Managing and operating equipment are also perennial problems, normally requiring considerable training of local employees, including technical, administrative and language aspects to facilitate effective communication between the home-based staff and site personnel. Furthermore, ill-trained local labour generally leads to extra maintenance on equipment because of the higher frequencies of breakdown and failures, etc., requiring in turn extra quantities of spares, consumables and materials, perhaps transported over difficult terrain and long distances and having to be co-ordinated without the aid of good telephone systems and postal services.

Other practical contracting problems concern the physical environment, for example locally sourcing bulk items such as cements, reinforcement, timber, pipes, bricks and blocks against demanding quality criteria, with only specific items being granted import licenses. Furthermore, high specifications enforced by the client or their representatives demand much effort in sourcing and back-up testing, necessitating generous allowances for defect replacement.

Climate will also have to be overcome, especially at the extremes, where typically concreting operations are problematical and the remedies, for example temperature control, are very expensive. Furthermore, construction equipment in general is affected especially by severe cold, requiring special heating and protection to facilitate cold starts, prevent solidification of diesels, oils, etc. Not least, adequate water supplies sometimes can only be reliably secured from site-based purification plant, electrical generating sets, etc.

The kind of project itself can cause difficulties for both the designer and the contractor in interpreting contract conditions, especially where international forms are involved, each party perhaps having little experience of the other house practices, compounded by neither being familiar with the overseas territory procedures.

Finally, the growing threat posed by politically motivated violence and international terrorism needs to be fully assessed and appropriate security/protective measures undertaken to safeguard both personnel and resources on site, and not least throughout the transport/delivery chain.

Seeking international work

Once a decision to seek work in a foreign country has been made, then pre-contract studies must be undertaken and a fact-finding visit made at an early stage.

Much information on calls for tender, agents seeking principals, notification of available projects, etc. can be obtained from the home base through the government export advisory offices, the major banks and private organisations. For a fee some of the following associated details may also be supplied:

- *Remuneration*—pay for expatriates, length of assignment, leave, etc.
- *Conditions of Service*—normal hours of work, minimum wage rates, severance pay, etc.
- *Personal Taxation*—rates of tax, specimen calculations, procedures on departures, etc.
- *Social Security*—national schemes, contributions and unemployment benefit.
- *Expatriates*—immigration requirements, health and medical facilities, remitability, etc.
- *Background information*—the economy, climate, history, religion, political stability, etc.

Furthermore, private agencies and firms are available to offer specialist services related to circumventing obstacles to deal-making, gaining business and securing contracts, supplying details on potential violence and terrorism, providing security and risk assessments, installing the appropriate security training and protection requirements, etc.

The Foreign Office and Diplomatic Service also presents opportunities through its high-level international business lobbying and associated overseas trade mission involvements, while substantial sums are dispensed as project aid by the Department for International Development.

Other important data to be sought out to assist in planning operations and estimating the costs of work would include:

- Contractors and subcontractors presently working in particular regions
- Costs of labour, both local and expatriates
- Cost of materials, history of shortages
- Details of plant and equipment agents, available spares, etc.
- Background details on agents with local knowledge
- Local firms and companies available and capable of becoming partners in a joint venture

Project funding

Funding generally falls into two categories, that needed by the contractor and that required to finance the project itself. The contractor would generally seek liquidity through its normal banking arrangements, retained profits, shareholders, etc. However, for the promoter, and often in overseas work the promoter may also be the contractor, especially where Turnkey and Design, Build and Operate projects are involved. The banks may finance the whole scheme but sometimes limit their financial exposure in project financing by sharing the risks with partners, i.e. other banks, financial organisations, government bodies or even the promoter.

The degree to which each bank participates depends on an assessment of the project, promoter/contractor involved, experience in the country concerned and the amount already allocated to that type of risk.

Funds are generally provided to suit the construction phasing and the money organised accordingly, reaching a maximum towards the end of the project. Repayment is generally required by instalments and not in a lump sum at the end of the loan period. The banks are often flexible on the loan period and understand that the variable nature of construction work has a direct bearing on the rate at which the loan can be repaid.

Participation in export credit guarantees, especially for work in developing countries, and normally arranged through the home government of the promoter(s) is crucial to the participation of the banks. The guarantor, however, has to be convinced that a project is viable before providing guarantees.

In this respect a common form of finance for purchases on overseas projects is to arrange for buyer (borrower) credits. The usual method involves a loan agreement between the buyer and the bank, the latter paying the money to the seller. The seller also has an insurance agreement with the export credit agency, with the bank normally requiring the guarantee from the export agency. In the event of failure of the borrower to repay, the bank receives back the advance from the guaranteeing agency. Up to 85% of the financial package can normally be guaranteed and, for overseas work, host countries sometimes add loans to cover the shortfall. Indeed, even an entire project may be funded by such methods, in which case the borrower generally has some influence on the choice of currency and the interest rate to be paid, thus requiring careful judgement on the likely changes expected in exchange rates. The combination of these alternatives is often a mix of aid, government soft loan and export guarantee agency preferential interest rate financing.

At the time of writing many countries are experiencing debt servicing difficulties on large international loans and as a consequence are unable to fund many new projects. The types of financial package offered by competing firms are thus becoming increasingly important. For example, arrangements requiring the host country to cover only the difference between the specific interest rate demanded by the bank and the generally commercial rate of the currency used is not uncommon.

Finance and payment

Resources purchased direct for overseas contracts usually require the organising of import licenses and, of course, the mechanisms of payment. In addition, rationing through a waiting list, sometimes tempered by importance of use of the currency, e.g. capital resources over consumer goods, commonly solves import priorities in developing countries.

Transactions in local currency may be available via the foreign exchanges, but for very poor or ill-managed economies, few opportunities exist to exchange 'hard' (i.e. dollars, £s, euros, etc.) for soft currency, which can really only be used to make purchases in the local economy. Some international bartering may be possible, and today organisations specialising in exchanging commodities for consumer and capital goods do exist.

Such difficulties are usually less problematic for projects funded by aid bodies like the World Bank Group (IBRD)/IFC/IDA, EBRD/EAR, UK-DFID, USAID and similar reliable government, private sector and pan-continental development banks and agencies. A soft loan is usually involved, i.e. a loan in a hard currency at a low interest rate or with easy payment terms, a stipulation being sourcing of supplies from a particular country. This may, therefore, produce less effective solutions than might otherwise have prevailed, especially when several foreign donors happen to focus on a particular country and the aid is unco-ordinated. This situation is fairly typical of current donor practice.

Finally, import duties should not be overlooked, i.e. they vary considerably from country to country, even though international agreements may have been negotiated through the World Trade Organisation (WTO). Some countries waive the import duty if the equipment is to be re-exported at the end of its duties.

Conditions of contract

Contracts in the home environment usually follow well-established procedures regarding duties of the promoter/client, project manager, designer, supervisor, quantity surveyor, main contractor, subcontractors, materials suppliers, etc., where payment systems, changes and variation orders, claims, arbitration, legal redress and so on are all usually laid out in the conditions of contract and can be upheld. Also the common methods of contracting—including traditional lump sum, management contracting, design and build, turnkey, build and operate, etc.—are all usually well known.

In foreign situations, however, such familiarity does not always translate so well. For example, should disputes arise, then difficulties may surface concerning entitlement to be heard in local, international, third-party or home-based courts. Even the use of model forms such as EU-PHARE, World Bank, ENAA of Japan, FAR, US Government Federal procurement procedures, FIDIC, etc. will not entirely replicate normal expectations, and clearly adequate insurance needs to be taken out to cover indemnification of both design and construction activities. To this end Engineers Against Poverty (EAP—www.engineersagainstpoverty.org) and Engineers Without Frontiers (EWF—www.ewb-international.org/) are jointly drawing up a draft new contract potentially for use by NGOs and others working with local firms in developing countries, principally by facilitating the inclusion of more appropriate clauses into existing contracts.

Bonds are a further complication and are usually demanded by clients to cover tender, performance, advance payment and retention aspects.

Documentation and bureaucracy

The bureaucracy of the importing country can be complex and certainly needs to be understood. For example, in addition to the specification and quality inspections, other barriers such as tendering and trading procedures, legislation relating to labour employment and the environment, road transport regulations and restrictions regarding particular imports can cause problems. Consequently overriding considerations may point to a joint venture with an indigenous company in smoothing the path. Furthermore, the host country may insist on this form of arrangement to enable the local partner to gain technical and management expertise. Ideally a 51% majority shareholding in the joint company or undertaking provides controlling ownership but foreign firms may be restricted to minority stake. Indeed, recent experience suggests that foreign firms, in avoiding questionable local practices, have achieved better financial results from sole ownership. Irrespectively, great care in the evaluation process and in establishing such relations would be essential; otherwise conflicts could eventually surface concerning strategy, decision-making and management style. Even straightforward issues covering accountancy practices, production management, personnel and industrial relations policies could pose potential areas of disagreement and, not least, confidential business development and research information may prove sensitive issues that threaten the partnership.

At the very minimum for the newcomer, the advice of a local agent would be essential, not least to deal with language interpretation. Indeed, a local agent of this sort may be necessary in each country through which the goods have to pass, where good practice also requires employing a shipping agent to expedite and co-ordinate orders and deliveries, and to be responsible for packing lists, bills of lading, shipping, customs forms, currency payments and so on. The shipping agent will normally encourage marking of items with part numbers and the provision of detailed descriptions on the documentation to aid both checking and inspection, but also to enable follow-up enquiries when matters become aggravated through pilferage and/or transport deterioration.

Construction labour

If local labour is used for the construction trades, careful vetting of subcontractors will be vital, especially where the work is fairly basic. In the case of specialist activities, these problems are usually less critical as expatriate managers and trades workers are commonly employed. However, foreign firms may be involved and hence specifications, working conditions, labour remuneration practices and subcontract legalities could be unfamiliar.

Equipment

Contractors working overseas usually have to establish their plant set-ups from scratch. The exceptions may be work undertaken in the more developed countries with sophisticated construction markets well supplied and serviced by manufacturers and agents.

Thus, equipment needs to be versatile and able to perform a range of different duties, e.g. tractor to carrier. Wherever possible, the opportunities for 'cannibalisation' through standardisation of make and components should be thought out in advance. Clearly, sensitive equipment such as instruments needs protecting from harsh conditions such as extremes of heat and cold, dust, humidity, harsh winds, etc. where the provision of a protective controlled environment may be essential, e.g. a cold store, air conditioning, etc.

Regular maintenance may be more essential than in the home territory, especially where operating in severe climates. Consequently stocks of spares, consumables, general items, etc. need to be plentiful and probably not less than about 15% of equipment value, unless the local market is very reliable. Indeed, the major equipment manufacturers may themselves only deal through local agents and therefore have very limited back-up facilities. In these circumstances, technicians, mechanics and general labour will have to be trained from the local population, which requires investments in time, resources and especially expatriate assistance until the new arrangements are operating satisfactorily.

Materials and equipment support items

In setting up a construction site overseas some general common items considered sensible might include:

- Concrete batching and mixing equipment including provision for dealing with hot or cold climates
- Transport vehicles, such as trucks, four-wheel-drive Land Rovers, utility vans and pickup trucks, etc.
- Welding and fabrication shops, equipped with a good choice of welding and burning gear, workshop equipment and grinders, etc.
- Maintenance workshops for equipment and vehicles, together with spare parts and stores.
- Power generators, including stand-by water storage tank(s), fuel dumps, etc. A water treatment capability may also be necessary.
- Waste disposal facilities for sewage, e.g. septic tanks, incinerators. On large long-term sites proper sewage treatment installations may be unavoidable.
- Furnished site offices, canteens, stores, compound and plant yard. Many projects will also require the specific provision of furnished accommodation for some of the work force, possibly with additional recreational facilities for expatriate staff and their families.

- Quarry operations and aggregate production equipment will almost certainly be necessary for remote areas.

Supply and delivery

While many of the above points affect the selection of materials, equipment and resources, other factors such as supply, packaging, transport, agents, customs, finance, insurance and so on are equally important.

Pre-shipment inspection

Materials and equipment ordered are commonly specified to certain standards, for example, ISO, EN, etc., and many developing countries insist on pre-shipment inspections charged to the supplier, to ensure the conditions are met before items leave the source country. The procedures requested may involve simply counting of items and dimensional checks, or more thorough inspections, including testing of materials, components, complete items of equipment, etc., in accordance with the particular specification standard. Other aspects could include monitoring the export price against like items sold on the home market.

Transport to site

Transport over long distances through unfamiliar territory and other countries en route clearly needs special attention and planning. Wherever transit periods exceed several months and different means of carriage, strong packaging and preferably containerisation is obviously necessary. The choice of transport method itself depends upon several factors—for example, roads, while generally economically competitive, usually involve long journey times and can be fraught with problems, such as poor road surfaces, lack of signposting, pilferage, banditry and unreliable drivers and/or haulage firms. Also when at journey's end, special temporary roads may need to be constructed, especially for the very remote construction site.

Alternatively, shipment over long sea routes, while generally cheap, is very slow, but is commonly preferred for large plant items to avoid disassembly demanded by container sizes. However, even with shipping, containerisation needs link-ups with trucks for ease of handling. A further consideration is the quality of facilities at the unloading port. Increasingly, rail transport is becoming competitive over long distances, especially where national rail systems are well organised — for example Trans-Siberia, Trans-Canada and systems in parts of Africa and Europe.

Airfreight is usually the quickest means of transport but also the most expensive. Heavy or bulky items are usually unsuitable, with large units often needing dismantling. Occasionally very specialist transport can be considered, such as helicopters, and even more exotic means such as snowmobiles, pack animals and so on.

Finally, the courier should not be overlooked for small items, facilitating considerable shortening of customs delays.

Conclusions

Clearly, when embarking on overseas contracts, obtaining knowledge about the likely conditions to be faced is vital. Much advice is available in the home country through Government export agencies, trade associations, R & D organisations, shipping agents, private consultancy firms, etc. However, reconnoitring the local environment is paramount and in addition for many countries, local agents, distributors, contacts, etc. are absolutely essential, while in some cases forming partnerships or joint companies may be justified in order to secure knowledge of:

- Business opportunities and deal-making
- Regulatory obstacles to hiring/firing workers, starting/closing a business, enforcing contracts, obtaining credit and protecting investments
- Financial and banking procedures
- Import legislation, quality specifications and standards
- Customs, shipping, delivery and distribution procedure
- Management expertise and labour skills availability
- Legal systems
- Risk, security and protection issues

17
Information resources and ICT systems

Summary

Information as a construction resource, management issues concerning construction industry information systems, email, web sites, intranets, on-line information data and transfer, bespoke and commercial software matters, data exchange and integration of systems.

Introduction

In order to stay competitive, construction organisations have to efficiently exploit and manage every resource they utilise for their operations. Executives in the industry implicitly accept that information is a key management resource and underpins the processes and operations of every construction company. However, the management of this resource rarely receives adequate attention from senior executives. This chapter focuses on the changing role of information resources within construction. It outlines the role of information in the construction process and presents some common sources of information resources that are utilised by construction companies. It also presents various information communication technology (ICT) tools that can enhance the productivity of construction companies.

The construction company's business

The construction company's business in principle is no different from that of any other company. Essentially, it involves four main aspects:

- Obtain sufficient workload or orders.
- Execute whatever workload has been acquired efficiently and profitably.
- Sustain the first two aspects against competition from other construction companies and changes imposed by its markets (clients, economic conditions, resource availability and environment issues, etc.).
- Provide the administrative mechanism and organisational structure that will ensure the attainment of the above three aspects.

Processes involved in construction business

To achieve the four aspects that make up their business, construction companies implement various processes that address the different functions required for the operational activities. The processes involved in the business activities of construction companies include:

- Marketing
- Estimating
- Tendering
- Design
- Construction
- Research and development
- Administration

The principal activities that make up each of these processes are generally covered in the other chapters of the book. Information and its associated technology provide the vehicle that links the activities within each process and between processes. The activities of each of the functional and operational processes presented in the list above can therefore be viewed as an information process. As an illustration of an information process, the next section looks at the design and construction phases of a project from an information process standpoint.

Information processes in construction

Design phase

The design phase of a construction project involves interaction between the client and other parties (e.g. architect, structural engineer, quantity surveyor) of the project team and at this stage a *brief* that captures the client's requirement in the form of defining the purpose of the project, its scope and limits of functional requirements is identified. This information is outlined in verbal and textual formats and is eventually captured and set down as a document that forms the basis of the design solution. The design solution itself will normally be furnished as a set of drawings and specifications. These activities essentially involve information transactions and culminate in an information-based solution. Equally, the other major activities of the pre-construction phase, such as developing a contract strategy, procurement strategy, financial strategy and a project implementation plan, which are normally influenced by the design solution, depend on information transactions between the several parties that form the project team, and beyond their respective organisations. The design phase of construction projects can therefore be viewed essentially as a transaction of information and knowledge resources. The output of the process feeds into the construction phase of the project.

Construction phase

The construction phase of projects involves converting the output from the design phase, along with other physical inputs such as materials, plant and manual labour, to generate the industry's products. Besides utilising information directly from the design phase, the construction phase also involves considerable information transactions. Examples include information exchanges between contractor and subcontractors, between contractor and material or plant suppliers, between contractor and regulatory bodies, and between subcontractors and designers. These information transactions help to facilitate the co-ordination and timely delivery for the construction phase of the project. The effective management of information resources within construction should therefore impact on both the success of projects and the overall performance of individual construction companies.

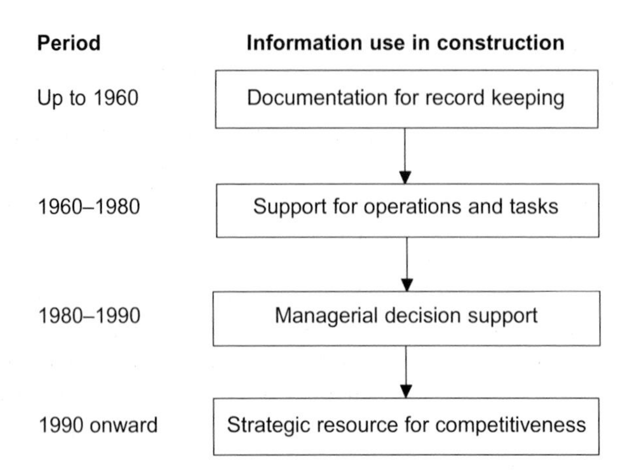

Fig. 17.1 Progressive change in role of information for contractors.

Changing role of information in construction

Technological developments in the last few decades have elevated the role that information plays in the management of companies and is causing a re-think in the way organisations in general treat information, information systems and its associated technologies. Until the 1980s, managers in the construction industry generally did not concern themselves with how information was collected, processed and distributed within their organisations. The reliance on paper-based communication formed an essential part of most construction organisations, and often got in the way of real productive work. The use of information within construction has seen a significant change from this position. Within the last three decades the concept of information for construction organisations has shifted from this role of general support for the contractor's operations to its use as a means for more effective managerial decision-making. The driving force for this shift in the role of information is to improve and speed up the decision-making processes of specific managers and executives in a broad range of tasks at both the project and company level. Figure 17.1 depicts this gradual shift in how construction companies have been deploying information over the last three decades.

From the early 1990s, information has assumed a different role for construction organisations, changing from serving as decision support to being a strategic resource. The effective deployment of information can affect the competitiveness of construction companies. This emerging role of information impacts on the way construction businesses have to conduct their functional operations and how they are structured. The strategic importance of this new role for information in construction derives from the simple fact that its activities at design, site, project and business level are dominated by information. The information is often in the form of documentation, such as drawings, specifications and conditions that are communicated between parties. As a major resource for sustaining competitiveness, information and associated technologies need effective management if contractors are to benefit from the deployment of this resource.

Information needs to support business processes

Construction organisations have to rely on information from various sources for their operations. These sources of information can be grouped into two broad categories of *internal* and *external* sources. Internal sources cover both the formal and informal reporting mechanisms employed by construction companies to manage and control their projects and other corporate activities. They range from documents that have company-wide impact, such as circulars and policy statements, to ones that address specific projects or issues. Internal information is often of a stable nature and requires less frequent revision. External information addresses the interaction between a construction company and its business environment. The sources of external information available to a construction company are diverse and the types of information they yield are of a less stable nature. This means that construction companies need a systematic approach to updating the information they use from these sources.

Sources

Information used by construction companies is acquired from several sources. Table 17.1 presents some of the sources of information that construction companies employ for their business and operational processes.

Construction managers are often confronted with situations whereby information required by them has to be sourced from different outlets. In the past such information requirements were limited because of the limitations in accessing large volumes of paper-based information. This situation is presently altering as a result of a shift in the mode of access to such information from paper format to electronic options. There is, therefore, a large volume of material that can be sourced for each type of information listed in Table 17.1. The electronic options include both *on-line* (intranet and internet) such as 'Building online', and *off-line* (CD-ROM) resources such as 'Construction Information Service'. These options potentially make available to the construction executive masses of information that previously could not be readily accessed.

The distinctions between the different functions in construction are increasingly becoming blurred as contractors' staff, designers, project managers and other stakeholders find themselves in an emerging teamworking environment often calling for knowledge and information beyond their traditional functional role. The way a construction company organises making such increasing information resources available in a systematic and structured manner will affect the effectiveness of decisions taken by its engineers and managers. This can be undertaken in an *extended enterprise* environment, whereby the timely flow of information can be from and between several sources within the company, and also to and from external sources. Operating within an extended enterprise environment should help to eliminate the isolation of the functional unit and bring about greater integration for all functional decisions. Figure 17.2 presents a schematic diagram of such an extended enterprise. The use of an extended enterprise approach for organising the contractor's information should help to eliminate situations in which estimators work without regard to other functions in the contractor's organisation, such as site management, and similarly for designers and other parties involved in the design phase of projects.

Table 17.1 Sources of information in construction.

Information Type	Primary Source	Secondary Source
New Developments	Research publications Corporate communications Commercial journals Technical abstracts	Industry magazines Academic journals
Processes	Corporate documentation Research publications Internal reports Cost data Output performance	External benchmarks Internal benchmarks Quality control unit Professional bodies BCIS
Competitiveness	Companies in the same strategic group Construction industry organisations outside strategic group Benchmarking reports	Construction Forecasting Group Statistical Digest for construction
Future workload	Forecasts by major client organisations Development plans and budgets for central government departments Development plans and budgets for local government departments *Official Journal of the European Commission*	Regional economic trends National economic forecasts Global economic trends
Standards	HMSO BSI HSE ISO	Research institutes TRADA CIRIA Professional Bodies (ICE etc.) Trade associations
Specifications	Trade associations Bodies concerned with regulatory compliance Electronic Libraries (Barbour Index, GENIAL)	National Building Specification Services EMAP Business Publishing

Management of contractors' information resources

The construction company's information resources can be grouped under two broad categories, which form the extremities of a continuum of information types appropriate for the execution of projects and also in managing at the corporate level. The two categories reflect information in the form of *documented artefacts and records* on the one hand and *knowledge* on the other hand.

Artefacts and records

Figure 17.3 presents examples of the different types of information that are classified under the artefacts and records category. The different types of information reflect three sub-categories of unstructured, semi-structured and structured information. Unstructured information comprises information that is frequently changing in content in an unsystematic manner. Information that is unstructured usually relies on substantial knowledge support to make it valuable for managing construction. Examples of unstructured information types include still images of conditions from a construction site or webcam clips of real-time progress of works. Structured

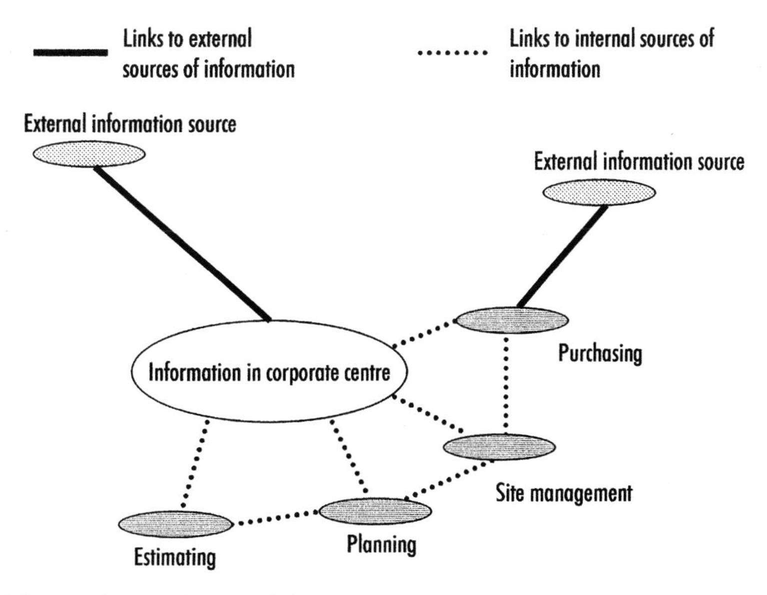

Fig. 17.2 Schematic diagram of an extended enterprise.

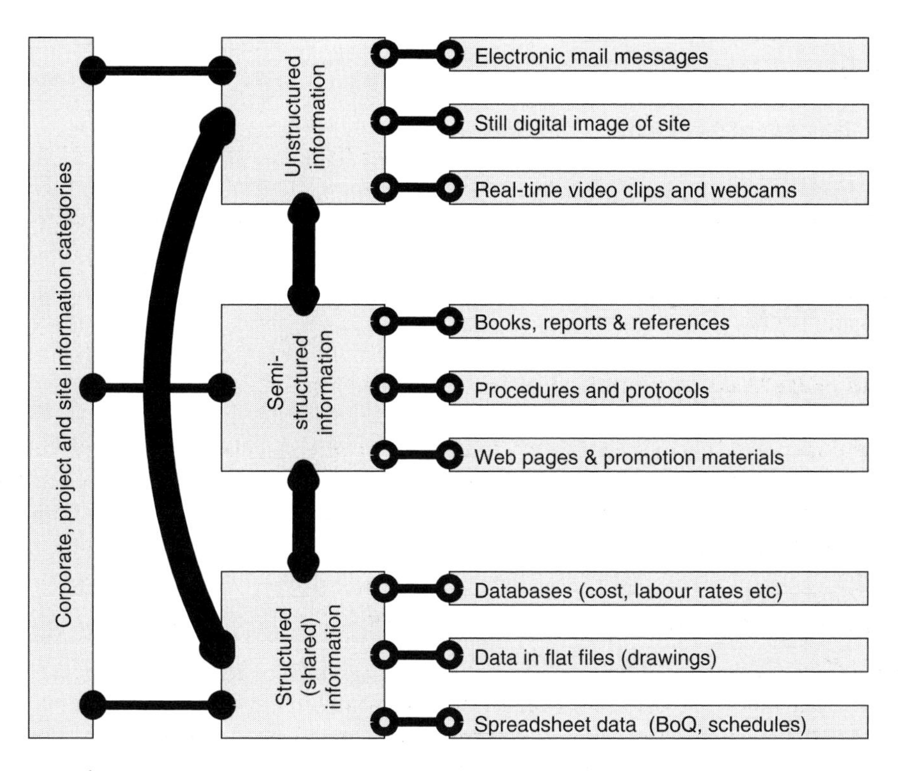

Fig. 17.3 Information genres requiring management in construction.

information comprises information types that are fairly stable in their content, and can often be appropriated without further manipulation or extensive knowledge support. At the half-way house between the unstructured and structured information typologies is semi-structured information. Examples of semi-structured information types include annual reports on corporate performance and project protocols. The essence of considering information types from these three perspectives is that a construction company requires a different tactical plan to address each type for effective exploitation. Whilst structured information can be addressed by clerical or operational input because of its routine nature, unstructured information requires more senior management input.

Knowledge management

Information can enable the effective integration of a contractor's operations, which are often widely spread geographically. This often involves transfer of knowledge from one part of the company to another. The term *knowledge* when used in relation to a construction company usually encompasses features such as *experience*, *concepts*, *values*, *beliefs* and *ways of working* that can be shared and communicated. Knowledge management means attending to the processes for creating, sustaining, applying, sharing and renewing knowledge to enhance a contractor's performance and create value. It involves developing appropriate strategy and processes that will enable the creation and flow of relevant knowledge throughout a contractor's organisation in order to create a value for both the company and its stakeholders (for example clients, designers and end-users). Knowledge management is therefore the broad process of acquiring, locating, organising, transferring and using the information and expertise within a construction organisation. Four key enablers influence the management of knowledge resources in construction companies: leadership, culture, technology and measurement. The embodiment of knowledge resources is therefore the executive and staff that make up the organisation. Knowledge for the construction company is not limited to information alone, but can also include awareness, experience, skill, insight, competence, know-how, practical ability, capability, learning, wisdom, certainty and so on. As such, knowledge for the construction company can be summed up as information that is relevant to its competitiveness and operational efficiency. The knowledge is normally actionable, and at least partially based on experience. For the construction contractor such knowledge transfers would normally occur between head office and a project, or between two projects. It could also involve information transactions between a contractor on the one hand and a supplier, subcontractor, the client, designer or other stakeholders and third parties to a project on the other. Real-time access to the knowledge resources and information enables the effective and efficient management of processes involved in the project. To continually improve themselves, construction organisations have to develop a systematic approach to capturing and applying such knowledge resources. Timely feedback of such information and knowledge, on for example the progress or performance of a project, should allow for incremental self-correction of processes. Similarly, access to comprehensive historical information should enable simulation, optimisation and modelling of the processes in major redesign efforts. The availability of such timely information for the contractor is captured in an *information system* (*IS*).

Although knowledge is recognised as a key resource and the need to manage it efficiently is well established, curiously the construction industry has not yet taken the step of appointing a Director of Knowledge to the company board. This role is still fragmented amongst several other functions. Given the increasing relevance of knowledge resources to the competitiveness of construction contractors, this position should become a reality within the foreseeable future.

Construction information systems

For most construction organisations, their use of information systems is normally internally focused and oriented towards the control or monitoring of operational activities. However, information systems can also facilitate the planning aspects of the construction organisations at both the business and project levels. A typical example relates to the use of output data from previous projects to establish the characteristics of future projects: estimating the cost; determining a programme for the works; and managing the project when won. Because of their internal orientation, such systems relate to relatively known and stable information resources. Construction companies need information systems because this can help them to augment their internally oriented information systems with an external perspective. This is often achieved in the form of an *executive information support* system (EIS). EIS provide construction executives with external information that enables strategic decisions to be taken. An EIS will form part of the *overall information system strategy* for the contractor's organisation.

Information system (IS) strategy

The effective exploitation of information resources within the construction company's management system is essential and can be achieved if the role of information is appreciated from a strategic viewpoint. This will involve an information strategy for the company with input from the contractor's IT manager and top executives, as well as contributions from all managerial positions that have significant impact on the provision and utilisation of information resources. Without an information strategy, construction contractors will suffer from the following effects:

- A mismatch of information resources as a result of devising different systems for the various functional activities of the company, for example estimating and planning
- Information overload
- Information losses due to inadequate capture of relevant information
- A general misunderstanding of the role of information management by site and project managers
- Poor co-ordination of information with decision-making needs at both the business and the operational levels

To prevent the above effects, construction companies need an information strategy agreed at and involving the highest level of its management. The changing trend of seeing information as a competitive resource for construction companies is a tacit acknowledgement that technology, in particular ICT, on its own does not automatically confer competitive advantage. As such any strategic considerations will not only have to recognise information as a resource, but also have to focus on the deployment and management of information, as well as accounting for its use. Such a strategy will consider all resources that facilitate the acquisition, creation, storage, processing and provision of information that generates knowledge or other values required for setting and achieving the goals and objectives of construction companies.

Information audit

Developing an effective IS strategy for a construction company requires determining what and where the information resources are and managing the relationships between them. To achieve such an effective IS strategy, a construction company has first to undertake an *information audit.* The audit is a process of discovering, monitoring and evaluating the company's information flows and resources in order to implement, maintain or improve its information management. In times past, such an audit would have been focused on the internal operations of construction companies. The rate of change in the IT industry, however, suggests that construction companies can no longer afford to undertake information audits on that basis alone. It will need to give consideration to the external environment of the company. The audit will identify the following key aspects at all the levels of the company, namely site, project, business unit and strategic:

- The construction company's information resources
- The construction company's information requirements
- The costs and benefits of the company's information resources
- Opportunities to use information resources for strategic advantage
- Information flows and processes both within the company and for its projects
- IT investment options that can facilitate the construction company's business initiatives

The identification of the construction company's information requirements and resources is best undertaken in a systematic framework matrix, as depicted in Fig. 17.4. The matrix can be developed to focus on various sections of the company and provide more detailed information, which will combine the *resources*, *requirements* and *ICT options*. Alternatively, a separate matrix can be developed for each of *resources*, *requirements* and *ICT options*. To establish a *requirements* profile for a construction company, a representative sample of the managerial staff in the

O – origin of information X – information is utilised by

Information type / **Functional Process**	Project management	Site engineering	Costing/estimating	Project planning	Contract/marketing	Finance	Administration	Purchasing
Site progress records	X	O		X	X			X
Programme of works		O	X	O	X			X
Safety procedures	X	X	X	X				
Project budgets	X		O	X		X		X
Project expenditure profile		X	O					
Quality conformance	X							
Materials utilisation		O						X
Claims	X	O	O					
Accounting						O	X	X
Labour and employment regulations							X	
Subcontractor progress	X	O	X	X	X	X		
Future orders			X	X	O	X	X	
Media profile/corporate image	X						O	
Audit reports						O	X	

Fig. 17.4 Contractors' information matrix.

organisation needs to be surveyed as to how and where they obtain information for their functional roles. The survey should also focus on the perceived constraints of their information environment, and what their future information requirements are likely to be. The results of the survey can then be summarised in a matrix such as Fig. 17.4. The information matrix can also be constructed for only a project, a site or a business unit within the construction organisation. The example in Figure 17.4 shows for selected information types the sources where they originate and where they are utilised within the company's processes (indicated as O and × respectively). When the matrix is developed to cover the whole organisation it will enable the organisation to map out common information requirements. This will help to eliminate unnecessary duplication in the storage, processing and acquisition of all the common information resources.

Information strategy options

The next phase of the contractor's IS strategy should develop options for an integrated information policy. This will cover the *what, why, how, where* and *when* for all information resources utilised by the company. Information sharing with respect to storage and content will play a crucial role in structuring such a policy. For example the considerable sharing of information between estimating and planning departments presents a strong case for common storage. Equally, the overlap of the two functions with site production, claims department, buying department and the activities of other functional units within a contractor's organisation provides scope for both common content and storage options. A decision on the option for an integrated solution to the construction company's IS system is an executive one. This will often be based on the information flows that are essential for the company's processes. Once this is resolved, a framework for implementing the strategy, along with guidelines for action within different sections of the company, will be drawn up.

Implementation and monitoring

The last phase of the IS strategy will outline a scheme for monitoring, evaluating and revising the strategy to ensure its continued relevance to the construction company's business and operations. This will normally take the form of a periodic information audit. The main drive for such an evaluation is to be able to establish a better return on information use consistent with the company's return on investment. This should address both the content and the processes involved in the provision of requisite information. This will entail addressing the following key requirements associated with the use of information outlined below by the company.

Timeliness

A review of the processes involved in making relevant information available to those who utilise it needs to reflect the level of urgency required for associated decisions. This will differ for different categories of information.

Ease of access

The procedures that users would have to go through to obtain requisite information to aid their decision or facilitate the tasks they have to undertake. This necessitates the structuring of information to enable different access protocols.

Ease of use

This entails the extent to which information is ready for use by end-user requirements, or the ease with which it can be manipulated by the user to match their requirements.

Utility

The extent to which the information provided can satisfy multiple requirements. Information that is very specific to only one user and not required by other users could often prove expensive to maintain.

Quality

The quality of the information primarily addresses the content. It involves assessing the accuracy, reliability and credibility of the information provided. This aspect of information use relates closely to the timeliness, particularly for information that changes in content at a rapid pace. A typical example is the progress or value of works on site.

The construction information manager

A new role emerging in construction among contractors, especially for large projects is that of a construction information manager. The functions performed by the information manager include the following:

- Advice on IT system for the project
- Developing an information management plan for the project
- Attending design co-ordination meetings
- Receiving information from design team and distributing
- Receiving all information from design subcontractors and distributing
- Monitoring and reviewing the flow of information
- Inspecting and commenting on details, obtaining a project team's input and relaying it back to designers
- Assisting in the preparation of subcontract enquiry packages
- Reviewing subcontract quotations for compliance with design
- Reviewing design alternatives
- Prioritising and processing information requests with designers
- Processing comments and clarifications
- Monitoring and collating information for the HSE file
- Co-ordinating design subcontractors' drawings
- Obtaining design subcontractors' risk assessments
- Information tracking
- Maintaining project archives

ICT in construction

Similar to the use of information, the introduction of technology within construction to improve the management of such information resources has progressed through three phases.

First phase of construction ICT

The first phase lasted up to the late 1970s. This entailed use of IT to achieve efficiency and cost savings in the processing of information. Improvements in this phase did not alter the way in which the information was organised but rather attempted a speed up of the manual processes. As such, many of the systems developed were directed at the operational level and found use in activities, such as accounting and book-keeping, that indirectly supported the management of construction companies and projects.

Second phase of construction ICT

The second generation of IT developed for the construction industry evolved during the late 1970s and continued through the late 1980s. The primary aim for most of the developments in this era was to align the technology with, or support, the functions of a construction company. The development of various application systems to tackle time-consuming or iterative operations took place within this period. The main achievement of this phase was isolated automation of tasks in the design and construction process. These included stand-alone systems for project planning, estimating, cost control, cash flow reporting and accounting and financial reporting. Many of the benefits from IT systems in this period proved to be less sustainable than originally predicted. Many construction organisations emerged from this era with large IT cost structures, raising doubtful opinions on the ascribed benefits of IT in the minds of senior management.

Third phase of construction ICT

The third generation of ICT dates from the early 1990s and is still evolving. This is addressing the integration of the stand-alone systems developed in the second phase, in order to maximise the use of IT as a strategic resource for construction. It also focuses on the use of ICT as a communication medium capable of establishing and maintaining favourable supply chain relationships.

ICT change within construction

Not too long ago, the term IT meant putting a computer on every worker's desk, i.e. computerisation. There is evidence that the emphasis is now switching from computerisation to business enhancement. Thus, there is greater concentration on automating the processes of the project and the construction organisation. The project management function is based on the timely delivery and documentation of construction information. The delivery of construction information has traditionally been dominated by paper documents like blueprints and specifications. These take the form of graphic, textual or verbal information formats exchanged between project team members. Typical information components include drawings, specifications, change directives, estimates, management logs and field reports. Because members of a project team share these information components, the exchange of construction information is seen as one of the critical tasks in the construction management process. In today's highly competitive world, meeting the need to compress project lead times and to achieve cost optimisation is vital. At the same time, the growing involvement of construction organisations in multi-partner projects is increasing the complexity of managing projects. For any construction enterprise involved in the running of such large and complex projects, being able to handle the volume of documentation more efficiently can have a major impact on reducing project time scales and costs. The potential for such cost and time savings exists in the form of integrated information technologies.

Enablers of current construction IT

The main benefits of the change brought about by IT are *speed* and *virtual proximity*. These two critical elements are essential for achieving success in managing modern construction projects. Large volumes of data can be processed faster and distributed to dispersed geographical locations much more quickly. This creates a virtual proximity for distant geographical locations. These benefits have been widely shared by industry because of the falling cost of computer hardware, an explosion in software development and cheaper telecommunication costs.

Hardware and software revolution

While construction computing was once the domain of large, well-capitalised construction firms, the diminishing cost of personal computers has put computing power within the reach of even the smallest contractors. At the same time, rapidly developing hardware performance, coupled with the development of storage drives with very large volumes, modems, scanners and back-up devices, has made the computer amenable to the storage and distribution of drawings and other data in electronic format. The evolution of servers, network cards, modems and routers has linked computers together, providing a forum for community collaboration. For example, the newest versions of the CAD programs Autocad and Micro-station include features that allow multiple users to post, view and mark up details on a drawing and collaborate through the use of an internet-based web browser. These technological advances have made possible a situation in which design work can proceed on a 24-hour basis, involving several design teams in different parts of the world. Uncompleted design drawings are transferred at the close of work from one time zone to another time zone, where the working day is about to commence. In addition, recent advances in wireless technologies have opened the door to achieving a truly mobile work environment for both the site and design/project office tasks. This extends the advantages of two-way radio frequency from simple audio communication to all types of information required for managing an extensive site such as a road or rail project.

Lower cost of communication

High telecommunication tariffs have long been a major stumbling block to exploiting technologies that rely on electronic communication. However, the implementation of the package of telecommunication liberalisation measures is already leading to lower prices and to more flexible pricing schemes. The World Trade Organisation (WTO) Agreement on Basic Telecommunication has contributed directly to the emergence of a global marketplace in electronic commerce. Similarly, international agreements to eliminate tariff (ITA) and non-tariff barriers should rapidly bring down the cost of key ICT products, encourage the take up of electronic commerce and reinforce competitiveness.

Using ICT resources in construction

There has been a growth in the use of ICT resources within construction. The effective exploitation of this resource can often lead to the following benefits:

- It saves employee time, lost phone messages and the three-day time delay often associated with surface mail.
- It avoids circuitous means of transferring data, for example printing a document, faxing it and then retyping the data at the receiving end in order to save it as an electronic file.
- It offers the company and individuals the opportunity to publish and distribute their work efficiently, while attaining a high and consistent quality in textual and graphical appearance.
- It provides access to information and allows communication and distribution of documents in a single, uniform fashion.

Besides acting as a means for general management and processing of project and company information, there are other ways in which ICT has been taken on by construction. These developments affect the construction process itself and can be categorised into four main areas. These are standardisation (examples include the use of EDI and bar coding), visualisation (comprising CAD, Virtual Reality and Augmented Reality), communication (including video/data conferencing and intranets) and integration (employing infobases and project-specific databases). The impact of these developments is leading to a new agenda for the construction industry.

Standardisation

Examples of construction work where standardisation has been applied include the use of bar-code technology and electronic data interchange.

Bar-coding

The use of bar-coding for inventory management in construction is gaining increasing acceptance. The application of bar-codes for managing materials on site is now commonplace, especially for the utilisation of unit items such as bricks. The use of bar-codes relies on an optical character recognition technology to maintain inventory records. A hand-held scanner is used to read the bar-code, which is printed on paper and attached to the stock items. Maximising the economic benefits of such technology requires considerable standardisation of the information on construction materials for the whole industry.

Electronic data interchange (EDI)

Essentially, EDI is the exchange of structured data between computer systems according to agreed message standards. The data transfer is achieved by electronic means. The structured nature of the data ensures an unambiguous method for presenting the data content of a document, thus ensuring the correct interpretation of the information by the computer system.

Examples of EDI implemented in construction include procurement of materials and other project procedures that employ document type processing, such as an invoice. In its truest sense EDI effectively means electronic exchange of the requisite information without human intervention. EDI messages are intended for, and are therefore structured for, automatic processing. EDI standards define the techniques for structuring data into the electronic message equivalents of paper-based documents. In other words, there are standardised methods for describing the components that make up any document associated with the project transaction, such as *product code*, *price*, *name* and *address*. The protocols for EDI are covered by the '*Standard for the Exchange of Product* model data', otherwise commonly referred to as STEP. Use of EDI in construction is common between the contractor, on the one hand, and materials suppliers and plant hire organisations, on the other. Any price revision by the supplier is then immediately available to the contractor.

Visualisation

The impact of developments in visualisation on construction processes has been in the area of digitising information, which hitherto had been in paper format. This not only improves the quality and visual effect of construction information, but also eliminates the redrafting associated with the design process.

Digitisation of construction data

In the past, design and construction relied on paper-based documents to exchange information. Developments in information technology are changing the way that construction teams generate, store, transmit and co-ordinate information. Over the past decade, many design firms have moved toward digital production of construction documents. Today, most design and engineering work is completed using computer-aided design software, and this is replacing traditional paper-based construction drawings. Traditional text-based documents are taking on a digital form as well. Specifications, standards, permits, licences and other text-based construction documents are increasingly produced using word-processing software or spreadsheet programs.

Computer-aided design (CAD)

Computer aided design is any interactive drawing system that permits a user to construct or edit a drawing on a graphics display screen. CAD systems will also support scanned paper-based drawings, which can then be manipulated for particular visual effects. Each drawing is stored as a disk file and CAD is only able to edit one drawing (or file) at a time. Most CAD software provides general purpose computer-aided drafting application programs designed for use on single-user, desktop personal computers as well as on graphics workstations. Originally CAD systems only featured CAD two-dimensional drawing capability. Nowadays, CAD systems combine 3D features, along with animation, to enable users to obtain the full visual impact of architectural and engineering developments prior to construction.

Virtual reality (VR)

Virtual reality is a generic term associated with computer systems that create a real-time experience of visual, audio and *haptic* effects.

Haptic effects relate to a sense of feel that is obtained through direct body contact. The essential components of virtual applications are:

- A three-degree of freedom virtual environment, in which a single-user or multiple-users can view objects using haptic displays. Haptic displays are manipulators used to provide force or tactile feedback to humans interacting with remote environments.
- A Shared Interactions Software Library, to build real-time interactive applications, such as shared virtual simulations, teleconferencing, etc.

Virtual reality presents a powerful tool for training, simulation and computer-aided design. The sensation of being in a real environment, while interacting with a VR simulation, is usually referred to as sense of presence or sense of immersion. For example, VR systems can offer the possibility of a sense of immersion for training construction workers on hazard-prone construction operations without risking casualties and also minimising the cost. In many site accident situations (for example, fire and electrical risk), a sense of feel for the conditions that lead to the occurrence is indispensable. Simple multimedia are unable to address such perceptual experience and a virtual world will enable trainees to gain a real sense of any dangerous conditions.

Construction operations often employ teams or gangs to achieve specific activities. For situations of that nature it is desirable to allow multiple users to interact together in a shared virtual environment. The potential of this technology is not limited to training. It can also be employed for simulating safe and productive work environments on site.

Augmented reality

Augmented reality refers to the combination of the real and the virtual to assist the user in his environment. Applications include tele-medicine, architecture, construction, devices for the disabled and many others. Several large augmented-reality systems already exist (for example, the Interactive Video Environment system), but a wearable computer with a small camera and digitiser opens a whole new set of applications.

Communication and data exchange tools

The most significant impact that technology has had on the management of information resources in construction is perhaps in the area of communications. Below follows a catalogue of communication tools that are finding increasing use within construction.

The internet

The internet evolved around the latter part of the 1970s from research originally funded by the US Advanced Research Projects Agency in the late 1960s and early 1970s. Today's internet is a global resource connecting millions of users in a labyrinth that comprises a network of computer networks. The internet is based on a set of rules for data exchange called the TCP/IP protocol suite, [which stands for *Transmission Control Protocol* (TCP) and *internet Protocol* (IP)], an agreed method of communication between all parties associated with the internet. In one sense, the internet represents a community of people who use and develop the isolated computer networks and provide a collection of resources that can be reached from those networks on a global basis. Common uses of the internet include information sharing, interaction and communication.

While the networks that make up the internet are based on a standard set of protocols, the internet also has gateways to networks and services that are based on other protocols. The value of the internet to construction companies derives from its ability to easily connect globally to a vast amount of data, which would otherwise have taken more time and money to organise. By exploiting the resources of the internet construction companies can gain the following benefits:

- Reduced communication costs
- Enhanced co-ordination and communication
- Acceleration in the distribution of knowledge resources within and outwith the company
- Promotion and marketing for the company

The take-up of the facilities available through the internet by construction companies has, however, been slow. This is due to a number of technological as well as social constraints often associated with the internet. They include security concerns, technology issues, legal uncertainties and social acceptance by construction executives.

Intranets

The intranet is a communication infrastructure that is based on the communication and content standards of the internet, but which is internal to an organisation. The tools employed to create intranets are identical to those used for the internet and its applications. The distinguishing feature of an intranet is that access to information is restricted to the construction company's personnel. The development of intranets was in direct response to the concerns of business regarding data security on the internet. Intranets evolved at about the same time as the internet revolution commenced. Construction companies can set up an intranet to allow project managers access to data from both central databanks and different projects.

Extranets

An extranet is a network that uses internet protocols and the public telecommunication system for communicating both privately and selectively with the contractor's clients and business partners. The technology allows a contractor to securely share part of their company information resources or operations with suppliers, subcontractors, project partners, clients or other construction companies. Extranets, therefore, introduce an additional functionality to intranets. The main benefit of this technology is the acceleration of construction business between different companies. Within construction, the advantage in deploying extranets derives from a cheaper and more efficient way for contractors to connect with their suppliers, subcontractors and other project partners. For example, suppliers can receive proposals, submit bids, provide documents and in some cases collect payments through an extranet site. Not only does this cut down on redundant ordering processes and keep suppliers up to date on future deliveries and design changes, but it also allows quicker response times to suppliers' problems.

Extranets can be used to exchange large volumes of data, including the sharing of product catalogues, providing design specifications and details, distribution of news to trading partners and collaborating with other contractors on joint project schemes. EDI is a special form of extranet that can be set up between a contractor and suppliers.

Using the internet

There are four protocols, or services, that are widely used over the internet today. These are electronic mail (email), File Transfer Protocol (FTP), Telnet and the world wide web.

Electronic mail

Electronic mail (email) is perhaps the most popular use of the internet. The basic concepts behind email are similar to that of regular mail. Documents are sent by mail to personnel in organisations at their particular addresses. In turn, they write back to a return mail address. Electronic mail, however, has a distinct advantage over regular mail: speed. Instead of several days, the message can reach the recipient in minutes or perhaps even seconds (depending on location and the state of the connection to the recipient). It can be argued that the telephone achieves a similar speed of transmission. However, the email provides the additional advantage of convenience and a record of communication. Messages are sent when it is convenient. Equally, the recipients respond at their convenience. Emails can also be used to transmit documents as attachments to messages. By means of email, details of work sections can be exchanged between designers and contractors.

File Transfer Protocol (FTP)

One of the first information delivery systems was the File Transfer Protocol (FTP). FTP enables the electronic transfer of a whole file from one computer to another without altering the file format. This is achieved by depositing the files in a repository described as an FTP site. Although email systems can also transfer files, the use of FTP options presents greater flexibility. First of all, FTP avoids restrictions on the type of file sent, in particular whether or not it is a binary (non-text) file, such as program software and pictures. Second, there are no restrictions on the size of file as there are with email. The transfer of engineering drawings, which are normally of large file sizes, from one section of a construction company to another or between project partners can readily be achieved with this technology.

Telnet

It is possible to log on to one computer and link into a remote computer to work with its software resources. The technology for doing this is called Telnet. The deployment of this technology within a construction company can allow estimators to use the resources of the planning department from a remote computer source.

World wide web

The world wide web (www or just 'web' for short) is a networked hypertext information system, originally devised at CERN in Geneva, but now in use all over the world. CERN is the European Laboratory for Particle Physics, founded in 1954. The basic idea of WWW was to merge the techniques of computer networking and hypertext into a powerful and easy to use global information system. Hypertext is text with links to further information, on the model of references in a scientific paper or cross-references in a dictionary. With electronic documents, these cross-references can be followed by a mouse-click and, within the world wide web environment, the references can lead to anywhere in the world. Electronic documents are similar to the pages of paper documents. A computer is used to display the pages on its screen. Inside each page, sensitive spots are exploited by the computer so that the user can switch automatically from one page to another by clicking on a sensitive spot. This navigation by wandering from one page to another is called *browsing*. WWW is 'seamless' in the sense that a user can see the whole web of information as one vast hypertext document. There is no need to know where information is stored, or any details of its format or organisation. Behind this apparent simplicity, of course, there is a set of ingenious design concepts, protocols and conventions that control how it works.

The world wide web is a hypermedia information and communication system popularly used on the internet computer network with data communications operating according to a client/server model. Web clients (browsers) can access multi-protocol and hypermedia information from servers (possibly using helper applications in conjunction with the browser) by way of an addressing scheme. Figure 17.5 presents a typical interface of a browser for accessing the WWW. The current uses of the internet in construction are dominated by its role as a publicity medium.

Other uses of the internet

Use of the internet is not restricted to the transfer of text and graphic information only. It is also used for communication between people in text, audio and visual format. This is achieved

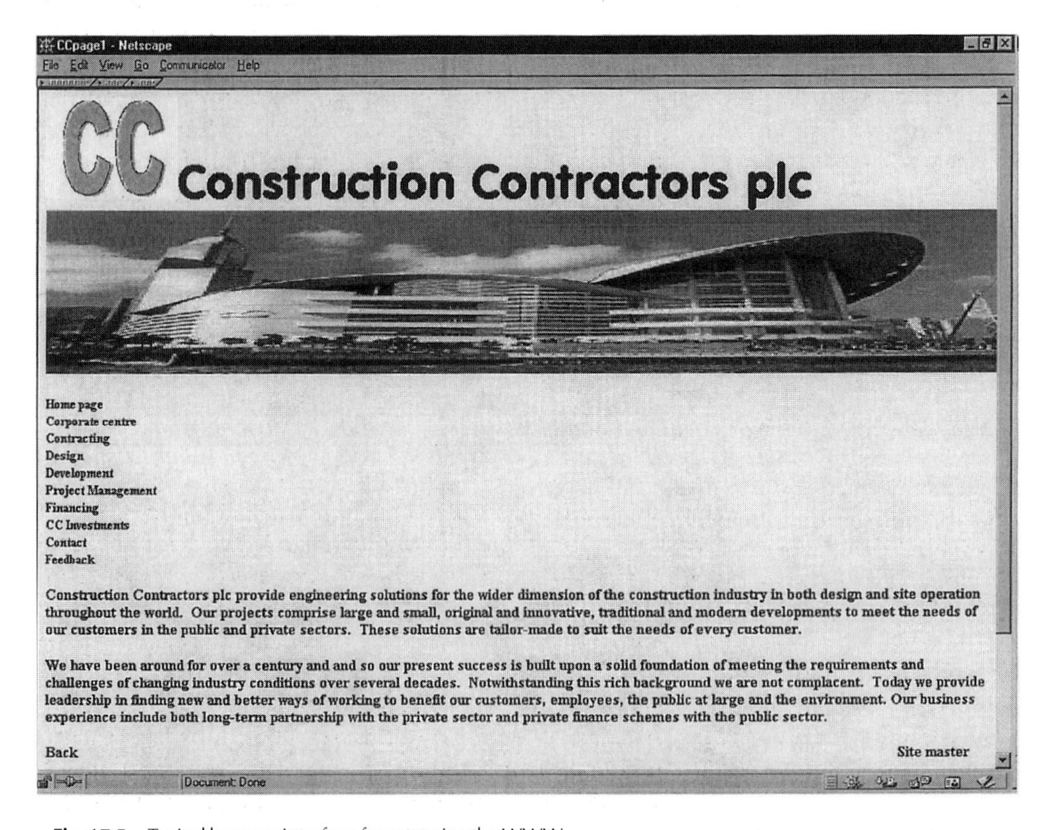

Fig. 17.5 Typical browser interface for accessing the WWW.

through teleconferencing. Teleconferencing is a technology that allows a group of people to confer simultaneously from different locations. The technology was originally based on the telephone. In recent times, however, most teleconferencing is undertaken through computer-based communication. Examples include data conferencing and video conferencing.

Data conferencing

Data conference resources allow interaction between parties so that they can confer only over text and graphic documents. This technology has already found some applications in construction, whereby designers located in different geographical regions work on the same drawing simultaneously.

Another form of data conferencing is internet Relay Chat (IRC). IRC provides a way of communicating in *real time* with people from all over the world. It consists of various separate networks of servers that allow users to connect to a particular IRC site. Generally, the user (such as you) runs a program (called a 'client') to connect to a server on one of the IRC nets. The server relays information to and from other servers on the same network. The main disadvantage associated with IRCs is that it only employs text for communication.

Video conferencing

In its simplest form, video conferencing is the live connection of two or more people using some combination of video, audio and data for the purpose of communication, with video being the only pre-requisite to fulfil the definition. Nevertheless, it is helpful to think solely of the visual impact of the medium from the outset because the roots of how we communicate with one another predate the development of language. Non-verbal actions often described as body language, essential for effective communication, are lost in the absence of visual interaction.

Rapidly developing video conferencing tools are also changing the way that construction projects are run. Virtual conferencing tools like CU-SeeMe, Microsoft's Net Meeting and Netscape's Collabra are now being bundled with standard computer software packages, and these tools will enable project teams to collaborate, redline drawings and solve problems without having to travel to the job site. These advances may help reduce travel costs and improve project communications.

Electronic commerce

Electronic commerce is employing information technology to improve transactions between companies. It involves the integration of email and similar technologies into a comprehensive electronic-based system of business functions. It is based on the electronic processing and transmission of data, including text, sound and video. It encompasses many diverse activities including electronic trading of goods and services, on-line delivery of digital content, electronic fund transfers, electronic invoicing, payments and receipts, electronic share trading, collaborative design and engineering, on-line sourcing of specifications, public procurement, direct client marketing and after-sales service. It involves both products (e.g. technical goods such as material components in construction) and services (e.g. information services, financial and legal services), traditional activities (e.g. healthcare, education) and new activities (e.g. virtual malls). Although electronic commerce is not an entirely new phenomenon, its potential impact on construction has yet to be realised and will lead to profound structural changes in construction. The take-up of electronic commerce in construction is being accelerated by the exponential growth of the internet. This growth allows transactions on a global scale between an ever-increasing number of participants, corporate and individual, known and unknown, on global open networks making up the internet. Electronic commerce, of course, is not limited to the internet. It includes a wide number of applications in the narrow-band (videotex), broadcast (teleshopping) and off-line (catalogue sales on CD-ROM) environments, as well as proprietary corporate networks (banking). Also, the internet is generating many innovative hybrid forms of electronic commerce by combining, for example, digital television infomercials with internet response mechanisms (for immediate ordering), CD-ROM product catalogues with internet connections (for content or price updates) and commercial web sites with local CD-ROM extensions (for memory-intensive multimedia demonstrations). Specific developments in construction include remote estimating, whereby cost estimates for a project can be obtained by accessing and processing relevant information through the web site on a consultancy basis.

Integration of systems

Integration of information and communication technologies has been achieved in the areas of project document management, establishment of project and company data warehouses and development of industry-wide infobases.

Project document management

For most construction companies, project documentation forms an essential element in the production process. In the case of engineering design and consultancy firms, this is often the only output of their processes. The management of project documentation is often essential for the schedule performance of the project, avoidance of extensive re-working and effective quality control of the project. The use of electronic options for management of such documentation is growing as a result of current technology. Earlier versions of electronic document management systems concentrated on storage and retrieval of project documentation, and were generally referred to as Document Filing Systems (DFS). Electronic document management systems (EDMS) provide a combined set of tools for full organisation of all aspects relating to project documents. They cover the control of their creation, revision, distribution, storage and retrieval throughout and beyond the project lifecycle. The tools employed by EDMS vary, but many systems include:

- Imaging for transforming paper to digital format using scanners
- Full-text retrieval for accessing archived textual documents
- Workflow for mapping and controlling the route of documents within the organisation
- Multimedia for managing audio and graphic information, particularly progress reports for site activities
- CAD for creating and editing documents

Data warehousing

Data warehousing is a process for assembling and managing data from various sources for the purpose of gaining a single detailed view of part or all of a business. This takes further the concept of databases to give consideration to access to and availability of the contents of the many databases that a construction company will have. The relevance of data warehousing to the contractor or designer comes from the need to achieve effective integration of all information resources in order to improve the company's projects and operations. Its purpose therefore is to achieve the following:

- Time savings for information suppliers and users by automating the reporting processes
- More and better information for managers, leading to better decisions
- Improve company and project processes
- Support strategic evaluation of a construction company

To implement a data warehouse scheme, there are a number of issues that a construction company will have to take into account:

- *Costs.* These in effect represent the size and scale of the warehouse, and need to be distributed between hardware, software, other tools and services.
- *Users.* Consideration has to be given to what the users currently need by way of information and what they are likely to require in future. The information needs of different users in a construction company will not be the same, and will influence access options for the data warehouse.
- *Development and maintenance.* A decision will have to be made on whether the system will be built in-house or involve external IT consultants. Similarly, a decision will be needed on whether it should employ standard hardware, software and tools or a bespoke system. Attention will also have to be given to the maintenance and possible replacement of the system.
- *Technical details.* Technical details of the warehouse include the following aspects:
    - The content of the system relates to the amount and type of data that will reside in the warehouse.
    - Metadata management involving defining and controlling all instructions and information that help to organise the contents of the data warehouse.
    - Access options to define the number of users, the mode of access and the display options for the warehouse.

In general, contractors will set up databases for each project they undertake. The use of a data-warehouse approach ensures integration of all the projects that a company has and so ensures more effective co-ordination.

Infobases

Infobases are commercial databases that cater for the information resource needs of specific interest groups. Their main benefit is the ready accessibility of information that is current. The pace of change requires that construction companies will have to continuously update themselves on issues such as Health and Safety regulations. Also, construction manufacturers are increasingly using the internet to publish information regarding their product lines. Traditional paper-based brochures, catalogues and specifications are seen as cumbersome to use, they take up room and their information goes out of date quickly. Digital catalogues are now being posted on a manufacturer's web site, where specifications and pricing can be updated regularly. Currently, several web sites provide access to manufacturer's data including FM Link, Construction Net, Canadian Engineering Network, Building On-Line and others. There are other projects at advanced stages of development to ensure effective exploitation of these technologies for the engineering sector. For example, the GENIAL project under the Global Engineering Network Initiative is an ESPRIT programme aimed at developing Intelligent Access Libraries to enable engineers, designers and others with a professional interest in construction to obtain electronically technical information resources relevant to their work. These include specifications, standards, design details, product descriptions and component details, as well as regulations.

Construction in a wireless world

The very nature of production in construction lends essence to the need for high levels of mobility for the organisations involved in the delivery of projects. Traditionally, this has been achieved through the setting up of mobile offices on site to manage and support the production operations and to act as a point of contact for general administrative representation of the contractor. The availability of technologies to enable real-time interaction and communication between staff located in disparate places (such as the project site and head office) is taking the mobility of individual workers and the siting of production offices to new heights in construction. This has been facilitated by the availability and deployment of various wireless technologies that are redefining the very nature of interaction in the work environment. Whilst it is not uncommon to find hard-wired connectivity on construction sites, the hazardous nature of most sites, as well as the temporary nature and spread of a project site, often means that there is a limitation in the level of communication points that can be set up. Setting up a wireless hot-spot that spans the full site eliminates the need for wired cables to connect the project's network with the benefit of flexibility so that operatives and staff can work wherever they need to be located. Construction companies deploy wireless and mobile technology solutions for mobile mail and messaging, and also to support automated control data collection to enable them to achieve time savings, productivity and collaboration throughout a project, as well as their wider organisation.

Wireless technologies utilise radio waves rather than wires to send data along a communication path. Devices that rely on wireless technologies can operate in different network environments. These include the wireless equivalent of the hard-wired networks such as wireless local area networks (WLAN). A WLAN enables wireless devices to be joined into a network without the constraint of their location. They achieve this by connecting to a small collection of wireless access points (or hotspots) that are wired to the network. The advantage presented by continued connection achieved through wireless technology is invaluable to interactions and meetings on the site as well as in the design office.

Information security

Information security has traditionally been a part of the general security of construction companies. This is because much of the information that had to be managed was in hard-copy formats such as drawings, bills of quantities and logbooks. The transition to a digital world presents a different risk for information access in construction projects and companies. An attendant feature of this transition is a shift of the responsibility for information security to the IT departments in construction companies. This is because most construction companies see information security as a technological problem calling for technological solutions, notwithstanding the reality that networks currently available cannot be made fully impenetrable. The need to plan against potential hacking and the continuous stream of viruses that have become the norm in digital environments has to be tackled not only from a technological standpoint, but also from operational and business perspectives.

The function of information security varies depending on the nature of the task and position of the staff or operatives in the industry. This is influenced by the value the construction company or project places on the information under question and the extent to which the information is structured. The role of information security is given added impetus when wireless networks and devices are deployed. In a hard-wired project office, there will be some combination of firewalls, anti-virus software, anti-spyware software, routers, proxy servers and secure login protocols to protect the corporate or project network as well as the individual project staff. While these security measures can be translated to the wireless network, the mobility of the devices often means straying away from the strict, corporate controls of firewalls, internal scanning of servers and external email sentries. The need to address the security of all project information needs to be managed with the same detail and degree of attention as other resources. This would normally form part of the role of the construction information officer.

18
Financial management

Summary

The purpose of this chapter is to provide an understanding of capital and its effective use in a construction company so that the manager is better equipped to deal with the financial crises that inevitably occur. The chapter describes sources, types and means of acquiring funds and capital and gives examples illustrating the capital workings. This is subsequently reinforced by a case study illustrating the accounts for a typical construction company. The study describes the build-up of the profit and loss account and the balance sheet, giving examples highlighting the important financial information that can be derived from such accounts.

Introduction

The construction industry is characterised by having a very great number of firms, from the tiny to large multinational companies, with contracts conventionally undertaken on the basis that payment for completed work is progressively supplied by the client throughout the life of the project. Therefore, on the assumption that the contractor is able to raise sufficient capital to finance possibly the first two months of the contract, steady interim payment thereafter provides the necessary cash flow. However, whenever the market takes a downward turn, competition for contracts becomes more intense and tenders need to be keener. Should any unforeseen difficulties thereafter arise during the course of the contract's execution, less finance is then available to pay creditors and bankruptcy could ensue. Furthermore, in times of rapid inflation companies carrying excessive stocks may suffer cash flow problems because of the purchase of the more expensive resources out of previous earnings. In many cases such companies could be in a profit-making situation in the long term, but not have the immediate cash to pay creditors and bankruptcy similarly may result. Indeed, for these reasons the level of bankruptcies in the industry tends to compare unfavourably with other sectors and needs to be understood in the context of the principles and procedures underpinning sound financial management. This is especially important for business activity in the design-and-build segment, where financing the whole project may fall on the principal contractor/developer.

Types of businesses

The limited company, which emerged during the nineteenth century, has proved successful and is the predominant type found in the construction industry today. The significant feature is that private individuals introduce capital into the company and in so doing become shareholders, but with personal liability limited to the amount unpaid on their shareholdings. Furthermore, under the same principle, individual companies are also able to invest in other firms by acquiring shareholdings, indeed even to the extent of assembling a majority stake holding (more than 50%) to produce the controlling interest. Such subsidiaries are normally grouped to form a pyramidal network headed by the holding company with separate consolidated accounts.

Notably, resident firms are liable for paying corporation taxes on trading profits. However, if they are owned by a holding company, the latter is commonly able to offset losses in one resident subsidiary against the profits of another similarly resident for taxation purposes in the consolidated parent (holding) company accounts. This principle, furthermore, may eventually also apply across the European Union but with firm rules on transfer pricing. Consequently, shareholders frequently continue to register the parent company in a convenient 'offshore tax haven' to limit general scrutiny of their group-holding affairs for both local and domicile tax advantage purposes and dividend regulations, particularly convenient in manipulating transfer pricing and financial transactions of intra-trading in a collection of internationally spread companies.

There are two types of limited company in the UK: private and public.

Private company

The legal procedure for forming a private company (Ltd) is fairly straightforward, requiring the persons wishing to form the company place with the Registrar of Companies:

- The memorandum
- The articles of association (if any)
- A list of persons who have consented to be directors
- A statement of the nominal share capital
- A statutory declaration of compliance (the purpose of which is to show that all legal requirements have been met)

Also, within 14 days of registering, they must provide:

- The address of the registered office
- Particulars of the director(s) and secretary

The memorandum of association contains five compulsory clauses:

(1) The name of the company (which must end with the word limited)
(2) The objectives of the company—these can be drawn so wide that they may include almost any legitimate activity
(3) The amount of authorised share capital
(4) The domicile of the company
(5) The limitation of liability of the members of the company (shareholders) signed by each subscriber in the presence of at least one witness; each subscriber must take at least one share

The articles of association

A company is not bound to have articles of association. However, if a company fails to register, articles will automatically be ruled by the model set to be found in the Companies Act.

The articles comprise the rules and regulations governing the internal workings of the company and cover matters such as the issue and transfer of shares, alterations to capital holdings, borrowing powers, shareholders' meetings and voting rights, the appointment and powers of directors and the presenting and auditing of accounts.

Other types of arrangement, such as separate Supervisory and Management boards, exist in some EU states.

Public limited company (PLC)

When a private company reaches a certain size, the problem of obtaining additional capital for further expansion is often best achieved by increasing the shareholdings in the company. However, the number of individuals willing to invest capital in an unquoted company is likely to be restricted since a realistic valuation of shares cannot be determined without the facilities of the market forces operating through the Stock Exchange or unlisted securities markets. To become a public company, therefore, involves raising capital via these or similar bodies and complying with specific legal requirements, i.e. the company must be registered, the number of directors must be at least two and the number of persons signing the memorandum must be at least seven, although there is no maximum. Additionally, as the company wishes to issue shares to the public, it must file a copy of the prospectus (a document inviting the public to subscribe to the company's share capital) with the Registrar General. (Sometimes the capital is subscribed privately by a large institution, in which case the company must file a statement in lieu of the prospectus.)

Multinational companies, for example, commonly secure separate registrations and listings in other major financial centres abroad in order to give access to a wider range of investors.

The prospectus

A company wishing to raise capital on the stock market usually approaches a broker, who then lodges a letter of application setting out the particulars for which a quotation is sought. At the Stock Exchange, various Committees on Quotations consider the application and judge whether the company should be allowed to advertise for subscribers. The advertisement contains all the information required by the prospectus plus other information demanded by the Stock Exchange, as follows:

(1) A statement of the company's previously issued capital, dividends and capital repayments, and the voting rights of the different classes of shareholders
(2) A list of names and addresses of the company directors, auditors, secretary, bankers, brokers and solicitors
(3) An outline history of the company
(4) A profit statement for the last ten years, together with a summary of the previous audited balance sheet
(5) Details from the articles of association
(6) A summary of contracts and other work entered into during the previous two years
(7) The views of the board of directors on the future trading and financial prospects for the company

Stock Exchange quotations

Whenever a share issue is to be made to the public, much advice is required regarding:

- The price at which to issue the shares for subscription
- The type and number of shares to issue
- Any legal rights and implications
- The viability of the issue

It is usual to engage the services of an experienced issuing house to carry out this work. The issuing house will also take on the administrative duties involved in placing the necessary advertisements and allocations of the shares. The considerable experience of issuing houses is called upon to judge the market, since if all the shares are not taken up and their market price subsequently falls, the plan to raise the necessary capital will not be fulfilled. For this reason, the issuing house either underwrites the risk itself or arranges elsewhere for the share issue to be underwritten. Only reputable companies with a solid record are therefore likely to be considered by such houses. Once an agreed minimum number of shares have been taken up, the company is given a Stock Exchange quotation. (A similar procedure applies in gaining separate listings on other stock exchanges.)

For the company not sufficiently large to gain a Stock Exchange listing but otherwise substantial in performance, there now exists in some well-developed economies the Unlisted Securities Market, which also gives the opportunity to raise equity finance from the general public and other investors, so freeing the firm from the limitations of private fund raising.

SE Company

Under new European Union legislation, a Societas Europaea (SE) company may be registered and regulated for pan-European Economic Area status, but compliance with local taxation and corporate governance rules of the particular nation state where the company is based continue to prevail. Relocation to a more favourable business situation in another member state without re-registration thus becomes an option.

Nevertheless, the legislation also allows a company to operate across the European Economic Area under the registration of any member state. Hence, firms may shop around to find the lowest set-up costs and capital required for registration as a PLC, Ltd, AG, etc.

In the USA, Delaware State registration proffers similar advantages.

Buying and selling shares

PLC

A simple holding of 51% of the shares with voting rights in effect provides a majority interest (similarly for a private company), so giving the holder of the shares control of the company. Indeed, possession of as little as 10% held may prove invaluable at an AGM when perhaps only a few of those eligible may actually be present or casting votes. In this respect the recent growth of shareholdings by institutional investors, e.g. pension funds, insurers, fund managers and unit trusts (mutual funds), is giving such investors more power to scrutinise the running of companies, but unfortunately they remain conservative, often preferring to sell shares rather than risk upsetting the market by openly declaring dissatisfaction with company performance. Also conflicts of interests sometimes arise, especially with the large corporations, insofar as the institutional investor may also be carrying out services such as looking after the firm's pension fund itself. Hence, investor/owners' ability to influence board management performance remains somewhat ambiguous.

The shares of a PLC are floated on the stock market at a previously decided price called the nominal price (par value) of the share, which represents the capital to be raised by the issue. However, subsequent buying or selling of the shares is subject to the market forces operating at the time, causing the price of the share to fluctuate wildly. The reasons for this are many, but mainly associated with the company's earnings yield and trend growth, asset values, money interest rates and the vitality of the economy. The gains or losses are independent of the company's financial accounts dealings, and are borne entirely by shareholders. The total traded price of the shares broadly represents the market value of the firm. On gaining possession of 75% or more of the shares in a PLC, UK company regulations allow the new majority shareholder under a special resolution voted at a general meeting to turn the company private and change the articles of association.

Limited company

Long-term inflationary distortions and business performance may have led to understated asset and other balance-sheet item values; thus, for the private firm not having a stock market listing, the purchase value has to be assessed, the owner persuaded to sell and sufficient funds raised by the buyer to secure the controlling proportion of shares. The amount paid for such a business in excess of the book value comprises monetary inflation on the tangible assets plus 'goodwill', i.e. assessed value of customers, network of contacts, skills built up, reputation, ability to earn future profits, etc. Commonly, goodwill is immediately totally depreciated on acquisition of the company since its deemed value may not stay constant over time. The tangible assets are usually restated at the fair rather than book value, and capital reserves (liabilities) accordingly adjusted to match.

In particular, private equity firms are becoming more active in construction take-overs using leveraged financing, the prospective company under acquisition scrutiny being valued on the basis of a multiple of historical profits guided by banker benchmarks; currently about 7X Ebitda —see price to equity (p/e) ratio discussed below. Such firms rely on debt funding to finance purchases of undervalued companies and, after carrying out appropriate re-engineering and restructuring, make a return on the investment by subsequently selling, floating or refinancing the acquisition.

Types of capital

The small company beginning operations will normally need to find the initial capital to start the business from private resources. Few others are willing to take the risks associated with an unknown enterprise. Gradually, though, as the firm prospers by ploughing back retained profits, the usual lenders of capital may then consider providing loans and investments.

Loan capital can be classified into short-term and long-term borrowings. Short-term loans usually carry lower interest rates, by virtue of the fact that repayment periods are short, of the order of months, and therefore open to much less risk than loans granted over a period of years.

The merits of the various finance alternatives are summarised as follows in Table 18.1:

Table 18.1 Financing alternatives.

Finance	Source	Advantages	Disadvantages	Costs
Overdraft	Bank	Usually cheapest source Quickly arranged Flexible No minimum Renewable Interest paid only on usage Sometimes available unsecured	Subject to changes in government economic policy Repayable on demand Subject to changes in bank policy Tempting to use for funding long-term purchases	Floating interest charge at base rate plus 1–4% may incur commitment fee
Short-term loan	Bank or finance house	Term commitment by loan institution Competition between lending houses, especially for hire purchase Relatively quickly arranged Can be used in conjunction with overdraft facility Sometimes available unsecured	Generally more expensive than overdraft Term commitment and funds may therefore be idle if forecast for funds inaccurate Tends to require security against other assets	Floating interest charge at base rate plus 2–5%
Medium-term loan	Clearing banks Investment banks	Term commitment by lender Capital and interest repayment can be arranged to suit borrower's future cash flow position Size of loan may be small or large, especially from clearing banks Inflation reduces real cost over time Fixed interest charge can sometimes be negotiated	Usually higher interest charge than shorter-term finance Long-term commitment which may require short-term borrowings to finance interest payment if cash flows became distorted from forecasts Negotiation fee likely Legal costs also often incurred	Fixed or variable interest charge set at 11/2–4% above 6 months inter-bank rate

Table 18.1 *continued*

Finance	Source	Advantages	Disadvantages	Costs
Hire purchase facility	Finance house	Inexpensive and specially arranged Payments fixed over term agreed Ideal for short-term requirements Normally overdraft facility not affected Capital allowances against corporation tax available immediately Not classed as borrowings	Expensive Subject to government economic policy changes, but never retrospectively Defaults usually rigorously prosecuted Interest rate quoted may be misleading, because of period payments and compound interest calculations Purchased assets not legally passed over to lender until final payment	Interest fixed at time of negotiations at finance house base rate plus 4–5%
Lease facility	Finance house	Similar advantages to hire purchase Over-geared firm can acquire resources without affecting Balance Sheet	Ownership does not pass to lessee and therefore capital allowances not available Values not reflected in balance sheet assets, and so may give distorted impression of firm's capabilities	Similar to hire purchase costs, but user must be aware of possible commitments to pay for mandatory maintenance and repairs Tax concessions foregone must be compared with HP alternative

Long-term/medium-term finance

Long-term finance is that capital required for five to ten years, either to start the business or to carry out expansion programmes. Broadly, the capital is used to purchase buildings, plant and equipment and to carry stocks of materials. The risks to the lender are high because of the time scale involved; consequently, only established firms are generally considered by the lending institutions. Some of the more important sources of long-term capital are shown in Fig. 18.1.

Short-term finance

The firm when established often needs short-term capital to overcome immediate cash flow problems. Materials have to be purchased, plant hired, labour and subcontractors paid and so on before payment is received for the finished product. Furthermore, capital may be required to smooth out the strains on cash flow resulting from rapid fluctuations in the market demand for the company's goods. Many sources of short-term finance are available to ease the situation, but naturally, the firm must be well managed and profitable before the lending institutions will consider any loan application. The main sources are shown in Fig. 18.2, the clearing bank overdraft facility being the most important source.

Capital sources and institutions

The more common types are described below.

Shares

Shares are called the equity in the company. The shareholder is entitled to the residual profits in the company after all other commitments have been met. The share also entitles the holder to voting rights and is termed an ordinary share. Other forms of shareholding include: those without voting rights; preference shares, which entitle the holder to a dividend in preference to the ordinary shareholders; cumulative preference shares, whereby a dividend is paid in full later to allow for those years when a dividend was not declared; and bearer shares, these being an unusual form in which there are no registered shareholders so that investors cannot readily be identified.

Debentures

Debentures and bonds are loans made to the company, but differ from conventional loans insofar as they are offered to the market at a fixed interest rate repayable at a set time, the loan being secured either by the mortgage on the firm's property or simply on the basis of the company's reputation. An independent secondary market exists where debentures and bonds can subsequently be bought and sold, the price fluctuating depending on general interest rates and credit-rating agency risk assessment of the company's prospects to pay fixed interest. As general interest rates fall, the bond price rises and thus the yield falls, and vice versa. Corporate bond yields are usually higher than less risky treasury bonds or gilts, and broadly track nominal GDP growth rate. So-called *junk* bonds are the lowest grade and offer very risky high yields.

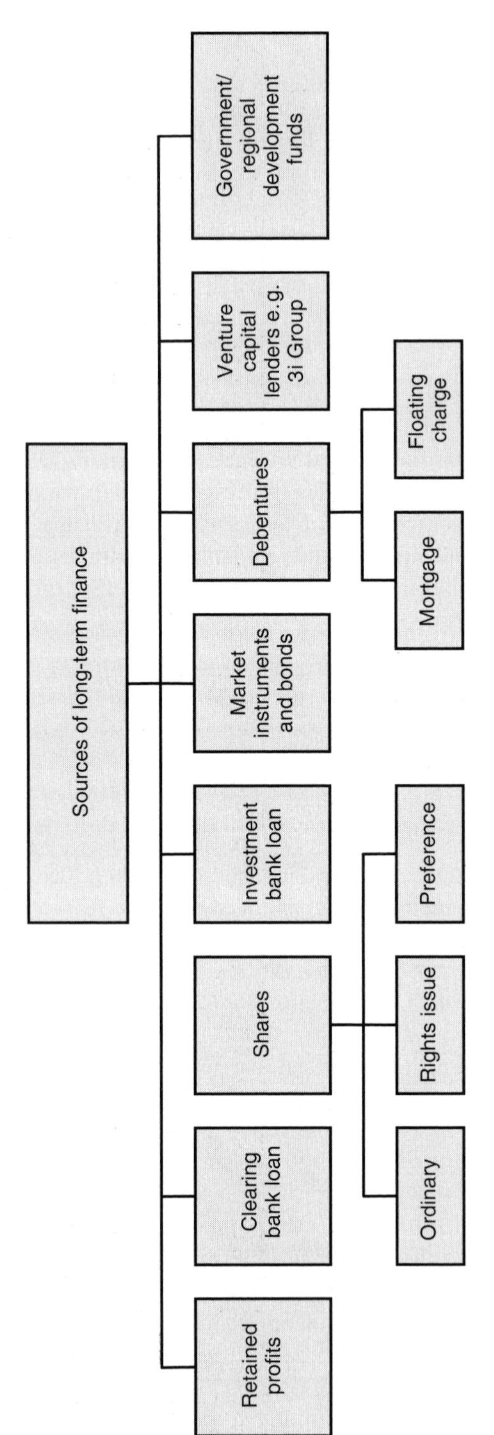

Fig. 18.1 Sources of long-term finance.

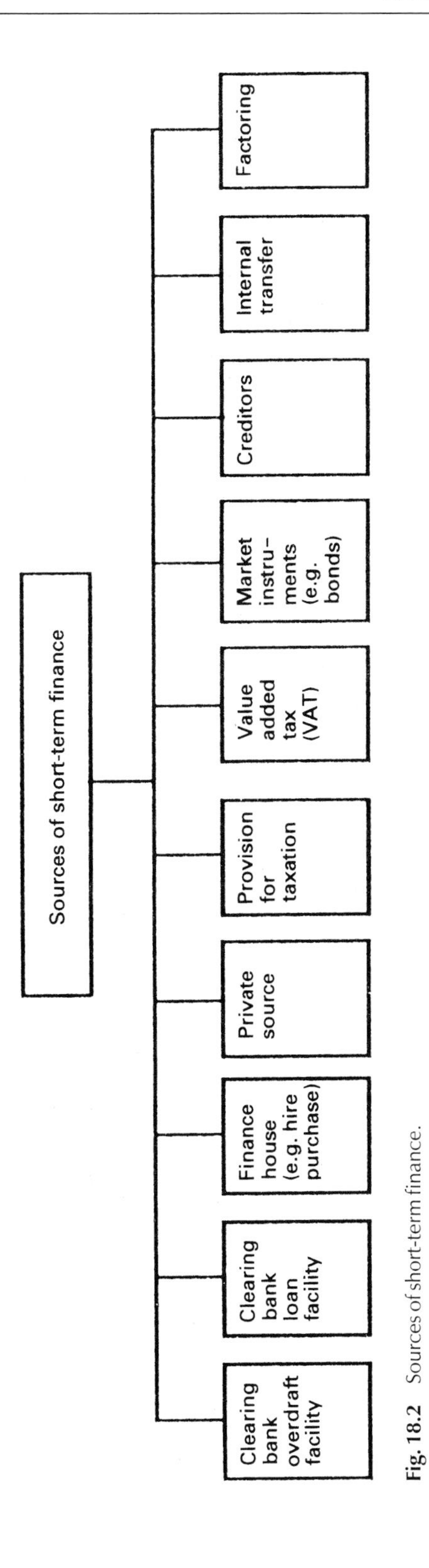

Fig. 18.2 Sources of short-term finance.

Debenture and bond holders rank ahead of almost all other creditors in the case of liquidation of the firm's assets. Like other loans, they represent a cost to the company and, as such, the interest payment made to the holders is deducted from profits before allowance is made for corporation tax payments. In addition, the interest payments rank ahead of any dividend declared to shareholders.

Loans

Banks and other finance providers are generally reluctant to lend long term to construction firms without collateral surety, and the borrower is commonly also requested to supply a proportion of the finance from internal sources. The (international) investment banks tend to demand higher rates of interest than the clearing banks since they are normally dealing with a large loan. Their role is not great in the construction industry, but such institutions are gradually becoming more involved with the large contractor interests in package deals and speculative ventures, both at home and increasingly overseas. Subordinate-ranked 'mezzanine' loans have less effect on Basel 2 credit ratings and, internally considered by the banks, are an alternative option. However, higher capital charge margins are incurred.

Various other inducements have been introduced by governments to encourage entrepreneurs to set up businesses, particularly in manufacturing and commerce. For example, business expansion schemes, venture capital from the major banks and pension fund investments are increasingly being made available to provide 'start-up' loans and even equity finance for small companies. A well-prepared case outlining the nature of the business, the potential of its activities and products, the experience of management, the skill of workforce, etc. is required.

Nevertheless, in the main, the bulk of construction firms must rely on good relationships developed with one of the 'high street' banks to secure at best an overdraft facility and perhaps later, when firmly established, loans of various kinds.

Rights issue

Shareholders are offered new shares in exchange for their present holdings. Any not taken up are put up for sale; for example, a company with shares holding a nominal value of, say, £2 offers its shareholders three new shares at £0.90 each in exchange. The market price of the new share is now likely to fall, but, of course, the shareholder has more shares to take account of this.

Provision for corporation tax

Payment of this tax is usually made in arrears; the cash therefore remains in the business during that time and acts as a valuable source of short-term funds.

Cash from creditors

Delayed payments to creditors and prompt ones from debtors, if handled with care, ease cash-flow problems. The construction industry is well suited for this sort of financial arrangement since completed work is paid for by the client in stages.

Retained profits

Retained profits and dividend payments are an important source of investment funds.

Financial instruments

Floating-rate notes, currency swaps, bills of exchange, promissory notes, etc. are used by international companies, particularly for trade purposes. In addition, derivatives, hedge funds and futures facilitate the option to buy or sell (put) an asset at a fixed price in the future; these include shares, foreign currency or a credit derivative to 'insure' against creditor default. Risks that are even more exotic are becoming tradable in the financial markets.

The control of capital

Liabilities

The capital in a business is the money invested by shareholders, money loaned by individuals or institutions and retained profits. The capital represents a liability on the company and is used to purchase its assets. Short-term liabilities, sometimes called *current liabilities*, include creditors, short-term loans and provisions for taxation and are not usually included as part of the capital employed for analysis purposes.

Assets

Fixed assets

The fixed assets are the buildings, land, plant and equipment, etc., permanently retained in the business for the purpose of facilitating the use of the working capital.

Current assets

Current assets include material stocks, work in progress, cash at the bank, debtors and short-term investments. They are the assets locked up in the working capital cycle.

Working capital

Working capital comprises the liquid or near-liquid assets needed to lubricate the daily transactions of the business. It is represented by the difference between current assets and current liabilities, and is locked up in a continuous cycle, as shown in Fig. 18.3. Working capital should not be confused with capital employed, described later.

Example—illustrating working capital requirements
A precast concrete manufacturer is considering whether to increase production. In order to do so, a new and fully equipped factory is required at an estimated cost of £450 000. The value of the increase in turnover is projected to be £2 000 000 p.a. broken down as 40% materials, 35% wages and other production costs, 15% administration, the remaining 10% representing profit. On average, the concrete units take two weeks to produce, after which they are put into store until

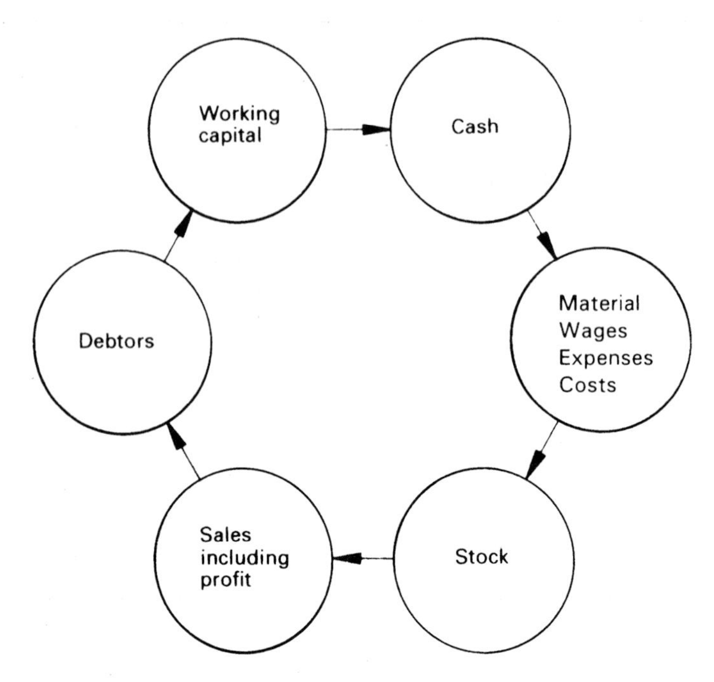

Fig. 18.3 The current assets and current liabilities cycle.

fully cured and ready for sale. The average period between completion of manufacture and sale is two months, and two months' credit is allowed for customers. Three months' supply of raw materials will be kept in stock at all times, for which suppliers customarily allow one and a half months' credit.

Problem
Determine the capital required to equip the new business, plus the working capital to finance the normal transactions.

New capital requirements		£
Buildings and equipment	450 000	
		450 000
Working capital requirements		
Materials (40% of £2 million = £800 000)		
Time factor		
Materials held in stock	3 months	
Materials held in production	1/2 month	
Finished goods held before sale	2 months	
Credit to purchasers	2 months	
	7 1/2 months	
Less credit from suppliers	1 1/2 months	
	6 months	

New capital requirements		£	
Thus,	$\frac{6 \times 800000}{12}$	400 000	
Labour (35% of £2 million = £700 000)			
Time factor			
Production period	½ month		
Selling period	2 months		
Credit to purchaser	2 months		
	4½ months		
Thus,	$\frac{4\frac{1}{2} \times 700\,000}{12}$	262 500	
Administration/selling (15% of £2 million = (£300 000)			
Thus,	$\frac{4\frac{1}{2} \times 300\,000}{12}$	112 500	
		775 000	
Total new capital requirement			1 225 000

The capital raised for the purchase of the building is needed on a long-term basis. However, that required to ease the cash-flow situation generated by the normal trading processes has to be financed with internal sources of capital or with short-term loans, for which a bank overdraft facility is the more usual source.

Capital gearing

Capital gearing is defined as the ratio of fixed return capital to ordinary share capital. If a company can confidently expect to make a reasonable profit in most years, then it is wise to raise some of its capital by means of debentures and/or preference shares at a fixed interest payment, so providing an improved yield for the ordinary shareholder, as demonstrated in the following examples.

Fixed return capital (FRC)—preference shares, debentures and loans

Company X is high-geared while Company Y is low-geared

	Company X (£)	Company Y (£)
10% debentures	10 000	100 000
5% preference shares	6 000	300 000
Ordinary shares	4 000	1 600 000
Total FRC	16 000	400 000
Total ordinary shares	4 000	1 600 000
Capital gearing	$\frac{16\,000 \times 100}{4000} = 400\%$	$\frac{400\,000 \times 100}{1\,600\,000} = 25\%$

Leverage

The question of capital gearing is extremely important when considering the injection of new finance into the company. The following example illustrates how high gearing can raise the yield on ordinary shares. (Assume 15% profit earned on capital employed.)

	Period 1 (£)	*Period 2* (£)
Ordinary shares at £1 each	2000	1000
1000 10% debentures at £1 each	–	1000
	£2000	£2000
Disposable profit	£300	£300
Interest on debentures	–	£100
Profit	£300	£200
Corporation tax (say 50%)	£150	£100
Profit available to ordinary shareholders	£150	£100
Earnings per share	**7.5p**	**10p**
Gearing ratio	– (low)	100 (high)

Note: A fall of £200 in profits would entirely eliminate the yield on the ordinary shares for the highly geared company. It is unwise to gear up highly if profits are likely to fluctuate wildly from year to year, since debentures have to be paid irrespective of the profits. Furthermore, when the average profit is good, but fluctuates over a period of years, the preference shareholder is entitled to only a fixed interest payment, while the ordinary shareholder gains in the good years. This is overcome by means of cumulative preference shares, on which a dividend is paid to cover all the years when full payment was not possible.

The company accounts

All UK limited companies are required to file their annual accounts with the Registrar of Companies and, in order to ensure that these give a true and fair view of the company's trading position, are subject to an annual independent audit carried out by regulated individuals or professional firms. In parts of the EU, auditing is performed by a government auditing body. Normal practice requires the company to present to its shareholders a *balance sheet* and *profit and loss account* derived from the trading transactions recorded in a book-keeping system. Also, operating and financial reviews of potential business risks together with possible recovery plans are provided to investors. Furthermore, a non-financial report may be presented, giving the firm's overview of its environmental and social impact, according to the guidelines set out under the Global Reporting Initiative (GRI 2006), which offers a checklist of issues.

The accounts clearly separate assets, liabilities and trading results for the period of one year just past. The balance sheet represents a 'photograph' of the company's trading position at a given point in time.

In practice, the exercise takes many weeks to complete and is therefore only the accountant's forward projections of the trading position to that time.

Book-keeping system

The company accounts are kept in a ledger, normally using the *double-entry book-keeping* convention. The figures record each and every transaction handled by the business, from which the profit and loss statement is finally derived. The process is commonly executed with the aid of a suitable computer program as part of a software package covering all aspects of financial management; nevertheless the methodology is quite straightforward and can be carried out manually as demonstrated in the following transactions and subsequently absorbed in the summary information given in Tables 18.2 and 18.3.

Table 18.2 Company trading position at 31 December.

Description	Contract (£)				
	A	B	C	D*	E*
Plant hired	2 500	1 500	1 000	500	500
Materials used	10 000	12 000	7 500	9000	5000
Wages and salaries incurred	7 500	6 000	2 500	2000	1000
Value of completed work	25 000	24 000	14 000	–	–
Work certified and paid for	25 000	24 000	8 000	–	–

* Contracts D and E are incomplete at the end of the year.

Table 18.3 Details of trading position.

Item	£
(1) Value of company fixed assets of land and buildings as at 31 December of the previous year	15 000
(2) Materials purchased and delivered to contracts	44 500
(3) Payments completed in respect of:	
(a) Overhead expenses on contracts	3 000
(b) Salaries and overheads at head office	6 000
(c) Wages and salaries on contracts	18 000
(4) Payments completed in respect of creditors	49 000
(5) Price of plant and equipment purchased during year (depreciation allowance on plant and equipment is 10% p.a.)	5 000
(6) Cash at bank as at 31 December	2 500
(7) Depreciation of buildings ($3\frac{1}{3}$% p.a.)	15 000
(8) Opening stock of materials at 31 December of the previous year (not yet paid for)	500

Double-entry book-keeping example

J Paris invests £10 000 as capital in the company on 1 January. On 1 February the company purchases an item of equipment for £50 cash from Dice Tractors and on 1 April purchases on credit £200 of materials from J Smith (Builders Merchants). The company builds a house extension for John Brown and invoices the following sums for payment as work is completed.

1 May	£100
10 May	£50
17 May	£75
30 May	£25
10 June	£50

John Brown sends a cheque for £100 on 1 June, and J Paris withdraws £3000 capital from the company on 30 June.

Book-keeping convention requires debits to be defined as resource flows *into* the accounts (i.e. the account receives), credits being resource flows *out* (i.e. the account gives). Set up in this manner the transactions may be recorded as follows:

(i) The first account established is a credit in the capital account with £10 000 investment from J Paris. Thus:

Capital account

Debit (+)	Credit (–)
	£10 000 from J Paris to Cash A/C on 1 Jan

Cashbook becomes

Debit (+)	Credit (–)
1 Jan from Capital A/C £10 000	

(ii) The first transaction involves the purchase of plant for £50 cash from Dice Tractors Ltd and three accounts—Purchase, Supplier and Cash—must be established. Thus:

Dice Tractors account gives out a resource valued at £50

Debit (+)	Credit (–)
	£50 to purchase A/C on 1 Feb for supplying an item of plant

Since the transaction was completed in cash, the Dice account is immediately balanced by a transfer of £50 from the Cash account.

(A) The Dice account then becomes:

Debit (+)	Credit (–)
1 Feb from Cash A/C £50	£50 to Purchase A/C on 1 Feb
£50	£50

and the Cashbook becomes

Debit (+)	Credit (–)
1 Jan from Capital A/C £10 000	£50 to Dice A/C on 1 Feb for plant item

The Purchase account receives the resource worth £50

Debit (+)		Credit (–)
1 Feb from Dice A/C for plant item	£50	

(iii) The purchase of £200 materials from J Smith Ltd can be treated in a similar manner, except that the Cash account is not involved in the transaction because the goods are obtained on credit. Thus:

(B) J Smith account gives out resources valued at £200.

Debit (+)	Credit (–)
	£200 to Purchase A/C on 1 Apr for materials
	£200

(C) Purchase account receives the materials

Debit (+)		Credit (–)
1 Feb from Dice A/C for plant item	£50	
1 Apr from J Smith A/C for materials	£200	
	£250	

(iv) The invoices for the work completed for John Brown require accounts for John Brown and Sales, and if cash is involved, the Cash account also. Thus:

(D) The Sales account gives out resources to John Brown, as follows:

Debit (+)	Credit (–)
	£100 to John Brown A/C 1 May for house extension
	£50 to John Brown A/C 10 May for house extension
	£75 to John Brown A/C 17 May for house extension
	£25 to John Brown A/C 10 Jun for house extension
	£300

The John Brown account receives resources from the company, as follows:

Debit (+)		Credit (–)
1 May from Sales A/C for house extension	£100	
10 May from Sales A/C for house extension	£50	
17 May from Sales A/C for house extension	£75	
30 May from Sales A/C for house extension	£25	
10 Jun from Sales A/C for house extension	£50	

(v) The cheque of £100 sent by John Brown is received in the Cash account and taken from the John Brown account. The Sales account is not required in this transaction, as returns of goods are not involved. Thus:

Cash account becomes:

Debit (+)		Credit (–)
1 Jan from Capital A/C	£10000	£50 to Dice A/C on 1 Feb for plant item
1 Jun from J Brown A/C	£100	

(E) The John Brown account now becomes:

Debit (+)		Credit (–)
1 May from Sales A/C for house extension	£100	£100 to Cash A/C on 1 Jun for 1st payment on house extension
10 May from Sales A/C for house extension	£50	
17 May from Sales A/C for house extension	£75	
30 May from Sales A/C for house extension	£25	
10 Jun from Sales A/C for house extension	£50	
	£300	£100

(vi) Finally, on 30 June J Paris withdraws (i.e. credit) £3000 from the Cash account, which is in turn received (debit) in the Capital account.

(F) The final Cash account becomes:

Debit (+)		Credit (–)
1 Jan from Capital A/C investment	£10000	£50 to Dice A/C on 1 Feb for plant item
1 Jun from J Brown A/C	£100	£3000 to Capital A/C on 30 Jun for J Paris
	£10100	£3050

(G) The Capital account becomes

Debit (+)		Credit (–)
30 Jun to J Paris from Cash A/C	£3000	£10000 from J Paris to Cash A/C on 1 Jan
	£3000	£10000

Trial balance

With all the transactions completed, the debits and credits in A–G accounts are each totalled and transferred to the trial balance sheet. If all the items have been correctly recorded then the debits and credits totals should exactly balance.

(£)	Debit (+)	Credit (–)
(G) Capital account	3000	10000
(F) Cash	10100	3050
(E) John Brown	300	100
(D) Sales	–	300
(C) Purchase	250	–
(B) J Smith	–	200
(A) Dice	50	50
	13700	**13700**

Note:

(1) Only accounts A–G are shown since the others were only intermediate steps for illustrative purposes.

(2) Also, by reference to account E for example it is evident that John Brown is in a £200 debt situation, having received £300 of the company's resources and paid only £100 by 30 June. John Brown is the debtor to the company and as such owes money.

(3) Account B, however, shows that J Smith is a creditor of the company and is due to be paid £200 for materials previously supplied.

At the end of the financial year, the books are closed, the above process concluded for all the transactions and the final profit and loss statement produced from the sales, purchase and wages (salaries) accounts, etc., with the trial balance forming the basis for the fuller and more informative balance sheet preparation.

Analysis of the P&L account

The information shown in Tables 18.2 and 18.3 is interpreted to give the profit and loss account (Table 18.4) and balance sheet (Table 18.5).

Notes:

(1) Work on site not yet completed is valued at 50% added to wages, materials and plant costs incurred.

(2) The capital employed in the business is represented by:

£21 500	ordinary shares at £1 each;
£5000	7% debentures at £1 each;
£10000	10000 5% preference shares at £1 each

(3) Corporation tax is set at 35% (corporation tax will be treated simply as a straightforward deduction from profits). In fact, corporation tax policy is quite complex and varies depending upon the country of residence.

Table 18.4 Profit and loss statement.

Tabulated revenue account year ended 31 Dec.	£	£	
Sales	25 000		
	24 000		
	14 000	63 000	
Work in progress	9 000		
	500		
	2 000		
	5 000		
	500		
	1 000		
	18 000 × 1.5	27 000	
Value of production			90 000
Materials used	10 000		
	12 000		
	7 500		
	9 000		
	5 000	43 500	
Wages incurred	7 500		
	6 000		
	2 500		
	2 000		
	1 000	19 000	
Plant hire	2 500		
	1 500		
	1 000		
	500		
	500	6 000	
Overhead expenses on sites		3 000	
Cost of production			71 500
Production profit			18 500
Head-office overheads and salaries		6 000	
Trading profit			12 500
Depreciation of plant and buildings		1 000	
Net profit (operating)			11 500
Interest on 7% debentures		350	
			11 150
Corporation tax (say 35%)		3 903	
Profit after tax and interest			7 247
Proposed dividend		2 500	
Transfer to general reserve		4 700	
Balance C/F			47

Analysis of the balance sheet

All companies are required to produce a yearly balance sheet showing: (1) the capital invested in the business and its classification, (2) how the capital is being employed by the business. The presentation is either in: (a) tabular form, showing the employment of the capital with deduc-

Table 18.5 Tabulated balance sheet at 31 December.

(a) Balance sheet as at 31 Dec.					As at 31 Dec. of the previous year
EMPLOYMENT OF CAPITAL			Dep'n		
Fixed assets		£	£	£	£
Buildings		15 000	500	14 500	
Plant and equipment		5 000	500	4 500	
		20 000	1 000	19 000	(15 000)
Current assets					
Cash at bank			2 500		(–)
Stock of materials	45 000				
Less used	43 500				
			1 500		(500)
Work in progress			27 000		(28 250)
Debtors: Contract C	14 000				
Less paid	8 000				
			6 000		(5 000)
				37 000	33 750
Current liabilities				56 000	48 750
Creditors: Equip.	5 000				
Plant hire	6 000				
Mat.	45 000				
	56 000				
Less paid	49 000		7 000		(5 500)
Wages and salaries on site	19 000				
Less paid	18 000		1 000		(500)
Proposed dividend			2 500		(150)
Interest due on loans			350		(700)
Bank overdraft			(–)		(250)
Corporation tax liability			3 903		(150)
				14 753	(7 250)
				£41 247	£41 500

(b) CAPITAL EMPLOYED	£	£
Share capital authorised and issued		
21 500 ord. shares at £1 each	21 500	(21 500)
10 000 5% pref. shares at £1 each	10 000	(10 000)
5 000 7% debentures at £1 each	5 000	(10 000)
Reserves		
General	4 700	(–)
Capital	–	(–)
Profit and loss account	47	(–)
	£41 247	(£41 500)

Note: In practice it is usual for corporation tax payments to be paid approximately one year in arrears, in which case the tax liability would appear as part of the capital employed and not as a current liability as shown.

tions made for current liabilities followed by the classification of the capital employed, or (b) liabilities shown on the left-hand side and the assets on the right of the page. In each case, the opposing sets of figures must balance.

Some regular terms used in the accounts

Debtor

Monies owed to the business by others who have received goods or services.

Creditor

Monies owed by the business to others for goods and services supplied.

Depreciation

The assets of the business wear out in producing the goods and finally have to be replaced. The loss in value each year represents a cost on the business to be taken out of the profits and in theory is set aside to replace the asset when worn out. Buildings, however, sometimes appreciate. At present, tax is not payable on the increase in value until the asset is sold, at which time the company is liable for capital gains tax. The usual way to cover such future payments is to create a capital reserve fund.

General reserve

Profits that are not distributed to shareholders are allocated to the general reserve and as such are reinvested into the business. If the general reserve grows large relative to other capital employed in the business, either the shareholders are given free shares or the nominal price of the existing shares is increased, whereby the general reserve is proportionately reduced. Such action removes any confusion that the general reserve can be declared as a dividend. (The general reserve, remember, has probably already been spent on assets, so remaining as a liability on the balance sheet.)

Stock options

Company executives are sometimes granted stock options as part of the salary package, an incentive that provides the right to buy an amount of company shares for a stated price—valuable if exercised when shares are rising. Stock options are expensed on the profit and loss account at the grant date, i.e. when awarded, as a cost item with an estimated fair-value price. Hence, the potential gain represents a contingent variable liability on the company until actually exchanged for real shares.

Dividend

The dividend is that proportion of profits distributed to shareholders after all other commitments have been met, including corporation tax, the proportion being set out in the tax policy of the country where the company is based. For example, the regulations may have different requirements for non-resident shareholders/owners by levying only minimum corporation tax and then giving deferrable tax opportunities on the larger dividend component unless repatriated. The precise conditions, however, vary depending on the country and need careful investigation beyond these general points.

Interpreting balance sheet information

The flow of funds

It is becoming more common for the presentation of the annual accounts to include, with the balance sheet and profit and loss account information, an analysis of the source and application of funds.

The example below shows details of (1) the source and application of funds and (2) the working capital changes, using the information given for the company in Tables 18.4 and 18.5 for the year ended 31 December.

Internal sources of funds	£
Profit before tax and interest	11 500
Add depreciation	1 000
Decrease in work in progress	1 250
Increase in creditors	1 500
Wages and salaries	750
	£15 750
Application	
Dividends, loan interest and tax paid for last year	1 000
Increase in debtors	1 000
Purchase of new equipment	5 000
Decrease in debenture holdings	5 000
Increase in stocks	1 000
Reduction of overdraft at bank	250
	£13 250
Increase in cash at bank	£2 500
Cash at bank	£2 500
Working capital =	current assets–current liabilities
Dec. this year =	£37 000 − £14 753 = £22 247
Dec. last year =	£33 750 − £7 250 = £26 500
Drop in working capital over one year =	**£5925**

Ratio analysis

Certain ratios selected from the accounts data can provide historical measures of the company's pattern of performance, which are useful as both internal and inter-firm comparison instruments for self-diagnosis, notably:

- Means of evaluating profitability, solvency and productivity
- Basis for setting targets of performance
- Measure of relative performance in changing trade conditions

Working capital ratios

Changes in working capital must be observed fairly regularly. A sudden demand for payment by creditors may cause severe cash-flow problems, even though an adequate profit is being made. Two tests are commonly applied:

(1) Current ratio (CR) $= \dfrac{\text{Current assets}}{\text{Current liabilities}}$

CR this year $= \dfrac{37\,000}{14\,753} = 2.51$ CR last year $= \dfrac{33\,750}{7\,250} = 4.65$

(2) Acid test (AT) $= \dfrac{\text{Cash and debtors}}{\text{Current liabilities}}$

AT this year $= \dfrac{8\,500}{14\,753} = 0.58$ AT last year $= \dfrac{5\,000}{7\,250} = 0.7$

Most accountants look for ratios of 2:1 and 1:1 respectively. During the previous year, the current assets were running too high, but the CR test was only corrected because of the large tax liability. A look at the acid test shows that little cash is available to meet this liability. A better policy would be to reduce the large amount of work in progress and thereby increase the cash balances.

Profitability and operating ratios

(1) Primary ratio $= \dfrac{\text{Net profit after tax and before interest}}{\text{Capital employed}} = \dfrac{7\,597 \times 100}{39\,575} = 19.2\%$

(2) Control ratio $= \dfrac{\text{Net profit before tax and interest}}{\text{Capital employed}} = \dfrac{11\,500 \times 100}{39\,575} = 29\%$

Net profit is profits before tax and interest. The control ratio is useful for comparing divisions within a company or a group.

(3) Profit/Sales ratio $= \dfrac{\text{Net profit bti}}{\text{Turnover}} = \dfrac{11\,500 \times 100}{90\,000} = 12.5\%$

(4) Turnover ratio

The *turnover ratio* can provide telling information about the capital structure of a company. For example, the firm that purchases most of its plant and equipment will tend to have a much lower ratio than the firm that has a hiring policy, and furthermore it possesses assets that are not easy to convert into a more liquid form in times of crisis.

$$\frac{\text{Turnover}}{\text{Capital employed}} = \frac{90\,000}{39\,575} = 2.3$$

(5) Debt ratio $= \dfrac{\text{Debt}}{\text{Ebitda}} < 5$ to avoid trend towards bankruptcy

Ebitda = Profit or earnings before, interest, tax, depreciation and amortisation of loans

Debt is mainly loan capital other than equity.

For the purposes of the example, debentures are taken to be debt, but would also include bank loans as appropriate, thus Ebitda is the trading profit:

$$\frac{\text{Debt}}{\text{Ebitda}} = \frac{£5\,000}{£12\,500} = 0.4, \qquad \text{i.e. OK}$$

Expressed inversely, earnings can easily meet likely interest and amortisation payments.

Other ratios

(1) $\dfrac{\text{Debtors} \times 12}{\text{Turnover}} = \dfrac{6\,000 \times 12}{90\,000} = 0.8$ months

Customers are given three weeks' credit.

(2) $\dfrac{\text{Creditors} \times 12}{\text{Purchases}} = \dfrac{7\,000 \times 12}{50\,000} = 1.7$ months

Suppliers give six weeks' credit.

The ratios indicate that the company is applying a sensible policy with respect to both its creditors and its customers, and is in fact using suppliers' credit as a short-term form of finance.

(3) A deeper investigation into a company's affairs may yield further important information regarding productivity, typically:

$$\frac{\text{Turnover}}{\text{Number of employees}}$$

$$\frac{\text{Profit}}{\text{Number of employees}}$$

$$\left[\frac{\text{Plant and equipment value}}{\text{Number of employees}}\right]$$

Investment ratios

(1) Earnings related to shareholdings = p/e ratio $= \dfrac{\text{Market share price}}{\text{Earnings per share}}$

(i) Earnings per share is calculated by deducting from gross profits: corporation tax, depreciation, interest on loans, payments to preference shareholders and other minority interests, i.e. the total amount available for distribution as dividend to ordinary shareholders and then dividing by the number of ordinary shares.

(ii) Market share price is the variable price of an ordinary share of the company traded daily by the hour on the stock market.

A company may retain a proportion of earnings for new ventures to help growth; hence dividend yield is not necessarily a satisfactory measure of the value of the stock for PLC investors. Consequently, alternative indicators have been developed. One of the most favoured methods adopted by analysts is the p/e ratio, which expressed inversely represents the total earnings yield on the market price of the share. Over the last century the UK market average p/e ratio has fluctuated widely around a mean of about 16:1, which corresponds to a nominal yield of between 6% and 7% and represents the premium over base interest rate as reward for the higher investment risk.

The principles for the determination of the p/e value are as follows:

- Simple calculation shows that to sustain the earnings yield, the share price will need to rise when earnings increase and vice versa.
- Notably, the company with a regularly rising share price (e.g. due to increasing earnings or asset values) will tend to have greater demand for its shares, thereby pushing the price even higher, so further raising its p/e ratio (and vice versa for less reliable earnings, e.g. construction).
- The prospective earnings situation for the share may also be factored into the price, for example the raising of nominal base interest rate will tend to depress future earnings expectations, and in anticipation the share price will fall—and vice versa.
- Rising inflation produces a lower p/e value (i.e. higher yield) because the central bank normally lifts the nominal base interest level as a combating measure, which has the effect of neutralising the increase in share price arising from the inflated earnings, except the element resulting from any real growth of earnings. Moreover, persistently high inflation creates uncertainty and weakens demand for shares; thus the p/e ratio may trend even further downwards. Significantly, the market average p/e has at times fallen below 10 during periods of intractable inflation.
- When inflation is judged to be sufficiently under control, the interest rate is allowed to fall; prospects for real growth of earnings and share price thereby improve and the p/e ratio therefore gradually recovers.

Example

For the example shown in Tables 18.4 and 18.5, the *this year* situation is:

Net profit after tax and interest		= £7247
Payment to preference shareholders		= £500
Earnings		£6747
Earnings per ordinary share =	$\frac{6747}{21500}$	= **£0.314**

Assuming the construction company has a p/e of say 9, then the current market price of the share would be 9 = p/0.314, i.e. p = £2.83

If next year's profits were 8% higher while inflation was 5% (accompanied by a rising interest rate trend), the price of the share would move upwards broadly in proportion to the real growth of earnings, i.e. 3%. However, earnings would increase the full 8%, hence the p/e would reduce as shown by the following calculation:

$$p/e = \frac{2.83 \times [1 + {}^{(8-5)}\!/_{100}]}{0.314[1 + {}^{8}\!/_{100}]} = \frac{2.91}{0.334} = \mathbf{8.7}$$

(2) Percentage growth in total earnings (i.e. retained and distributed) at least matching the nominal GDP rate of growth is commonly sought by investors. Furthermore, any profit undistributed raises the capital of the company, thereby potentially leading to a greater increase of future total earnings. Hence, a less than maximum dividend would be compensated by a rising p/e value.

(3) $$\text{Dividend yield} = \frac{\text{Earnings per ordinary share minus retained earnings}}{\text{market share price}}$$

$$= \frac{(6747 - 4747)}{2.83 \times 21500} \times 100 = \mathbf{3.28\%}$$

Theoretically, the share price should equal the discounted net present value of all prospective dividends; however, the equation components lie in the future and are therefore obviously indeterminate.

Bankruptcy prediction by Z score analysis

Discriminating variables derived from the published accounts of the company are mathematically combined using multivariate discriminant analysis (MVDA) to produce a single Z score value of the form:

$$Z = 14.6 + 82V_1 - 14.5V_2 + 2.5V_3 - 1.2V_4 + 3.55V_5 - 3.55V_6 - 3V_7 \qquad [18.1]$$

where

V_1 = profit after tax divided by net capital employed
V_2 = current assets divided by net assets
V_3 = turnover divided by net assets
V_4 = short-term loans divided by earnings before tax and interest charges
V_5 = tax trend
V_6 = profit after tax trend
V_7 = short-term loan trend

Using the formula, a plot of the Z score for the company over a period is compared to historical Z scores determined for failed and solvent companies and any potential bankruptcy trend threat identified. Further details of the use of Z scores have been provided by Abidali & Harris (1995) and Mason & Harris (1979).

Non-financial indicators

The important non-financial indicators of company performance for navigating long-term sustainability and regularly reported to stakeholders include:

- Customer satisfaction
- Product and service quality
- Operational performance
- Employee commitment
- Social and environmental performance data

Financial reporting standards for public companies

The listed companies of about 90 countries, including the EU and the USA, agreed at the beginning of 2005 to adopt the International Financial Reporting Standards (IFRS) developed under the auspices of the International Accounting Standards Board (IASB) and US Financial Accounting Standards Board (FASB). The focus of the new regime concentrates on 'fair value accounting' rather than historic costs, consequently affecting debt and profit statements by the requirement to provide current market valuations for fluctuating items like property portfolios, loans, financial instruments and similar—presently mainly important in the banking and insurance sectors, while others such as stock options and pension commitments are potentially more pertinent to construction.

In contrast, company tax rates levied on profits vary widely depending on the company's country of residence or the domicile of group holdings. Significantly, common tax, dividend and corporate governance regulations have yet to be universally agreed, even between the major regional trading blocks. Hence, governments, depending on economic priorities and constraints, are able to self-interestedly set tax rates and controls to be advantageous for attracting foreign direct investment, trade, etc.

Regulatory authorities

Standards for business practices such as the issue and exchange of publicly traded shares, dealing in securities, brokerage, investment management and advice, banking and lending, auditing, etc. are broadly guided by IASB, FASB and IOSCO and similar organisations. For example, for banking supervision and capital risks management, the accords agreed by the Basel Committee (referred to as Basel 1 and 2) specifically deal with the soundness and stability of the international banking system, implemented in the EU under the Capital Requirements Directive (CRD).

Regulatory responsibility, however, varies widely internationally and, in the UK, the Financial Services Authority (FSA) promotes efficient, orderly and fair financial markets; the Financial Reporting Council (FRC) regulates corporate reporting and governance, while the Accountancy Investigation and Discipline Board (AIDB) is the independent, investigative and disciplinary body for accountants and auditors.

SEC/FRB operates in the USA and as a result of the Sarbanes-Oxley Bill; SEC registered companies listed in the USA are further required to comply with PCAOB reporting standards on company internal accounting controls. Direct regulation by government is more typical for many EU countries.

References

Abidali, M. & Harris, F.C. (1995) A methodology for predicting company failure in the construction industry. *Construction Management & Economics*, **13(3)**, 189–196.

Mason, R.J. & Harris, F.C. (1979). Predicting company failure in the construction industry. *Proceedings of the Institution of Civil Engineers*, Part 1, May, **66**, 301–307.

Section four

Self-learning exercises

19
Questions and solutions

Questions

Question 1 Method study exercise

The following exercise illustrates the use of a flow diagram, flow process chart and multiple activity chart to study crane activity at a sewage works involving the simultaneous construction of various structures. They include:

- Digestion tanks
- Consolidation tanks
- Heater house
- Ducts

The structures are constructed in reinforced concrete and the tanks' walls formed in 1 m high lifts, the shutters being raised manually into position. The heater house comprises reinforced concrete foundations and a ground slab, the ducts also being of concrete.

The construction method chosen by the contracting firm adopts a crawler crane with a concrete skip attachment to place the concrete. To achieve the required levels of production, the crane must travel from structure to structure as each pour of concrete is completed. The objective is to work on each structure during the day and return to the plant compound overnight as illustrated by the Machine Flow Diagram Fig. Q1.1 and Process Chart Fig. Q1.2. As can be observed the sequence is interrupted, in particular by the resident engineer insisting on checking all the work after the concrete placing activity and before allowing the crane to move on. Also, the crane experiences severe delays en route caused by soft ground. Consequently, the required number of concrete pours cannot be achieved every day and the project programme begins to fall behind schedule.

***Task* (i)** Using the ASME symbols given in Fig. Q1.3 determine the type and number of events for a daily cycle.

***Task* (ii)** Given the information below, prepare a *multiple activity chart* of the present method of working.

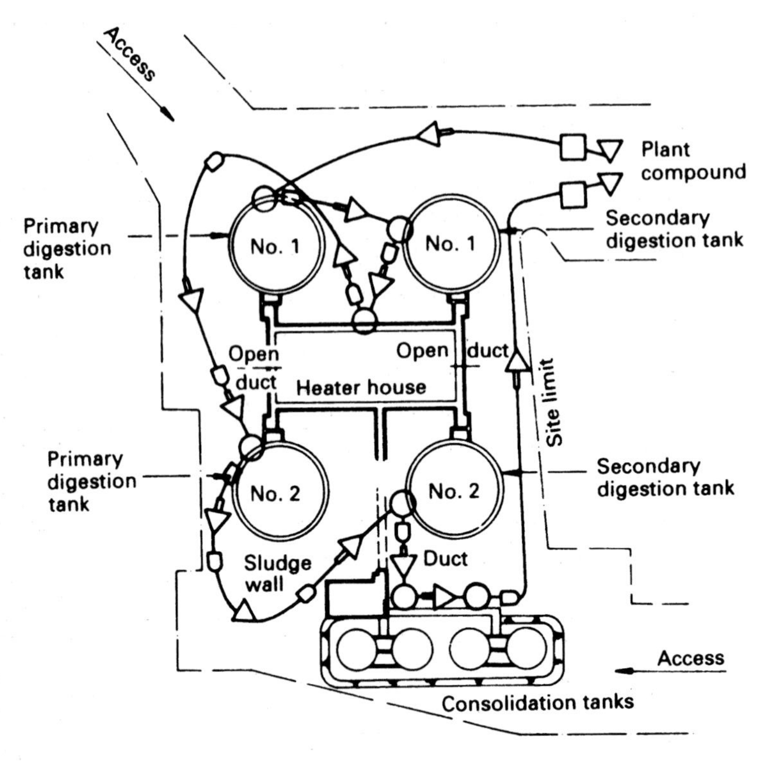

Fig. Q1.1 Flow diagram for Bogtown Sewage Works site layout.

The flow process chart indicates that concreting time at the Primary Digestion Tank No. 1 was rather prolonged. The detailed site layout is shown in Fig. Q1.4 and the activities performed by the concrete mixer, dumpers and crane when placing concrete are as follows:

(1) Concrete batching plant

(a)	Sand, aggregate, cement are loaded into the hopper	2 standard minutes
(b)	After completion of task (a), the mixer driver lowers the hopper and discharges the contents into the mixing drum and adds water	0.5 standard minutes
(c)	The hopper is then raised ready for refilling	0.5 standard minutes
(d)	The materials are mixed immediately after task (b) is complete	2 standard minutes
(e)	The concrete is discharged from the mixing drum into dumpers	0.5 standard minutes

Flow process chart for a crane				
Time	Dis-tance	Symbol	Activity	Type of activity
		▽ 1	Crane in plant compound	Non-productive
		□ 1	Crane inspected for service	Non-productive
	60m	⇨ 1	Travel to Primary Digestion Tank No. 1	Non-productive
90 min		○ 1	Pick up skips and pour concrete in shutters	Productive
		D 1	Delay while engineer checks work	Non-productive
	20m	⇨ 2	Travel to Secondary Digestion Tank No. 1	Non-productive
60 min		○ 2	Pick up skips and pour concrete in shutters	Productive
		D 2	Delay while engineer checks work	Non-productive
		⇨ 3	Travel to heater house	Non-productive
20 min	20m	○ 3	Pick up skips and pour concrete	Productive
		D 3	Delay while engineer checks work	Non-productive
	40m	⇨ 4	Travel and turn	Non-productive
		D 4	Delay as crane manoeuvres on the soft ground	Non-productive
	40m	⇨ 5	Travel to Primary Digestion Tank No. 2	Non-productive
		D 5	Delay as crane negotiates obstructions	Non-productive
	10m	⇨ 6	Complete travel to digestion tank	Non-productive
60 min		○ 4	Pick up skips and pour concrete	Productive
		D 6	Delay while engineer checks work	Non-productive
	20m	⇨ 7	Travel to Secondary Digestion Tank No. 2	Non-productive
		D 7	Delay as crane meets soft ground	Non-productive
	20m	⇨ 8	Travel and turn to tank	Non-productive
		D 8	Further delay on soft ground	Non-productive
	20m	⇨ 9	Complete travel to tank	Non-productive
60 min		○ 5	Pick up skips and pour concrete	Productive
		D 9	Delay while engineer checks work	Non-productive
	20m	⇨ 10	Travel to consolidation tanks	Non-productive
20 min		○ 6	Pick up skips and pour concrete in first two tanks	Productive
	10m	⇨ 11	Travel to next two tanks	Non-productive
20 min		○ 7	Pick up skips and pour concrete in final two tanks	Productive
		D 10	Delay while engineer checks work	Non-productive
	100m	⇨ 12	Travel back to plant compound	Non-productive
		□ 2	Inspect crane after day's work	Non-productive
		▽ 2	Store crane in compound	Non-productive

Fig. Q1.2 Process chart for a crane.

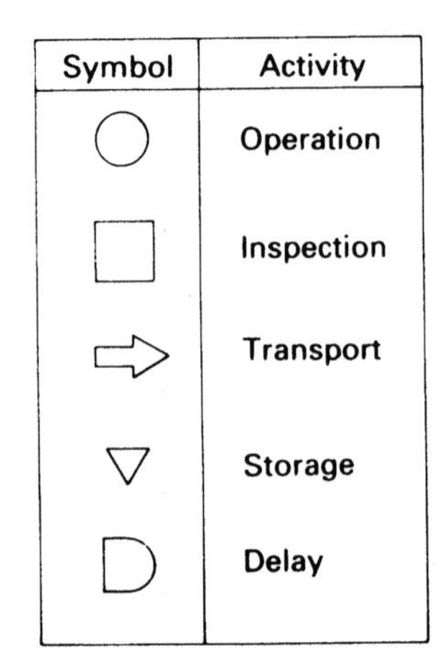

Fig. Q1.3 ASME symbols.

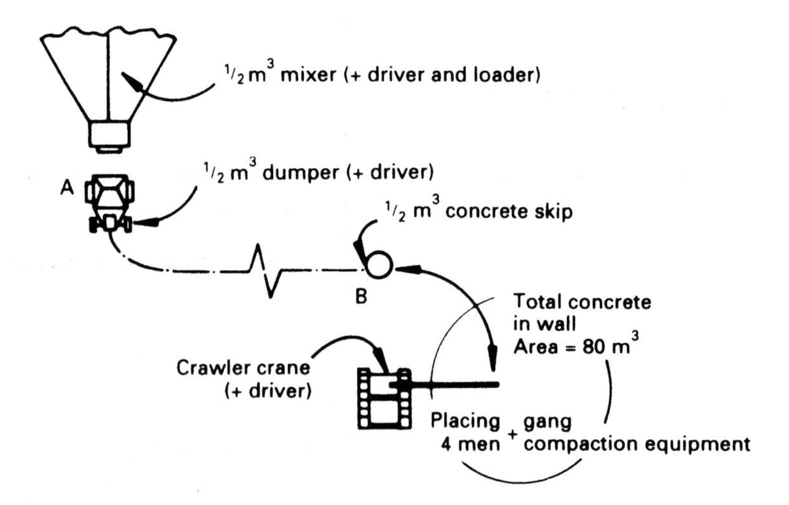

Fig. Q1.4 Site layout at primary digestion tank no. 1.

(2) Concrete placing
Immediately after discharging the concrete into the dumper the mixing cycle is restarted.

(a) Dumper travels from batching plant to the tank and positions ready to discharge concrete into skip — 2 standard minutes
(b) Dumper discharges concrete into skip and driver attaches to crane hook — 0.5 standard minutes
(c) Dumper returns to batching plant — 1.5 standard minutes
(d) Crane lifts full skip, travels, slews and discharges concrete from skip into shutters — 3 standard minutes
(e) Crane lifts empty skips, slews and travels back to position — 2.5 standard minutes

This sequence of activities reveals a cyclical *interaction* between the crane, dumper and concrete mixer as shown in the Outline process diagram Fig. Q1.5.

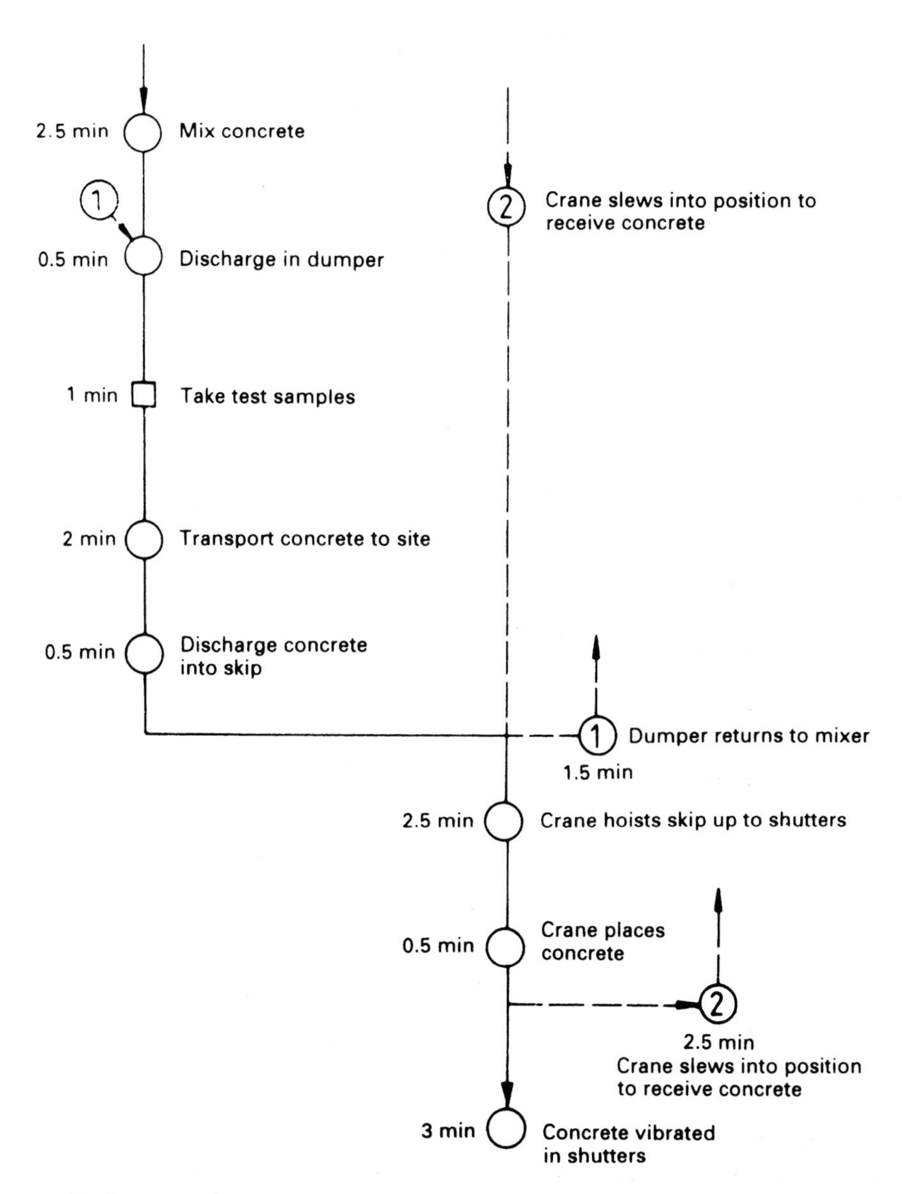

Fig. Q1.5 Outline process diagram.

***Task* (iii)** Determine the cost per m^3 of placing the concrete given the following prices:

(a)	½ m^3 concrete batching plant	£120	per week
(b)	½ m^3 dumper	£40	per week
(c)	½ m^3 concrete skip	£8	per week
(d)	Crawler crane	£400	per week
(e)	Labourer	£2.5	per hour
(f)	Compaction equipment	£12	per week

The site works a 5-day, 40-hour week.

***Task* (iv)** Prepare an improved solution and reduced placing cost.
***Task* (v)** Determine the utilisation factors of the resources used.

Question 2 Time study exercise

As part of the previous method study exercise on the sewage works, the standard times for mixing and placing of concrete in the walls of the Primary Digestion Tank were obtained by direct observations separately taken at the batching plant, the dumper truck route and at the Digestion Tank itself, shown in elemental form on the time study sheet in Chapter 3, Fig.3.7, beginning at 8.00 am. Figure Q2.1 is an abstract sheet and shows subsequent records of the elements for 24 cycles.

***Task* (i)** Prepare a time study sheet and calculate the Basic Time for each element.
***Task* (ii)** Using the typical relaxation allowances shown in Fig. 3.11 and realistic contingencies calculate the Standard Time for each element.
***Task* (iii)** Determine the overall performance rating.
***Task* (iv)** Comment on the relationship between the standard Time values and the Multiple Activity Chart data.

TIME STUDY ABSTRACT SHEET (24 observation rounds only are shown)

STUDY No.

REF. No.

DATE

ELEMENTS	BASIC TIMES (min)																								TOTAL BT (min)	FREQUENCY		UNIT BASIC TIMES
																										PER	QTY	
Fill $\frac{1}{2}$ m^3 concrete skip from $\frac{1}{2}$ m^3 dumper	0.2	0.2	0.3	0.1	0.2	0.3	0.3	0.2	0.2	0.2	0.2	0.1	0.1	0.2	0.4	0.2	0.2	0.2	0.2	0.1	0.1	0.3	0.3	0.2	5.0	SKIP	24	0.21 BM/SKIP
Attach/detach skip to/from crane hook	0.2	0.2	0.1	0.2	0.2	0.3	0.3	0.2	0.1	0.1	0.2	0.2	0.2	0.2	0.2	0.3	0.3	0.1	0.1	0.2	0.2	0.2	0.2	0.2	4.7	SKIP	24	0.20 BM/SKIP
Crane lifts, slews, travels	1.9	2.0	1.9	2.0	2.2	1.8	1.7	1.9	1.9	2.0	2.9	2.1	2.1	2.0	1.9	1.8	1.7	1.9	1.8	2.0	2.0	2.1	2.0	1.9	46.6	SKIP	24	1.94 BM/SKIP
Concrete gang, vibrates concrete	8.0	7.9	8.9	7.4	8.1	8.1	8.2	8.0	8.0	8.0	8.2	7.9	8.0	8.1	8.2	8.2	8.1	8.0	8.0	8.0	7.9	7.9	8.1	8.0	192.3	SKIP	24	8.01 BM/SKIP
Crane returns to position	2.1	2.0	2.1	2.1	1.9	1.8	1.9	2.0	2.0	2.0	2.3	2.3	1.8	1.8	2.1	2.1	2.0	2.0	2.0	2.1	2.1	2.1	2.0	2.0	48.6	SKIP	24	2.02 BM/SKIP
Discharge concrete	1.7	1.7	1.6	1.6	1.4	1.5	1.8	1.9	1.5	1.3	1.7	1.7	1.7	1.7	1.8	1.8	1.6	1.6	1.9	1.5	1.7	1.7	1.7	1.7	39.8	SKIP	24	1.66 BM/SKIP

Fig. Q2.1 Time study abstract sheet.

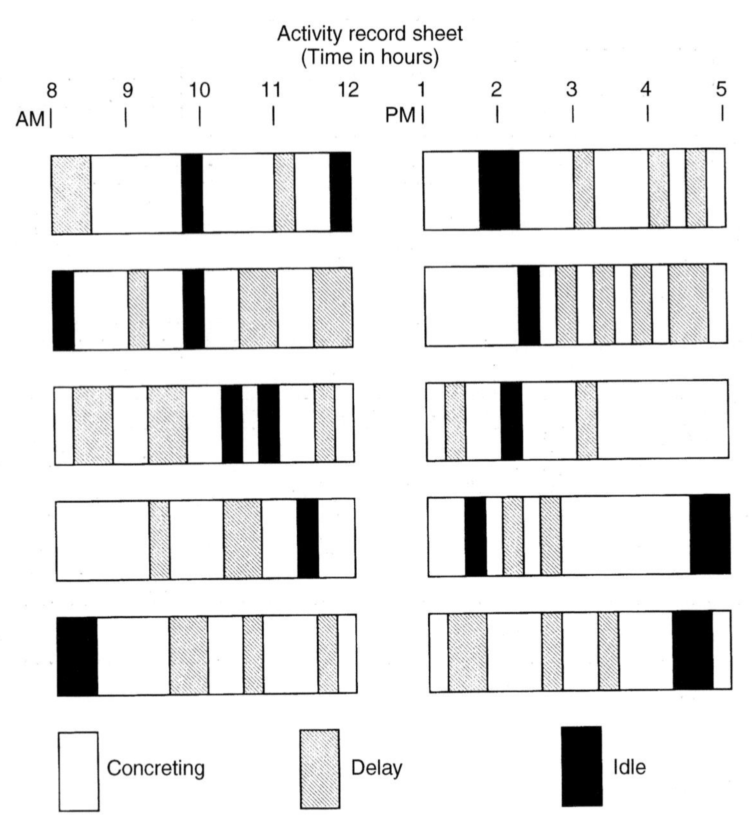

Fig. Q3.1 Activity sampling exercise.

Question 3 Time study and activity sampling comparison

Figure Q3.1 shows the time study pattern of work over a 5-day period for a concreting gang on a large civil engineering contract. The results revealed 66.25% placing concrete, 22.5% delays and 11.25% idle.

Use the activity sampling technique to derive similar results.

Question 4 Site layout exercise

The construction site shown in Fig. Q4.1 illustrates the layout of equipment, building materials and access roads for the erection of a low-rise building.

(1) Comment on and criticise the present layout in relation to the positioning of both the materials and hoists.
(2) Suggest an improved site layout.

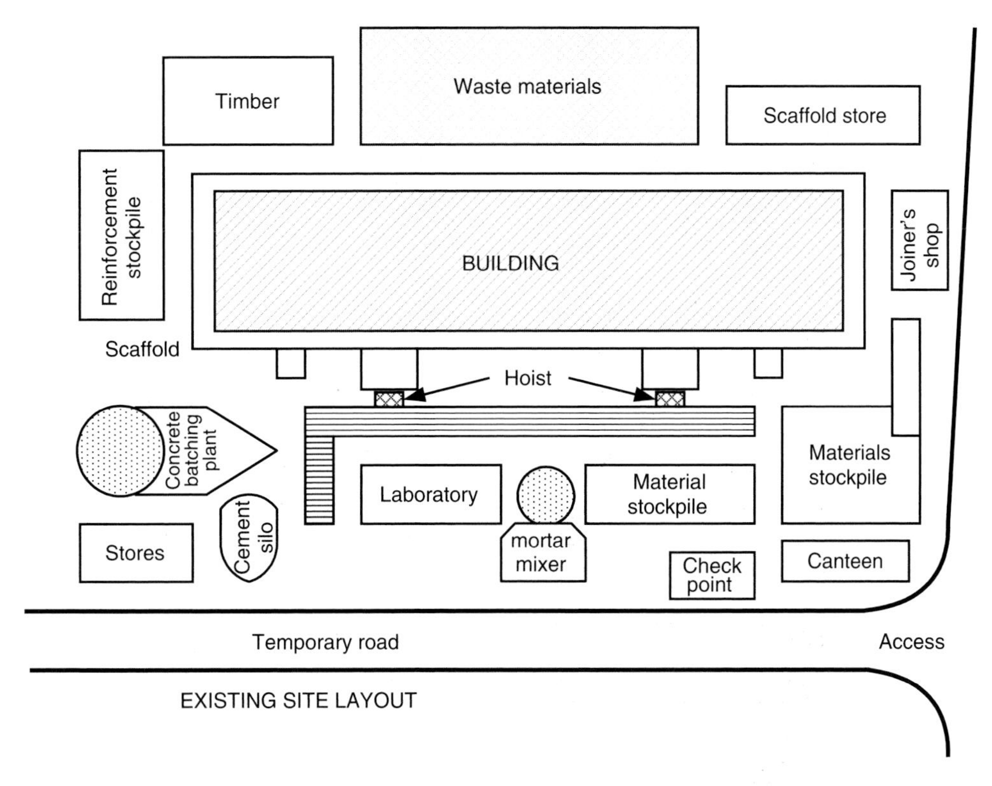

Fig. Q4.1 Existing site layout.

Question 5 Activity sampling of crane operations

An activity sampling study is taken to determine the performance of a crawler crane on a construction site. Four hundred observations were made and separated into their different elements as shown in Table Q5.1.

Table Q5.1 Craneage activity sampling.

Operation	No. of observations
Crane lifting or lowering	160
Crane moving to place of work	80
Unloading or loading crane hooks	60
Crane idle	100

(1) From the above observations determine the proportion of the time the crane was idle and the degree of accuracy of the result. Ninety-five per cent confidence is required.
(2) If the accuracy required is to be ±2%, how many further observations are needed?

Question 6 Activity sampling of labour

A site engineer casually observing a gang of carpenters notes that only 20% of the workers appear to be actually erecting formwork during this single observation. On reporting back the results of this field count to the site manager, it is decided to take a much fuller sampling study. Calculations show that, if the results are to be accurate to ±10%, with 95% confidence a total of 64 observations are needed. The engineer decides to subdivide the operations to be observed into the following categories:

- Fixing formwork (FS)
- Stripping formwork (RF)
- Cleaning formwork (CL)
- Worker idle (I)

After completion of the fuller sampling study of 64 observations it is revealed that 24 observations showed the workers to be engaged in fixing formwork and 16 observations clearing formwork; in eight observations the workers were idle and in the remaining 16 observations stripping formwork.

(1) Calculate the accuracy of the results for each of the observations.
(2) How many additional observations need to be taken to give 10% accuracy, assuming 95% confidence is required?

Question 7 Crashed activities

Application of extra resources on an activity, and/or overtime working, can assist in producing a shorter time, but at extra cost, and indeed congestion is a likely outcome causing a non-linear shortening to the crash time.

In the example shown in Table Q7.1 below, the standard cost and estimated crashed cost/duration of each activity are given:

(1) Determine the earliest crashable project finish time and associated project direct crashed cost.
(2) If site overheads are £40 per day determine the optimum earliest crashed project finish time and associated total crashed cost.

Table Q7.1

Activity	Activity duration	Standard cost (£)	Standard time	Crashed cost (£)	Crashed cost per day (£)
1–2	7	140	5	170	15
2–3	3	90	1	140	25
3–5	4	120	3	140	20
2–5	5	100	4	117.5	17.5
2–4	4	40	3	70	30
4–5	2	20	1	60	40
5–7	6	180	3	270	30
1–6	5	100	5	100*	–
6–7	2	40	2	40*	–
					£830

No opportunity to crash activities marked with an asterisk (*).
For simplicity, the cost per day of crashing is shown as a linear amount, whereas in practice the true rate is likely to be curvilinear.

Question 8 Networks

Figure Q8.1 presents an arrow network without event numbers but showing the duration for each activity in days.

(1) Number the events.
(2) Calculate the earliest and latest event times.
(3) Identify the activities that are critical.
(4) Calculate the total float and free float of the non-critical activities.

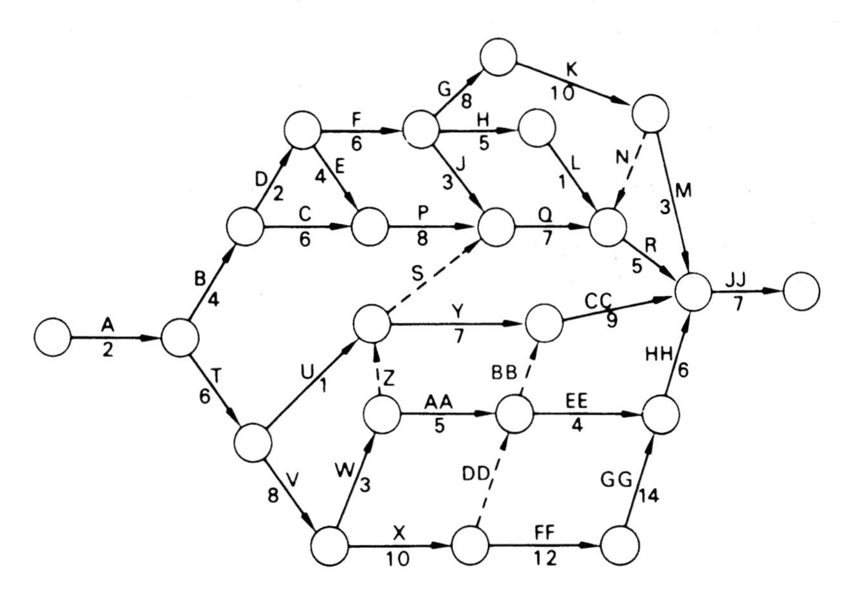

Fig. Q8.1 Arrow network showing duration in days.

Question 9 Planning exercise

The construction of a submerged culvert to cross a waterway

Your company has been successful in obtaining the contract for the construction of a submerged concrete culvert to connect up with existing culverts on each side of the lock walls at the entrance to a dock. General details of the scheme are given in stages 1–6 of Fig. Q9.1 and Fig. Q9.2.

You are asked by the project manager to prepare a project plan of construction work in the form of a critical path network showing the sequence of operations, the earliest and latest times and total float for each activity. The plan is to comply with the method statement for the construction procedure submitted as part of the tender documents.

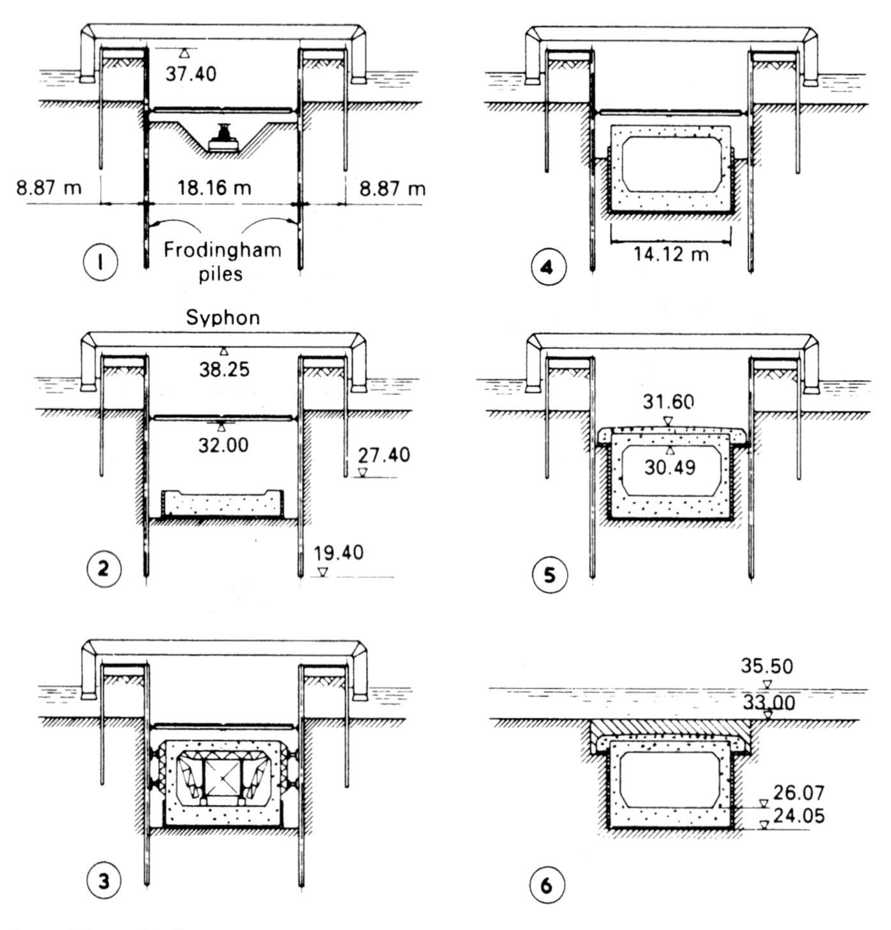

Fig. Q9.1 Scheme details.

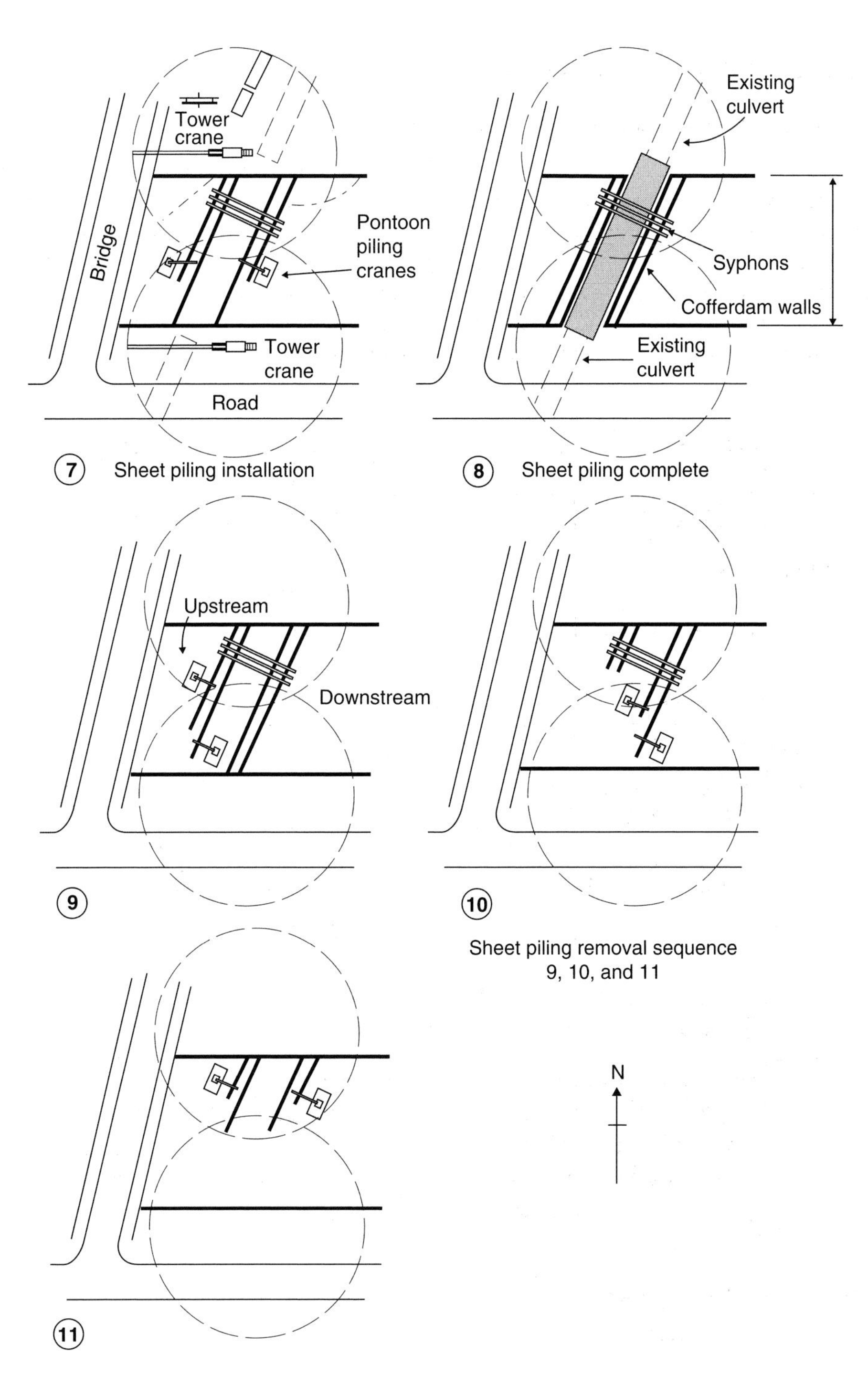

Fig. Q9.2 Construction sequence details.

Method statement

Assumptions

(1) The culverts on both the northern and southern banks are complete and can be conveniently coupled to the new culvert.
(2) During the construction period access to shipping is closed and an alternative route is available.
(3) The northern shore is to be used for setting up site offices, site yard and a tower crane—and for storing sheet piles.
(4) The southern shore is to be used to store sheet piles and formwork, and a tower crane is also to be erected.
(5) Site traffic may use the docks road system and can cross the lock via a nearby bridge.
(6) All floating plant that exceeds the minimum bridge height can move from one side of the construction works to the other by means of the network of connecting locks and channels.
(7) All concrete will be supplied ready mixed.

Constructing the temporary cofferdam

On each side of the line of the culvert, a double wall of sheet piling is temporarily installed in order that construction work may be carried out in the dry. The existing lock walls are then broken out and made good on completion of the culvert. The final arrangement is shown in stage 8 of Fig. Q9.2. Three pipes 1200 mm in diameter are erected above the line of the culvert and must be in place before the sheet pile wall cuts off the free flow of water. If this is not done, the changing tide levels will cause a differential pressure head between the two sides of the cofferdam. The syphons are located near to, and fixed in place using, the tower crane on the northern shore. As the syphons are located near the northern lock wall, piling will commence on this side of the lock.

The driving of the sheet piles is done with two crawler-mounted cranes, each fitted with a leader and each mounted upon a pontoon. One will work on the upstream cofferdam walls and the other on the downstream walls. The tower cranes erected on each shore are able to reach all piling work and will serve to handle all lifting of materials. As soon as a section of double wall is completed it is cross-tied, braced and filled with impermeable material using the tower cranes. On completion of this work the cofferdam is pumped out and continuously de-watered thereafter.

Excavation for the culvert

The two inner walls of the cofferdam are braced just below the bed of the lock as shown in stage 1. In order to minimise the possibility of soil slip, the excavation is carried out in two phases. After removal of the first-phase excavation, the main bracing across the cofferdam is erected. The second phase will be excavated starting on the centre line of the culvert working out to the sides of the cofferdam. A bulldozer is used to push the material towards the nearest lock wall and load it on to trays, which are lifted out by the tower cranes and transported away in muck wagons.

Construction of the culvert

The sequence of operations is shown in stages 1–6. The bottom of the excavation is sealed with a thin layer of blinding concrete. The culvert itself is constructed in two parts. The base section is cast in a bituminous-sealed layer of brickwork, which acts as formwork. The walls and roof of the culvert are concreted at the same time and formed using travelling shuttering. This formwork is removed at an inspection chamber in the existing culvert on the northern quay. The quay walls are now made good. The tower cranes are used for all concrete placing.

The walls are sealed externally with bitumen-impregnated paper and then surrounded with a facing of engineering brickwork. As soon as these operations are complete, the backfill is placed from dumpers, which ride along the top of the cofferdam walls, and the bracing is removed. Sufficient space is now available to place the final roof cover of concrete. Above this level a hardcore protective covering is placed up to the existing lock floor level.

Removal of the sheet piling

The cofferdam is flooded using pumps and the water level inside the cofferdam raised to the level of that on the outside. One of the pontoon-mounted grabs then begins removing material from the upstream wall of the cofferdam, working from the south shore towards the syphons. The other pontoon crane, fitted with pile extraction equipment, follows close behind removing both sides of the wall. The process is then repeated on the downstream wall of the cofferdam. The syphons can now be removed, as the water is once again free flowing. This work is carried out by the tower crane on the north shore. The rest of the excavation is then completed and the piles withdrawn. Finally the site is cleared and all ground made good.

Equipment available

- One compressor
- Two tower cranes
- Two pontoons
- One bulldozer
- Two piling gangs
- One formwork gang
- One concrete gang/general gang
- Two crawler cranes, capable of conversion to pile driver
- One pile extractor
- Two 150 mm diameter centrifugal pumps

All brickwork and culvert waterproofing is carried out under subcontract.

The critical path network

(1) Draw the logic of the network using the information given in Table Q9.1.
(2) By considering the availability of resources and equipment determine:
  (a) The completion time for the project;
  (b) The period of working below water level;
  (c) The period the subcontractor is needed;
  (d) The periods that the tower cranes, the pontoons and the crawler cranes are required.

Table Q9.1 Activities and resources information.

Activities	Duration (days)	Description	Resources used		
	5	Prepare site south shore	1 general gang	1 bulldozer	
	5	Prepare site north shore	1 general gang	1 bulldozer	
	5	Set-up north shore	1 general gang		
	6	Erect Crane No. 1 on pontoon	1 pontoon/crane	1 pile gang	
	6	Erect Crane No. 2 on pontoon	1 pontoon/crane	1 pile gang	
	14	Pile cofferdam upstream walls	1 pontoon/crane	1 pile gang	1 tower crane
	14	Pile cofferdam downstream walls	1 pontoon/crane	1 pile gang	1 tower crane
	8	Fix ties U/S cofferdam wall	1 pontoon/crane	1 pile gang	1 tower crane
	8	Fix ties D/S cofferdam wall	1 pontoon/crane	1 pile gang	1 tower crane
	5	Place fill U/S c/dam walls	Tower crane	1 general gang	
	5	Place fill D/S c/dam walls	Tower crane	1 general gang	
	2	Erect tower crane south	Tower crane	1 general gang	
	2	Erect tower crane north	Tower crane	1 general gang	
	5	Install syphons	Tower crane	1 pile gang	
	5	Pump out cofferdam	2 pumps	1 general gang	
	4	Break out lock walls south	1 compressor	1 general gang	
	4	Break out lock walls north	1 compressor	1 general gang	
	3	Convert Crane No. 1 to grab	1 pontoon/crane	1 pile gang	
	3	Convert Crane No. 2 to extractor	1 pontoon/crane	1 pile gang	
	8	Excavate Stage 1 culvert	1 bulldozer	1 general gang	1 tower crane
	18	Excavate Stage 2 culvert	1 bulldozer	1 general gang	1 tower crane
	10	Install braces	1 general gang	1 tower crane	
	5	Place blinding concrete seal	1 concrete gang	1 tower crane	
	21	Lay culvert base brickwork	Subcontract		
	18	Concrete base	1 concrete gang	1 tower crane	
	9	Erect travelling formwork	Formwork gang		
	65	Construct walls and roof to culvert	1 concrete gang	1 formwork gang	1 tower crane
	34	Apply culvert waterproofing	Subcontract		
	5	Backfill around culvert	1 general gang	1 tower crane	
	8	Remove bracing	1 general gang	1 tower crane	
	10	Place s/con culvert roof	1 concrete gang	1 tower crane	
	5	Place hardcore	1 bulldozer	1 general gang	
	5	Make good lock walls north	1 formwork gang	1 concrete gang	1 tower crane
	5	Make good lock walls south	1 formwork gang	1 concrete gang	1 tower crane
	5	Flood cofferdam	1 general gang		
	4	Excavate U/S wall to syphon	1 pontoon/crane	1 general gang	
	8	Remove piles U/S wall to syphon	1 pontoon/crane	1 tower crane	1 pile gang
	4	Excavate D/S wall to syphon	1 pontoon/crane	1 general gang	
	8	Remove piles D/S wall to syphon	1 pontoon/crane	1 tower crane	1 pile gang
	4	Remove syphons	1 tower crane	1 general gang	
	2	Complete excavation of wall	1 pontoon/crane	1 general gang	
	4	Complete removal of pile wall	1 pontoon/crane	1 general gang	
	5	Dismantle tower/crane north	1 general gang		
	5	Dismantle tower/crane south	1 general gang		
	2	Dismantle pontoons	1 general gang		
	5	Clear site	1 general gang		

Question 10 Resource smoothing

The network shown in Fig. Q10.1 represents a section of work being undertaken by a subcontractor. The subcontractor's labour requirements are shown in Table Q10.1

Because this work has to phase in with the work of the main contractor, this section must be completed within 27 weeks. Nevertheless, the subcontractor wishes to carry out some resource smoothing in order that there are no excessive 'peaks' or 'troughs' in the labour aggregation chart.

Prepare two labour charts, one based on all activities starting as early as possible and one based on all activities starting as late as possible.

By inspection indicate on one of these charts which adjustments you would make in order to meet the subcontractor's wishes.

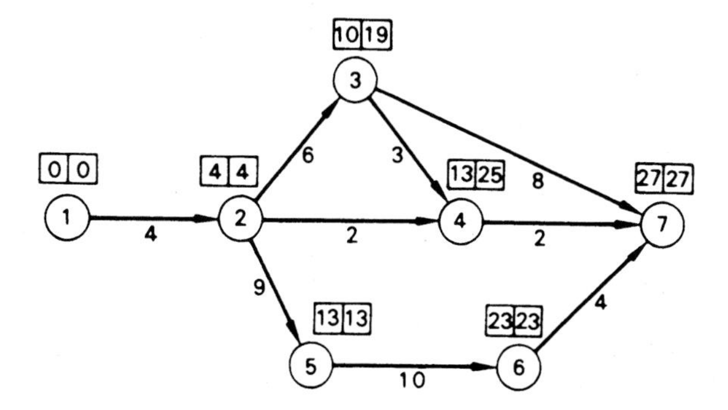

Fig. Q10.1 Network for subcontractor's operations. *Note:* Duration in weeks alongside each arrow.

Table Q10.1 Subcontractor's labour requirements.

Activity	Labour required in numbers of operatives
1–2	2
2–3	3
2–5	4
2–4	4
3–4	3
3–7	4
5–6	2
6–7	2
4–7	1

Question 11 Resource allocation

Figure Q11.1 shows the precedence diagram for a small project. Table Q11.1 shows the resources required for each activity. Both resources X and Y are limited to 6. Using a priority listing of activities based on an early start–total float sort, prepare resource loading diagrams for both X and Y so that neither exceeds the limit. List the activities and their scheduled start dates.

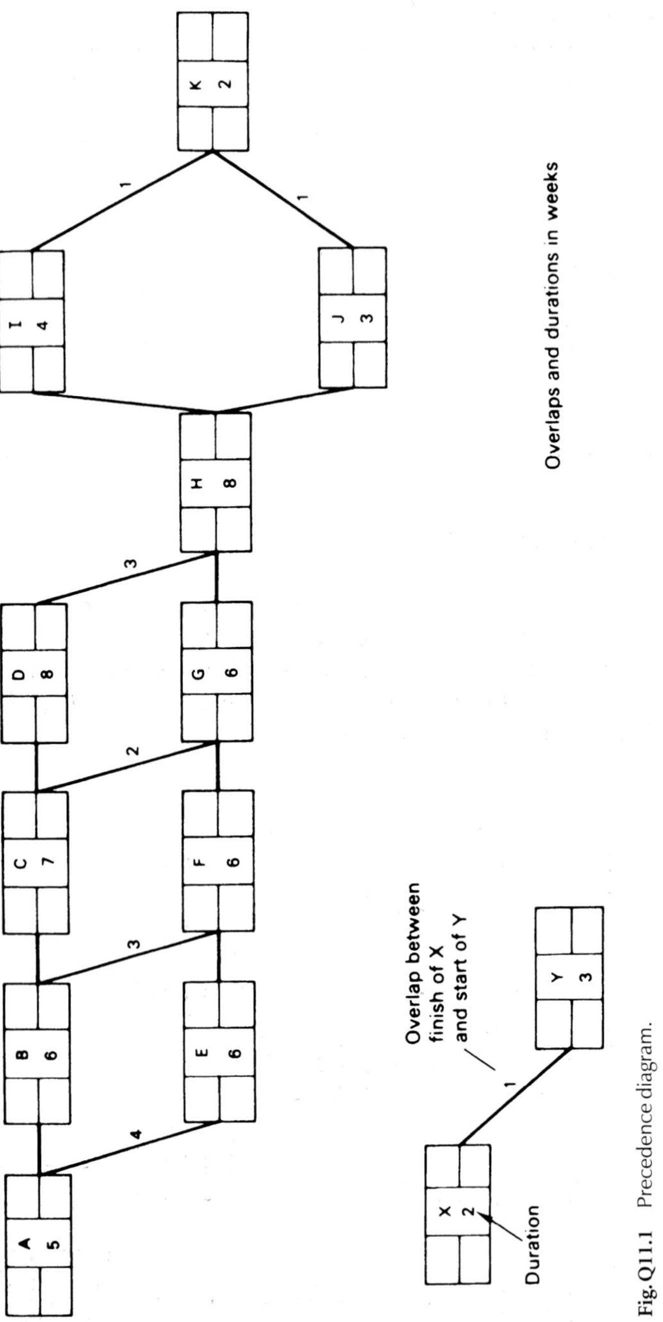

Fig. Q11.1 Precedence diagram.

Table Q11.1 Resource requirements.

Activity	Duration	Resource requirement X	Resource requirement Y
A	5	3	
B	6		4
C	7	4	
D	8		4
E	6	4	
F	6		2
G	6	2	
H	8		2
I	4	5	
J	3	4	
K	2	2	

Question 12 Resource allocation

Figure Q12.1 shows the network for a project. Table Q12.1 shows the resources required for each activity. Resource M is limited to 5, resource L is limited to 8.

Produce a scheduled start date for each activity based on resource loading diagrams for both M and L that do not exceed the resource limitations. Use a priority sorting of early start–total float in preparing the resource loading diagrams.

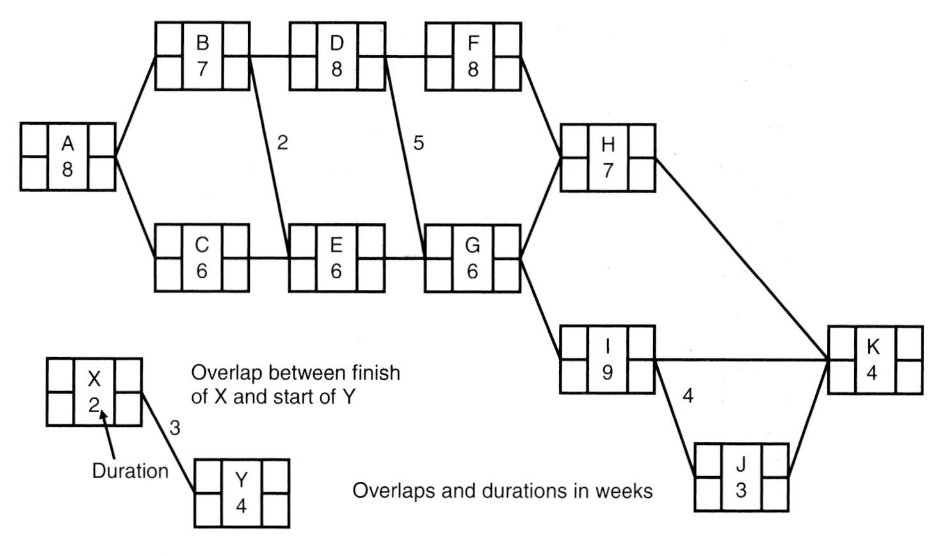

Fig. Q12.1 Network for a project.

Table Q12.1 Resource requirements.

Activity	Duration (weeks)	Resource requirements	
		M	L
A	8	2	2
B	7	4	2
C	6	3	2
D	8	4	2
E	6	4	–
F	8	–	5
G	6	3	–
H	7	–	4
I	9	4	–
J	3	–	4
K	4	–	4

Question 13 Line of balance

Your company has been awarded a contract to erect 124 pylons for the electricity board. Table Q13.1 shows the sequential operations involved in the construction of each pylon, together with the estimated labour-hours and optimum number of operatives for each operation. The handover rate specified is six pylons per week and this can be taken as the target rate of build.

Prepare a line of balance schedule assuming that each gang works at its natural rate. State clearly the contract duration. Assume a five-day working week, eight hours per day and a minimum buffer time of two days.

Table Q13.1 Operations, labour-hours and number of operatives.

Operation	Labour-hours	Optimum number of operatives per operation
A Excavate	55	4
B Concrete foundations	64	4
C Erect tower	145	8
D Fix cantilever cable arms	90	8
E Fix insulators	25	5

Question 14 Line of balance

The construction plan for a house is shown in Fig. Q14.1. Table Q14.1 gives the labour-hours required and the team size for each operation.

Prepare a line of balance schedule for a contract of 30 houses using a target rate of build of four houses per week and each team working at its natural rate. Assume a minimum buffer time of five days between operations and five eight-hour days per week.

What is the overall duration of the project and when will the first team of bricklayers (superstructure operation) leave the site?

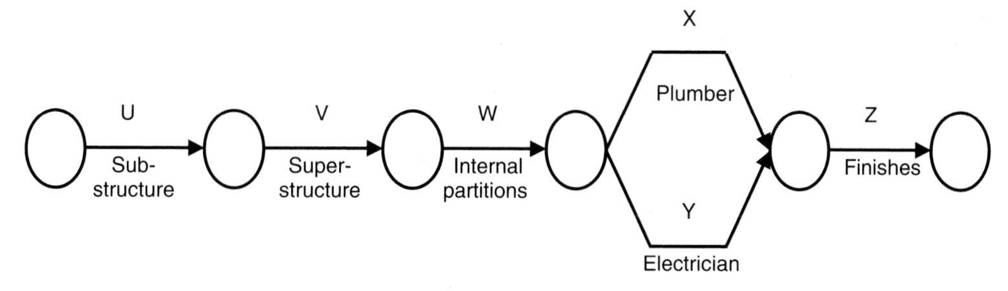

Fig. Q14.1 Construction plan for one house.

Table Q14.1 Labour-hours and team size.

Operation	U	V	W	X	Y	Z
Labour-hours per house	120	290	250	40	30	220
Men per team	3	6	4	3	2	5

Question 15 Line of balance

Prepare a line of balance schedule for a small contract of 15 houses based on a rate of build of three houses per week, assuming five eight-hour days per week. A minimum buffer of five days should be assumed. Table Q15.1 shows the operations together with the estimated labour-hours and optimum number of operatives for each operation.

The sequence of operations is shown in Fig. Q15.1. Make two suggestions as to how the overall duration of the project could be reduced.

Table Q15.1 Operations, labour-hours and number of operatives.

Operation	Labour-hours	Optimum number of operatives per operation
A Substructure	180	6
B Brickwork	320	4
C Joiner, first fix	200	4
D Tilers	60	2
E Glazing	40	2
F Joiner, second fix	120	3
G Electrician	80	2
H Plumber	100	2
I Painter	40	3

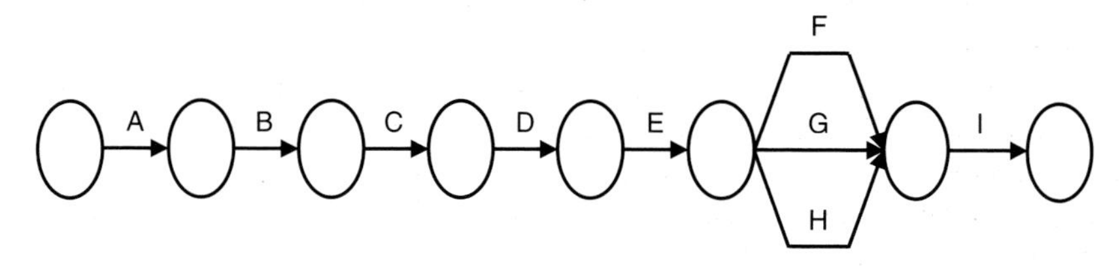

Fig. Q15.1 Construction plan for one house.

Question 16 PERT

Your company STS & Associates has been awarded a negotiated subcontract to undertake specialist work as part of a much bigger civil engineering project. The completion of the subcontract works is crucial to the main contractor's progress and, as such, your contract conditions employ a disproportionate penalty for time overrun. Your Planning Engineer has decided that the chances associated with the timely completion of the subcontract should be established in order to negotiate a more realistic project duration. Table Q16.1 presents a breakdown of the activities in the subcontract works prepared by the planning department along with the Site Manager's estimates for times based on existing internal resources of the company.

Table Q16.1 Activity times based on internal resources of company.

Activity	Predecessor	Duration (weeks)		
		Optimistic	Most likely	Pessimistic
A		4	5	9
B	A	4	8	9
C	A	6	7	11
D	A	5	9	10
E	D	4	6	8
F	B, E	2	8	9
G	C	5	9	10
H	G	3	6	8
K	F, H	5	6	9

(1) Calculate the duration of the project using the following activity times:
  (a) optimistic
  (b) most likely
  (c) pessimistic
(2) What are the chances that the project will be completed in 35 weeks?
(3) What contract period should be negotiated to give the project a 95% chance of completion on time?

Question 17 Estimating the cost of a formwork panel

Details of a formwork panel are given in Fig. Q17.1 Calculate the cost to a contractor of making, erecting, striking, cleaning and oiling one panel based on the following data:

- Cost of softwood timber = £295/m^3
- Cost of 19 mm ply = £12.65/m^2
- Wastage allowance = 17.5%
- Cost allowance for nails = £1.50/m^2
- Carpenter all-in rate = £12.00/h
- Labourer all-in rate = £9.00/h

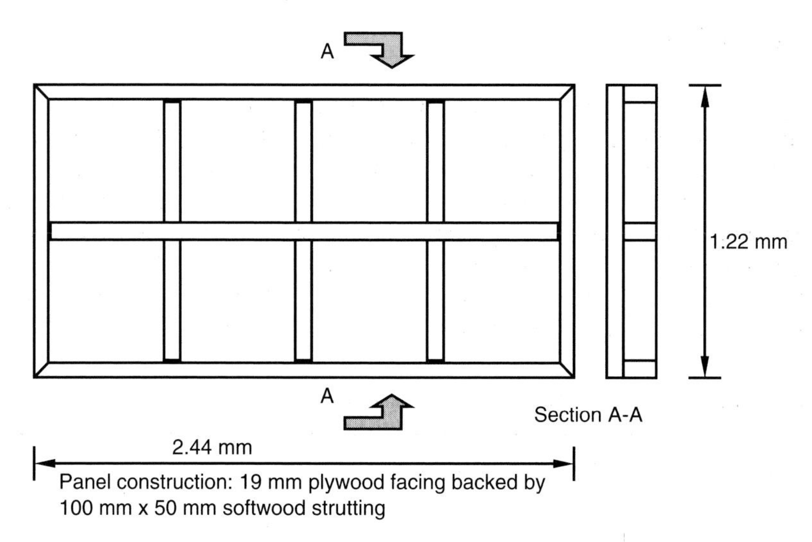

Fig. Q17.1 Formwork panel.

- Composite labour rates
  (composite labour gang = two carpenters + one labourer)
  Labour hours to make panel = 1.25 h/m^2
  Labour hours to erect panel = 1.00 h/panel
  Labour hours to strike panel = 0.55 h/panel
  Labour hours to clean and oil panel = 0.04 h/m^2
- Mould oil cost = £1.20/litre
- Mould oil usage = 0.36 litres/m^2

Question 18 Estimating the cost of reinforcement

Two bills of quantity items from a bridge contract are:

- Supply and fix mild steel re-bar less than or equal to 16 mm diameter 14 tonne
- Supply and fix mild steel re-bar equal to or greater than 20 mm diameter 23 tonne

Calculate the contractor's cost for these items from the following data.

The bar-bending schedule for the bridge gives the quantities of reinforcement steel shown in Table Q18.1.

- Cost of steel-fixers is £12.00 per hour
- Usage rates for steel-fixers are shown in Table Q18.2
- Include a wastage allowance of 2.5%
- Allow 1.5% of the material cost for chairs and spacers
- Allow binding wire as shown in Table Q18.3
- The cost of binding wire is £1.56/kg

Table Q18.1 Bar diameter and weight.

Bar diameter (mm)	Weight (tonnes)
12	1.54
16	12.46
20	6.67
25	4.14
32	8.28
40	3.91

Table Q18.2 Usage rates for steel-fixers.

Bar diameter (mm)	Steel-fixer hours per tonne for offloading transport and fix	Material cost per tonne cut, bent and delivered (£)
12	31	329
16	24	326
20	20	326
25	19	327
32	17	325
40	15	326

Table Q18.3 Binding wire allowance.

Bar diameter (mm)	Binding wire (kg/tonne)
12	11
16	9
20	7
25	5
32	5
40	5

Question 19 Estimating the cost of placing concrete in foundations

To calculate the cost of concreting in any location a decision on the plant and labour to be employed must be made. In this case it is decided to use the following:

Plant

- One crane and two skips at £175 per day
- Vibrators and small plant at £30 per day

Labour

- Labour gang of one ganger and five labourers
- Labourer cost = £72 per day
- Ganger cost = £78 per day
- Crane driver = £88 per day

The cost of providing concrete by ready mix trucks is £70.09 per m^3. Allow wastage of 10%.

Calculate the cost to a contractor of placing concrete in foundations using the above labour and plant configuration. The average daily pour size is 70 m^3.

Question 20 Estimating cost of placing concrete—operational estimating

An example of a bill item is:
The total quantity of concrete in the project is 11 250 m^3.

Item	Description	Unit	Quantity	Rate	Amount
F723	Placing of RC concrete class 22.5/20 to bases thickness 300–500 mm	m^3	35		

Calculate the cost to the contractor of placing concrete and give the 'cost' element that should be included in the rate for Item F723. Base the cost calculation on the following data.

- The average rate of placing is 250 m^3 per week. The concrete placing is planned to take 45 weeks
- The plant provided for the placing of concrete is shown in Table Q20.1
- The labour required is one ganger and four labourers
- Ganger cost = £390.00 per week
- Labourer cost = £360.00 per week

Table Q20.1 Plant provided for the placing of concrete.

Number	Plant item	Dedication	Cost per week (£)
2	22 RB cranes	1–100% concrete placing 1–50% concrete placing	800.00
4	Skips	100%	22.00
5	Dumpers	100%	84.00
6	Vibrators	100%	42.00

Question 21 Estimating cost of laying a 300 mm concrete pipe

The following data are extracted from a pipe-laying contract:

- Concrete pipe diameter: inside 300 mm, outside 334 mm
- Overall pipe length = 925 m
- Width of trench = 750 mm
- The trench depths are:

Depth (m)	Length (m)
1.50	631
2.00	274
2.50	20

Calculate the total cost and cost for each depth to a contractor of excavation, planking and strutting, laying pipes, surrounding concrete as shown in Fig. Q21.1 and backfilling using the following data.

Labour and plant for pipe-laying (up to 300 mm diameter):

- One JCB 3C excavator @ £18.64 per hour
- Small plant, pumps and compactor @ £2.85 per hour
- One pipe-layer @ £12.00 per hour
- Two labourers @ £9.00 per hour each

Assume a working week is 50 hours. Also, assume an average output of 50 linear metres of pipe laid per week, the output varying with the depth of trench. The output includes excavating, trimming, pipe laying, concreting and backfilling.

Labour allowances for planking and strutting are shown in Table Q21.1. Allowance for provision of planking and strutting materials, all of which are recoverable, is £0.95 per m^2.

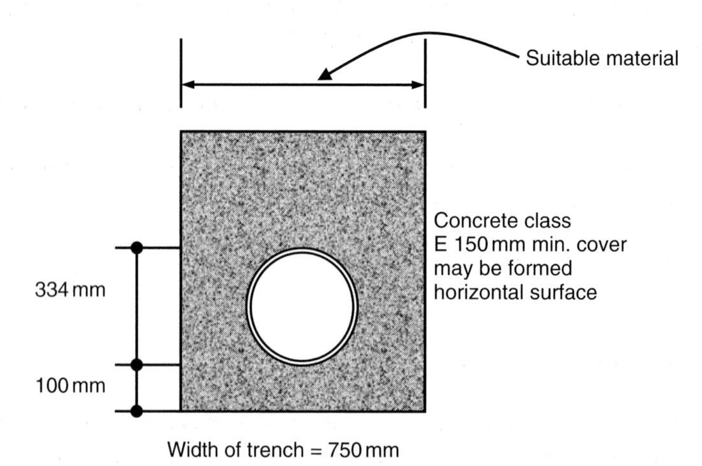

Fig. Q21.1 Concrete surround to pipe.

Table Q21.1 Labour allowances for planking and strutting.

Depth (m)	Labour hours per metre
1.50	0.20
2.00	0.30
2.50	0.70

- Ready-mix concrete for bedding = £70.09 per m^3
  (Assume wastage of 5%)
- Pipe, concrete class L 300 mm diameter = £23.56 per m
  (Assume wastage of 3% for pipes)

Allowance for disposal of surplus material:

- Assume all excavated material is suitable as backfill and compacts to original volume
- The excavator is used to load tipper trucks with surplus material. The loading, travel, discharge and return cycle time of one 4 m^3 tipper-truck is 21 min. The cost of a 4 m^3 tipper-truck including driver is £32.16 per hour.

Question 22 Hours-saved incentive scheme

A carpenter is engaged on shuttering work for the construction of a reinforced concrete pumping chamber. The total area of formwork involved is 24 m^2. The bonus target set by the site manager is:

- Erect formwork
- Remove formwork, 1 m^2 per labour-hour

This target includes time for cleaning and making ready the formwork panels.

The manager has based the target on a 75–100 scheme, and a bonus of one-third the basic wage can be earned for performance at standard (100P).

(1) Show how the carpenter's earnings and the cost of the job vary when performance is 50P, 75P, 100P and 120P.
(2) Calculate how many hours should be allowed for the job, and the payment for each hour saved if a 50–100 scheme is to be used.

Assume the basic rate of pay is £1.00 per hour.

Question 23 Bonus scheme with plant

The site manager in charge of the construction of the cooling towers for a new power station decides to install financial incentive schemes for all groups of workers. The excavation of the deep trench in which to bury 2 km of 1 m diameter concrete pipe supplying the cooling towers with water from the nearby river is to be treated as an independent operation. The activity involves a dragline excavator, its driver and two labourers bottoming up the trench.

The total amount of excavation involved is 24 000 m^3 and the gang is set the output target excavation rate of 50 m^3 per hour to excavate, place the soil to the side of the trench and bottom up the trench base. At the end of the first week, which includes weekend working, the time booked for excavating 400 m^3 of earth is as shown in Table Q23.1.

The basic rate of payment is £1.00 per hour and bonus is to be paid at 50% of each hour saved. The machine driver is to receive 1.5 shares and the labourers 1 share each.

Calculate the bonus earned by each worker during this first week.

Table Q23.1 Week No. 1: Summary of hours worked.

Description	M	T	W	Th	F	Sa	Su	Total
Driver	10	8	10	10	8	8	4	58
Labourer No. 1	10	8	10	10	8	8	4	58
Labourer No. 2	10	4	–	–	8	8	4	34
Servicing machine	1	1	1	1	1	1	1	7
Total	31	21	21	21	25	25	13	157

Question 24 Gang bonus payments

A gang of carpenters on a civil engineering contract is set a target of 100 labour-hours to erect the formwork for the walls of a pump house. The manager decides to pay bonus using a direct incentive scheme. The gang consists of a charge-hand paid 1.5 shares, two carpenters paid 1.25 shares each and two labourers paid 1 share each. On completion of the job the bonus clerk analyses the time allocation sheets submitted by the charge-hand. The records show the time booked to each worker, including stoppages, to be as follows:

- Charge-hand 2 days at 10 h each day
- Carpenter No. 1 2 days at 10 h each day
- Carpenter No. 2 2 days at 8 h each day
- Labourer No. 1 2 days at 10 h each day
- Labourer No. 2 1 day at 10 h (Monday only)

Work was held up for four hours on the second day due to inclement weather.

If the basic rate of pay is £1.00 per hour for an 8-hour day, and overtime is paid at time and a quarter, calculate the bonus earned by each worker in the gang together with total earnings.

Question 25 Control of project costs

The subcontractor responsible for the construction of the substructure of the turbine house for a new coal-fired power station decides to install a cost control system. In order to facilitate the clerical procedures so that feedback is possible each week only labour and plant are considered; materials control is dealt with in a separate system. The main site functions, which can be carried out by stable gangs of workers, are given code numbers against which actual costs are recorded as follows:

- Code 10 Earthmoving gang and plant
- Code 20 Reinforcement fixing gang X
- Code 30 Reinforcement fixing gang Y
- Code 40 Carpenter gang X
- Code 50 Carpenter gang Y
- Code 60 Concrete gang X
- Code 70 Concrete gang Y
- Code 80 Project overheads

The contract plan of construction is shown in Table Q25.1 and the budget for this work in Table Q25.2. At the end of Week 37, the current progress is recorded as shown in Table Q25.3. Actual costs to the end of Week 37 have been recorded and are shown in Table Q25.4.

(1) Plot the budget curve for the full contract, and the value and cost curves to the end of Week 37.
(2) The recorded costs for each of the cost codes are shown in Table Q25.5.

Draw up a table of variances and make recommendations to the site manager for future action.

Table Q25.1 Contract programme.

Activity description	Start date week beginning	Activity duration in weeks
A Excavate turbine hall	1	15
B Turbine hall foundations	4	15
C Turbine hall retaining walls	15	18
D Turbine plinths	19	22
E Floor slab to turbine hall	19	23
F Cooling-water pump house	9	25
G Cooling-water culverts	9	33

Table Q25.2 Budget values (£000).

Activity description	Cost code								Total (£000)
	10	20	30	40	50	60	70	80	
A Excavate turbine hall	112	–	–	–	–	–	–	29	141
B Turbine hall foundations	–	30	–	72	–	34	–	35	171
C Turbine hall retaining walls	–	48	–	79	–	28	–	40	195
D Turbine plinths	–	20	–	42	–	30	–	6	98
E Floor slab to turbine hall	30	30	–	54	–	43	–	34	161
F Cooling-water pump house	10	–	32	–	20	–	20	20	122
G Cooling water culverts	–	–	10	–	20	–	16	10	66
Budget totals (£000)	152	128	42	247	40	135	36	174	954

Table Q25.3 Progress at end of Week 37.

Activity description	Date started week beginning	Projected completion week ending
A Excavate turbine hall	1	15
B Turbine hall foundations	5	22
C Turbine hall ret. walls	16	36
D Turbine plinths	23	46
E Floor slab to turbine hall	24	48
F Cooling-water pump house	13	36
G Cooling water culverts	13	48

Table Q25.4 Actual costs to end of Week 37.

Weeks	Cost in period (£000)
0–10	114
11–20	266
21–30	300
31–37	150
Total	830

Table Q25.5 Actual costs up to end of Week 37 (£000).

Cost code								Total (£000)
10	20	30	40	50	60	70	80	
155	100	40	210	31	104	33	157	830

Question 26 Budgetary control

A construction company engaged in heavy civil engineering work prepares at the beginning of each year a financial budget for its contract department. The situation existing at 1 January is as follows:

Contract	Fixed tender price	Period of work
1	£180 000	1 Jan to 31 Dec
2	£480 000	1 Jan to 31 Dec
3	£120 000	1 Jan to 31 Dec

However, the company is at present negotiating a further contract valued at £180 000 to commence on 1 April with a duration of 12 months. Each of these contracts includes an allowance to cover head-office overheads and profit as indicated in Table Q26.1.

Table Q26.1 Budgeted sum for overheads and profit.

Contract	HO overheads (£)	Profit (£)	Total
1	8 000	10 000	(Represents 10% of sales)
2	20 000	28 000	
3	6 000	6 000	
4	6 000	12 000	
Total	£40 000	£56 000	£96 000

The budget is reviewed six months later, to reveal the following position at 30 June:

Contract	Value of work done (£)	Direct cost to date (£)
1	90 000	84 000
2	240 000	220 000
3	50 000	40 000
4	Contract did not materialise	

The actual cost of head-office overheads to date on each contract is £5000.

(1) Assuming that the value of the work in each project can be spread uniformly throughout its duration, prepare a budget for the year.
(2) Contract 4 did not materialise; thus HO overheads and profit will be under-recovered as a result of this shortfall in obtaining contracts. Calculate the company sales and overhead variances, and the total variance on each contract at 30 June, and thereby obtain the actual profit and compare it with that budgeted.
(3) What action should management take on the basis of the figures?

Question 27 Cash flow

The programme for the construction of a small workshop building is displayed in the form of a precedence diagram, shown in Fig Q27.1. The value of the work contained in each activity has been calculated from the rates contained in the bill of quantities and is listed in Table Q27.1.

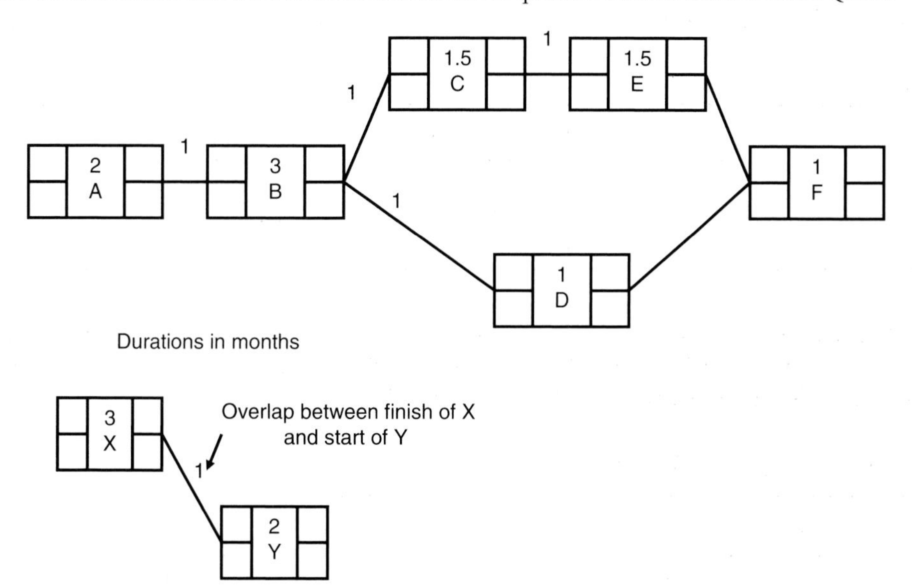

Fig. Q27.1 Precedence diagram for construction of a workshop.

The contractor undertaking the project would like you to prepare graphs of cumulative cost and monies received to date against time for activities starting as early as possible, in order to establish a clearer picture of the financial implications of this contract. From these graphs calculate the net profit if interest is charged at 8% p.a. on outstanding monies. The gross profit margin is 10% of contract value and retention is 5% up to a maximum limit of £3000.

Table Q27.1 Value of activities.

Activities	Duration (months)	Value (£)
A Excavate	2	9000
B Concrete base	3	12000
C Erect frames	1.5	18000
D Concrete floor slabs	1	15000
E Fix cladding	1.5	6000
F Install plant	1	20000

Note: Rate of work throughout any activity is uniform.

Measurement is made monthly with a payment delay of one month. Half of the retention is paid on practical completion and the remaining half six months later. To simplify the calculation you may assume that all costs must be met at the instant they are incurred. What is the maximum amount of cash the contractor needs to execute this contract and when does the contractor require this amount?

Question 28 Cash flow

Table Q28.1 is the contractor's budgeted cost liabilities for a contract. The contractor will add a contribution mark-up of 8% to the estimated cost of each item in the bill of quantities. The contract conditions allow for monthly measures to be made and payment of the amount certified less 10% retention is made one month later. The retention money is repaid six months after the practical completion of the project. The weighted average delay between incurring a cost liability and making payment may be calculated from the information given in Table Q28.1, which shows the cost and the delay associated with each cost element.

If the current interest rate on borrowed capital is 14% p.a., what reduction in the contractor's budgeted contribution will occur due to interest changes? What is the maximum amount of cash needed to execute the contract? (Each month may be assumed to be 4⅓ weeks.)

Table Q28.1 Data for cash flow forecasting exercise.

Cost element	Total cost (%)						Payment delays (weeks)		
Labour	25						1.12		
Plant	25						8.20		
Materials	50						4.00		
Month no	1	2	3	4	5	6	7	8	9
Cost of work executed in months (£000)	3	3	4	6	6	8	6	4	2

Question 29 Cash flow

Table Q29.1 shows a contractor's project budget and profit distribution for a newly awarded contract. The conditions of contract allow interim measurements to be made monthly and payment of the amount certified less 10% retention to be paid to the contractor one month later. Half the retention is included in the final certificate on practical completion and the other half is released six months after practical completion.

Determine the monthly net cash flows assuming an average delay of one month between the contractor incurring a cost liability and the outward cash flow.

Calculate the interest charges on locked-up capital for an annual interest rate of 12%.

Table Q29.1 Project budget and profit distribution.

Month no.	1	2	3	4	5	6	7	8	9	10
Value of work each month (£000)	2	3	4	8	9	9	8	5	4	2
Profit (% of value)	6	6	6	6	6	6	6	10	10	10

Question 30 Cash flow

Table Q30.1 shows a contractor's project budget and profit distribution for a newly awarded contract. The conditions of contract allow interim measurements to be made monthly and payment of the amount certified less 10% retention paid to the contractor one month later. Half the retention is released on practical completion and the other half is released six months later. The client is anxious to reduce administrative costs and has proposed to the contractor that measurements be made every two months, again with a delay of one month before the contractor receives payment. The contractor is equally anxious to reduce administrative costs but wishes to assess the financial implications. To assist the contractor to do this, prepare graphs of cumulative cash out and receipt of both monthly and two-monthly measurements. (All graphs to be on the same sheet.) An average payment delay of one month between the contractor's cost liability and the outward cash flow may be assumed.

From these graphs calculate:

(1) The maximum amount of capital needed to execute the project with monthly and two-monthly interim measurements.
(2) The interest charge of the extra finance needed, given that the cost of capital is 15% p.a.

Table Q30.1 Budgeted value and profit distribution.

Month No	1	2	3	4	5	6	7	8	9	10
Value of work each month (£000)	3	4	5	8	8	8	7	6	5	2
Profit (% of value)	15	15	10	10	10	10	10	10	5	5

Question 31 Cash flow

Table Q31.1 gives the monthly estimated cost for a construction project. The profit margin added to these costs was 8%. The retention is 10%, repaid six months after the practical completion of the project. The payment conditions are that measurements are to be made monthly and payment less retention will be made one month later. A payment delay of one month between the contractor incurring a cost liability and paying may be assumed.

Table Q31.1 Monthly estimated costs for a contract.

Month	1	2	3	4	5	6	7	8	9
Estimated costs (£000)	2	2	3	3	4	4	4	3	2

Based on past experience, this contractor in preparing the cash flow forecast will assume that the margin achieved will be reduced by increased costs and is assuming that costs will be 4% higher than estimated. The contractor also expects to recover another 4% of the actual incurred total costs as a claim that, for cash flow forecasting, it is assumed will be paid three months after practical completion.

Prepare a cash flow forecast for this contract.

Question 32 Present worth comparison

A pumping scheme being developed has three different possible systems of pumps and pipe work. If the life of the scheme is 20 years, which scheme should be recommended as the most economic?

Scheme	Pipe diameter (mm)	Installation cost (£)	Annual running cost (£)
A	500	18 250	7 250
B	600	20 200	4 600
C	750	24 000	4 000

Use 15% to represent the cost of capital.

If the cost of capital was 10%, would the recommendation alter?

Question 33 Present worth comparison

A town built on a river is considering building an additional bridge across the river. Two proposals have been put forward for bridges at different sites. The costs of each proposal are summarised as follows:

	Bridge A	*Bridge B*
Initial cost of bridge	£65 000	£50 000
Initial cost of road works	£35 000	£30 000
Annual maintenance of bridge	£500	£900 for first 15 years, £1100 thereafter
Annual maintenance of roads	£300	£250
Life of bridge	60 years	30 years
Life of roads	60 years	60 years

With the cost of capital at 9%, which proposal should be adopted? Assessment of the proposals to be carried out by a comparison of their present worth.

Question 34 Equivalent annual cost comparison

A flood control pumping station is being designed. Three possible pumping schemes are proposed and the relevant costs are shown in Table Q34.1.

What is the most economical range of pumping time in hours/year for each scheme? Figure Q34.1 shows the frequency of pumping demand. What is the most economical scheme? The cost of capital may be taken as 19%.

Table Q34.1 Cost of alternative schemes.

	Scheme		
	A	B	C
Cost of pumps (£)	12 000	18 000	28 000
Life (years)	15	15	20
Maintenance per annum (£)	1 000	1 500	1 500
Cost of pipes (£)	22 000	18 000	12 000
Life (years)	30	30	30
Cost of pumping (p per h)	120	90	80

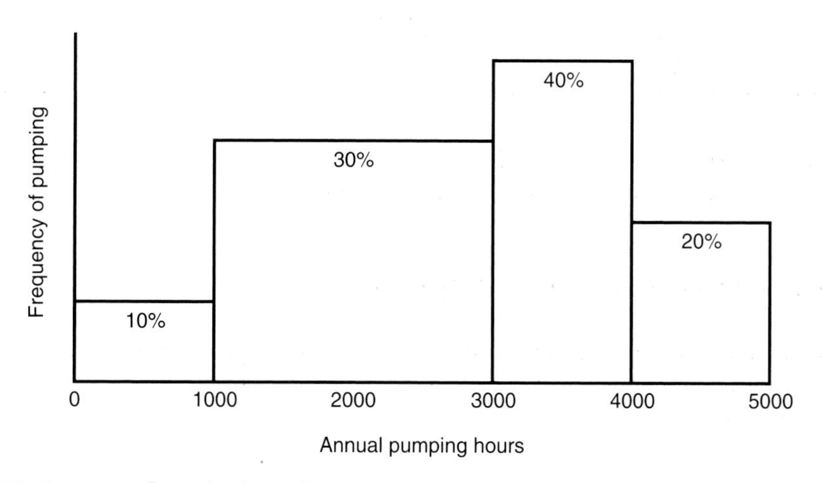

Fig. Q34.1 Frequency of pumping demand.

Question 35 Discounted cash flow yield

A company is considering the installation of new process plant equipment at a cost of £59 500 spread over three years, as follows:

- Year 0 £17 500
- Year 1 £22 000
- Year 2 £20 000

The net annual income from this plant will be £10 500 starting from Year 3. The suppliers of the plant recommend a life of only 10 years for the equipment, but the company intends to run the plant for 13 years after completion of the installation, and this means incurring heavy replacement and maintenance costs of £6500, £7500 and £9500 during the last three years. Calculate the rate of return for this project.

Question 36 Dual rate of return

A material supplier has bought quarrying rights and is free to quarry stone provided the farmland affected is reinstated afterwards. The reinstatement work involves the company in heavy capital expenditure towards the end of the project. Table Q36.1 shows the capital expenditure and the net revenue.

Calculate the paying rate given that the earning rate is 5% p.a. Why is the 'paying rate' a measure of the project's profitability?

The estimate of £180 000 for reinstatement is thought to be the least reliable of all these estimates. Explain how you would deal with such an unreliable estimate in an assessment of the project's rate of return.

Table Q36.1 Capital expenditure and net revenue.

Year	Capital Expenditure (£)	Net Revenue (£)
0	500 000	
1	300 000	400 000
2		200 000
3		200 000
4		200 000
5		200 000
6	180 000	

Table Q37.1 Projected revenues.

Year	Net revenue (£)
1	15 000
2	18 000
3	20 000
4	20 000
5	10 000
6	10 000

Question 37 Corporation tax and development grants

A company with annual pre-tax profits of approximately £2 million proposes to purchase an item of plant for £50 000. There is no expected resale value.

Capital allowances at the rate of 25% writing down, and a development grant of 20%, are available.

Corporation tax is 30% and a time lag of one year exists before tax is paid and the grant is received.

Calculate the cash flows net of tax based on the projected revenues given in Table Q37.1 and determine whether the net-of-tax yield is more than 15%.

Question 38 Corporation tax and different allowances

A company proposes to invest in the project for which cash flows are shown below. Corporation tax is 30% and taxation allowances are shown below, any balancing allowance being taken in Year 8. A tax time lag of one year occurs before tax is paid. No development grants are available. It can be assumed that the company has other profitable projects in operation.

The company's net-of-tax criterion rate of return is 18%; does this project meet this requirement?

Project details

- Investment

Industrial building	£250 000
Plant	£120 000

- Projected resale value at Year 8

Building	£200 000
Plant	Nil

- Taxation allowances

Building	4% of total cost, straight-line method
Plant	25% writing down

Net revenue

Year	£
1	80 000
2	90 000
3	100 000
4	100 000
5	90 000
6	80 000
7	70 000
8	70 000

Question 39 Replacement age

The purchase price of a small electricity generating plant is £20 000. The operating costs based on the annual average estimated hours of operation are £800 in the first year, when manufacturers' warranties operate, and £1200 in the second year, rising by £300 each year thereafter. The resale value of the plant can be assumed to be as predicted in Table Q39.1. The cost of capital is 15%. Calculate the optimum replacement age.

Table Q39.1 Predicted resale values.

Year	Predicted resale values (£)
1	18 000
2	16 000
3	15 000
4	12 000
5	8 000
6	5 000
7	2 000

Question 40 Replacement age

A precast concrete company is considering purchasing a small concrete mixing plant to be used for on-site remedial work. The equipment suppliers claim that, because of the limited use, the resale value will remain high and the equipment will operate efficiently for 20 years or more.

However, the precast concrete company has asked you to assess the optimum replacement age assuming that there will be a continuous demand and that when the equipment is retired replacement will be made by a similar piece of equipment capable of the same performance. The initial cost of the equipment is £15 000.

The operating costs of the plant, based on an average estimate of hours used each year, is £500 in the first year when manufacturers' warranties operate and £1000 in the second year, increasing by £300 per annum thereafter. The resale value is estimated to decrease by 12.5% per annum compound.

The cost of capital can be assumed to be 12%.

Question 41 DCF yield and inflation

Ten years ago an investment project was entered into by a company because the calculated rate of return on the predicted costs and revenues was a satisfactory 14% in real terms. The project did not prove to be as lucrative as expected, and the net annual revenues obtained during the project are shown below.

If over the last ten years inflation was on average 5% per annum, what was the real rate of return to the company on the capital invested in this project?

Year		
0	Investment	£24 450
1	Net revenues	£4 500
2		£4 700
3		£5 000
4		£5 100
5		£4 900
6		£5 100
7		£5 300
8		£4 900
9		£4 800
10		£4 300

Question 42 DCF yield and inflation

The predicted cash flow for a project as forecast at the start of the project in Year 0 prices was:

Year 0	Investment	–£12 000
Year 1–7	Operating costs	–£3 000
	Operating revenue	+£5 000

However, both inflation and price restraints affected this project and the cash flows actually experienced are given in Table Q42.1.

Calculate the apparent rate of return. If inflation on average was 12% over this period, what was the real rate of return?

Table Q42.1 Cash flows experienced.

Year	Cash Flow (£)	
	In	Out
0		–12 000
1	+6000	–3 000
2	+6000	–3 500
3	+6500	–4 000
4	+6500	–5 250
5	+7000	–5 000
6	+7000	–5 250
7	+7500	–5 500

Question 43 Present worth with increasing annual costs

The estimates for the cost of constructing two alternative small electricity generating stations are as follows:

	Scheme A (£M)	Scheme B (£M)
Capital costs	132	50
Fuel: annual cost	20	25
Maintenance:		
—annual cost estimated at Year 0 —budgeted to rise at 2.75% per year	5	8
Operating staff	0.5	0.8
Administration	0.5	0.8

The proposal also includes the upgrading of the station in the tenth year of operation to increase its output. The estimates for the upgrade are:

	Scheme A (£M)	Scheme B (£M)
Capital costs of upgrade in Year 10 (estimated at Year 0 prices)	15	15
Fuel: annual costs increase (at Year 0 prices)	+2	+2
Maintenance: annual cost increase (at Year 0 prices)	+1.5	+1
Operating staff and administration costs to remain the same.		

Using an interest rate of 9% to represent the value of money, compare these schemes on a present worth basis and determine which is the more economic over a 20-year period.

Question 44 Present worth with increasing annual costs and inflation

Repeat the present worth comparison of the two schemes described in Question 43 and determine whether the choice of schemes will alter if cost inflation is predicted as follows:

Element	Annual inflation rate (%)
Capital equipment	2
Fuel	8
Maintenance	5
Operating staff	2.75
Administration staff	2.75

Question 45 Valuation of a plant item

An item of plant had an original purchase price of £10 000 and has been in use for two years. It has a remaining useful life of four years and the estimated running costs, at present prices, for the next four years are £661.25, £766.44, £874.50 and £1005.68. The estimated resale value at the end of these four years is £2000, when the plant is six years old.

To purchase a similar item of plant would now cost £11 000 with running costs as follows:

Year	Running costs (£)
1	400.00
2	460.00
3	529.00
4	608.35

The resale at the end of four years is estimated at £4000.

Calculate the value to the owner of the existing item of plant by comparing the costs of keeping the plant item with the costs of replacing it. The comparison should be based on the next four years of use and the value of money may be taken at 10%.

Question 46 Valuation of a plant item with longer life

Repeat the calculation of value for the existing plant item in comparison with the new plant item as in Question 45 but base the comparison on a 20-year period. The new item of plant may be kept for six years; the running costs for Years 5 and 6 are £699.60 and £804.54, respectively. The resale values for the new plant item are:

Year	Resale (£)
2	8000
4	4000
6	2000

Question 47 Valuation of a plant item whose life is in perpetuity

Repeat the calculation of value as in Questions 45 and 46 but base the comparison on replacements to perpetuity.

Question 48 Selection of mutually exclusive projects using incremental analysis

An appraisal of three mutually exclusive projects is being carried out. The criterion rate of return for the company is 9%.

The cash flow forecasts for the three projects are as follows:

Project	A	B	C
Initial investment	£6000	£8000	£9000
Net annual receipts	£1100 (for years 1–3 and £1400 thereafter)	£1500	£1900
Life	8 years	8 years	8 years
Salvage value	£800	£1200	£800

Which project should be recommended? Support your recommendation by calculation.

Question 49 Analysis of plant purchase

A crane hire company, after carrying out a marketing study of its operations for the next five years, decides to expand its truck-mounted hydraulic crane services. A meeting between the managing director, financial director and plant manager is called to consider the financial and commercial figures uncovered during the study. Past records indicate that a machine in the 25–30 tonne range would have the most likely prospect of achieving suitable utilisation targets. The national economic indicators for future work look favourable and the investment study shows that an adequate return on capital is possible if the machine can be purchased for £80 000 and have a five-year working life. The managing director considers that, to take full advantage of the market, the new machine must be added to the fleet within the next three months. The plant manager insists that the machine (a) has a boom capable of extending 20 m, (b) has an extension fly jib that is hydraulically positionable and (c) conforms fully to International (ISO) standards. It is further agreed that the manufacturer of the machine must be located in the home country or, if foreign, then the machine must be purchased through a franchised agent. Past experience of directly imported purchases has proved unreliable regarding the supply of spare parts.

The purchasing manager obtains details of all the truck-mounted cranes currently on sale. Several are unsuitable regarding price and capacity and are quickly eliminated, leaving four to be more carefully examined: the Harris 30T, the McCaffer L25, the Dunsby 25-tonne Hydracon and Brock's 25-tonne Superlift made in the USA. Details of these trucks are given in Table Q49.1.

Table Q49.1 Specifications of four truck-mounted cranes.

Description	Harris 30T	McCaffer L25	Hydracon	Superlift
Price	£79 000	£75 000	£80 000	£78 000
Delivery period	3 months	4 weeks	6 months	2 months
Maximum lift	27.5 tonnes	25 tonnes	26 tonnes	30 tonnes
Boom length	3–24 m	8.63–25.97 m	8.9–28.7 m	5.2–26.3 m
Fly jib	Power assisted	Power assisted	Power assisted	Manual
One-operator	Yes	Yes	Yes	Yes
Fully telescopic boom	Yes	Yes	Yes	Yes
Simple operation	Good	Poor	Excellent	Poor
Variable hoisting speeds	20–35 m/min	30 m/min only	22–37 m/min	20–45 m/min
Fail-safe brakes	Yes	Yes	Yes	Yes
Four-wheel drive	Yes	Yes	Yes	Yes
Travelling speed	65 km/h	70 km/h	60 km/h	65 km/h
Engine horsepower	135 hp Turbins	138 hp Rumbings	210 hp Harland	160 hp Hertz
Turning radius	11.5 m	11.3 m	11.00 m	12 m
Safe load indicator	Enco Auto	Wicht Audio	Enco Auto	Dale 424
Overall length	11.85 m	10.64 m	10.59 m	11.00 m
Overall width	2.5 m	2.5 m	2.44 m	2.48 m
Travelling weight	29 000 kg	33 000 kg	29 500 kg	30 000 kg
Cab comfort	Poor	Fair	Good	Very good
Outriggers	Hydraulic	Hydraulic	Manual	Hydraulic
Gearbox	12 forward 3 reverse	6 forward 2 reverse	7 forward 2 reverse	6 forward 2 reverse
Power-assisted steering	Curman	Rico	Curman	Curman
Accessories	Reasonable	Good	Poor	Poor
Slewing speed	2.25 rev/min	2.35 rev/min	2.40 rev/min	2.0 rev/min
Boom telescoping speed	20 m/min	23 m/min	18 m/min	20 m/min
Cost of spares	Reasonable	High	Reasonable	Low
Back-up services	Poor	Good	Good	Reasonable
Low running cost	Good	Poor	Good	Reasonable

The plant manager favours the Harris truck, because several similar machines made by this company already exist in the fleet and some spares would be interchangeable. However, back-up services from Harris had been poor in the past, relative to experience with equipment held from the other manufacturers.

The Harris 30T is now marketed with the newly developed displacement pumps to provide the hydraulic control. Although reliable in general, there have been several instances of these pumps seizing up in the first few weeks of usage. However, replacement is fairly straightforward, but the machine is put out of action for a day at a cost of £150. This phenomenon occurs in one in ten cases, and the new pump costs £300. The McCaffer L25 sometimes suffers gearbox failure early on in life; the probability of this is small, 0.05 at a cost of £3000 for replacement.

Exercise brief

Advise the plant manager on which truck to purchase. Use the Kepner and Tregoe method of analysis. Consider the MUSTS outlined in the brief, choose your own WANTS and apply rankings and ratings appropriate to the information given.

Question 50 Setting a hire rate for plant

(1) Determine the cost per hour of owning a 1 cubic metre capacity hydraulic backhoe excavator during the third year of its life, given the following information. The declining balance method of depreciation is to be used for the machine, and straight-line depreciation for the tyres.

(2) If the return on capital is to be 15% p.a., calculate the hourly cost of owning and operating the machine using the discounted cash flow (DCF) method of analysis.

• Initial cost excluding tyres	£15 000
• Resale value	£2 000
• Useful life of machine	6 years
• Interest charged on borrowed capital	15% p.a.
• Fuel consumption	10 litres/h
• Cost of fuel	10p/litre
• Oil and grease	10% of fuel cost
• Repairs to machine	10% of purchase price
• Insurance and tax	2% of purchase price
• Average working hours of machine in year	2 000 h
• Purchase price of a set of tyres	£1 000
• Useful life of a set of tyres	2 years
• Tyre maintenance and repairs	5% of purchase price

Question 51 Plant workshop budget

The annual budget for a plant workshop is £200 000, of which 80% comprises direct costs of consumable materials, spare parts, lubricants and labour such as mechanics, fitters, welders and electricians. The remaining 20% is a fixed charge for overheads made up of staff salaries, rent, rates, insurance, administration, power and depreciation on workshop equipment.

At the end of the year the firm's activities were higher than expected with the plant fleet operated 10% more hours. As a consequence the direct costs increased to £170 000 and the actual costs of the overheads were £45 000.

Calculate the direct, overhead and total variances.

Question 52 Sales variance on plant hire

The budgeted hire revenue for a plant division over the next 12 months is £800 000. Profit, overheads and depreciation each represent 20% of hire revenue, the latter two being fixed costs, and the remaining 40% is made up of variable costs of workshop maintenance and other running costs. At the end of the period, actual revenue was £600 000 and variances recorded for direct and overhead costs were respectively +£6000 and +£2000. In addition, several items of plant were disposed of prematurely leading to a negative variance of £3000 on depreciation.

Determine the budgeted and actual levels of profit.

Question 53 Variances on a plant item

(1) During the month an item of plant was hired out for 280 hours at a hire rate of £2.40 per hour. The budget anticipated only 200 hours of work at a hire rate of £2.50 per hour.

Calculate the utilisation and price variances for the item during the month, where

- Utilisation variance is the financial effect of using the plant either more or less than the hours budgeted. Thus:

UV = (actual hours × budgeted hire rate) – (budgeted hours × budgeted hire rate).

- Price variance is the financial effect of charging more or less than the budgeted hire rate. Thus:

PV = (actual hours × actual hire rate) – (actual hours × budgeted hire rate).

(2) Profit, fixed overheads and depreciation are each set at 20% of the budgeted hire rate at the beginning of the month, the remainder being direct costs allowed for maintenance.

If the actual costs incurred for overheads, depreciation and maintenance during the month were, respectively, £120, £125 and £250, calculate the variances and actual profit.

Question 54 Capital gearing

The capital structures for two construction companies are given in Table Q54.1, and the profits available for interest payments and dividends for a range of company performance levels in Table Q54.2.

(1) Calculate and comment upon the capital gearing ratios for each of the companies.
(2) Calculate the dividend available to the ordinary shareholders and comment upon the earnings per share at each level of profit.

Corporation tax is 50%.

Table Q54.1 Capital structures.

	Company A (£000)	Company B (£000)
Ordinary shares (50p)	155	35
7% preference shares	45	45
10% loan stock (debentures)		120

Table Q54.2 Performance levels.

	1	2	3	4	5	6
Profit (£)	10000	12000	18000	20000	25000	30000

Question 55 Working capital

The managing director of a construction company, after carrying out a thorough market investigation of the opportunities for work in a development area, decides to set up a division in the region. The work will be obtained by tender only, and a small management organisation is to be set up in an establishment purchased for £45 000. As an approximate rule, the estimate for tenders is built upon the following figures:

(1) Labour, 20%
Plant, 15%
Materials, 40%
Overheads, 15%
Profit, 10%.
(2) All materials purchased and plant hired are given one month's credit by suppliers and three months' stock of materials is held on sites at all times.
(3) The average time between starting a section of work and receipt of the interim payment is two months. The working capital is £180 000.

Calculate the annual turnover required to justify the director's decision to go ahead if the working capital is to be fully utilised, and give a summary of the new capital requirements.

Question 56 Balance sheet

The figures shown in Table Q56.1 are balances extracted from the accounts of a construction company as at 31 December.

(1) Prepare a tabulated profit and loss statement in respect of the figures.
(2) Prepare a tabulated balance sheet at 31 December, given the following information:
(a) Value of land and buildings as at 31 December is £50 000;
(b) Cash at bank as at 31 December is £40 000;
(c) Value of work certified but not yet paid is £25 000;
(d) Payment completed in respect of:

Materials, £295 000;
Plant and equipment, £95 000;
Wages and salaries, £345 000.

(e) Authorised and issued share capital is in the form of 150 000 ordinary shares at £1 each;
(f) The general reserve stood at £20 000 as at 31 December the previous year;
(g) Corporation tax is paid at 50% of profits;
(h) Only 50% of profit after tax is to be paid as dividend to shareholders. The remainder is to be transferred to the general reserve.

Table Q56.1 Account balances of a construction company.

	£000
Materials purchased and delivered to site	300
Closing stock as at 31 Dec. the previous year	15
Materials on site as at 31 Dec. this year	10
Wages incurred	250
Salaries incurred	100
Plant and equipment purchased during year (annual depreciation 20%)	100
Work certified (as at 31 Dec. this year)	695
Work in progress	10

Question 57 Financial ratios

Calculate the important financial ratios for Question 56.

Question 58 Interpretation of the balance sheet

The balance sheet for a medium-size construction company engaged in general building as at 31 December is shown in Table Q58.1.

During the following 12 months normal trading continued, involving the following transactions:

- Net profit made before tax = £67 500;
- Dividend paid = £18 000;
- Tax paid to Inland Revenue = £30 000;
- Interest paid to debenture holders = £6000;
- Debtors increased by £30 000;
- 10 000 ordinary shares of £1 issued at a price of £1.25 per share, all cash is received;
- 7500 8% debentures are redeemed;
- Creditors are increased by £45 000;
- £100 500 of new plant is purchased;
- Annual depreciation on plant holdings is 10%;
- Work in progress is decreased by £7500;
- Stock increased by £15 000;
- Bank overdraft is increased by £10 500:

Table Q58.1 Tabulated balance sheet as at 31 December the previous year.

Employment of capital	£	£
Fixed assets		
Land and premises		220 500
Plant		229 500
		450 000
Current assets		
Debtors	112 000	300 000
Cash	31 000	750 000
Stock	62 000	
Work in progress	95 000	
Current liabilities		
Creditors	90 000	
Bank overdraft		90 000
	–	660 000
Capital employed		450 000
450 000 ordinary shares at £1 each		75 000
75 000 8% debentures at £1 each		135 000
General reserve		–
Profit and loss account		660 000

(1) Prepare a tabulated profit and loss account and balance sheet for 31 December.
(2) Show the source and application of funds during the year.
(3) Calculate for the period up to 31 December:
 (a) Current ratio;
 (b) Liquidity ratio (acid test).
 Compare these ratios with those of the previous year.

Question 59 Mass haul problem

A firm of civil engineering contractors undertakes to carry out as a subcontract the evaluation work on a road construction project. The work involves forming cuttings and embankments described in the bill of quantities as follows:

- Excavate materials on site and form embankments
- Excavate materials on site and cart away
- Excavate materials in borrow pit and form embankment on site

The quantities for the above items are calculated from longitudinal sections and cross sections, and are shown in Table Q59.1.

The subcontractor plans to use tractors and scrapers, and estimates the cost of doing the work from the following rates.

- Dig and form banks up to 200 m haul, 30p per m^3
- Dig and form banks with a haul over 200 m and up to 300 m, 80p per m^3
- Dig and form banks with a haul over 300 m and up to 400 m, 160p per m^3
- Dig and form banks with a haul over 400 m and up to 500 m, 280p per m^3
- Dig in cutting and cart to tip, 50p per m^3
- Dig in borrow pit, bring to site and form embankments, 75p per m^3 (struck volume)

Table Q59.1

Length along road 100 m	Cross-area cut	Sectional fill	Volumes (m³) from drawings cut	Measured fill	Fill needed = fill × 100/90.9	Cumulative volume (m³)
0	0		0			0
2	400		+40 000			+40 000
4	300		+70 000			+110 000
6	0	0	+30 000			+140 000
8		100		−10 000	−11 000	+129 000
10		300		−40 000	−44 000	+85 000
12		200		−50 000	−55 000	+30 000
14	0	0		−20 000	−22 000	+8 000
16	200		+20 000			+28 000
18	0	0	+20 000			+48 000
20		200		−20 000	−22 000	+26 000

All fill material compacts to 90.9% of its struck volume.

(1) Calculate the most efficient method of carrying out the work if the cost is to be minimised.
(2) State the final cost.

Question 60 Development budget

Increasingly contractors are becoming associated with the lending institutions in developing schemes from inception to completion and in some cases even operation of the facility such as with BOOT and DBFO projects. Given the following cost information concerning the development of an office block in freehold:

(1) Evaluate the development value against the costs of the scheme in order to ensure that an adequate yield is available.
(2) Explain the effects of an enforced lettable void on rents and developer's profit.

Office Block–Freehold	
• Gross floor area	1000 m²
• Rent	£50/m²
• Construction costs	£400/m²
• Yield required	5%
• Interest rate on finance	10%
• Utilisation factor	0.8
Development period	
• Pre-contract	6 months
• Contract	18 months
• Design fees	10%
• Developer's profit	15% Gross Development Value
• Management costs	10% of rent p.a.

Question 61 Effects of a lease on development budget

(1) If the building described in Q.60 were let on a 52-year lease, determine the revised development budget calculations and comment on the comparative effects.
(2) Comment on the comparative effect of the alternative of freehold land, although the building is planned to have a life of only 52 years.

Solutions

Solution 1 Method study exercise

Task (i)

The crane's daily cycle is summarised in Table S1.1. Note that, because the process is directed towards recording the crane, inspection of the concrete represents a delay.

Had a material type chart been adopted to study the flow of concrete, then this probably would be represented by an inspection symbol. Such charts may take two forms:

Table S1.1 Summary of craneage activity of a daily cycle.

Summary of craneage activity of a daily cycle				
○	□	⇨	▽	D
7	2	12	2	10
$5^1/_2$ h		380 m		

- Labour/machine type to record what happens to the machine or operator
- Material type to record what happens to the materials handled by the operator

The chart has highlighted the excessive travel and delays experienced by the crane, together with a record of where the main problems are occurring.

Task (ii)

Multiple activity chart. It is observed in the multiple activity chart (Fig. S1.1) that the complete sequence of operations has a minimum 6 minute cycle time.

Task (iii)

(1) Concrete production and placing
   (a) Cycle time = 6 min
   (b) Hours worked per day = 8
   (c) ½ m^3 concrete placed every 6 min
   (d) Output per day = $\dfrac{60 \times 8 \times 0.5}{6} = 40\ m^3$
   (e) Time to place 80 m^3 = 2 days

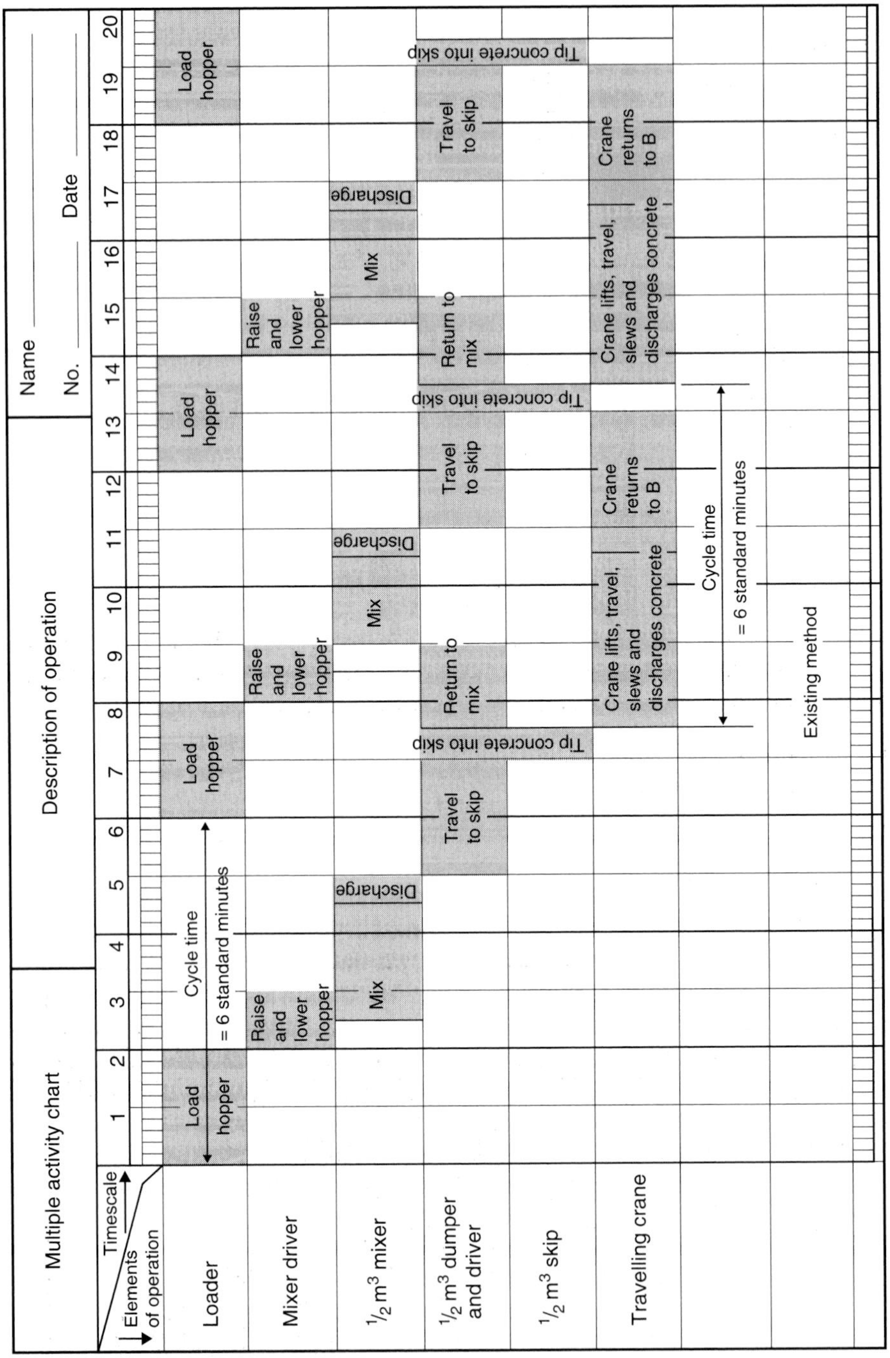

Multiple activity chart

Description of operation

Name

No. Date

Timescale

Elements of operation

1 2 3 4 5 6 7 8 9 10 11 12 13 14 15 16 17 18 19 20

Loader: Load hopper; Load hopper; Load hopper; Load hopper

Cycle time = 6 standard minutes

Mixer driver: Raise and lower hopper; Raise and lower hopper; Raise and lower hopper

$^1/_2$ m^3 mixer: Mix; Discharge; Mix; Discharge; Mix; Discharge

$^1/_2$ m^3 dumper and driver: Travel to skip; Return to mix; Travel to skip; Return to mix; Travel to skip

$^1/_2$ m^3 skip: Tip concrete into skip; Tip concrete into skip; Tip concrete into skip

Travelling crane: Crane lifts, travel, slews and discharges concrete; Crane returns to B; Crane lifts, travel, slews and discharges concrete; Crane returns to B

Cycle time = 6 standard minutes

Existing method

Fig. S1.1 Multiple activity chart for existing method.

(2) Costs

(a)	½ m^3 batch	0.4 × 120	= £48
(b)	½ m^3 dumper	0.4 × 40	= £16
(c)	½ m^3 skip	0.4 × 8	= £3.2
(d)	Crane	0.4 × 400	= £160
(e)	Labour	16 × 8 × 2.5	= £320
(f)	Compaction equipment	0.4 × 12	= £4.8
	Total Costs		£552.0

(3) Labour

Mixer loader and driver	= 2
Dumper driver	= 1
Concrete gang	= 4
Crane driver	= 1
Total labour	= 8 workers

Therefore cost of placing concrete $= \dfrac{552}{80} = £6.9$ per m^3

Task (iv) Proposed improved solution

It can be seen from Fig. S1.2 that the batching plant could easily produce a batch of concrete every 3 minutes. A possible improved solution would be to use two skips of 1 m^3 capacity and to increase the number of ½ m^3 dumpers to two as shown in Fig. S1.3.

(1) Revised concrete production and placing
- (a) 1 m^3 concrete placed every 6 min
- (b) Output per day $= \dfrac{60 \times 8}{6} = 80\,m^3$
- (c) Time to place 80 m^3 = 1 day

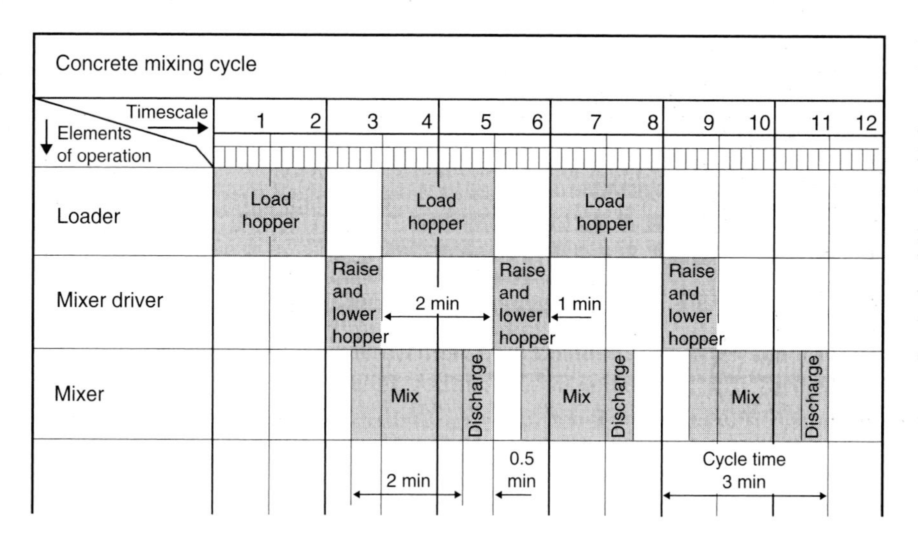

Fig. S1.2 Multiple activity chart for concrete batching plant.

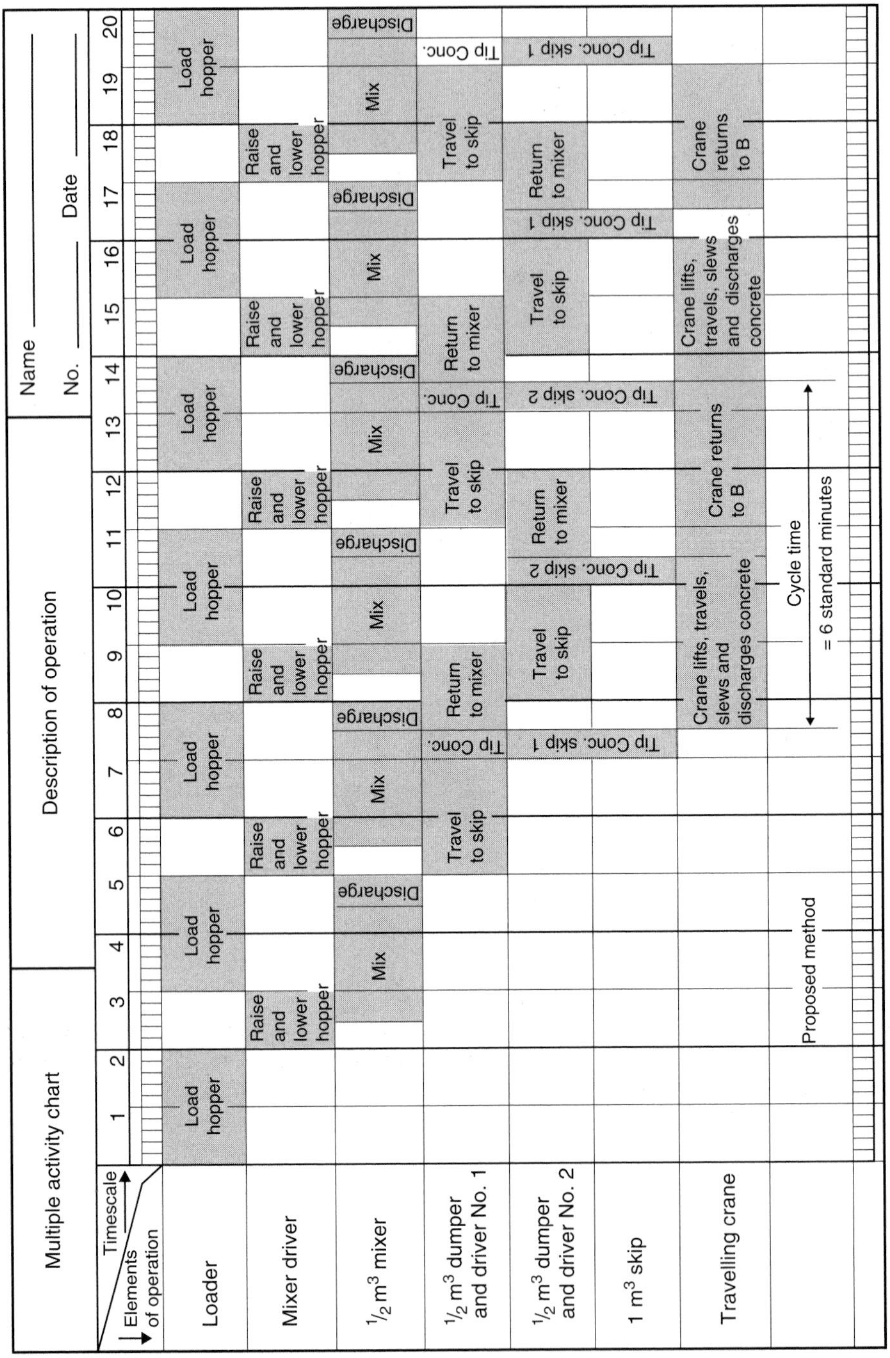

Fig. S1.3 Multiple activity chart for revised method.

(2) Costs

(a)	½ m^3 mixer	0.2×120	= £24
(b)	$2 \times$ ½ m^3 dumpers	$2 \times 0.2 \times 40$	= £16
(c)	2×1 m^3 concrete skips	$2 \times 0.2 \times 12$	= £4.8
(d)	Travelling crane	0.2×400	= £80
(e)	Labour	$8 \times 10 \times 2.5$	= £200
(f)	Construction equipment	0.2×12	= £2.4
			£327.2

(3) Labour

(a)	Mixer loader and driver	2
(b)	Dumper drivers	2
(c)	Concrete gang 5 (extra worker)	5
(d)	Crane driver	1
Total labour		10 workers

(e) Proposed cost of placing concrete $= \dfrac{327.2}{80} = £4.09/m^3$

Other alternative solutions

The reader should now examine other possible solutions before deciding on the most economic alternative. For example:

Alternative 3	$2 \times$ ½ m^3 dumpers
	$2 \times$ ½ m^3 skips
Alternative 4	1×1 m^3 dumper
	1×1 m^3 skip
Alternative 5	$3 \times$ ½ m^3 skips
	1 m^3 skip + ½ m^3 skip

Task (v)

Utilisation factors. The utilisation factor for each resource is calculated by extracting the working time of the resource and dividing by the cycle time. In this way, alternative schemes can be compared immediately. This is shown in Table S1.2.

Table S1.2 Summary of utilisation factors of multiple activity charts.

Activity		Present		Proposed		Difference	
		Work	Idle	Work	Idle	Work	Idle
Loader	Cycle time (min)	2	4	4	2	+2	−2
	(%)	33	67	67	33	+34	−34
Mixer driver	Cycle time (min)	1	5	2	4	+1	−1
	(%)	17	83	33	67	+16	−16
Mixer	Cycle time (min)	2.5	3.5	5	1	+2.5	−2.5
	(%)	42	58	83	17	+41	−41
Dumper No.1	Cycle time (min)	4	2	4	2	0	0
	(%)	67	33	67	33	0	0
Dumper No. 2	Cycle time (min)	–	–	4	2	–	–
	(%)	–	–	67	33	–	–
Crane	Cycle time (min)	5.5	0.5	5.5	0.5	0	0
	(%)	92	8	92	8	0	0

Solution 2 Time study exercise

Tasks (i) and (ii)

Figure S2.1 is the summary sheet on which the basic times and assessed allowances are collected against the measured quantity of work done.

Task (iii)

It can be seen from the summary sheet that the performance data indicates a low level of activity if idle time is included.

Task (iv)

With reference to Fig. S1.1, where the standard times are used to establish a multiple activity chart, a cycle time of 6 minutes is required to place ½ m^3 of concrete, so producing 5 m^3 per hour. The potential output can be considerably improved as demonstrated in Fig. S1.3, where it should be noted that the same standard times have been adopted in the revised method.

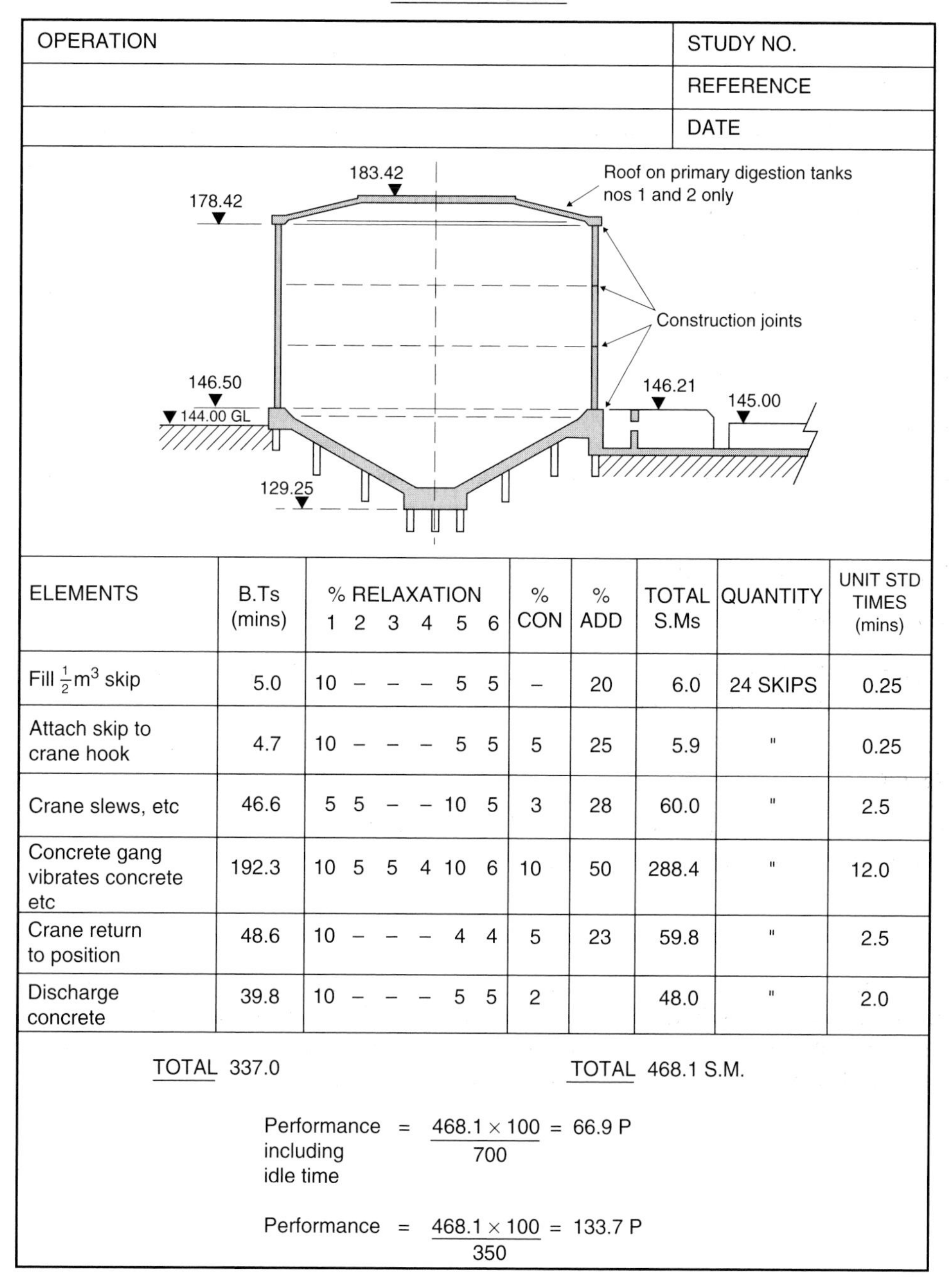

SUMMARY SHEET

OPERATION	STUDY NO.
	REFERENCE
	DATE

ELEMENTS	B.Ts (mins)	% RELAXATION 1	2	3	4	5	6	% CON	% ADD	TOTAL S.Ms	QUANTITY	UNIT STD TIMES (mins)
Fill $\frac{1}{2}$ m^3 skip	5.0	10	–	–	–	5	5	–	20	6.0	24 SKIPS	0.25
Attach skip to crane hook	4.7	10	–	–	–	5	5	5	25	5.9	"	0.25
Crane slews, etc	46.6	5	5	–	–	10	5	3	28	60.0	"	2.5
Concrete gang vibrates concrete etc	192.3	10	5	5	4	10	6	10	50	288.4	"	12.0
Crane return to position	48.6	10	–	–	–	4	4	5	23	59.8	"	2.5
Discharge concrete	39.8	10	–	–	–	5	5	2		48.0	"	2.0

TOTAL 337.0 TOTAL 468.1 S.M.

Performance including idle time $= \dfrac{468.1 \times 100}{700} = 66.9$ P

Performance $= \dfrac{468.1 \times 100}{350} = 133.7$ P

Fig. S2.1 Time study summary sheet.

Solution 3 Time study and activity sampling comparison

As a trial, take four observations each hour, selected randomly. Mark these in pencil above each daily strip on Fig. Q3.1, e.g. a number between 0 and 24 means that the observation took place on the quarter hour, 25 to 49 on the half hour, etc.

Note what is happening at each pencil mark.

Calculate the percentage time for each operation and the resulting accuracy of each using the activity sampling formula given previously in the text.

Compare the sampled results with the time study results.

Repeat the exercise respectively with two and eight observations each hour and compare the levels of accuracy.

Solution 4 Site layout exercise

(1) Criticism of existing site layout (Fig. Q4.1):
  (a) Both hoists have separate scaffold staging, causing increased costs.
  (b) Materials are not stockpiled near hoists.
  (c) Entrance to the site is too narrow for trucks to pass.
  (d) Stores are located behind the batching plant so obscuring storekeeper's view and the checkpoint is separated from the stores.
  (e) Concrete and mortar mixers are located too far from the hoists, and some distance from the temporary road.
  (f) Stockpiles are dispersed and hinder unloading and policing.
  (g) Temporary roads are long and narrow.
  (h) Some stores areas are difficult to reach and cannot be seen from stores.

(2) Suggested improved layout (Fig. S4.1):
  (a) Both hoists are housed in a common scaffold.
  (b) Batching plants have direct discharge into dumpers onto the hard-standing.
  (c) The temporary road is shorter and wider.
  (d) The access is widened near the site entrance.
  (e) The stores are located to give a good view of all materials stockpiles and are sited near the temporary road. The checkpoint is also near the stores.
  (f) Aggregate bins can be directly accessed from the hard-standing.
  (g) A compound is provided to police non-bulk materials.

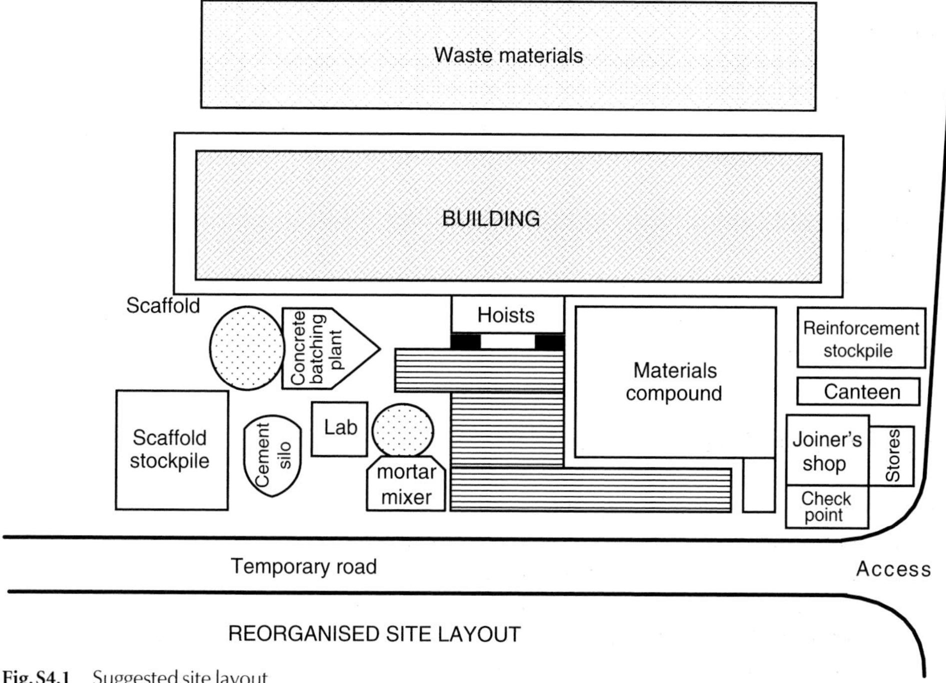

Fig. S4.1 Suggested site layout.

Solution 5 Activity sampling of crane operations

(1) $N = Z^2 \times P(1-p)/L^2$

For 95% c onfidence $Z=2$, $N=400$, the total number of observations taken. P = 100/400, the proportion of the observations that the crane was idle. L = degree of accuracy of the result

$$L = \frac{2^2 \times 0.25 \times 0.75 \times 100}{400}$$

$$= 4.33\%$$

Therefore proportion of time crane was observed idle = 25 ± 4.33% with 95% confidence.

(2) $L = 2\%$; $P = 0.25$.

$$N = \frac{2^2 \times 0.25(1-0.25)}{0.02^2}$$

$$= 1875 \text{ observations}$$

Four hundred observations have already been made.

Therefore 1875 – 400 = 1475 further observations to be recorded.

Solution 6 Activity sampling of labour

(1)

$$L=\sqrt{\frac{Z\times Z\times P(1-P)}{N}}$$

L = accuracy required. N = total number of observations.
P = observed proportion of time spent on an operation.

- Fixing formwork (FS)

$$P=0.375; Z=2; N=64.$$

$$\mathrm{L}=\sqrt{\frac{4\times 0.375\times 0.625}{64}}$$

$$=0.121\times 100$$
$$=\pm 12.1\%.$$

- Stripping formwork (RF)

$$P=0.25; Z=2; N=64.$$

$$\mathrm{L}=\sqrt{\frac{4\times 0.25\times 0.75}{64}}$$

$$=0.108\times 100$$
$$=\pm 10.8\%$$

- Cleaning formwork (CL)

$$P=0.25; Z=2; N=64.$$

$$L=\sqrt{\frac{4\times 0.25\times 0.75}{64}}$$

$$=0.108\times 100$$
$$=\pm 10.8\%$$

- Idle time (I)

$$P=0.125; Z=2; N=64.$$

$$L=\sqrt{\frac{4\times 0.125\times 0.875}{64}}$$

$$=0.827\times 100$$
$$=\pm 8.27\%$$

(2) The results on the next page show that only the idle time can be defined with the required 10% accuracy and clearly more observations are required to be certain of the other operations. It can be seen from the formula

Table S6.1

Observations			Percentage of total time		
Operation	Number	% of total (P)	Accuracy (L)	Min	Max
FS	24	37.5	12.1	25.4	49.6
RF	16	25.0	10.8	14.2	35.8
CL	16	25.0	10.8	14.2	35.8
I	8	12.5	8.27	4.23	20.77
	64 *(N)*	100.0			

$$N = \frac{Z^2\,(1-P)}{L^2}$$

that the closer P is to 0.5 the larger the value of N. Therefore, taking $P = 0.375$ and $L = 0.1$

$$N = \frac{Z^2 \times 0.375 \times 0.625}{0.1 \times 0.1}$$

$$= 93.75$$

A further 30 observations are needed to bring the results within the 10% accuracy range.

Solution 7 Crashed activities

(i)

Examination of the crashed cost daily rates indicates that some activities are cheaper per day than others and would normally be selected first. Also, closer inspection suggests that crashing all the activities at the same time would be an inefficient use of resources since those not on the critical path would not reduce the overall time of the project; hence careful analysis is necessary, starting first by crashing activities on the critical path, progressing to those with float that will subsequently change the route of the critical path. The activity with the least expensive crash rate is the first action point, followed by the next expensive as follows:

Step 1

Action	Duration of project	Rate	Amount	Crashed cost
Crash 1–2 by 2 days	18	15	30	£860
Crash 3–5 by 1 day	17	20	20	£880
Crash 5–7 by 3 days	14	30	90	£970

It is observed that although activity 2–3 has a cheaper crashed rate than 5–7, the effect of crashing embraces the need to crash 2–4, 4–5 and 2–5 in that order since they will also form identical critical path durations, and thus produce combined high crashed cost rates.

Step 2

Action	Duration of project	Rate	Amount	Crash cost
Crash 2–3 by 1 day and Crash 2–4 by 1 day	13	25 30	55	£1025

Step 3

Action	Duration of project	Rate	Amount	Crash cost
Crash 2–3 by 1 day and Crash 4–5 by 1 day and Crash 2–5 by 1 day	12	25 40 17.5	82.5	£1107.5

The maximum direct cost of the project with a critical path time of 12 days is £1107.5 compared to £8300 at normal duration; there is also a potential saving in site overheads.

(ii) Effect of indirect costs

Site overheads are £40 per day; thus the total costs at the different crash durations are:

Project time costs	Direct costs	Indirect cost	Total
20	830	800	1630
18	860	720	1580
17	880	680	1560
14	970	560	1530
13	1025	520	1545
12	1107.5	480	1587.5

The optimum occurs at a crash cost rate of £30 per day on activity 5–7 to produce an earliest finish time of 14 days, direct costs of £970 and total project cost of £1530.

Solution 8 Networks

Calculate event times

Calculation of floats based on Fig. S8.1 is set out in Table S8.1.

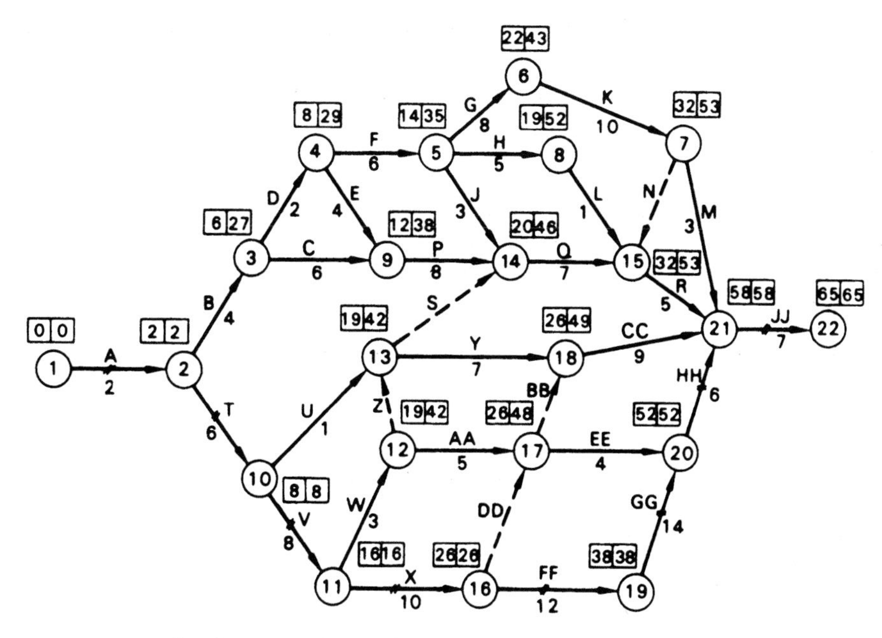

Fig. S8.1 Arrow network with event numbers, earliest and latest event times and critical activities identified.

Table S8.1 List of non-critical activities with their total and free float calculated.

Letter	Activity i–j	Duration (days)	Earliest time of start event	Latest time of start event	Earliest time of finish event	Latest time of finish event	Total float	Free float
B	2–3	4	2	2	6	27	21	0
C	3–9	6	6	27	12	38	26	0
D	3–4	2	6	27	8	29	21	0
E	4–9	4	8	29	12	38	26	0
F	4–5	6	8	29	14	35	21	0
G	5–6	8	14	35	22	43	21	0
H	5–8	5	14	35	19	52	33	0
J	5–14	3	14	35	20	46	29	3
K	6–7	10	22	43	32	53	21	0
L	8–15	1	19	52	32	53	33	12
M	7–21	3	32	53	58	58	23	23
P	9–14	8	12	38	20	46	26	0
Q	14–15	7	20	46	32	53	26	5
R	15–21	5	32	53	58	58	21	21
U	10–13	1	8	8	19	42	33	10
W	11–12	3	16	16	19	42	23	0
Y	13–18	7	19	42	26	49	23	0
AA	12–17	5	19	42	26	48	24	2
CC	18–21	9	26	49	58	58	23	23
EE	17–20	4	26	48	52	52	22	22

Total float = latest time of finish event – earliest time of start event – duration.
Free float = earliest time of finish event – earliest time of start event – duration.

Solution 9 Planning exercise

A suggested network for the construction of the submerged culvert is shown in Fig. S9.1.

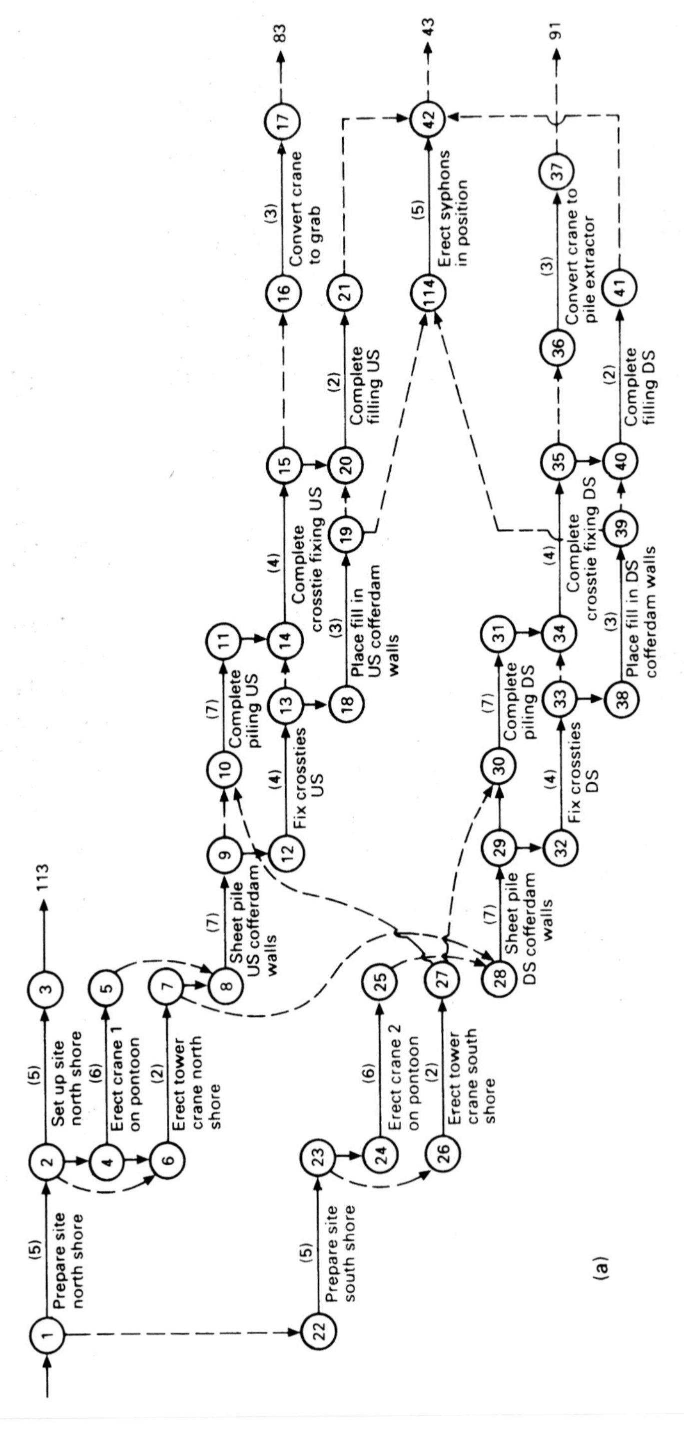

Fig. S9.1 Suggested network for submerged culvert (without resource restraints).

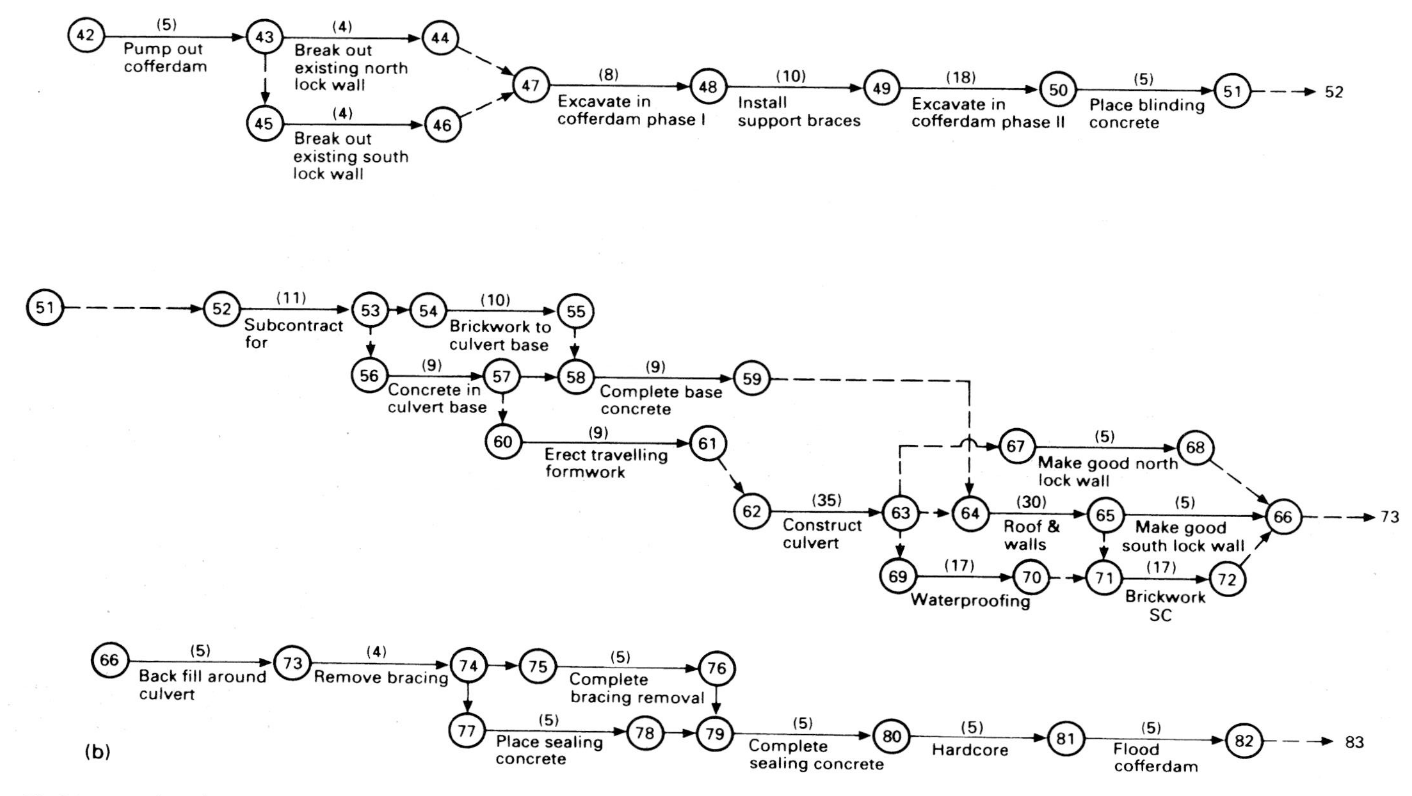

Fig. S9.1 *continued*

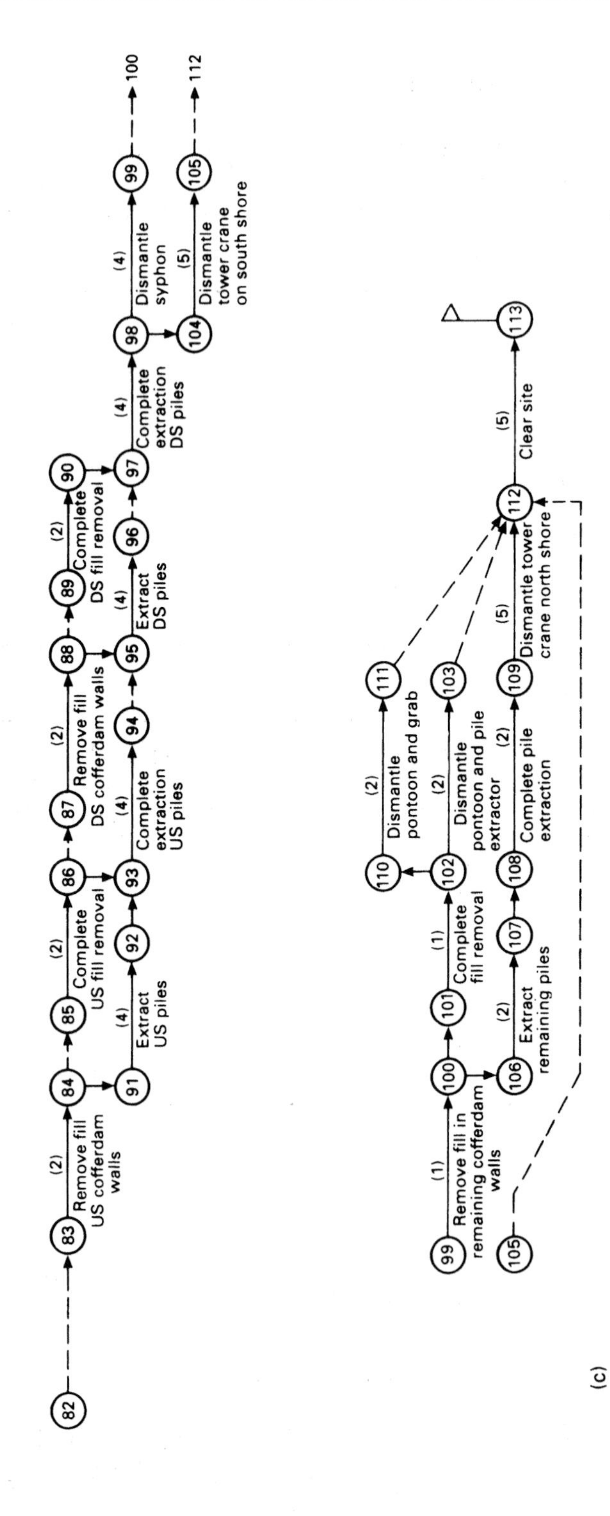

(c)

Fig. S9.1 *continued*

Solution 10 Resource smoothing

(1) Calculate float and earliest and latest start from Fig. Q10.1 (Table S10.1).
(2) Rank activities in a priority order.
(3) Prepare resource loading charts based on early starts and latest starts (Figs. S10.1–S10.3).

Table S10.1 Total float, earliest start and latest start.

Activities	Total float	Earliest start	Latest start calculation	Latest start
1–2	0	0		0
2–3	9	4	19–6	13
2–4	19	4	25–2	23
2–5	0	4		4
3–7	9	10	27–8	19
3–4	12	10	25–3	22
4–7	12	13	27–2	25
5–6	0	13		13
6–7	0	23		23

Table S10.2 Activities in a priority order.

Earliest start—total float		Latest start—total float	
Activities	Earliest start	Activities	Latest start
1–2	0	1–2	0
2–5	4	2–5	4
2–3	4	5–6	13
2–4	4	2–3	13
3–7	10	3–7	19
3–4	10	3–4	22
5–6	13	6–7	23
4–7	13	2–4	23
6–7	23	4–7	25

Total float = latest time of finish event – earliest time of start event – duration.
Latest start = latest time of finish event – duration.

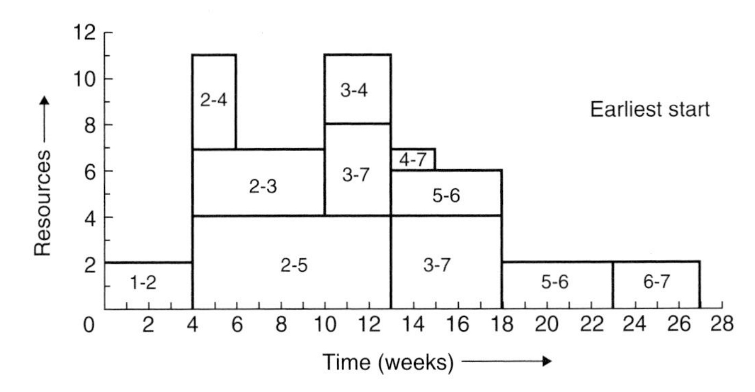

Fig. S10.1 Resource profile—early starts.

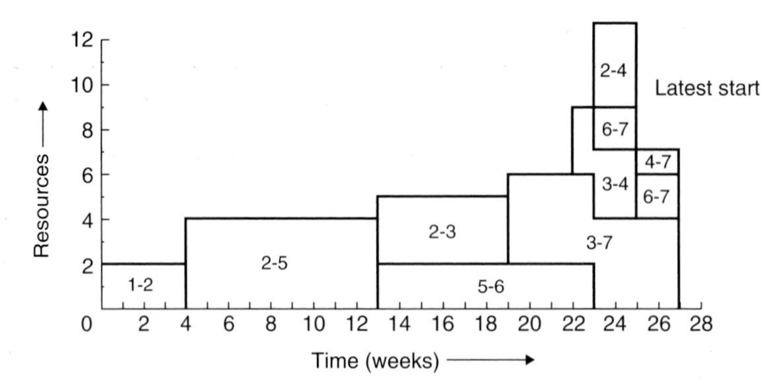

Fig. S10.2 Resource profile—latest starts.

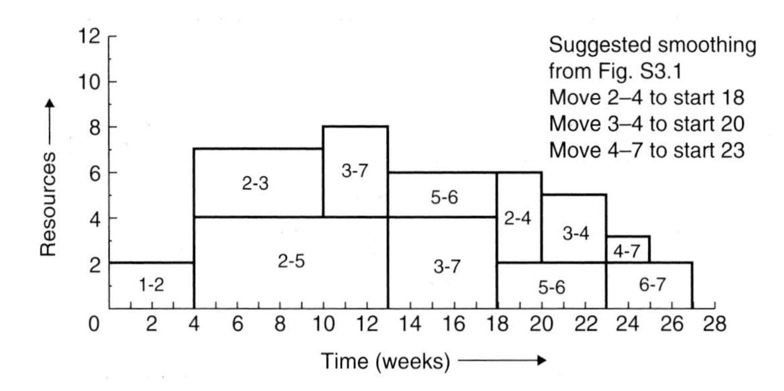

Fig. S10.3 Resource profile after smoothing.

Solution 11 Resource allocation

(1) Calculate activity times (Fig. S11.1).
(2) Calculate float and rank activities in priority order (Table S11.1).
(3) Prepare resource loading charts (Fig. S11.2).
(4) Extract list of scheduled start dates (Table S11.2).

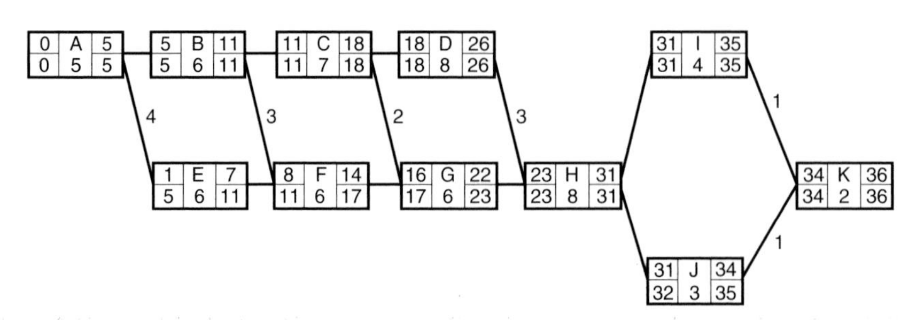

Fig. S11.1 Calculation of activity times.

Table S11.1 Float and activities ranked in order.

Activity	Earliest start	Total float	Activities earliest start – total float order	Earliest start
A	0	0	A	0
B	5	0	E	1
C	11	0	B	5
D	18	0	F	8
E	1	4	C	11
F	8	3	G	16
G	16	1	D	18
H	23	0	H	23
I	31	0	I	31
J	31	1	J	31
K	34	0	K	34

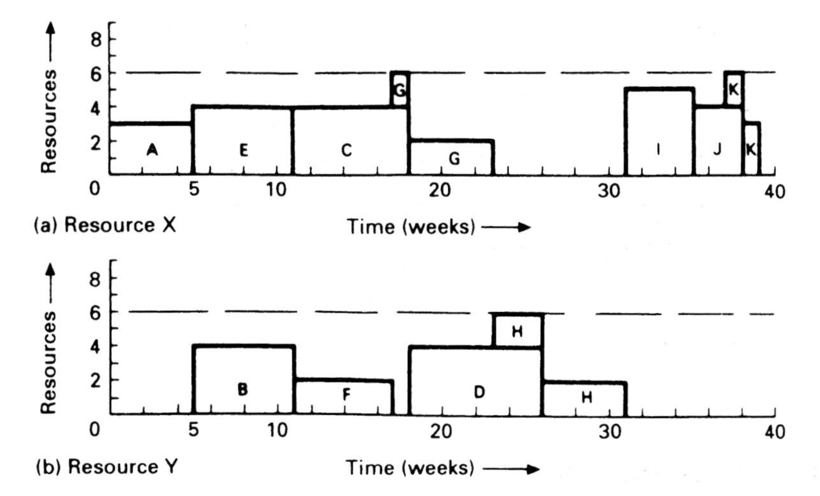

Fig. S11.2 Resource loading charts.

Table S11.2 List of scheduled start dates.

Activity	Scheduled start date
A	0
E	5
B	5
F	11
C	11
G	17
D	18
H	23
I	31
J	35
K	37

Scheduled finish = 39.

Solution 12 Resource allocation

(1) Calculate activity times (Fig. S12.1).
(2) Calculate earliest start and float and rank in priority order (Tables S12.1 and S12.2).
(3) Prepare resource loading charts (Fig. S12.2).
(4) Extract list of scheduled start dates (Table S12.3).

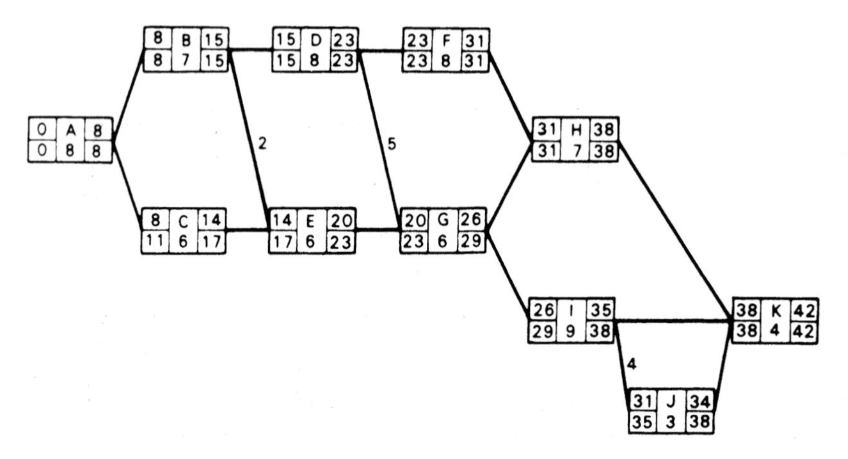

Fig. S12.1 Network from Question 5 with activity times calculated.

Table S12.1 Activity list with earliest start and total float.

Activity	Earliest start	Total float
A	0	0
B	8	0
C	8	3
D	15	0
E	14	3
F	23	0
G	20	3
H	31	0
I	26	3
J	31	4
K	38	0

Table S12.2 Activities ranked in priority early start—total float.

A	F
B	I
C	H
E	J
D	K
G	

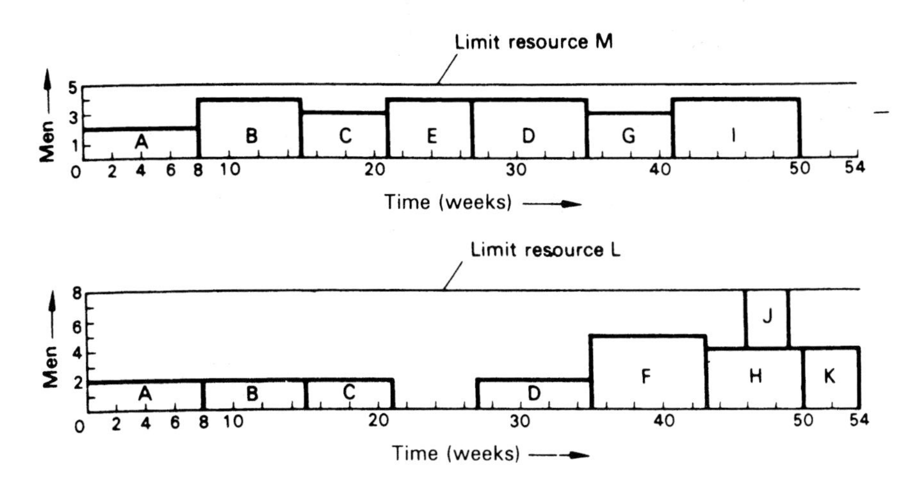

Fig. S12.2 Resource loading charts.

Table S12.3 Scheduled start dates.

Activity	Scheduled start
A	0
B	8
C	15
D	27
E	21
F	35
G	35
H	43
I	41
J	46
K	50

Overall project duration = 54 weeks.

Solution 13 Line of balance

(1) Calculate gang sizes and rates of build (Table S13.1).
(2) Prepare a line of balance schedule (Fig. S13.1).

Table S13.1 Calculation of gang sizes and rates of build (target rate of build = six pylons per week).

Operation	Labour-hours (M)	Theoretical gang size $\left(G = \frac{6 \times M}{40}\right)$	Men per operation (Q)	Actual gang size (in multiples of Q) (g)	Natural rate of build $\left(U = \frac{g}{G} \times 6\right)$	Time per operation (days) $\left(T = \frac{M}{8} \times \frac{1}{Q}\right)$	Elapsed time between start on first pylon and last pylon $\left(S = \frac{(n-1) \times 5}{U}\right)$
A Excavation	55	8.25	4	8	5.8	1.72 → 2	106.03 → 106
B Foundations	64	9.6	4	8	5.0	2.00 → 2	123.00 → 123
C Tower erection	145	21.75	8	24	6.6	2.26 → 3	93.18 → 94
D Cable arms	90	13.5	8	16	7.1	1.41 → 2	86.62 → 87
E Insulators	25	3.75	5	5	8.0	0.63 → 1	76.88 → 77

Note: Buffer = Two days.
→ = Rounding.

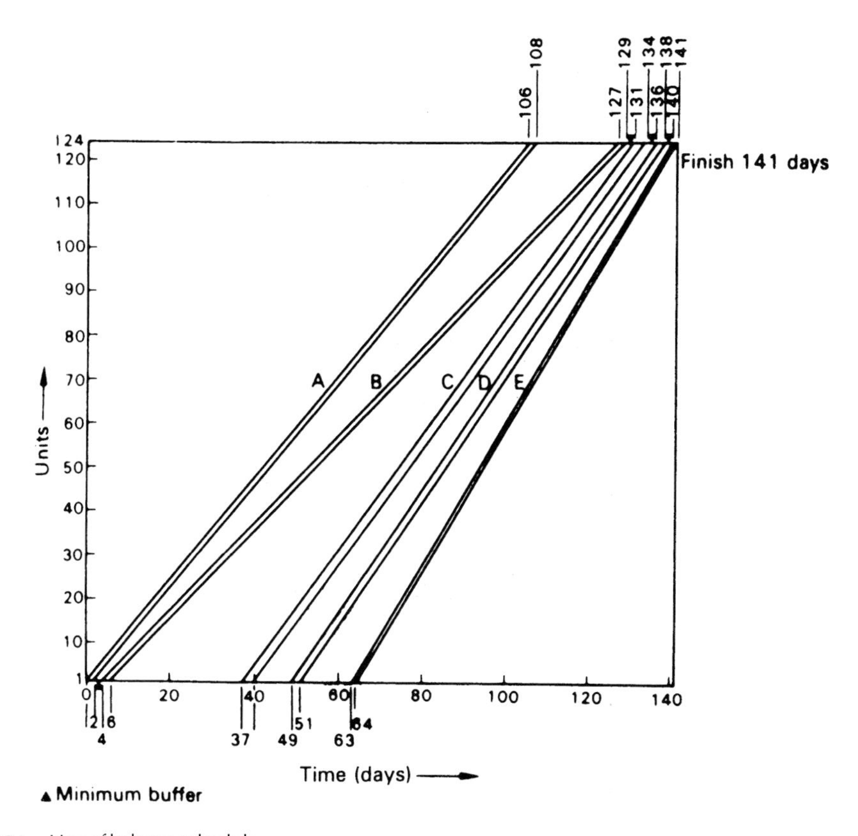

Fig. S13.1 Line of balance schedule.

Solution 14 Line of balance

(1) Calculate gang sizes and rates of build (Table S14.1).
(2) Prepare a line of balance schedule (Fig. S14.1).

Table S14.1 Calculation of gang sizes and rates of build (target rate of build = 4).

Operation	Labour-hours per house (M)	Theoretical gang size $\left(G = \frac{4 \times M}{40}\right)$	Men per house (Q)	Actual gang size (in multiples of Q) (g)	Actual rate per week $\left(U = \frac{g}{G} \times 4\right)$	Time in days for one house $\left(T = \frac{M}{8 \times Q}\right)$	Time between start of first and last houses (days) $\left(S = \frac{(n-1) \times 5}{U}\right)$
U	120	12	3	12	4.0	5.0 → 5	36.25 → 37
V	290	29	6	30	4.14	6.04 → 6	35.02 → 35
W	250	25	4	24	3.84	7.81 → 8	37.76 → 38
X	40	4	3	3	3.0	1.67 → 2	48.33 → 49
Y	30	3	2	2	2.67	1.87 → 2	54.31 → 55
Z	220	22	5	20	3.64	5.5 → 6	39.84 → 40

→ = Rounding.

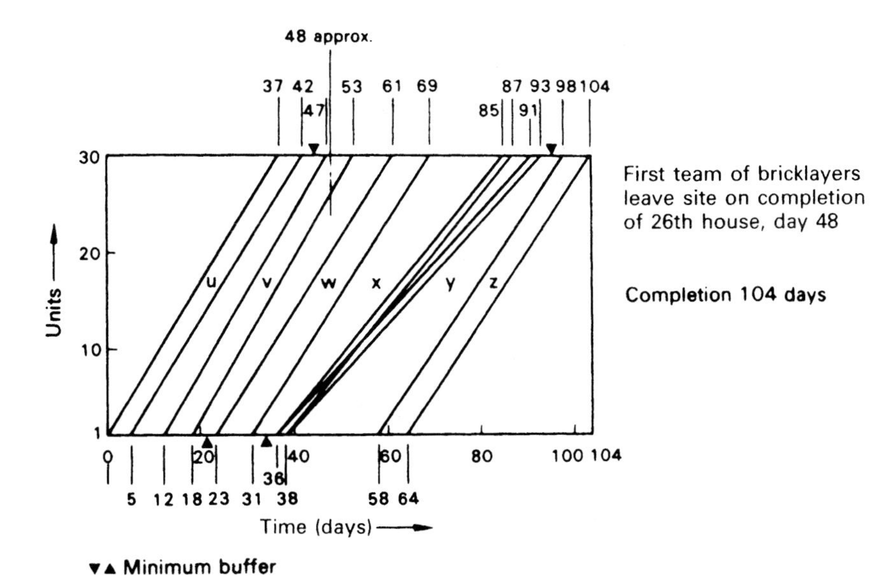

Fig. S14.1 Line of balance schedule.

Solution 15 Line of balance

(1) Calculate gang sizes and rates of build (Table S15.1).
Suggestions to reduce overall project duration:

(a) Use parallel scheduling (may involve overmanning), that is all operations scheduled and manned to operate at the same rate;
(b) Speed up Operation E to a rate of build of 3 (involves overmanning on E only);
(c) Select more than one operation, such as E and D, and bring their rate of build to 3.0 (involves overmanning on the selected operations only).

Table S15.1 Calculation of gang sizes and rates of build (target rate of build = 3).

Operation	Labour-hours per house (M)	Theoretical gang size $\left(G = \frac{3 \times M}{40}\right)$	Men per house (Q)	Actual gang size (in multiples of Q) (g)	Actual rate per week $\left(U = \frac{g}{G} \times 3\right)$	Time in days for 1 unit $\left(T = \frac{M}{8} \times \frac{1}{Q}\right)$	Time between start of first and last houses $\left(S = \frac{(n-1) \times 5}{U}\right)$
A	180	13.5	6	12	2.67	3.75 → 4	26.22 → 27
B	320	24.0	4	24	3.0	10.0 → 10	23.33 → 24
C	200	15.0	4	16	3.2	6.25 → 7	21.88 → 22
D	60	4.5	2	4	2.67	3.75 → 4	26.22 → 27
E	40	3.0	2	2	2.0	2.50 → 3	35.00 → 35
F	120	9.0	3	9	3.0	5.0 → 5	23.33 → 24
G	80	6.0	2	6	3.0	5.0 → 5	23.33 → 24
H	100	7.5	2	8	3.2	6.25 → 7	21.88 → 22
I	40	3.0	3	3	3.0	1.67 → 2	23.33 → 24

→ = Rounding.

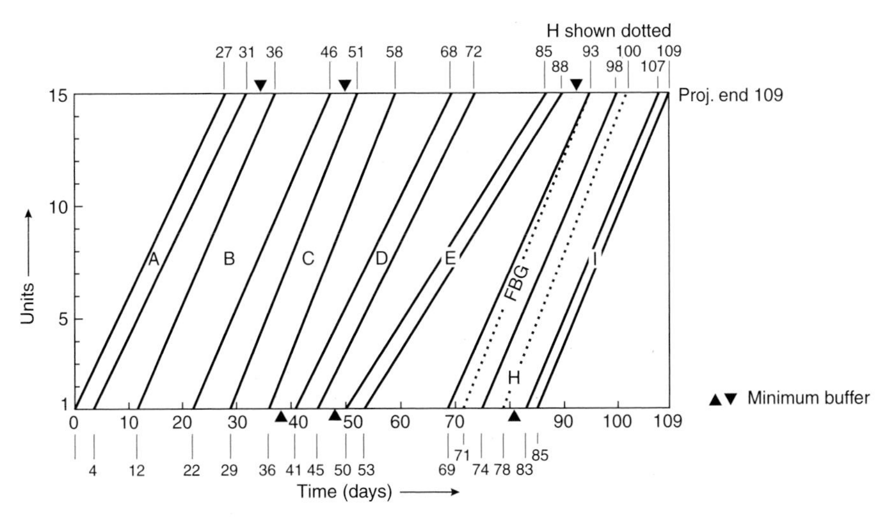

Fig. S15.1 Line of balance schedule.

Solution 16 PERT

Table S16.1 presents the computed expected activity duration and variance for the project generated by the equation below. Figure S16.1 shows the associated network for the project.

$$\text{Expected activity duration } d_e = (d_o + 4d_m + d_p)/6 \quad (1)$$

$$\text{Variance} \quad v = [(d_p - d_o)/6]^2 \quad (2)$$

(1) By undertaking a forward and backward pass for each of the durations for optimistic, pessimistic and most likely, the following is produced:

Duration type	Critical path	Overall project duration
(a) Optimistic	A–C–G–H–K	23 weeks
(b) Pessimistic	A–C–G–H–K	47 weeks
(c) Most likely	A–D–E–F–K	34 weeks

Table S16.1 Expected duration (weeks) and variances for activities.

Activity	Predecessor	Duration (weeks)				
		d_o	d_m	d_p	d_e	σ^2
A		4	5	9	5.50	0.69
B	A	4	8	9	7.50	0.69
C	A	6	7	11	7.50	0.69
D	A	5	9	10	8.50	0.69
E	D	4	6	8	6.00	0.44
F	B, E	2	8	9	7.17	1.36
G	C	5	9	10	8.50	0.69
H	G	3	6	8	5.83	0.69
K	F, H	5	6	9	6.33	0.44

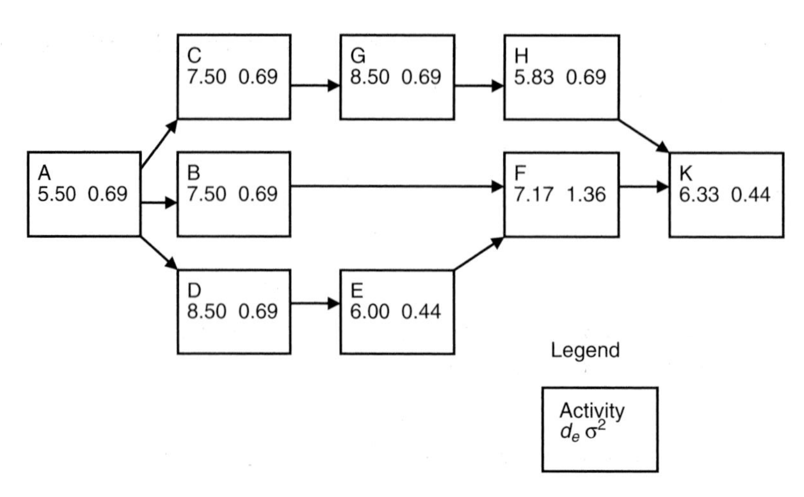

Fig. S16.1 Network of activities for project.

(2) Similarly, by doing a forward and backward pass for the expected duration the following is obtained.

Duration type	Critical path	Overall project duration
Expected	A–C–G–H–K	34.17 weeks

The expected timely completion of the project is therefore 34.17 weeks. Assume that the activities of this project are statistically independent of each other. The probability that the project will be completed within the 35 weeks is established with the equation:

$$Z = (\text{Target duration} - \text{Expected project duration})/(\sqrt{v})$$

$$Z = (\text{TD} - \text{EPD})/(\sqrt{v})$$

Where:

- Target duration (TD) is the desired or negotiated project completion time
- Expected project duration (EPD) is the critical time of the project (defined as the sum of expected activity duration d_e for those activities on the critical path)
- Z is the number of standard deviations of a normal deviation and
- v is the variance of the critical path (the sum of the variances vd_e for all activities on the critical path)

v for the activities on the critical path for the expected duration is $0.69+0.69+0.44+1.36+0.44=3.53$.

For a 35-week contract period based on the programmed project duration of 34.17 weeks with a variance of 3.53 weeks, this yields:

$$Z = (\text{TD} - \text{EPD})/(\sqrt{v})$$

$$Z = (35 - 34.17)/(\sqrt{3.53}) = 0.4418 \text{ standard deviations.}$$

The equivalent probability can be read off a normal probability distribution table.

$P(Z=0.4418)=0.672$, which equals a 67.2% chance of completing the project on time.

(3) A 95% chance of timely completion yields

$$P(Z=\text{TD}-34.17)=0.95$$

Reading off the normal probability tables, this gives

$$\text{TD}-34.17=1.6452$$

$$\text{TD}=34.17+1.6452=35.82 \text{ weeks}$$

The minimum duration to be negotiated for in order to give the project a 95% chance of timely completion should be 36 weeks.

Solution 17 Estimating the cost of a formwork panel

Calculate the cost of material

Volume of softwood timber:

$$=(3\times 2.44\times 0.1\times 0.05)+(5\times 1.22\times 0.1\times 0.05)$$
$$=0.04+0.03=0.07\,\text{m}^3$$

Cost of timber = £295.00/m^3
Wastage = 17.5%
Therefore, cost of timber in one panel:

$$=0.07\times 1.175\times 295.00=£24.26$$

Area of ply (19 mm plywood facing)

$$=2.44\times 1.22=2.98\ (\text{say } 3.0\,\text{m}^2)$$

Cost of ply = £12.65/m^2
Wastage = 17.5%
Therefore, cost of ply in one panel

$$=3.0\times 1.175\times 12.65=£44.59$$

Cost of timber + ply = 24.26 + 44.59 = £68.85
Allowance for nails £1.50/m^2 = £4.50

Therefore, total material cost for one panel = £73.35

Material cost of panel/one use	= £73.35
Material cost of panel/m^2/one use	= £24.45
Material cost of panel/five uses	= £14.67
Material cost of panel/m^2/five uses	= £4.89

Labour cost to make panel

Carpenter gang	= 2 carpenters and 1 labourer
Carpenter cost	= £12.00 per hour
Labourer cost	= £9.00 per hour

Therefore, composite labour rate = $(2 \times 12.00 + 1 \times 9.00)/3 = £11.00$/hour

Labour usage per m^2 of panel = 1.25/h

Therefore, total labour cost per panel = $1.25 \times 3.00 \times £11.00 = £41.25$

Labour cost of panel/one use	= £41.25
Labour cost of panel/m^2/one use	= £13.75
Labour cost of panel/five uses	= £8.25
Labour cost of panel/m^2/five uses	= £2.75

Labour cost to erect panel

Labour cost per panel = $1.00 \times £11.00$	= £11.00
Labour cost per panel/one use	= £11.00
Labour cost of panel/m^2/one use	= £3.67

(*Note*: only one use per erection)

Labour cost to strike panel

Labour cost = $0.55 \times £11.00$	= £6.05
Labour cost per panel/one use	= £6.05
Labour cost of panel/m^2/one use	= £2.02

(*Note*: only one use per strike)

Clean and oil panel

Labour cost per panel = $0.040 \times 3.00 \times £11.00$	= £1.32
Material cost per panel = $0.36 \times 3.00 \times £1.20$	= £1.30
Total cost per panel = 1.32 + 1.30	= £2.62
Total cost per panel/one use	= £2.62
Total cost per panel/m^2/one use	= £0.87

(*Note*: only one use per clean and oil)

Total cost of panel

Total cost per panel for one use of material and labour to make, erect and strike, and clean and oil:

$$= 73.35 + 41.25 + 11.00 + 6.05 + 2.62 = £134.27$$

Total cost per m^2 for one use = £44.76

Total cost per panel for five uses:

$$= 14.67 + 8.25 + 11.00 + 6.05 + 2.62 = £42.59$$

Total cost per m^2 of panel for one use:

$$= 24.45 + 13.75 + 3.67 + 2.02 + 0.87 = £44.76$$

Total cost per m^2 of panel for five uses:

$$= 4.89 + 2.75 + 3.67 + 2.02 + 0.87 = £14.20$$

Solution 18 Estimating the cost of reinforcement

Labour cost for 12 mm diameter is 31 h tonne × £12.00 = £372.00
Repeating the calculation for each diameter:

Bar diameter (mm)	Labour cost per tonne (£)
12	372
16	288
20	240
25	228
32	204
40	180

16 mm diameter and less

The percentage of re-bar for each bar diameter for 16 mm diameter or less is:

Bar diameter (mm)	Weight (tonnes)	Percentage of total
12	1.54	11
16	12.46	89

Weighted average labour rate:

$$= (0.11 \times 372) + (0.89 \times 288)$$
$$= 40.92 + 256.32$$
$$= £297.24 \text{ per tonne}$$

Weighted average material rate:

$$= (0.11 \times 329 \times 1.025) + (0.89 \times 326 \times 1.025)$$
(*Note*: 1.025 is the wastage allowance)
$$= 37.09 + 297.39 = £334.48 \text{ per tonne}$$

Allow for chairs and spacers at 1.5%

$$= 0.015 \times 334.48$$
$$= £5.02 \text{ per tonne}$$

Weighted average cost of binding wire:

$$= (0.11 \times 11 \times 1.56) + (0.89 \times 9 \times 1.56)$$
$$= 1.89 + 12.50 = £14.39 \text{ per tonne}$$

Therefore, the total cost of material per tonne:

$$= 334.48 + 5.02 + 14.39$$
$$= £353.89 \text{ per tonne}$$

Total labour and material rate:

$$= £297.24 + £353.89$$
$$= £651.13$$

Thus, the cost rate for the 16 mm re-bar (or less) is £651.13 per tonne giving a total cost for the item of 14 × 651.13 = £9115.82.

20 mm diameter and greater

The percentages of re-bar for each bar diameter for 20 mm re-bar and greater is shown in the table below.

Bar diameter (mm)	Weight (tonnes)	Percentage of total
20	6.67	29
25	4.14	18
32	8.28	36
40	3.91	17

Weighted average labour rate:

$$\begin{aligned}&= (0.29 \times 240) + (0.18 \times 228) + (0.36 \times 204) + (0.17 \times 180)\\&= 69.60 + 41.04 + 73.44 + 30.60\\&= £214.68 \text{ per tonne}\end{aligned}$$

Weighted average material rate:

$$\begin{aligned}&= (0.29 \times 326 \times 1.025) + (0.18 \times 327 \times 1.025)\\&\quad + (0.36 \times 325 \times 1.025) + (0.17 \times 326 \times 1.025)\\&= 96.90 + 60.33 + 119.93 + 56.81\\&= £333.97 \text{ per tonne}\end{aligned}$$

Allow for chairs and spacers at 1.5%

$$\begin{aligned}&= 0.015 \times 333.97\\&= £5.01 \text{ per tonne}\end{aligned}$$

Weighted average cost of binding wire:

$$\begin{aligned}&= (0.29 \times 7 \times 1.56) + (0.71 \times 5 \times 1.56)\\&= 3.17 + 5.54\\&= £8.71 \text{ per tonne}\end{aligned}$$

Therefore, the total cost of material per tonne:

$$\begin{aligned}&= 333.97 + 5.01 + 8.71\\&= £347.69 \text{ per tonne}\end{aligned}$$

Total labour and material rate:

$$\begin{aligned}&= £214.68 + £347.69\\&= £562.37\end{aligned}$$

Thus, the cost rate for the 20 mm re-bar (or greater) is £562.37 per tonne giving a total cost for the item of 23 × 562.37 = £12 934.51.

Solution 19 Estimating the cost of placing concrete in foundations

Labour costs

Concrete ganger	= £78.00 per day
Labourers (five)	= £72.00 × 5 = £360.00 per day
Total labour cost per day	= £438.00

Therefore, average labour cost per m^3 = £437.00 ÷ 70
= £6.26 per m^3

Plant cost

Crane and skips	= £175.00 per day
Crane driver	= £88.00 per day
Vibrators and small plant	= £30.00 per day
Total plant cost	= £293.00 per day

Therefore, average plant cost per m^3 = £293.00 ÷ 70
= £4.19 per m^3

Material cost

Cost of supplying concrete per m^3	= £70.09
Allowance for wastage at 10%	= £7.01

Therefore total material cost per m^3 = £77.10

Total cost per m^3

Total cost of placing concrete:

= labour cost + plant cost + material cost
= £6.26 + £4.19 + £77.10
= £87.55 per m^3

Solution 20 Estimating cost of placing concrete—operational estimating

Plant costs

These are shown in Table S20.1.
Therefore plant cost per m^3 = 88 200.00/11 250 = £7.84 per m^3

Labour costs

These are shown in Table S20.2
Therefore labour cost per m^3 = 82 350.00/11 250 = £7.32 per m^3

Table S20.1 Plant costs.

Plant item	No.	Dedication (%)	Cost per week (£)	No. of weeks	Cost (£)
22 RB cranes	1	100	800.00	45	36 000.00
22 RB cranes	1	50	800.00	45	18 000.00
Skips	4	100	22.00	45	3 960.00
Dumpers	5	100	84.00	45	18 900.00
Vibrators	6	100	42.00	45	11 340.00
			Total plant cost		88 200.00

Table S20.2 Labour costs.

Item	No.	Cost per week (£)	No. of weeks	Cost (£)
Ganger	1	390.00	45	17 550.00
Labourers	4	360.00	45	64 800.00
		Total labour cost		82 350.00

Therefore total cost of placing 11 250 m^3 of concrete

$$= 88\,200.00 + 82\,350.00 = £170\,550.00$$

and total cost per m^3 of placing concrete

$$= 7.84 + 7.32 = £15.16$$

Hence, £15.16 would be the cost element of the rate entered against Item F723.

Solution 21 Estimating cost of laying a 300 mm concrete pipe

Volume of concrete in bedding

Width of trench = 750 mm
External diameter of pipe = 334 mm
Depth of bedding = 100 mm (minimum)
Cover to pipe = 150 mm (minimum)

Volume of concrete per metre run = overall depth × width − volume of pipe

$$= (0.150 + 0.334 + 0.1) \times 0.75 - 3.142(0.334)^2/4$$
$$= 0.584 \times 0.75 - 0.088 = 0.35\ m^3$$

Material costs

Cost of pipe = £23.56 per metre run with 3% wastage.

Therefore, cost of pipe per metre = 1.03 × 23.56 = £24.27.

Concrete used per metre = 0.35 m^3/metre with 5% wastage. Cost of concrete = £70.09/m^3.

Therefore, cost of concrete surround per metre run

$$= 1.05 \times 0.35 \times 70.09 = £25.76.$$

Total cost of material per metre run = cost of pipe + cost of concrete

$$= 24.27 + 25.76$$
$$= £50.03$$

Excavation volumes and outputs

These are shown in Table S21.1.

Average volume per metre run = 1158.38/925 = 1.25 m^3/m

Average output excavation, pipe-laying and backfilling = 50 linear metres per week. Therefore, the outputs per depth are as shown in Table S21.2.

Table S21.1 Excavation volumes and outputs.

Width (m)	Depth (m)	Length (m)	Volume (m^3)	Volume/m (m^3/m)
0.75	1.50	631	709.88	1.13
0.75	2.00	274	411.00	1.50
0.75	2.50	20	37.50	1.88
Totals		925	1158.38	

Table S21.2 Outputs per depth.

Depth (m)	Length (m)	Vol./metre (m^3/m)	Average vol./metre	Adjusted average output (m/wk)
1.50	631	1.13	1.25/1.13 = 1.11	55.50
2.00	274	1.50	1.25/1.50 = 0.83	41.50
2.50	20	1.88	1.25/1.88 = 0.66	33.00

Durations of pipe-laying

These are shown in Table S21.3.

Table S21.3 Durations of pipe-laying.

Depth (m)	Length (m)	Output per week (m/week)	Duration (weeks)
1.50	631	55.50	631/55.50 = 11.37
2.00	274	41.50	274/41.50 = 6.60
2.50	20	33.00	20/33.00 = 0.61
		Total	18.58 weeks

Labour and plant costs

Gang cost per hour:

Excavator	= £18.64
Small plant	= £2.85
Pipe layer	= £12.00
2 Labourers	= £18.00
Total	= £51.49

Assuming 50 hours per week then weekly gang cost

$$= 50 \times 51.49 = £2574.50$$

Therefore, labour and plant cost for full pipe run

$$= 18.58 \text{ weeks} \times £2574.50 \text{ per week}$$
$$= £47\,834.21$$

Labour and plant cost per metre pipe run

$$= £47\,834.21/925$$
$$= £51.71/\text{m}$$

Labour and plant cost for each depth

These are shown in Table S21.4.

Table S21.4 Labour and plant cost for each depth.

Depth (m)	Duration (weeks)	Cost per week (£)	Total cost (£)	Length (m)	Cost/m (£/m)
1.50	11.37	2574.50	29 272.07	631	46.39
2.00	6.60	2574.50	16 991.70	274	62.01
2.50	0.61	2574.50	1 570.45	20	78.52
			47 834.22		

Labour costs for planking and strutting

These are shown in Table S21.5.
Labour cost is £9.00 per hour.
Therefore, labour costs for planking and strutting $= £9.00 \times 222.40$

$$= £2001.60$$

Average cost per metre run $= 2001.60/925 \qquad = £2.16$

Table S21.5 Labour costs for planking and strutting.

Depth (m)	Length (m)	Labour (h/m)	Total labour (h)
1.50	631	0.20	126.20
2.00	274	0.30	82.20
2.50	20	0.70	14.00
		Total	222.40

Labour costs for planking and strutting by depth

These are shown in Table S21.6.

Table S21.6 Labour costs for planking and strutting by depth.

Depth (m)	Labour (h)	Cost (£/hr)	Total cost (£)	Length (m)	Cost/metre (£/m)
1.50	126.20	9.00	1135.80	631	1.80
2.00	82.20	9.00	739.80	274	2.70
2.50	14.00	9.00	126.00	20	6.30
	222.40		2001.60		

Material costs at £0.95/m² planking and strutting

These are shown in Table S21.7.

Table S21.7 Material costs at 0.95/m² planking and strutting.

Depth (m)	Length (m)	Area (m²)	Total cost (£)	Cost per metre (£/m)
1.50	631	1893	1798.35	2.85
2.00	274	1096	1041.20	3.80
2.50	20	100	95.00	4.75
		3089	2934.55	

Disposal of surplus material

Assume all excavated material is suitable fill and compacts to original volume, then surplus material is equivalent to that displaced by concrete and pipe.

Volume of surplus material for disposal per metre run
$= (0.15 + 0.334 + 0.1) \times 0.75 \times 1\ m^3/m = 0.438\ m^3/m$

Therefore total volume $= 0.438 \times 925 = 405.15\ m^3$

Capacity of tipper truck $= 4\ m^3$

Therefore, number of loads $= 405.15/4 = 102$

Cycle time for loading, travel, discharge and return = 21 min.
Therefore total time required of tipper trucks $= 102 \times 21$ min
$= 2142$ min
$= 35.7$ h

Cost of tipper truck including driver = £32.16/h. Therefore cost of disposing of surplus material $= 35.7 \times 32.16 =$ £1148.11.

Cost of disposing surplus material per metre = 1148.11/925
= £1.24/m

The total cost per metre run and total cost are shown in Tables S21.8 and S21.9.

Table S21.8 Total cost per metre run.

Depth (m)	Length (m)	Pipe and concrete (£/m)	Labour and plant (£/m)	Labour, planking and strutting (£/m)	Material, planking and strutting (£/m)	Disposal of surplus material (£/m)	Total (£/m)
1.50	631	50.03	46.39	1.80	2.85	1.24	102.27
2.00	274	50.03	62.01	2.70	3.80	1.24	119.74
2.50	20	50.03	78.52	6.30	4.75	1.24	140.80

Table S21.9 Total costs.

Depth (m)	Length (m)	Pipe and concrete (£)	Labour, and plant (£)	Labour, planking and strutting (£)	Material, planking and strutting (£)	Disposal of surplus material (£)	Total (£)
1.50	631	31 568.93	29 272.07	1135.80	1798.35	782.44	64 557.59
2.00	274	13 708.22	16 991.70	739.80	1041.20	339.76	32 820.68
2.50	20	1 000.60	1 570.45	126.00	95.00	24.80	2 816.85
Totals		46 277.75	47 834.22	2001.60	2934.55	1147.00	100 195.12

Solution 22 Hours-saved incentive scheme

(1) 75–100 scheme (direct scheme)

Standard performance = 100P. Bonus = 33⅓% of basic wage for working at standard performance. Total time allowed for doing the job = 1 hour × 24 = 24 hours when working at 75P.

Let the time actually taken at standard performance = y hours.

Figure S22.1 shows the earnings saved will be straight proportional to effort and thus the gradient is 1, i.e. (33.3/100)/(25/75).

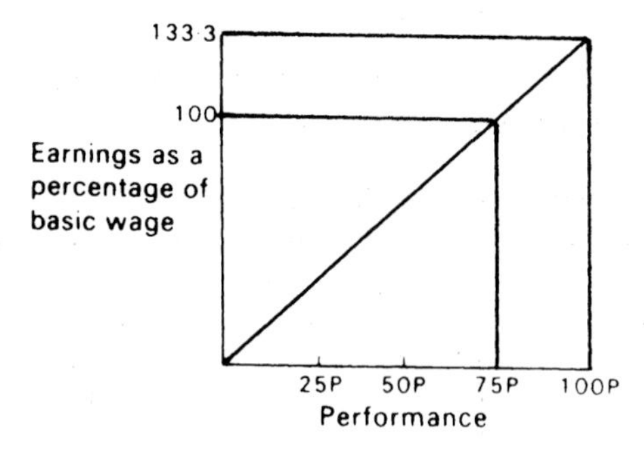

Fig. S22.1 75–100P scheme.

Table S22.1 Effects on earnings and the cost of the job.

Rate of working	Actual time taken (h)	Hours saved	Bonus (£)	Hourly earnings	Cost of job (£)
120P	15	9	9	1.60	24
100P	18	6	6	1.33	24
75P	24	0	0	1.00	24
50P	36	0	0	1.00	36

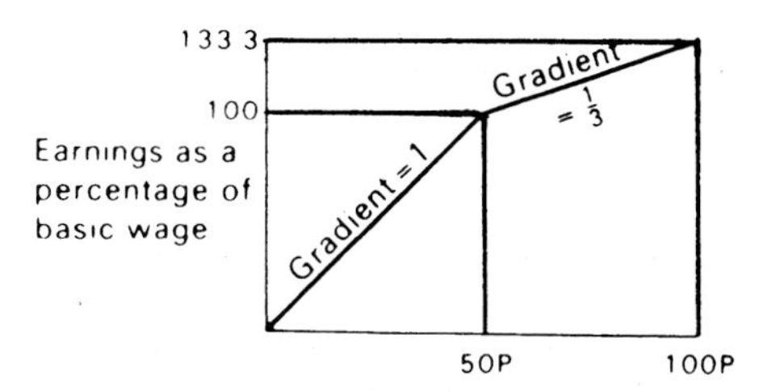

Fig. S22.2 50–100P scheme.

Time allowed = 24

Thus $24 = y + (⅓ \times y \times 1)$

$4y = 3 \times 24$

$y = 18\,\text{h}$

Therefore, bonus starts at $100\text{P} \times (18 \div 24) = 75\text{P}$.

The site manager estimates that if the carpenter works at standard performance the whole job will be completed in 18 labour-hours. In so doing the carpenter will earn six hours' bonus. Table S22.1 shows the effects on earnings and the cost of the job for performances of 50P, 75P, 100P and 120P.

(2) 50–100 scheme (geared scheme)
It is clear from Fig. S22.2 that the earnings are not straight proportional to effort. If bonus is to remain at £6 for working at standard performance, but is to commence at 50P, each hour saved must be paid at $6/(36 - 18) = £0.33$. (The gradient of the graph is 33/100.)
The allowed hours are proportionately increased as follows:

Allowed hours = actual hours × ⅓ × actual hours × 1/0.33
Allowed hours = 18 × ⅓ × actual hours × 1/0.33
= 36 hours

Table 22.2 shows the effects on earnings and cost of the job in the new situation.
Hours saved at standard performance = 18 hours.

Table S22.2 Effects on earnings and the cost of the job in the new situation.

Rate of working	Actual time taken (h)	Hours saved (h)	Bonus (£)	Hourly earnings (£)	Cost of job (£)
120P	15	21	7	1.46	22
100P	18	18	6	1.33	24
75P	24	12	4	1.16	28
50P	36	–	–	1.00	36

Solution 23 Bonus scheme with plant

Production during week = 4000 m^3. Therefore target hours for this amount of production = 4000/50 = 80 h or 3 × 80 = 240 labour-hours.

The actual time spent by the workers = 157 − 7 = 150 labour-hours.

Target = 240 h
Actual = 150 h
Saving = 90 h at 0.5 × £1.00 = £45.00

Shares

Machine driver	58 × 1½	= 87
Labourer No. 1	58 × 1	= 58
Labourer No. 2	34 × 1	= 34
Total shares		= 179

Earnings per share = 45.0 ÷ 179 = £0.25

Bonus paid

Machine driver	87 × 0.25	= £21.8
Labourer No. 1	58 × 0.25	= £14.6
Labourer No. 2	34 × 0.25	= £ 8.6
Total bonus		= £45.00

Solution 24 Gang bonus payments

In a good bonus system, the worker should not be penalised for stoppages due to inclement weather or other unforeseen delays. Therefore to arrive at the actual time spent by the gang on the shuttering operations, four hours must be deducted from the hours booked against each worker, for the stoppage on the second day.

Charge-hand	actual hours 10 + 6	= 16
Carpenter No. 1	actual hours 10 + 6	= 16
Carpenter No. 2	actual hours 8 + 4	= 12
Labourer No. 1	actual hours 10 + 6	= 16
Labourer No. 2	actual hours 10	= 10
Total		= 70 hours

Hours saved = 100 − 70 = 30 h. Basic rate of payment is £1.00 per hour.
Bonus earned by the gang = 30 × 1.00 = £30.00.

Total shares = 16 × 1½ + 16 × 1¼ + 12 × 1¼ + 16 × 1 + 10 × 1
= 24 + 20+ 15 + 16 + 10
= 85

Bonus per share = 30 ÷ 85 = £0.353

Final bonus

Charge-hand	$0.353 \times 24 =$	£8.472
Carpenter No. 1	$0.353 \times 20 =$	£7.060
Carpenter No. 2	$0.353 \times 15 =$	£5.290
Labourer No. 1	$0.353 \times 16 =$	£5.648
Labourer No. 2	$0.353 \times 10 =$	£3.530
Total		= £30.000

Basic earnings

	Paid at basic rate	*Paid at overtime rate*		
Charge-hand	$2 \times 8 \times 1.00 +$	$2 \times 2 \times 1.00 \times 1\frac{1}{4} =$	$16 + 5$	= £21.0
Carpenter No. 1	$2 \times 8 \times 1.00 +$	$2 \times 2 \times 1.00 \times 1\frac{1}{4} =$	$16 + 5$	= £21.0
Carpenter No. 2	$2 \times 8 \times 1.00$		$= 16$	= £16.0
Labourer No. 1	$2 \times 8 \times 1.00 +$	$2 \times 2 \times 1.00 \times 1\frac{1}{4} =$	$16 + 5$	= £21.0
Labourer No. 2	$1 \times 8 \times 1.00 +$	$1 \times 2 \times 1.00 \times 1\frac{1}{4} =$	$8 + 2.5$	= £10.5
	Total			= £89.5

Total earnings

Charge-hand	$8.472 + 21.000 =$	£29.472
Carpenter No. 1	$7.060 + 21.000 =$	£28.060
Carpenter No. 2	$5.290 + 16.000 =$	£21.290
Labourer No. 1	$5.648 + 21.000 =$	£26.648
Labourer No. 2	$3.530 + 10.500 =$	£14.030
	Total =	£119.500

Therefore

Average earnings per hour = $119.5 \div 86$ = £1.39
Average bonus per hour = $30 \div 86$ = £0.35

Solution 25 Control of project costs

Table Q25.2 shows the budgeted values of each activity, which is further subdivided among each of the cost headings. If it can be assumed that the planned total expenditure on each activity is uniform with time, the summation of the weekly values for each activity will yield the budget curve. Figure S25.1 shows the contract programme in bar-chart form with the planned expenditure allocated. The weekly budgeted totals are plotted graphically on a cumulative basis as shown in Fig. S25.2.

(1) Table Q25.3 gives the state of progress up to the end of Week 37, together with the projected completion times for each activity. By interpreting this table in the form of a bar-chart as shown in Fig. S25.3, the percentages complete are determined. Once again it is assumed that expenditure is uniform with time. The value of work done can be calculated by using the

updated bar-chart and the budgeted values. The value to date curve is shown in Fig. S25.2. The value of work done in each of the individual cost headings for each activity is computed on a similar basis and summarised in Table S25.1. This table should compare with Fig. S25.2.

(2) The plot of the value and cost curves shows that the contract was doing better than planned up to Week 24. Since then the project has been losing money. Clearly from the table of variances, Cost Code 10, the excavation work, needs attention, but unfortunately this work is almost complete. Only Code 40 has sufficient work left on which to take corrective action. The adverse variance of £3270 on site overheads is probably due to the project being six weeks behind schedule; therefore an attempt to shorten this overrun may reduce the potential loss.

Bar chart for contract programme showing budgeted expenditure

W/E / Act'y	4	8	12	16	20	24	28	32	36	40	44	48	52
(A)					£ 9400/week								
(B)						£ 11400/week							
(C)									£ 10833/week				
(D)											£ 4454/week		
(E)											£ 7000/week		
(F)									£ 4880/week				
(G)											£ 2000/week		
Monthly	37600	83200	103840	112153	116506	116668	116668	116668	74409	53816	22454		
Cumul.	37600	120800	224640	336793	453299	569967	686635	803303	87712	931528	954000		

Fig. S25.1 Bar-chart for contract programme showing budgetal expenditure.

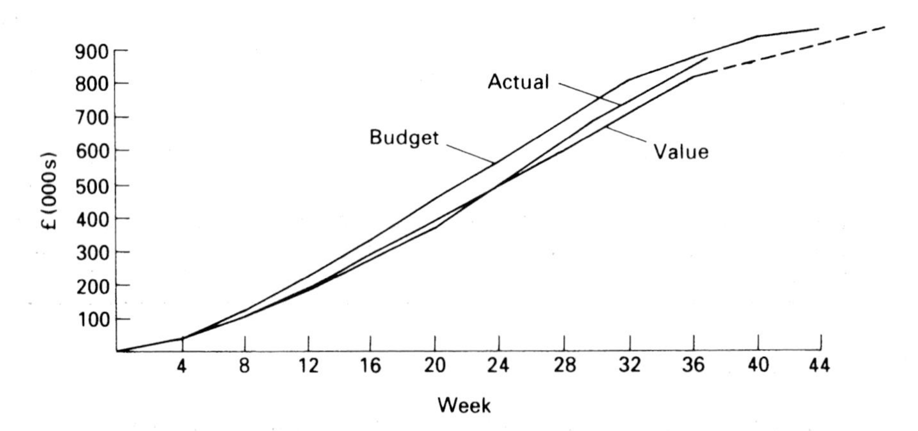

Fig. S25.2 Budget value and cost curves.

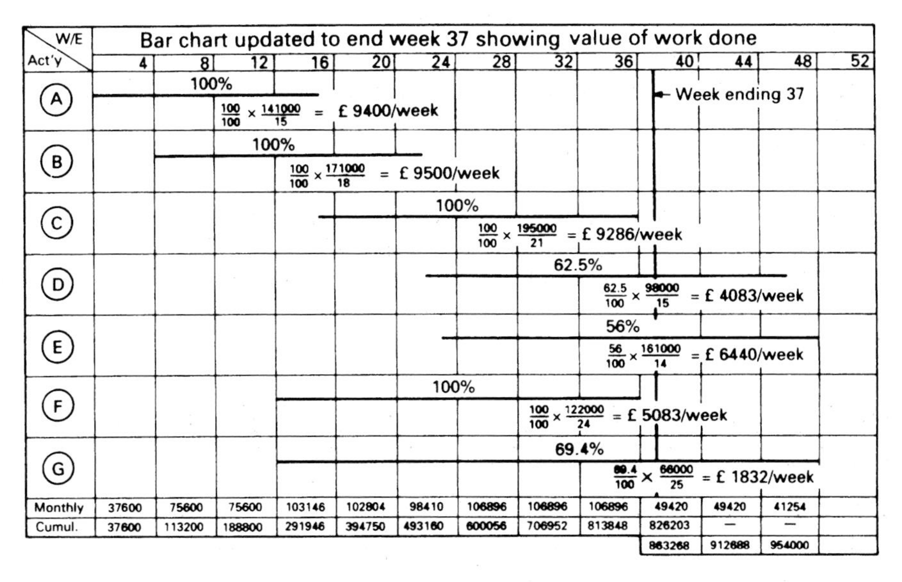

Fig. S25.3 Bar-chart updated to end of Week 37 showing value of work done.

Table S25.1 Value of work done up to the end of Week 37 (£000).

Activity	Complete (%)	Cost code 10	20	30	40	50	60	70	80	Total (£000)
A	100.0	112.00	–	–	–	–	–	–	29.00	141.00
B	100.0	–	30.0	–	72.00	–	34.00	–	35.00	171.00
C	100.0	–	48.0	–	79.00	–	28.00	–	40.00	195.00
D	62.5	–	12.5	–	26.25	–	18.75	–	3.75	61.25
E	56.0	–	16.8	–	30.24	–	24.08	–	19.04	90.16
F	100.0	30.00	–	32.00	–	20.00	–	20.00	20.00	122.00
G	69.4	6.94	–	6.94	–	13.88	–	11.10	6.94	45.80
	Value	148.94	107.30	38.94	207.49	33.88	104.83	31.10	153.73	826.21
	Actual	155.00	100.00	40.00	210.00	31.00	104.00	33.00	157.00	830.00
	Variance	−6.06	+7.30	−1.06	−2.51	+2.88	+0.83	−1.90	−3.27	−3.79

Solution 26 Budgetary control

(1) At the beginning of the year, the budget for receipts is as shown in Table S26.1.

Budgeted profit = £10 000 + £28 000 + £6 000 + £9 000
= £53 000

Budgeted HO overhead = £8 000 + £20 000 + £6 000 + £4 500
= £38 500

} £91 500 at end of Dec

(2) Table S26.2 shows the position at 30 June on each contract.

Table S26.1 Budgeted receipts.

Period	Contract 1 (£)	Contract 2 (£)	Contract 3 (£)	Contract 4 (£)	Total (£)
1 Jan	15000	40000	10000	–	65000
Feb	15000	40000	10000	–	65000
Mar	15000	40000	10000	–	65000
Apr	15000	40000	10000	15000	80000
May	15000	40000	10000	15000	80000
Jun	15000	40000	10000	15000	80000
Jul	15000	40000	10000	15000	435000
Aug	15000	40000	10000	15000	80000
Sep	15000	40000	10000	15000	80000
Oct	15000	40000	10000	15000	80000
Nov	15000	40000	10000	15000	80000
Dec	15000	40000	10000	15000	80000
Total	180000	480000	120000	135000	915000
1 Jan the following year				15000	
Feb				15000	
Mar				15000	
Total				180000	

Table S26.2 Table of variances in £s.

	Budgeted for this date (£)	Value of work done (£)	Actual cost (£)	Variance (£)
Contract 1				
Direct	81000	81000	84000	−3000
HO overhead	4000	4000	4000*	0
Profit	5000	5000	5000*	0
Total	90000	90000	93000	−£3000
Contract 2				
Direct	216000	216000	220000	−4000
HO overhead	10000	10000	10000*	0
Profit	14000	14000	14000*	0
Total	240000	240000	244000	−£4000
Contract 3				
Direct	54000	45000	40000	+5000
HO overhead	3000	2500	3000*	−500
Profit	3000	2500	3000*	−500
Total	60000	50000	46000	+£4000

* Fixed costs allocated by Head Office.

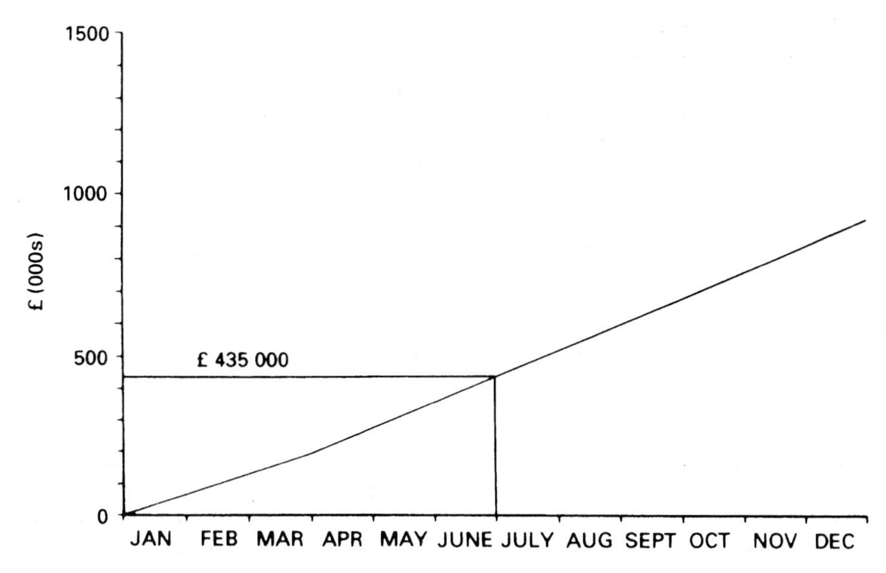

Fig. S26.1 Monthly cumulative budgeted turnover.

Figure S26.1 shows the monthly cumulative budgeted turnover.

The budgeted sales to the end of June are £435 000 (from Fig. S25.1).
However, Contract 4 did not materialise; therefore the budget that can be realised by this date is only £390 000, i.e. (90 000 + 240 000 + 60 000).

Shortfall = £45 000

This shortfall will mean that head office overheads and profits will be under-recovered by:

Sales variance = 10% of £45 000
= £4500

Total costs to date:

Direct costs	= 84 000 + 220 000 + 40 000	= £344 000
HO overheads	= 5000 + 5000 + 5000	= £15 000
		= £359 000
Total receipts	= 90 000 + 240 000 + 50 000	= £380 000
Company profit to date		= +£21 000

Reconciliation

Budgeted profit to date from Table Q26.1
(i.e. (10 000 × ½ + 28 000 + 6000 × ½ + 12 000 × ¼)) + £25 000

Total contract variances from Table S26.2:

Contract 1 = −3000
Contract 2 = −4000
Contract 3 = +4000
= −£3000

Therefore, total variance for all contracts	= −£3 000
Sales variance	= −£4 500
Budgeted HO overheads (4000 + 10 000 + 3000 + 1500)	= £18 500
Actual HO overheads	= £15 000
Therefore, HO overhead variance	= +£3 500
Company profit at 30 June	= +£21 000

Alternative summary of variances

Contract variance	Operating variance	Volume variance	Total
1	−3000	0	−3000
2	−4000	0	−4000
3	+5000	−1000	+4000
	−£2000	−£1000	−£3000

(3) Management action should concentrate on obtaining a further contract as soon as possible. In addition, Contracts 1 and 2 show losses on direct costs with half the work completed. Clearly some attention is required. Further observation reveals that considerably less overheads have been expended by head office with respect to Contract 2 than was budgeted; perhaps more vigorous control from head office might improve the profitability of this project. Contract 3 is slightly behind schedule.

Solution 27 Cash flow

(1) Calculate time of each activity (Fig. S27.1).
(2) Calculation of interest charges.

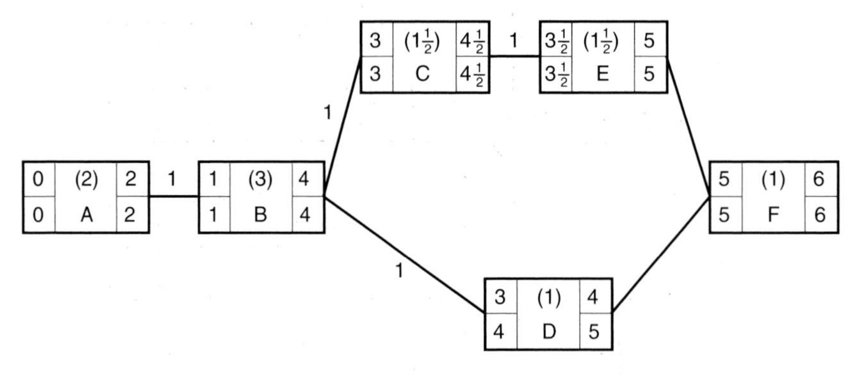

Fig. S27.1 Precedence diagram of small workshop with activity times calculated.

Table S27.1 Cash flow calculations.

Months	1	2	3	4	5	6	7	12
Monthly value	4500	8500	4000	3 3000	10 000	20 000		
Cumulative value	4500	13 000	17 000	50 000	60 000	80 000		
Cumulative profit	450	1 300	1 700	5 000	6 000	8 000		
Cumulative cost	4050	11 700	15 300	45 000	54 000	72 000		
Cumulative value less retention	4275	12 350	16 150	47 500	57 000	77 000		
Cumulative monies received		4 275	12 350	16 150	47 500	57 000	78 500	80 000

Activities	Months 1	2	3	4	5	6
A	4500	4500				
B		4000	4000	4000		
C				12000	6000	
D				15000		
E				2000	4000	
F						20000
Value	4500	8500	4000	33000	10000	20000

Fig. S27.2 Bar-chart of activities and monthly value calculations.

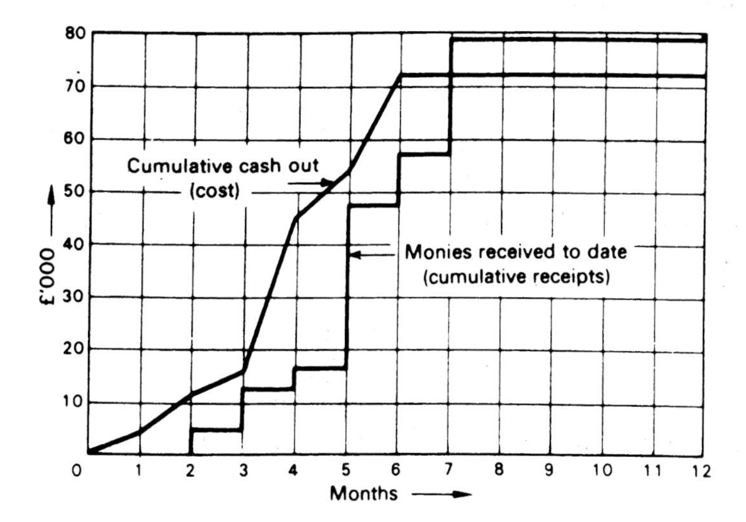

Fig. S27.3 Cash in and cash out.

Finance area (area between cash in and cash out curves) = 10.0* × (10 000 × 1 month).
Interest = 8%.

$$\text{Interest charge} = 10.00 \times 10\,000 \times {}^{1}/_{12} \times 0.08 = £666.67$$

* 10.0 represents the number of squares in this area. Each square is £10 000 × 1 month.

(3) Maximum cash needed

Maximum cash needed = £37 850 immediately before payment is received at Month 5.

Solution 28 Cash flow

Table S28.1 and Fig. S28.1 show the calculation of cash in and out.

Table S28.1 Calculation of cash in and out.

Months	1	2	3	4	5	6	7	8	9	10	15
Costs (£000)	3	3	4	6	6	8	6	4	2	–	–
Cumulative costs	3	6	10	16	22	30	36	40	42		
Cumulative contribution	0.24	0.48	0.80	1.28	1.76	2.40	2.88	3.20	3.36		
Cumulative value	3.24	6.48	10.80	17.28	23.76	32.40	38.88	43.20	45.36		
Cumulative value less retention	2.91	5.83	9.72	15.55	21.38	29.16	34.99	38.88	40.82		
Cumulative receipts		2.91	5.83	9.72	15.55	21.38	29.16	34.99	38.88	40.82	45.36
Cumulative payments†		3	6	10	16	22	30	36	40	42	

Note: †Calculation of payment delay:

Payment delay: 1 × 1.12 = +1.12
+1 × 8.20 = +8.20
+2 × 4.00 = +8.00
Total = 17.32

$$\text{Weighted mean} = \frac{17.32}{4} = 4.33 \text{ weeks} = 1 \text{ month.}$$

Cost of locked-up capital

Finance area (area between cash in and cash out curves)
= 6.40* × (£5000 × 1 month) = 6.40 × 5000 £months.
Interest = 14%.

$$\therefore \text{Interest charges} = 6.40 \times 5000 \times {}^{1}/_{12} \times 0.14 = £373.33$$

$$\therefore \text{Net profit} = \text{Gross profit} - \text{interest charge} = £3360 - £373.33 = £2986.67$$

* 6.40 is the number of squares of £5000 × 1 month taken from the graph (Fig. S28.1).

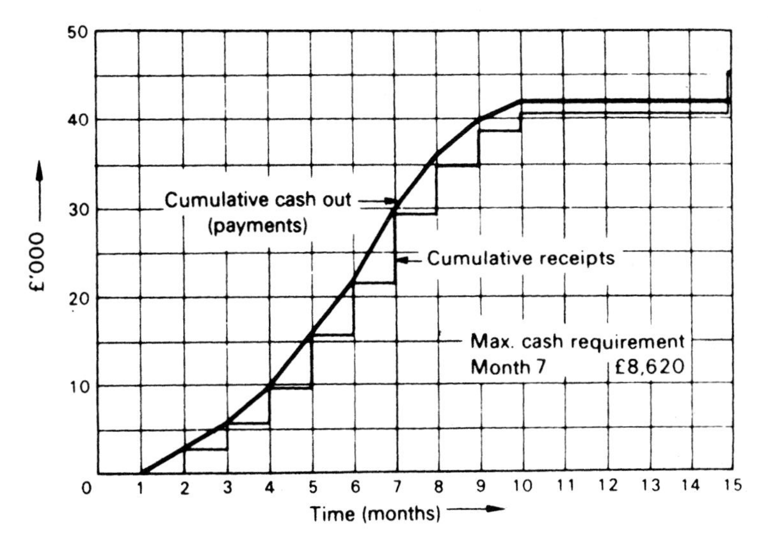

Fig. S28.1 Cash in and cash out.

Solution 29 Cash flow

Table S29.1 shows the calculation of cash out and cash in, and Fig. S29.1 shows the cumulative cash flows.

Calculation of interest charges

Borrowings (area under negative part of cash flow curve in Fig. S29.1)

$$= 0.08 \times 1 + 0.2 \times 1 + 0.36 \times 1 + 0.68 \times 1 + 1.04 \times 1 + 1.4 \times 1 + 1.72 \times 3$$

$$= 8.92 \times 1000 \text{ £months}$$

$$\text{Interest charges} = 8.92 \times 1000 \times 1/12 \times 12\%$$

$$= £89.20$$

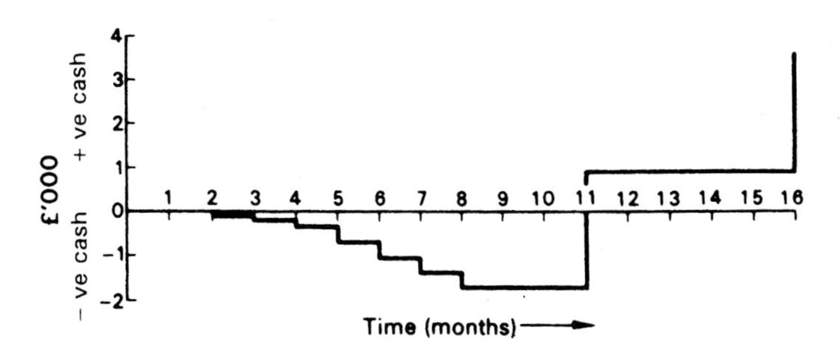

Fig. S29.1 Cumulative cash flows.

Table S29.1 Calculation of cash out and cash in.

Month	1	2	3	4	5	6	7	8	9	10	11	16
Value (£000)	2	3	4	8	9	9	8	5	4	2		
Cumulative value	2	5	9	17	26	35	43	48	52	54		
Cumulative value less retention	1.8	4.5	8.1	15.3	23.4	31.5	38.7	43.2	46.8	48.6		
Cumulative monies received		1.8	4.5	8.1	15.3	23.4	31.5	38.7	43.2	46.8	48.6	48.6
Cumulative retentions repaid											2.7	5.4
Cost (£000)*	1.88	2.82	3.76	7.52	8.46	8.46	7.52	4.5	3.6	1.8		
Cumulative cost	1.88	4.7	8.46	15.98	24.44	32.90	40.42	44.92	48.52	50.32		
Cumulative cash out		1.88	4.7	8.46	15.98	24.44	32.90	40.42	44.92	48.52	50.32	50.32
Cumulative cash flow†		−0.08	−0.2	−0.36	−0.68	−1.04	−1.4	−1.72	−1.72	−1.72	0.98	3.68

* Value less profit %.
† Cumulative monies received + cumulative retentions repaid − cumulative cash out.

Solution 30 Cash flow

Table S30.1 shows the cash in and cash out calculations. Figure S30.1 illustrates these calculations based on one- and two-month measurement intervals.

Maximum cash needed for monthly measurement is £6850 at Months 6 and 7 immediately before payment is received.

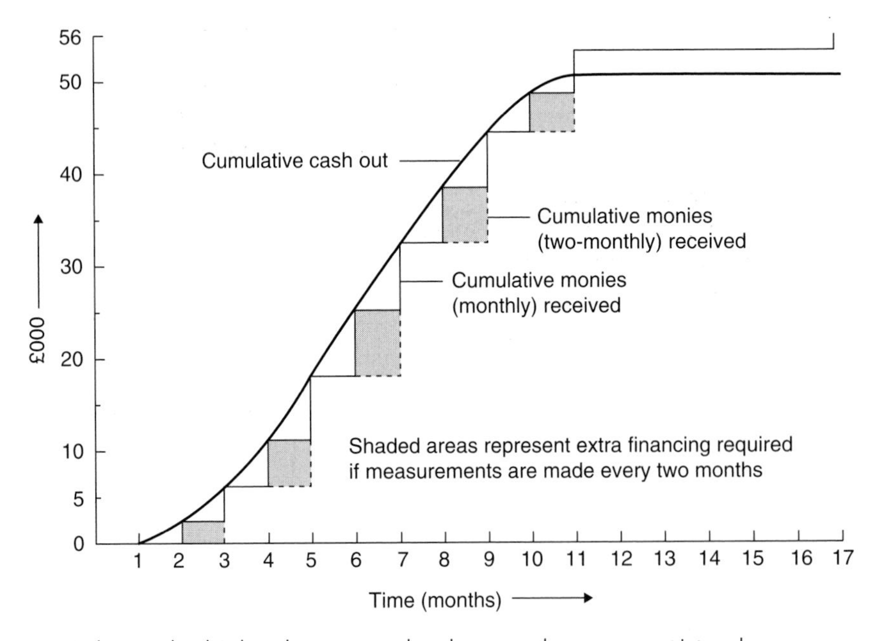

Fig. S30.1 Cash out and cash in based on one-month and two-month measurement intervals.

Table S30.1 Cash in and cash out calculations.

Month	1	2	3	4	5	6	7	8	9	10	11	17
Value (£000)	3	4	5	8	8	8	7	6	5	2		
Cumulative value	3	7	12	20	28	36	43	49	54	56		
Cumulative value	2.70	6.30	10.80	18.00	25.20	32.40	38.70	44.10	48.60	50.40		
less retention		2.70	6.30	10.80	18.00	25.20	32.40	38.70	44.10	48.60	50.40	
Cumulative monies received on monthly measurement											+2.8* 53.20	+2.8* 56.00
Cumulative monies received on a two-monthly measurement			6.30		18.00		32.40		44.10		53.20	56.00
Profit distribution (%)	15	15	10	10	10	10	10	10	5	5		
Cost/month	2.55	3.40	4.50	7.20	7.20	7.20	6.30	5.40	4.75	1.90		
Cumulative cost	2.55	5.95	10.45	17.65	24.85	32.05	38.35	43.75	48.50	50.40		
Cumulative cash out		2.55	5.95	10.45	17.65	24.85	32.05	38.35	43.75	48.50	50.40	

* Retention payments.

Maximum cash needed for two-month measurements is £14 050 at Month 7 immediately before payment is received.

Calculating the interest charge on the extra finance needed to fund a two-monthly measurement:

$$\begin{aligned}\text{Finance area (shaded on Fig. S30.1)} &= 2.55\times 1\\ &+(10.80-6.30)\times 1\\ &+(25.20-18.00)\times 1\\ &+(38.70-32.40)\times 1\\ &+(48.60-44.10)\times 1\\ &=2.55+4.50+7.20+6.30+4.50\\ &=25.05\ (\text{£months}\times 1000)\end{aligned}$$

Interest rate = 15% p.a.

$\therefore$ Interest charges on extra funding

$= 25.05\times 1000\times {}^{1}\!/_{12}\times 0.15$

$= £313.13$

Solution 31 Cash flow

Table S31.1 shows the cash in and cash out calculations.

Month	1	2	3	4	5	6	7	8	9	10	12	15
Monthly estimated costs	2	2	3	3	4	4	4	3	2			
Cumulative estimated costs	2	4	7	10	14	18	22	25	27			
Cumulative costs increased by 4%	2.08	4.16	7.28	10.40	14.56	18.72	22.88	26.00	28.08			
Cumulative cash out		2.08	4.16	7.28	10.40	14.56	18.72	22.88	26.00	28.08		
Measured monthly value	2.16	2.16	3.24	3.24	4.32	4.32	4.32	3.24	2.16			
Cumulative monthly measured value	2.16	4.32	7.56	10.80	15.12	19.44	23.76	27.00	29.16			
Cumulative monthly measured value less retention	1.94	3.89	6.80	9.72	13.61	17.50	21.38	24.30	26.24			
Cumulative monies received		1.94	3.89	6.80	9.72	13.61	17.50	21.38	24.30	26.24		
Retention repaid												2.92
Claims paid											1.12	
Cumulative cash flow*	0	−0.14	−0.27	−0.48	−0.68	−0.95	−1.22	−1.50	−1.70	−1.84	−0.72	+2.20

* Cumulative cash flow = cumulative monies received + retention repaid + claims paid − cumulative cash out.

Solution 32 Present worth comparison

Calculate the present worth of each scheme using 15%

Scheme A

Present worth of installation cost = £18 250.00
Present worth of £7250 each year
for 20 years = 7250(6.2593) = £45 379.93
Total present worth = £63 629.93

(*Note*: 6.2593 is the present worth for a uniform series of 20 years taken from tables or calculated as $\frac{(1+i)^n-1}{i(1+i)^n}$, where $i = 15\%$ and $n = 20$ years.)

Scheme B

Present worth of installation cost = £20 200.00
Present worth of £4600 each year
for 20 years = 4600(6.2593) = £28 792.78
Total present worth = £48 992.78

Scheme C

Present worth of installation cost = £24 000.00
Present worth of £4000 each year
for 20 years = 4000(6.2593) = £25 037.20
Total present worth = £49 037.20

Therefore at 15% Scheme B is the most economical because it has the smallest present worth.

Repeating the calculations at 10%

Scheme A

Present worth of installation cost = £18 250.00
Present worth of £7250 each year
for 20 years = 7250(8.5135) = £61 722.88
Total present worth = £79 972.88

(*Note*: 8.5135 is the present worth factor for a uniform series as explained above with $i = 10\%$.)

Scheme B

Present worth of installation cost = £20 200.00
Present worth of £4600 each year
for 20 years = 4600(8.5135) = £39 162.10
Total present worth = £59 362.10

Scheme C

Present worth of installation cost = £24 000.00
Present worth of £4000 each year
for 20 years = 4000(8.5135) = £34 054.00
Total present worth = £58 054.00

At 10% Scheme C is the most economical because it has the smallest present worth.

The difference between these two calculations is the interest rate. Scheme C requires an extra investment of £3800 in comparison with B. The saving in running costs is £600 per year. When the interest rate is at 15% the savings in running costs are not sufficient to justify the extra expenditure, i.e. the extra expenditure can earn more than £600 per year. When the interest rate is only 10% the extra expenditure cannot earn more than £600 per year and would therefore be better employed earning this saving.

Solution 33 Present worth comparison

Calculation of present worth of the costs of the bridges

Bridge A

	Present Worth
Initial cost of bridge	= £65 000.00
Initial cost of roads	= £35 000.00
Maintenance for 60 years (500 + 300) × (11.0479) =	£8 838.32

(*Note*: both road and bridge last 60 years, 11.0479 is the present worth factor for a uniform series of 60 years with an interest rate of 9%, taken from tables or calculated as $\frac{(1+i)^n - 1}{i(1+i)^n}$, where $i = 9\%$ and $n = 60$ years.)

Present worth of Bridge A = £108 838.32

Bridge B

	Present Worth
Initial cost of bridge	= £50 000.00
Initial cost of roads	= £30 000.00
Maintenance for 30 years	
900×10.2736	= £9246.24
$200 \times 8.0606 \times 0.27453$	= £442.58
250×10.2736	= £2568.40

(*Note*: 11.0479 is the present worth factor for a uniform series of 30 years calculated as above. 8.0606 is the present worth factor for a uniform series of 15 years calculated as above. Since the £200 (i.e. 1100 – 900) used in this calculation runs from Years 16 to 30 the factor gives a present worth value in Year 15, at an interest rate of 9%; 0.27453 is the present worth factor for Year 15 taken from tables or calculated as $\frac{1}{(1+i)^n}$, where $i = 9\%$ and $n = 15$ years.)

Present worth of Bridge B lasting 30 years = £92 257.22

This present worth of £92 257.22 for Bridge B cannot be compared to the present worth of £108 838.32 for Bridge A because 'A' was calculated for a 60-year life and 'B' was calculated for a 30-year life. Therefore to make the comparison a replacement for 'B' must be included to make the life of 'B' and its replacement 60 years.

Replacement

	Present Worth
Initial cost of bridge	= £50 000.00
Initial cost of roads (roads from first 30 years had a 60-year life)	= £0
Maintenance of bridge and roads for 30 years	
= £9246.24	= £9246.24
£442.58 } as before	= £442.58
£2568.40	= £2568.40
Present worth for replacing Bridge B	= £62 257.22

This £62 257.52 refers to the present worth of the replacement at the start of the second 30 years.

∴ Present worth of the replacement at the beginning
$= 62\,257.22 \times 0.07537 = £4692.33$

(*Note*: 0.07537 is the present worth factor for Year 30 at an interest rate of 9%, taken from tables or calculated as $\frac{1}{(1+i)^n}$, where $i=9\%$ and $n=30$ years.)

$\therefore$ Present worth of Bridge B and its replacement
$= 92\,257.22 + 4692.33 = \underline{£96\,949.55}$

Thus Bridge B is the most economic because it has the smallest present worth.

Solution 34 Equivalent annual cost comparison

The solution is to plot the equivalent annual costs of each scheme for different pumping demands and determine the range of pumping demands that are cheapest for each scheme.

Scheme A

The equivalent annual cost of installation and maintenance

$$= 1000 + 12\,000 \times (0.20509) + 22\,000 \times (0.19103) = \underline{£7663.74}$$

(*Note*: 1000 is already in annual costs, 12 000 is converted to an annual cost by the capital recovery factor 0.20509 taken from the tables or calculated as $\frac{i(1+i)^n}{(1+i)^n - 1}$, where $i=19\%$ and $n=15$ years; for 0.19103, $n=30$ years.)

Scheme B

The equivalent annual cost of installation and maintenance costs

$$= 1500 + 18\,000 \times (0.20509) + 18\,000 \times (0.19103) = \underline{£8630.16}$$

(*Note*: comments on factors as above.)

Scheme C

The equivalent annual cost of installation and maintenance costs

$$= 1500 + 28\,000 \times (0.19604) + 12\,000 \times (0.19103) = \underline{£9281.48}$$

(*Note*: comments on factors as above, 0.19604 for 20 years.)

The equivalent annual costs calculated above are annual 'fixed' costs and are independent of the number of hours pumping.

The 'variable' cost depending on the number of hours pumping for each scheme are:

	Scheme A	Scheme B	Scheme C
Zero hours pumping	0	0	0
5000 hours pumping	(120p × 5000) = £6000	(90p × 5000) = £4500	(80p × 5000) = £4000

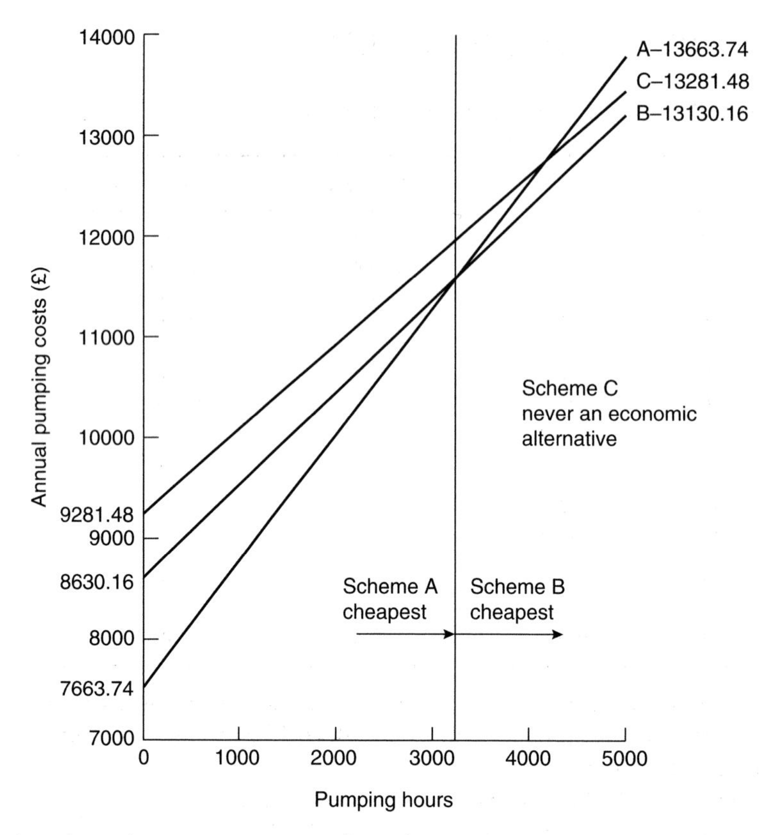

Fig. S34.1 Annual pumping costs versus pumping demand.

These pumping costs vary linearly between zero and 5000 hours. Taking the 'fixed' equivalent annual cost and these variable costs Fig. S34.1 can be plotted.

Examine Fig. Q34.1. Since almost 50% of pumping demand falls in the zone where 'A' is cheaper and 50% of pumping demand falls in the zone where 'B' is cheaper there is no economic difference between 'A' and 'B'.

Solution 35 Discounted cash flow yield

First determine the net cash flow for each year of the project.

Year	Cash in (£)	Cash out (£)	Net cash flow (£)
0		17 500	−17 500
1		22 000	−22 000
2		20 000	−20 000
3	10 500		+10 500
4	10 500		+10 500
5	10 500		+10 500
6	10 500		+10 500
7	10 500		+10 500
8	10 500		+10 500
9	10 500		+10 500
10	10 500		+10 500
11	10 500		+10 500
12	10 500		+10 500
13	10 500	6 500	+4 000
14	10 500	7 500	+3 000
15	10 500	9 500	+1 000

Find by trial and error the interest rate which will give a net present value of zero when used to discount (or reduce to present worth) the net cash flows. This interest rate is the DCF yield (or the rate of return). See Table S24.1.

At an interest rate of 11% the net present value is −£1442.22.

At an interest rate of 10% the net present value is +£1452.91.

The interest rate that gives a net present value of zero (i.e. DCF yield) can be found by interpolation (Fig. S35.1):

$$\text{DCF yield} = 10.00 + 1.00\left[\frac{1452.91}{1452.91-(-1442.22)}\right]$$

$$= 10.50\%$$

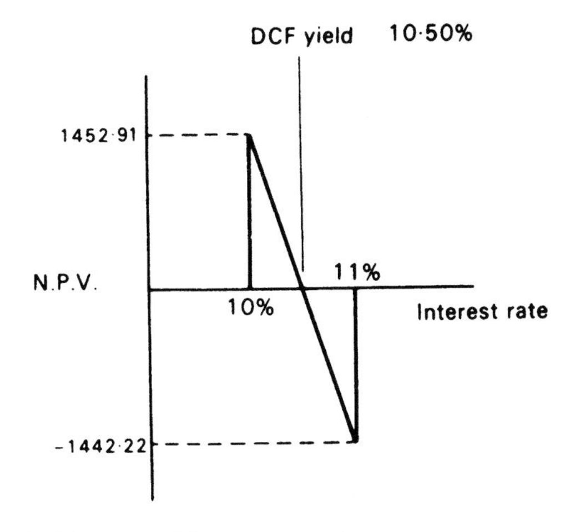

Fig. S35.1 Interpolating to find the DCF yield.

Table S35.1 Present worth factors.

Year	a Net cash flow	b Present worth factors or discount factors* for 15%†	c Present worth at 15% (a × b)	d Present worth factors or discount factors* for 11%‡	e Present worth at 11% (a × d)	f Present worth factors or discount factors* for 10%§	g Present worth at 10% (a × f)
0	−17 500	1.0	−17 500.00	1.0	−17 500.00	1.0	−17 500.00
1	−22 000	0.869	−19 118.00	0.900	−19 800.00	0.909	−19 998.00
2	−20 000	0.756	−15 120.00	0.811	−16 220.00	0.826	−16 520.00
3	+10 500	5.018 × 0.756 ‖	39 832.88	5.889 × 0.811 ‖	50 147.78	6.155 × 0.826 ‖	53 286.91
4	+10 500						
5	+10 500						
6	+10 500						
7	+10550						
8	+10 500						
9	+10 500						
10	+10 500						
11	+10 500						
12	+10 500						
13	+4 000	0.162	648.00	0.257	1028.00	0.289	1156.00
14	+3 000	0.141	423.00	0.231	693.00	0.263	789.00
15	+1 000	0.122	122.00	0.209	209.00	0.239	239.00
		Net present value	−£10 712.12		−£1442.22		£1452.91

* The present worth factors or the discount factors are taken from tables or calculated as $1/(1 + i)^n$ where i is the interest rate and n is the year number.

† 15% is a first guess.

‡ 11% is a second guess based on the negative net present value for 15% which implied 15% was larger than the DCF yield.

§ 10% is a third guess based on the negative net present value for 11% which implied 11% was larger than the DCF yield.

‖ The cash flows from Years 3 to 12 are uniform, therefore they can be treated as a uniform series. The factors of 5.018, 5.889 and 6.114 are the uniform series present worth factors for 15%, 11% and 10%, respectively. These are taken from tables or calculated as $\frac{(1+i)^n - 1}{i(1+i)^n}$ where i is the interest rate and n is the number of years. The uniform series of cash flows from 3 to 12 is 10 years, thus it is 10 in this case. The present worth of this series is given at the beginning of the series which is Year 2 of the cash flows. The factors of 0.756, 0.811 and 0.826 are the present worth factors for Year 2 for 15%, 11% and 10%, respectively. The present worth factors are used to convert the present worth of the uniform series from Year 2 to Year 0.

Solution 36 Dual rate of return

Determine the net cash flows for the project.

Year	Cash in (£)	Cash out (£)	Net cash flow (£)
0		500 000	–500 000
1	400 000	300 000	100 000
2	200 000		200 000
3	200 000		200 000
4	200 000		200 000
5	200 000		200 000
6		180 000	–180 000

Determine by trial and error the 'paying rate' (i.e. the maximum rate that *could* be paid for borrowed capital if the project was funded on an overdraft).

The calculation process involves a year-by-year determination of the amount of money outstanding at the end of each year and the interest paid or earned on that outstanding sum (Table S36.1). The dual rate DCF yield is the 'paying rate' which leaves a balance of zero.

Since the balance remaining is –£3708.12 the 'paying rate' of 15% is greater than the maximum possible paying rate. Therefore the calculations are repeated using a smaller paying rate (Table S36.2).

Table S36.1 Calculation process.

Year	Net cash flow (£)	Paying rate 15%* (£)	Earning rate 5% (£)	Borrowing account (£)
0	–500 000			–500 000.00
1	100 000	–75 000.00†	8394.85‡	–475 000.00
2	200 000	–71 250.00		–346 250.00
3	200 000	–51 937.50		–198 187.50
4	200 000	–29 728.13		–27 915.63
5	200 000	–4 187.34		+167 897.03
6	–180 000			–3 708.12

* 15% is a first guess.
† The interest paid on previous years' negative balance.
‡ The interest earned on previous years' positive balance.

With a paying rate of 14%, the balance remaining is +£19 938.40. Therefore the paying rate will give a zero balance (i.e. the dual rate DCF yield) is between 14% and 15%. By interpretation:

Dual rate DCF yield with an earning rate of 5%

$$= 14.00 + 1.00\left[\frac{19\,938.40}{19\,938.40 - (-3708.12)}\right]$$

$$= 14.84\%$$

This is a measure of profitability because it measures the maximum interest rate that could be paid for borrowed capital. Thus it is measuring how much the project is producing.

Table S36.2 Calculation process using a smaller paying rate.

Year	Net cash flow (£)	Paying rate 14% (£)	Earning rate 5% (£)	Borrowing account (£)
0	−500000			−500000.00
1	100000	−70000.00	9520.88	−475000.00
2	200000	−65800.00		−335800.00
3	200000	−47012.00		−182812.00
4	200000	−25593.68		−8405.68
5	200000	−1176.80		+190417.52
6	−180000			+19938.40

Unreliable estimates of reinstatement costs

Any variation in the reinstatement costs would vary the rate of return. The method of dealing with such variation would be to determine how 'sensitive' the rate of return is to changes in the reinstatement costs. One method would be simply to calculate several rates of return using different estimates of the reinstatement costs.

Solution 37 Corporation tax and development grants

Calculating the cash flows net of tax is shown in Table S37.1.

Capital expenditures are added into a company 'pool' for the purposes of calculating allowances. With no resale to be deducted from this pool there are, in effect, some allowances continuing which have been ignored in this example.

Table S37.1 Calculating the cash flows net of tax.

	a	b	c	d	e	f	g	h
Year	Investment in plant (£)	Net revenue (£)	Corporation tax at 30% payable on previous year's profits (£)	Plant tax allowances (£)	Tax saved by allowances (d × 30%) (£)	Tax paid (c − e) (£)	Dev't grant	Net cash flow (baf + g) (£)
0	50000.00			12500.00	3750.00	−3750.00		−46250.00
1		15000.00	0.00	9375.00	2812.50	−2812.50	10000.00	27812.50
2		18000.00	4500.00	7031.25	2109.38	2390.63		15609.38
3		20000.00	5400.00	5273.44	1582.03	3817.97		16182.03
4		20000.00	6000.00	3955.08	1186.52	4813.48		15186.52
5		10000.00	6000.00	2966.31	889.89	5110.11		4889.89
6		10000.00	3000.00	2224.73	667.42	2332.58		7667.42
7		0.00	3000.00	1668.55	500.57	2499.44		−2499.44

Table S37.2 Determining if the cash flows net of tax yield at least 15%.

Year	Net cash flow (£)	PW factors at 15%* (£)	PW (£)
0	−46 250.00	1.000	−46 250.00
1	27 812.50	0.870	24 184.78
2	15 609.38	0.756	11 802.93
3	16 182.03	0.658	10 639.95
4	15 186.52	0.572	8682.94
5	4889.89	0.497	2431.14
6	7667.42	0.432	3314.84
7	−2499.44	0.376	−939.63
		Net present value	**13 866.95**

* The present worth factors are taken from tables or calculated as $1/(1 + i)^n$, where i is the interest rate and n is the year number.

At an interest rate of 15% the net present value is £13 866.95. Therefore the cash flows net of tax yield more than 15%.

Solution 38 Corporation tax and different allowances

Calculating the cash flows net of tax is shown in Table S38.1. Table S38.2 shows how to determine if the net tax of cash flows yield is at least 18%.

As the net present value is negative at 18%, the cash flows net of tax do not yield 18%.

Table S38.1 Calculating the cash flows net of tax.

Year	a	b	c	d	e	f	g	h	j
	Building investment (£)	Investment in plant (£)	Net revenue (£)	Corporation tax at 30% payable on previous year's profits (£)	Tax allowances for building investment (£)	Tax allowances for plant investment (£)	Tax saved by allowances ($e + f \times 30\%$) (£)	Tax paid ($d - g$) (£)	Net cash flow ($c - a - b - h$) (£)
0	250 000.00	120 000.00			10 000.00	30 000.00	12 000.00	−12 000.00	−358 000.00
1			80 000.00	0.00	10 000.00	22 500.00	9750.00	−9750.00	89 750.00
2			90 000.00	24 000.00	10 000.00	16 875.00	8062.50	15 937.50	74 062.50
3			100 000.00	27 000.00	10 000.00	12 656.25	6796.88	20 203.13	79 796.88
4			100 000.00	30 000.00	10 000.00	9492.19	5847.66	24 152.34	75 847.66
5			90 000.00	30 000.00	10 000.00	7119.14	5135.74	24 864.26	65 135.74
6			80 000.00	27 000.00	10 000.00	5339.36	4601.81	22 398.19	57 601.81
7			70 000.00	24 000.00	10 000.00	4004.52	4201.36	19 798.64	50 201.36
8	−200 000.00		70 000.00	21 000.00	10 000.00	3003.39	3901.02	17 098.98	252 901.02
9				21 000.00	−40 000.00		−12 000.00	33 000.00	−33 000.00

Table S 38.2 Determining if the net tax cash flows yield at least 18%.

Year	Net cash flow (£)	Present worth factors* at 18%	PW (£)
0	−358000.00	1.000	−358000.00
1	89750.00	0.847	76059.32
2	74062.50	0.718	53190.53
3	79796.88	0.609	48566.84
4	75847.66	0.516	39121.38
5	65135.74	0.437	28471.43
6	57601.81	0.370	21337.53
7	50201.36	0.314	15759.46
8	252901.02	0.266	67281.32
9	−33000.00	0.225	−7440.05
		Net present value =	**−15652.23**

* The present worth factors are taken from tables or calculated as $1/(1 + i)^n$, where i is the interest rate (18%) and n is the year number.

Solution 39 Replacement age

The calculations involve determining the equivalent annual cost of keeping the generator for one year, for two years and for three years, etc.

The calculations in Table S39.1 are seeking the balance between rising running costs and declining resale values. The calculations are explained in the headings in the table opposite.

The minimum equivalent annual cost occurs at an operating life of three years. This is taken to be the optimum replacement age.

Table S39.1 Determining the equivalent annual cost for different operating lives.

Year	*a*	*b*	*c*	*d*	e	*f*	*g*	*h*	*j*	*k*	*l*	*m*	*p*
	Purchase price (£)	Capital recovery factors* at 15%	Equivalent annual costs of purchase price ($a \times b$) (£)	Running costs (£)	Present worth factors† at 15% (£)	Present worth of running costs ($d \times e$) (£)	Sum of present worth of running costs (£)	Equivalent annual cost of present worth of running costs ($b \times g$) (£)	Equivalent annual costs of purchase price and running costs ($c + h$) (£)	Resale value (£)	Present worth of resale ($k \times e$) (£)	Equivalent annual costs of present worth of resale ($l \times b$) (£)	Equivalent annual costs of purchase running costs and resale ($j - m$) (£)
0	20000				1.0								
1		1.150	23 000.00	800	0.869	695	695	799.25	23 799.25	18 000	15 642.00	17988.30	5810.95
2		0.615	12 300.00	1200	0.756	907	1602	985.3	13 285.30	16 000	12 096.00	7 439.04	5846.19
3		0.437	8 740.00	1500	0.657	985	2587	1130.51	9 870.51	15 000	9 855.00	4 306.64	5563.88
4		0.350	7 000.00	1800	0.571	1027	3614	1264.90	8 264.90	12 000	6 852.00	2 398.20	5866.70
5		0.298	5 960.00	2100	0.497	1043	4657	1387.78	7 347.78	8 000	3 976.00	1 184.85	6162.93
6		0.264	5 280.00	2400	0.432	1036	5693	1 502.95	6 782.95	5 000	2 160.00	570.24	6212.71

* The capital recovery factors are taken from tables or calculated as $\frac{i(1+i)^n}{(1+i)^n - 1}$, where i is 15% and n is the number of years.

† The present worth factors are taken from tables or calculated as $\frac{1}{(1+i)^n}$, where i is 15% and n is the number of years.

Solution 40 Replacement age

The calculations involve determining the equivalent annual cost of keeping the plant item for one year, two years, three years, etc. (see Table S40.1).

The minimum equivalent cost occurs at Year 6, this is taken as the optimum replacement. However, note how little difference exists between Years 5, 6, 7, 8, 9 and 10.

Table S40.1 Determining the equivalent annual cost for different operating lives.

Year	*a*	*b*	*c*	*d*	*e*	*f*	*g*	*h*	*j*	*k*	*l*	*m*	*p*
	Purchase price (£)	Capital recovery factors* at 12%	Equivalent annual costs of purchase price ($a \times b$) (£)	Running costs (£)	Present worth factors† at 12% (£)	Present worth of running costs ($d \times e$) (£)	Sum of present worths of running costs (£)	Equivalent annual cost of sum of present worth of running costs (£)	Equivalent annual costs of purchase price and running costs ($c + h$) (£)	Resale value (£)	Present worth of resale value ($k \times e$) (£)	Equivalent annual costs of present worth of resale value ($l \times b$) (£)	Equivalent annual costs of purchase price, running costs and resale ($j - m$) (£)
0	15000												
1		1.120	16 800.00	500	0.892	446.00	446.00	499.52	17 299.52	13 125.00	11 707.50	13 112.40	4187.12
2		0.591	8865.00	1000	0.797	797.00	1243.00	734.61	9599.61	11 484.38	9153.05	5409.45	4190.16
3		0.416	6240.00	1300	0.711	924.30	2167.30	901.60	7141.60	10 048.83	7144.72	2972.20	4169.40
4		0.329	4935.00	1600	0.635	1016.00	3183.30	1047.31	5982.31	8792.72	5583.38	1836.93	4145.38
5		0.277	4155.00	1900	0.567	1077.30	4260.60	1180.19	5335.19	7693.63	4362.29	1208.35	4126.84
6		0.243	3645.00	2200	0.506	1113.20	5373.80	1305.83	4950.83	6731.93	3406.36	827.74	4123.09
7		0.219	3285.00	2500	0.452	1130.00	6503.80	1424.33	4709.33	5890.44	2662.48	583.08	4126.25
8		0.201	3015.00	2800	0.403	1128.40	7632.20	1534.07	4549.07	5154.13	2077.11	417.50	4131.57
9		0.187	2805.00	3100	0.360	1116.00	8748.20	1635.91	4440.91	4509.87	1623.55	303.60	4137.31
10		0.176	2640.00	3400	0.321	1091.40	9839.60	1731.77	4371.77	3946.13	1266.71	222.94	4148.83

* The capital recovery factors are taken from tables or calculated as $\frac{i(1+i)^n}{(1+i)^n - 1}$, where i is 12% and n is the number of years.

† The present worth factors are taken from tables or calculated as $\frac{1}{(1+i)^n}$, where i is 12% and n is the number of years.

Solution 41 DCF yield and inflation

Calculate the DCF yield of the rate of return on the cash flows actually experienced, this is called the apparent rate of return (Table S41.1).

Apparent rate of return (DCF yield) is obtained by interpolation, i.e.:

$$\text{Apparent rate of return} = 14 + 1.0\left[\frac{823.60}{823.60-(-140.50)}\right]$$
$$= 14.85\%$$

The relationship between real rate of return (R_r), apparent rate of return (R_a) and the inflation rate (R_d) is

$$(1+R_a) = (1+R_r)(1+R_d)$$

R_a is 14.85% by calculation, R_d is given as 5%. Thus R_r can be found

$$(1+0.1485) = (1+R_r)(1+0.05)$$
$$(1+R_r) = \frac{(1.1485)}{(1.05)}$$
$$R_r = 1.0938 - 1$$
$$= 0.0938$$

Therefore the real rate of return was only 9.38%.

Table S41.1 Calculating the apparent rate of return.

Year	a	b	c	d	e
	Net cash flows (£)	Present worth factors or discount factors* at 14%†	Present worth (a × b) (£)	Present worth factors or discount factors* at 15%‡	Present worth (a × d) (£)
0	−24450	1.000	−24450.00	1.000	−24450.00
1	4500	0.877	3946.50	0.869	3910.50
2	4700	0.769	3614.30	0.756	3553.20
3	5000	0.674	3370.00	0.657	3285.00
4	5100	0.592	3019.20	0.571	2912.10
5	4900	0.519	2543.10	0.497	2435.30
6	5100	0.455	2320.50	0.432	2203.20
7	5300	0.399	2114.70	0.375	1987.50
8	4900	0.350	1715.00	0.326	1597.40
9	4800	0.307	1473.60	0.284	1363.20
10	4300	0.269	1156.70	0.247	1062.10
		Net present value	+£823.60		−£140.50

* Taken from tables or calculated as $\frac{1}{(1+i)^n}$, where i is the interest rate and n is the number of years.

† 14% is a first guess.

‡ 15% is a second guess, given that 14% produced a positive net present value.

Solution 42 DCF yield and inflation

(1) Calculate the net cash flows actually experienced (Table S42.1).
(2) Calculate the apparent rate of return (Table S42.2).

The apparent rate of return (DCF yield) is obtained by interpolation, i.e.:

$$\text{Apparent rate of return } = 6 + 1.0 \times \left[\frac{195.00}{195.00 - (-189.50)}\right]$$
$$= 6.51\%$$

The relationship between the real rate of return (R_r), the apparent rate of return (R_a) and the inflation rate (R_d) is given by

$$(1 + R_a) = (1 + R_r)(1 + R_d)$$

R_a is 6.51% by calculation, R_d is given as 12%. Thus R_r can be found

$$(1 + 0.0651) = (1 + R_r)(1 + 0.12)$$
$$(1 + R_r) = \frac{(1.0651)}{(1.12)}$$
$$R_r = 0.9509 - 1.0$$
$$= -0.049$$

Therefore the real rate of return is *negative* at −4.9%. That is, this project had an apparent rate of return that was smaller than the inflation rate.

Table S42.1 Cash flows experienced.

Year	Cash in (£)	Cash out (£)		Net cash flow (£)
0		−12 000		−12 000
1	+6000	−3 000	=	+3 000
2	+6000	−3 500	=	+2 500
3	+6500	−4 000	=	+2 500
4	+6500	−5 250	=	+1 250
5	+7000	−5 000	=	+2 000
6	+7000	−5 250	=	+1 750
7	+7500	−5 500	=	+2 000

Table S42.2 The apparent rate of return.

Year	a	b	c	d	e
	Net cash flows (£)	Present worth factors or discount factors* for 6%†	Present worth (a × b) (£)	Present worth factors or discount factors* for 7%‡	Present worth (a × d) (£)
0	−12 000	1.000	−12 000.00	1.000	−12 000.00
1	+3 000	0.943	2 829.00	0.934	2 802.00
2	+2 500	0.889	2 222.50	0.873	2 182.50
3	+2 500	0.839	2 097.50	0.816	2 040.00
4	+1 250	0.792	990.00	0.762	952.50
5	+2 000	0.747	1 494.00	0.712	1 424.00
6	+1 750	0.704	1 232.00	0.666	1 165.50
7	+2 000	0.665	1 330.00	0.622	1 244.00
		Net present value	£195.00		−£189.50

* Taken from tables or calculated as $\frac{1}{(1+i)^n}$, where i is the number of years.
† 6% is a first guess.
‡ 7% is a second guess, given that 6% produced a positive net present value.

Solution 43 Present worth with increasing annual costs

Calculate the present worth of Scheme A (Table S43.1) and that of Scheme B (Table S43.2).

The present worth of Scheme A is £397.08M.

The present worth of Scheme B is £398.83M.

Thus on the basis of this present worth comparison, Scheme A is marginally less expensive than Scheme B. However, the present worths of each scheme are so close that no real distinction can be made.

Present worth factors and adjustments

(1) Present worth of a uniform series calculated using 9%, taken from tables or calculated as:

$$\frac{(1+i)^n - 1}{i(1+i)^n}$$

where i is 9% and n is 10; factor is 6.4176.

(2) Present worth of a lump sum calculated using 9%, taken from tables or calculated as:

$$\frac{1}{(1+i)^n}$$

where $i = 9\%$ and n is 10; factor is 0.42241.

Table S43.1 Calculation of present worth of Scheme A.

Cost heading	Cost	Present worth factors	Present worths (£M)
Capital cost	£132M	1.0	132.00
Fuel costs	£20M per year from 1 to 10.	6.4176(1)	128.35
	£22M per year from 11 to 20.	6.4176(1) × 0.42241(2)	59.64
Maintenance	£5M per year from 1 to 10 increasing at 2.75% per year.	7.3322(3)	36.66
	£5M per year rising at 2.75% per year, increases to £6.56M by year 10. Adding the £1.5M increase, the cost of maintenance for Years 11 to 20 is £8.06(5)M increasing at 2.75% per year	7.3322(3) × 0.42241(2)	24.96
Operating staff and administration	£1.0M per year for Years 1 to 20.	9.1285 (4)	9.13
Capital cost of upgrade	£15M in Year 10.	0.42241(2)	6.34
		Total present worth	–£397.08M

The numbers in parentheses in Tables S43.1 and S43.2 are references to the notes given on the previous page and at the end of this solution.

Table S43.2 Calculations of present worth of Scheme B.

Cost heading	Cost	Present worth factors	Present worths (£M)
Capital cost	£50M	1.0	50.00
Fuel costs	£25M per year from 1 to 10.	6.4176(1)	160.44
	£27M per year from 11 to 20.	6.4176(1) × 0.42241(2)	73.19
Maintenance	£8M per year from 1 to 10 increasing at 2.75% per year.	7.3322(3)	58.66
	£8M per year rising at 2.75% increases to £10.49M in Year 10. Adding the £1.0M increase, the cost of maintenance for Years 11 to 20 is £11.49(5)M increasing at 2.75% per year.	7.3322(3) × 0.42241(2)	35.59
Operating staff and administration	£1.6M per year from 1 to 20.	9.1285(4)	14.61
Capital cost of upgrade	£15M in Year 10.	0.42241(2)	6.34
		Total present worth	–£398.83M

(3) The money rate is 9%; the costs are increasing at the rate of 2.75% per year. This can be accommodated by adjusting the interest rate representing the value of money to determine the 'effective' rate 'e':

$$e = \frac{(1+i)}{(1+c)} - 1$$

where i is 9% and c is 2.75%. Thus $e = 6.08\%$.

The present worth factor is then calculated using 6.08% and is the present worth of a uniform series for ten years, taken from tables or calculated as:

$$\frac{(1+e)^n - 1}{e(1+e)^n}$$

where e is 6.08% and n is 10; factor is 7.3322.

(4) Present worth is a uniform series calculated using 9%, taken from tables or calculated as:

$$\frac{(1+i)^n - 1}{i(1+i)^n}$$

where i is 9% and n is 20.

(5) The series runs from Years 11 to 20. To get the starting value for the series it is necessary to take the initial Year 0 value and add ten years of increasing costs at 2.75% per year (given by initial value $\times (1+d)^n$ where d is 2.75% and n is 10).

Solution 44 Present worth with increasing annual costs and inflation

Calculate the present worth of Scheme A (Table S44.1) and that of Scheme B (Table S44.2).

The present worth of Scheme A including allowances for inflation is £631.80M. The present worth of Scheme B including allowances for inflation is £698.26M.

Whereas in Solution 32 there is little difference between the present worths of each scheme, when inflation is included Scheme B, the scheme with the larger running costs, is shown to be considerably more expensive.

Present worth factors and adjustments

(1) Interest rate representing the value of money is 9%, the annual inflation rate for fuel costs is 8%. The 'effective' rate, allowing for inflation is given by:

$$e = \frac{(1+i)}{(1+d)} - 1$$

where i is 9% and d, the inflation rate, is 8%.

Table S44.1 Calculation of the present worth of Scheme A.

Cost heading	Cost	Present worth factors	Present worth (£M)
Capital cost	£132M	1.0	132.00
Fuel costs	£20M per year from 1 to 10.	9.5095(1)	190.19
	£47.50(6)M per year from Years 11 to 20 inflating at 8% per year.	9.5095(1) × 0.42241(2)	190.80
Maintenance	£5M per year from 1 to 10 increasing at 2.75% per year and inflating at 5%.	9.4560(3)	47.28
	£13.12(7)M per year from Years 11 to 20 increasing at 2.75% and inflating at 5%.	9.4560(3) × 0.42241(2)	52.41
Operating staff and administration	£1M per year inflating at 2.75%.	11.3958(4)	11.40
Capital upgrade	£15M in Year 10 with 2% inflation.	0.5149(5)	7.72
		Total present worth	−£631.80M

Table S44.2 Calculation of the present worth of Scheme B.

Cost heading	Cost	Present worth factors	Present worth (£M)
Capital cost	£50M	1.0	50.00
Fuel costs	£25M per year from 1 to 10.	9.5095(1)	237.74
	£58.29(8)M per year from Years 11 to 20 inflating at 8% per year.	9.5095(1) × 0.42241(2)	234.15
Maintenance	£8M per year from 1 to 10 increasing at 2.75% and inflating at 5%.	9.4560(3)	75.65
	£18.72(9)M per year from Years 11 to 20 increasing at 2.75% and inflating at 5%.	9.4560(3) × 0.42241(2)	74.77
Operating staff and administration	£1.6M per year from 1 to 20 inflating at 2.75%.	11.3958(4)	18.23
Capital upgrade	£15M in Year 10 with 2% inflation.	0.5149(5)	7.72
		Total present worth	−£698.26M

The numbers in parentheses in Tables S44.1 and S44.2 are references to the notes given on the previous page and at the end of this solution.

$$e = \frac{(1.09)}{(1.08)} - 1 = 0.925\%$$

The present worth of a uniform series calculated using 0.925% is calculated as:

$$\frac{(1+e)^n - 1}{e(1+e)^n}$$

where e is 1% and n is 10; factor is 9.5095.

(2) The present worth of a lump sum calculated using 9%, taken from tables or calculated as:

$$\frac{1}{(1+i)^n}$$

where i is 9% and n is 10; factor is 0.42241.

(3) The interest rate representing the value of money is 9%; costs rise at 2.75% (not due to inflation but as a result of increasing real costs). The effective rate allowing for this increasing cost is given by:

$$e' = \frac{(1+i)}{(1+c)} - 1$$

where i is 9% and c, the cost rise, is 2.75%.

$$e' = \frac{(1.09)}{(1.0275)} - 1 = 6.08\%$$

The rise in maintenance costs due to inflation is 5%. The effective rate adjusted to allow for inflation is e and is given by:

$$e = \frac{(1+e')}{(1+d)} - 1$$

where e' is the previous effective rate, 6.08%, calculated above and d is the inflation, 5%.

$$e = \frac{(1.0608)}{(1.05)} - 1 = 1.03\%$$

The present worth of a uniform series calculated using 1.03% is taken from tables or calculated as:

$$\frac{(1+e)^n-1}{e(1+e)^n}$$

where e is 1% and n is 10; factor is 9.4560.

(4) The interest rate representing the value of money is 9% and the annual inflation rate for operating staff and administration costs is 2.75%. The 'effective' rate allowing for inflation is given by:

$$e=\frac{(1+i)}{(1+d)}-1$$

where i is 9% and d, the inflation rate, is 2.74%.

$$e=\frac{(1.09)}{(1.0275)}-1=6.08\%$$

The present worth of a uniform series calculated using 6.08% is taken from tables or calculated as:

$$\frac{(1+e)^n-1}{e(1+e)^n}$$

where e is 6.08% and n is 20; factor is 11.3958.

(5) The interest rate representing the value of money is 9% and the annual inflation rate for capital costs is 2%. The 'effective' rate allowing for inflation is given by:

$$e=\frac{(1+i)}{(1+d)}-1$$

where i is 9% and d, the inflation rate, is 2%.

$$e=\frac{(1.09)}{(1.02)}-1=6.86\%$$

The present worth of a lump sum calculated using 6.86 is calculated as:

$$\frac{1}{(1+e)^n}$$

where e is 6.86% and n is 10; factor is 0.5149.

(6) £20M per year inflating at 8% inflates in Year 10 to £43.18M. This is given by £20M × $(1+d)^n$ where d is 8% and n is 10. The £2.0M increase will inflate at 8% to £4.32M by Year 10. The cost of fuel for Years 11 to 20 is £43.18M + £4.32M = £47.50M per year inflating at 8%.

(7) £5M per year increasing at 2.75% per year and inflating at 5% per year, increases and inflates to £10.68M by Year 10. This is given by £5M × $(1+c)^n(1+d)^n$ where c is 2.75%, d is 5% and n is 10. The £1.5M increase will inflate at 5% to £2.44M by Year 10. The cost of maintenance for Years 11 to 20 is £10.68M + £2.44M = £13.12M per year increasing at 2.75% and inflating at 5% per year.

(8) £25M per year inflating at 8% inflates to £53.97M. This is given by £25M × $(1+d)^n$ where d is 8% and n is 10. The £2.0M increase will inflate at 8% to £4.32M by Year 10. The cost of fuel for Years 11 to 20 is £53.97M + £4.32M = £58.29M per year inflating at 8%.

(9) £8M per year increasing at 2.75% per year and inflating at 5% per year, increases and inflates to £17.09M by Year 10. This is given by £8M × $(1+c)^n(1+d)^n$ where c is 2.75%, d is 5% and n is 10. The £1.0M increase will inflate at 5% to £1.63M by Year 10. The cost of maintenance for Years 11 to 20 is £17.09M + £1.63M = £18.72M per year increasing at 2.75% and inflating at 5%.

Solution 45 Valuation of a plant item

Existing plant item

The cash flows for the existing item of plant are given in Table S45.1. Note there is no cash flow in Year 0 because the plant item is already owned.

The present worth of these cash flows using an interest rate of 10% is £1212.45 calculated as shown in Table S45.2.

Alternative: Buying new plant item

The cash flows estimated at present prices for purchasing, operating and reselling after four years for a similar item of equipment are given in Table S45.3. The present worth of these cash flows using an interest rate of 10% is given in Table S45.4.

Value calculation

The difference between the present worth of keeping the existing plant item and replacing immediately with a new item of plant is:

£9824.71 − £1212.45 = £8612.26

Thus if the owner could sell the existing plant item for £8612.26 and offset this against the cost of the new equipment the present worth of acquiring the new equipment would be the same as

Table S45.1 Cash flows for existing plant item.

Year	Cash flows for existing plant item (£)
0	–
1	−661.25
2	−766.44
3	−874.50
4	−1005.68 + 2000

Table S45.2 Present worth using an interest rate of 10%.

Year	Cash flows for existing plant item (£)	Present worth factors	Present worth (£)
0	–	–	–
1	−661.25	0.90909	−601.14
2	−766.44	0.82644	−633.42
3	−874.50	0.75131	−657.02
4	−1005.68 + 2000	0.68301	+679.13
		Total present worth	−1212.45

Table S45.3 Cash flows for new plant item.

Year	Cash flows for purchase, operating and resale for new item of plant (£)
0	−11000.00
1	−400.00
2	−460.00
3	−529.00
4	−608.35 + 4000

Table S45.4 Present worth using an interest rate of 10%.

Year	Cash flows for new item of plant (£)	Present worth factors	Present worth
0	−11 000.00	1.0	−11 000.00
1	−400.00	0.90909	−363.64
2	−460.00	0.82644	−380.16
3	−529.00	0.75131	−397.44
4	−608.35 + 4000	0.68301	+2 316.53
		Total present worth	−9 824.71

keeping the existing item of plant. If the owner could sell the item of plant for more than £8612.26 then it would be more economic to sell and replace: if £8612.26 could not be realised from the sale of the equipment it would be more economic to retain the equipment.

Solution 46 Valuation of a plant item with longer life

Immediate replacement

The cash flows for immediate replacement with a new item of plant and replacements for 20 years are given in Table S46.1. Calculation of present worth, using an interest rate of 10%, is given in Table S46.2. Thus the cash flows for the immediate and subsequent replacements can be represented as shown in Table S46.3. The £12 316.35 at Years 0, 6 and 12 represents all the cash flows for the replacements purchased in those years. The present worth for the immediate and subsequent replacements up to 20 years is £24 115.92, calculated as shown in Table S46.4.

Keeping existing plant item

The cash flows for keeping the existing item of plant and subsequent replacements for 20 years are given in Table S46.5. The present worth of keeping the existing plant item until the end of its useful life and its subsequent replacement can be calculated first by representing the cash flows as shown in Table S46.6.

Table S46.1 Cash flows and replacements for 20 years.

Year	Cash flows		
	Purchase price (£)	Running costs (£)	Resale
0	−11 000		
1		−400.00	
2		−460.00	
3		−529.00	
4		−608.35	
5		−699.60	
6	−11 000	−804.54	+2000
7		−400.00	
8		−460.00	
9		−529.00	
10		−608.35	
11		−699.60	
12	−11 000	−804.54	+2000
13		−400.00	
14		−460.00	
15		−529.00	
16		−608.35	
17		−699.60	
18	−11 000	−804.54	+2000
19		−400.00	
20		−460.00	+8000

Table S46.2 Present worth using an interest rate of 10%.

Year	Cash flow (£)	Present worth factors	Present worth
0	−11 000.00	1.0	−11 000.00
1	−400.00	0.90909	−363.64
2	−460.00	0.82644	−380.16
3	−529.00	0.75131	−397.44
4	−608.35	0.68301	−415.51
5	−699.60	0.62092	−434.40
6	−804.54 + 2000.00	0.56447	+674.80
		Total present worth	−12 316.35

Table S46.3 Cash flows for replacements.

Year	Cash flow (£)
0	−12 316.35
6	−12 316.35
12	−12 316.35
18	−11 000.00
19	−400.00
20	−460.00 + 8000

Table S46.4 Present worth for replacements.

Year	Cash flow (£)	Present worth factors	Present worth (£)
0	−12 316.35	1.0	−12 316.35
6	−12 316.35	0.56447	−6 952.21
12	−12 316.35	0.31863	−3 924.36
18	−11 000.00	0.17985	−1 978.35
19	−400.00	0.16350	−65.40
20	−460.00 + 8000	0.14864	+1 120.75
		Total present worth	−24 115.92

The £1212.45 taken from Solution 45 represents all the cash flows for the existing item of plant.

The £12 316.35 in Years 4 and 10 is used to represent all the cash flows for the replacement in those years. Consequently, the present worth of keeping the existing plant item until the end of its useful life and subsequent replacements up to 20 years is £16 511.14. This is shown in Table S46.7.

Table S46.5 Cash flows for keeping existing plant item and replacements.

Year	Cash flows				
	Running costs existing plant (£)	Resale existing plant	Purchase of replacements (£)	Running cost of replacements (£)	Resale of replacements (£)
0	—				
1	−661.25				
2	−766.44				
3	−874.50				
4	−1005.68	+2000.00	−11 000.00		
5				−400.00	
6				−460.00	
7				−529.00	
8				−608.35	
9				−699.60	
10			−11 000.00	−804.54	+2000
11				−400.00	
12				−460.00	
13				−529.00	
14				−608.35	
15				−699.60	
16			−11 000.00	−804.54	+2000
17				−400.00	
18				−460.00	
19				−529.00	
20				−608.35	+4000

Table S46.6 Cash flows for Years 0 to 20.

Year	Cash flow (£)
0	−1212.45
4	−12316.35
10	−12316.35
16	−11000.00
17	−400.00
18	−460.00
19	−529.00
20	−608.35 + 4000

Value calculation

The difference between these two present worths is:

$$£24\,115.92 - £16\,511.14 = £7604.78$$

If the owner could sell the existing plant item for £7604.78 and offset this amount against the cost of the immediate replacement, the present worth of the immediate and subsequent replace-

Table S46.7 Present worth of keeping existing plant item and subsequent replacements.

Year	Cash flows (£)	Present worth factors	Present worth (£)
0	–1212.45	1.0	–1212.45
4	–12316.35	0.68301	–8412.19
10	–12316.35	0.38554	–4748.45
16	–11000.00	0.21762	–2393.82
17	–400.00	0.19784	–79.14
18	–460.00	0.17985	–82.73
19	–529.00	0.16350	–86.49
20	–608.35 + 4000	0.14864	+504.13
		Total present worth	–16511.14

ments would be the same as keeping the existing item. If the owner could sell the existing plant item for more than £7604.78 the acquiring of an immediate replacement would be more economic. If £7604.78 could not be realised by the sale of the existing plant item, then keeping the existing plant item would be more economic.

Solution 47 Valuation of a plant item whose life is in perpetuity

Immediate replacement

The cash flow for immediate and subsequent replacements to infinity is represented in Table S47.1.

The £12 316.35 is the present worth for purchasing, operating and reselling the new plant item. This present worth was calculated previously as part of Solution 46. If this plant item is replaced every six years then this amount recurs every six years to infinity.

The present worth of a sum recurring every year starting at the end of the first year is given by the factor:

$$\frac{(1+i)^n - 1}{i(1+i)^n}$$

which is the uniform series present worth factor.

The present worth of a sum recurring not every year but every y years starting in y years is given by the factor:

$$\frac{1}{(1+i)^y - 1}$$

Thus, if the sum recurring every y years was x, the present worth would be:

Table S47.1 Cash flow for immediate and subsequent replacements.

Year	Cash flows representing replacements to infinity
0	−12 316.35
6	−12 316.35
12	−12 316.35
18	−12 316.35
24	−12 316.35
–	–
–	–
–	–
–	–
∞	∞

$$x \times \left[\frac{1}{(1+i)^y - 1} \right]$$

and if the sum x occurred in Year 0 the total present worth of the whole series would be:

$$x + x \times \left[\frac{1}{(1+i)^y - 1} \right]$$

which reduces to:

$$\frac{x}{1 - \dfrac{1}{(1+i)^y}}$$

Thus, substituting £12 316.35 for x, six years for y and 10% for i, the present worth of the series starting in Year 0 and recurring to infinity is:

$$\frac{£12\,316.35}{1 - 0.56447} = £28\,278.99$$

The 0.56447 can be taken from tables since the element $1/(1+i)^y$ is the expression for the present worth factor.

The amount calculated, £28 278.99, represents the amount that would be required today which, if invested at 10%, would produce £12 316.35 every six years forever. This can be checked as follows. First, take away the initial £12 316.35 and this leaves £15 962.64 for investment. In six years this £15 962.64 would increase to £15 962.64 × 1.77157 + £28 278.99. This £28 278.99 is made up of the original capital, £15 962.64, and the interest earned in those six years of £12 316.35. The factor 1.77157 is the compound amount factor for 10% and is taken from tables. Thus, the sum of £12 316.35 can be used every six years to provide a replacement plant item. This

£12 316.35 is the interest earned on the capital of £15 962.64 and is entirely used up every six years. The capital of £15 962.64 remains undepleted and can go on producing £12 316.35 every six years indefinitely.

The £12 316.35 is the present worth of purchasing, operating and reselling one plant item, and £28 278.99 is the present worth of purchasing, operating and reselling every six years to perpetuity. This present worth is sometimes called the capitalised cost.

Keeping existing plant item

Taking the cash flows from Solution 45 for keeping the existing plant and its replacements, these can be represented as in Table S47.2.

The cash flows in Years 0 to 4 representing the running costs of the existing plant and the resale value have a present worth of £1212.45 as calculated in Solution 45.

The £12 316.35 every six years has a present worth of £28 278.99 as calculated. However, in this set of cash flows the first of the £12 316.35 occurs in Year 4 and the cash flows for keeping the existing plant item and its replacements to infinity can be represented as in Table S47.3.

Table S47.2 Cash flows for keeping existing plant and its replacements.

Year	Cash flows for keeping existing plant and its replacements		
	Running costs existing plant (£)	Resale (£)	Replacements (£)
0	–		
1	–661.25		
2	–766.44		
3	–874.50		
4	–1005.68	+2000	–12316.35
10			–12316.35
16			–12316.35
22			–12316.35
28			–12316.35
–			–
–			–
–			–
–			–
∞			∞

Table S47.3 Cash flows for keeping the existing plant item and its replacements up to Year 4.

Year	Cash flows for keeping the existing plant item and its replacements (£)
0	–1212.45
4	–28278.99

The present worth of these cash flows is £20 527.28, calculated thus:

$$£1212.45 + (£28\,278.99 \times 0.68301) = £20\,527.28$$

Given an interest rate of 10%, £20 527.28 is the amount required for investment today to give £1212.45 in Year 0, enough for the existing plant, and £28 278.99 in Year 4, enough for the replacements at £12 316.35 every six years until infinity. Thus £20 527.28 is the capitalised cost of keeping the existing plant and then replacing it in perpetuity.

Value calculation

The difference between these two present worths is:

$$£28\,278.99 - £20\,527.28 = £7751.71$$

Thus £7751.71 can be taken to represent the value of the plant item to the owner as in Solutions 45 and 46.

Solution 48 Selection of mutually exclusive projects using incremental analysis

The cash flows of the projects are given in Table S48.1.

Ranking the projects in ascending order of capital investment gives the order: A, B, C.

Selecting the smallest project in terms of capital investment to begin the analysis gives Project A.

Calculate the net present value of Project A using the criterion rate of return 9% (Table S48.2).

The positive net present value of +£1385.20 indicates that the yield of Project A is well above the criterion rate of return of 9%. Project A is therefore acceptable and can be used as a basis for comparison with the next project.

Selecting the next smallest project for analysis gives Project B. Calculate the incremental cash flow from Project A to Project B (Table S48.3).

Table S48.1 Cash flows of Projects A, B and C.

Year	Project A (£)	Project B (£)	Project C (£)
0	–6000	–8000	–9000
1	+1100	+1500	+1900
2	1100	1500	1900
3	1100	1500	1900
4	1400	1500	1900
5	1400	1500	1900
6	1400	1500	1900
7	1400	1500	1900
8	2200	2700	2700

Table S48.2 Net present value of Project A using the criterion rate of return 9%.

Year	Cash flow (£)	Present worth factors[1]	Present worth (£)
0	−6000	1.0	−6000
1	+1100	0.917	+1008.70
2	1100	0.841	925.10
3	1100	0.772	849.20
4	1400	0.708	991.20
5	1400	0.649	908.60
6	1400	0.596	834.40
7	1400	0.547	−65.80
8	2200	0.501	1102.20
		Net present value	+1385.20

[1] Taken from tables or calculated as: $\frac{1}{(1+i)^n}$ where i is 9% and n is the appropriate year.

Table S48.3 Incremental cash flow from Project A to Project B.

Year	Project A cash flow (£)	Project B cash flow (£)	Incremental cash flow (B – A) (£)
0	−6000	−8000	−2000
1	+1100	+1500	+400
2	1100	1500	400
3	1100	1500	400
4	1400	1500	100
5	1400	1500	100
6	1400	1500	100
7	1400	1500	100
8	2200	2700	500

The incremental cash flows represent the difference between Projects B and A. Project B requires an extra £2000 of capital investment. The cash flows that this extra £2000 yields, over and above what can be achieved in Project A, are shown in the incremental cash flow column above.

Calculate the net present value of the incremental cash flow (B – A) (Table S48.4).

The negative net present value of −£487.50 indicates that the yield on the incremental cash flows on the extra investment of £2000 is less than the criterion rate of return of 9%. Thus the extra investment does not earn an adequate rate of return and cannot be justified.

Project B is therefore rejected and Project A is retained as a basis of comparison.

Selecting the next project gives C. Calculate the incremental cash flow from Project A and Project C (Table S48.5).

The incremental cash flows represent the difference between Project C and Project A.

Project C requires an extra £3000 of capital investment. The cash flows that this extra £3000 yields over and above that which can be achieved in Project A are shown in the incremental cash flow column in Table S48.5.

Table S48.4 Net present value of the incremental cash flow (B – A).

Year	Incremental cash flow (B – A) (£)	Present worth factors[1]	Present worth (£)
0	–2000	1.0	–2000
1	+400	0.917	366.80
2	400	0.841	336.40
3	400	0.772	308.80
4	100	0.708	–0.80
5	100	0.649	64.90
6	100	0.596	59.60
7	100	0.547	54.70
8	500	0.501	250.50
		Net present value	487.50

[1] Taken from tables or calculated as: $\frac{1}{(1+i)^n}$ where i is 9% and n is the appropriate year.

Table S48.5 Incremental cash flow from Project A and Project C.

Year	Project A cash flow (£)	Project C cash flow (£)	Incremental cash flow (C – A) (£)
0	–6000	–9000	–3000
1	+1100	+1900	+800
2	1100	1900	800
3	1100	1900	800
4	1400	1900	500
5	1400	1900	500
6	1400	1900	500
7	1400	1900	500
8	2200	2700	500

Calculate the net present value of the incremental cash flow (C – A) (Table S48.6).

The positive net present value of +£524.50 indicates that the yield on the extra £3000 investment is greater than the criterion rate of return of 9%. Thus project C is acceptable as it gives all the cash flows and yield of Project A and an adequate return on the extra investment of £3000 required by Project C.

The project recommended for investment is Project C.

Table S48.6 Net present value of the incremental cash flow (C – A).

Year	Incremental cash flow (C – A) (£)	Present worth factors[1]	Present worth (£)
0	−3000	1.0	−3000.00
1	+800	0.917	+733.60
2	800	0.841	672.80
3	800	0.772	617.60
4	500	0.708	354.00
5	500	0.649	324.50
6	500	0.596	298.00
7	500	0.547	273.50
8	500	0.501	250.50
		Net present value	+£524.50

[1] Taken from tables or calculated as: $\frac{1}{(1+i)^n}$ where i is 9% and n is the appropriate year.

Solution 49 Analysis of plant purchase

A suggested analysis to act as a guide is shown in the following evaluation tables.

Re-evaluation based on adverse possibilities

Harris 30T

Seized pump, probability of occurrence	= 0.1
Seriousness represented by cost of pump	= £300
Delay cost	= £150
Total	= £450
Therefore, degree of threat = 0.1 × 450	= £45

McCaffer L25

Gearbox failure, probability of occurrence	= 0.05
Total replacement cost	= £3000
Therefore degree of threat = 0.05 × 3000	= £150

Harris 30T is still the better choice. However, the analysis produces quite a close result and in the circumstances several of the objectives might need careful examination as they may be of overriding importance.

Table S49.1. Evaluation table for craneage selection exercise—Crane No 1.

Objectives	Alternative	
MUST	Harris 30T	GO/NO GO
A Max lift 25 tonnes	27.5 tonnes	✓
B 20 m boom length	3–24 m	✓
C Max purchase price £80 000	£79 000	✓
D Delivery less than 3 months	3 months	✓
E Hydraulic fly jib	Power assisted	✓
F Purchase on home market	Yes	✓
G Conforms to ISO regulations	Yes	✓

Description	E/T	Ranking		Information	Rating	Weighting	
Price	E	10		£79 000	8	80	
Delivery Period	E	10		3 months	7	70	
Good lifting capability	T		10	27.5 tonnes	8		80
Boom length	T		10	3–24 m	5		50
Fly jib	T		10	Power assisted	6		60
Low running costs	E	9		Good	6	54	
Cost of spares	E	5		Reasonable	5	25	
One-driver operation	E	6		Yes	7	42	
Simple operation	E/T	8	8	Good	9	72	72
Fully telescopic boom	T		10	Yes	9		90
Cab comfort	T		7	Poor	3		21
Outriggers	T		8	Hydraulic	6		48
Power-assisted steering	T		6	Curman	1		6
Gearbox	T		7	12F + 2R	6		42
Engine HP	T		6	135HP Turbins	6		36
Variable hoisting speed	T		4	20–35 m/min	8		32
Steering speed	T		3	2.25 rev/min	7		21
Boom telescopic speed	T		5	20 m/min	7		35
Back-up services	E/T	6	6	Poor	1	6	6
Turning radius	T		2	11.5 m	7		14
Four-wheel drive	T		2	Yes	7		14
Safe load indicator	T		10	Enco Auto	8		80
Travelling weight	T		1	29 000 kg	9		9
Overall width	T		9	2.5 m	9		81
Overall length	T		1	11.85 m	6		6
Accessories	E/T	4	4	Reasonable	5	20	20
Fail-safe brakes	T		10	Yes	7		70
Travelling speed	T		5	65 km/h	6		30
E Economic element	29%	58				369	923
T Technical element	71%		144				
						1292	

Remarks: All MUSTS are OK

Table S49.2 Evaluation table for craneage selection exercise—Crane No 2.

Objectives	Alternative	
MUSTS	McCaffer L25	GO/NO GO
A Max lift 25 tonnes	25 tonnes	✓
B 20 m boom length	8.63–25.97 m	✓
C Max purchase price £80 000	£75 000	✓
D Delivery less than 3 months	4 weeks	✓
E Hydraulic fly jib	Power assisted	✓
F Purchase on home market	Yes	✓
G Conforms to ISO Standards	Yes	✓

Description	E/T	Ranking		Information	Rating	Weighting	
Price	E	10		£75 000	10	100	
Delivery Period	E	10		4 weeks	9	90	
Good lifting capability	T		10	25 tonnes	6		60
Boom length	T		10	8.63–25.97 m	7		70
Fly jib	T		10	Power assisted	4		40
Low running costs	E	9		Poor	2	18	
Cost of spares	E	5		High	2	10	
One-driver operation	E	6		Yes	6	36	
Simple operation	E/T	8	8	Poor	4	32	32
Fully telescopic boom	T		10	Yes	8		80
Cab comfort	T		7	Fair	5		35
Outriggers	T		8	Hydraulic	6		48
Power-assisted steering	T		6	Rico	4		24
Gearbox	T		7	6F + 2R	5		35
Engine HP	T		6	138HP Rumbings	8		48
Variable hoisting speed	T		4	30 m/min only	1		4
Steering speed	T		3	2.35 rev/min	8		24
Boom telescopic speed	T		5	23 m/min	8		35
Back-up services	E/T	6	6	Good	5	30	30
Turning radius	T		2	11.3 m	8		16
Four-wheel drive	T		2	Yes	7		14
Safe load indicator	T		10	Wicht Audio	8		80
Travelling weight	T		1	33 000 kg	6		6
Overall width	T		9	2.5 m	9		81
Overall length	T		1	10.64 m	7		7
Accessories	E/T	4	4	Good	7	28	28
Fail-safe brakes	T		10	Yes	7		70
Travelling speed	T		5	70 km/h	7		35
E Economic element	29%	58				344	907
T Technical element	71%		144				
						1251	

Remarks: All MUSTS are OK

Table S49.3 Evaluation table for craneage selection exercise—Crane No 3.

Objectives	Alternative	
MUSTS	McCaffer L25	GO/NO GO
A Max lift 25 tonnes	26 tonnes	✓
B 20 m boom length	8.9–28.7 m	✓
C Max purchase price £80 000	£80 000	✓
D Delivery less than 3 months	6 months	✗
E Hydraulic fly jib	Power assisted	✓
F Purchase on home market	Yes	✓
G Conforms to ISO Standards	Yes	✓

Description	E/T	Ranking		Information	Rating	Weighting	
Price	E	10					
Delivery Period	E	10					
Good lifting capability	T		10				
Boom length	T		10				
Fly jib	T		10				
Low running costs	E	9					
Cost of spares	E	5					
One-driver operation	E	6					
Simple operation	E/T	8	8				
Fully telescopic boom	T		10				
Cab comfort	T		7				
Outriggers	T		8				
Power-assisted steering	T		6				
Gearbox	T		7				
Engine HP	T		6				
Variable hoisting speed	T		4				
Steering speed	T		3				
Boom telescopic speed	T		5				
Back-up services	E/T	6	6				
Turning radius	T		2				
Four-wheel drive	T		2				
Safe load indicator	T		10				
Travelling weight	T		1				
Overall width	T		9				
Overall length	T		1				
Accessories	E/T	4	4				
Fail-safe brakes	T		10				
Travelling speed	T		5				
E Economic element	29%	58					
T Technical element	71%		144				
Remarks: Failed MUST objective D and is eliminated.							

Table S49.4 Evaluation table for craneage selection exercise—Crane No 4.

Objectives	Alternative	
MUSTS	Superlift	GO/NO GO
A Max lift 25 tonnes	30 tonnes	✓
B 20 m boom length	5.3–26.3 m	✓
C Max purchase price £80 000	£78 000	✓
D Delivery less than 3 months	2 months	✓
E Hydraulic fly jib	Manual	✗
F Purchase on home market	Imported from USA	✗
G Conforms to ISO Standards	Yes (extra)	✓ (?)

Description	E/T	Ranking		Information	Rating	Weighting	
Price	E	10					
Delivery Period	E	10					
Good lifting capability	T		10				
Boom length	T		10				
Fly jib	T		10				
Low running costs	E	9					
Cost of spares	E	5					
One-driver operation	E	6					
Simple operation	E/T	8	8				
Fully telescopic boom	T		10				
Cab comfort	T		7				
Outriggers	T		8				
Power-assisted steering	T		6				
Gearbox	T		7				
Engine HP	T		6				
Variable hoisting speed	T		4				
Steering speed	T		3				
Boom telescopic speed	T		5				
Back-up services	E/T	6	6				
Turning radius	T		2				
Four-wheel drive	T		2				
Safe load indicator	T		10				
Travelling weight	T		1				
Overall width	T		9				
Overall length	T		1				
Accessories	E/T	4	4				
Fail-safe brakes	T		10				
Travelling speed	T		5				
E Economic element	29%	58					
T Technical element	71%		144				
Remarks: Failed MUST objective D and is eliminated.							

Table S49.5 Summary of evaluation of crane data.

Element	% of Total	Harris 30T	McCaffer L25
Cost and Service (E)	29%	369 (28%)	344 (28%)
Technical (T)	71%	923 (72%)	907 (72%)
Total:	100%	1292	1251

First choice — Harris 30T.

Solution 50 Setting a hire rate for plant

(1) Declining balance depreciation

L = resale value; P = purchase price; n = life of asset; d = percentage depreciation.

$$d = 1 - \sqrt[n]{\frac{L}{P}}$$

$$d = \left(1 - \sqrt[6]{\frac{2000}{15\,000}}\right) \times 100 = 28.52\%$$

Table S50.1 shows the declining balance at 28.52% and Table S50.2 shows a cost analysis.

$$\text{Cost of owning and operating the machine} = \frac{8315}{2000} = £4.16 \text{ per hour}$$

The cost only represents the marginal cost. The true cost to the owner must of course include an additional amount to cover overheads and profit over and above that required to service the interest charges on borrowed capital.

Table S50.1 Declining balance at 28.52%.

End of year	% of dep.	Dep. for year (£)	Book value (£)
0	28.52	0	15000
1	28.52	4275	10725
2	28.52	3050	7675
3	28.52	2190	5485
4	28.52	1570	3915
5	28.52	1115	2800
6	28.52	800	2000
		13000	–

Table S50.2 Cost analysis.

	£ per year
Depreciation in third year	2190
Interest on borrowed capital expressed in terms of an annual mortgage type payment and calculated from interest tables*	
$15000 \times 0.0264 \times 6 - 15000 \div 6$	1460
Insurance and tax –2% × 15 000	300
Ownership cost	3950
Fuel = 2000 × 10 × 0.10	2000
Oil and grease = 10% × 2000	200
Repairs = 10% × £15 000	1500
Operating cost	3700
Straight-line depreciation on tyres	500
Interest on borrowed capital † (100 × 0.615 × 2 – 2000) ÷ 2	115
Tyre maintenance	50
Tyre cost	665
Total cost	8315

Note:

* Uniform series capital recovery factor $= \dfrac{i(1+i)^n}{(1+i)^n - 1}$

$$\frac{0.15(1+0.15)^6}{(1+0.15)^6 - 1} = 0.264$$

† Uniform series capital recovery factor $= \dfrac{i(1+i)^n}{(1+i)^n - 1}$

$$\frac{0.15(1+0.15)^2}{(1+0.15)^2 - 1} = 0.615$$

where n = period of borrowing in years, i = annual rate of interest.

(2) Annual income

Let the annual income, which is assumed to be generated on the last day of the year, be £X. The cash flows can now be represented in Table S50.3.

The present worth factor $= 1/(1+i)^n$. The factors for Years 1–6 based on an interest rate of return at 15% are:

Period	$\dfrac{1}{(1+i)^n}$
Year 1	0.869
Year 2	0.750
Year 3	0.658
Year 4	0.572
Year 5	0.497
Year 6	0.432

In order that the desired rate of return can be achieved the cash outflows, and cash inflows, reduced to a common base year, must be in balance. If this common base year is taken as beginning Year 1, then the above present worth factors are applicable. Column (f) in Table S50.3 can now be expressed in the following equation.

Table S50.3 Cash flows.

Period (years)	a Capital outflow (£)	b Tax and insurance (£)	c Tyres (£)	d Operating costs (£)	e Income (£)	Cash flow $f = e - (a + b + c + d)$ (£)
0	−15 000	−300	−1000	−3750	0	−20 050
1		−300	−1000	−3750	+X	X − 4050
2		−300	−1000	−3750	+X	X − 5050
3		−300	0	−3750	+X	X − 4050
4		−300		−3750	+X	X − 5050
5		−300		−3750	+X	X − 4050
6	+2 000	0		0	+X	X + 2000

Note: Operating costs include tyre repairs and maintenance and are assumed to occur at the beginning of the year.

$$-20\,050+(X-4050)\times 0.869+(X-5050)\times 0.750+(X-4050)\times 0.658+(X-5050)\times 0.572+(X-4050)\times 0.497+(X+2000)\times 0.432=0 \quad (1)$$

$$3.78X = 3519+3787+2664+2888+2012-864+20\,050 \quad (2)$$

$$X=\frac{34\,056}{3.78}=£9009$$

$$\text{Hourly cost}=\frac{9009}{2000}=£4.5 \text{ per hour}$$

Solution 51 Plant workshop budget

The direct, overhead and total variances relating to the workshop budget are given in Table S51.1

Although the level of activity had increased the direct costs were kept under good control and achieved a favourable variance of £6000. Unfortunately extra costs were incurred over and above those budgeted for overheads, resulting in an adverse variance of £5000. However, the net benefit was still £1000.

Table S51.1 Budget variances.

	Budgeted cost for year	Adjusted budget for work done	Actual cost	Variance
Direct costs	160000	176000	170000	+6000 (Fav)
Overheads	40000	40000	45000	−5000 (Adv)
Total	200000	216000	215000	+1000 (Fav)

Solution 52 Sales variance on plant hire

Profit and fixed costs are 60% of revenue, thus with actual revenue only £600 000 compared to £800 000 budgeted, these will be under-recovered by

$$(800\,000 - 600\,000) \times \frac{60}{100} = £120\,000$$

The variances are analysed as follows:

	£
Sales variance	−120 000
Overhead costs variance	+2 000
Direct costs variance	+6 000
Depreciation variance	−3 000
Shortfall on profits	£115 000

Budgeted profit on £800 000 turnover was 20%, or £160 000, but as the budgeted fixed costs remain the same irrespective of turnover, profit is overstated by the shortfall figure. Thus actual profit is

$$160\,000 - 115\,000 = £45\,000$$

Reconciliation

	£	£
Actual revenue		600 000
Actual costs (40% of £600 000)	240 000	
Fixed costs (40% of £800 000)	320 000	
	560 000	560 000
Revised profit		40 000
Actual variances		+5 000
Profit		45 000

Solution 53 Variances on a plant item

(1)

$$\begin{aligned}
\text{Utilisation variance} &= (280 \times 2.5) - (200 \times 2.5) = £200 \\
\text{Price variance} &= (280 \times 2.4) - (280 \times 2.5) = -£28 \\
\text{Total} &\quad +£172 \\
&\text{(favourable)}
\end{aligned}$$

Increasing the operating hours raised the potential revenue by £200, but was actually reduced to £172 by the low hire rate obtainable. If both variances were consistently negative over a period of months then the option to sell the plant should be considered.

(2)

£	Budgeted costs for month	Adjusted budget for work done	Actual cost	Variance
Direct costs	200	200 × 134.4%* = 268.8	250	+18.8
Overheads	100	100	120	−20
Depreciation	100	100	125	−25
Total	400	468.8	495	−26.2

* Note: Actual revenue is 280 × 2.4 = £672, budgeted revenue was 200 × 2.5 = £500, i.e. 172/500 = 34.4% more.

Analysis of variance

$$\begin{aligned}
\text{Sales variance} &= (672 - 500) \times \text{fixed cost percentage} \\
&= 172 \times 60\% = £103.2
\end{aligned}$$

Thus:

	£
Sales variance	= +103.2
Direct costs variance	= +18.8
Overhead variance	= −20
Depreciation variance	= −25
Actual increase of profit	= +77
Budgeted profit	= 20% × 500 = £100

Thus the firm has made £177 profit during the month.

Reconciliation

Revenue	£672
Actual cost	£495
Profit	£177

Solution 54 Capital gearing

In solutions to Questions 54, 56 and 58, corporation tax liability is affected by capital allowances and other factors, the complexities of which are outside the scope of this book.

Capital gearing = ratio of fixed return capital to ordinary share capital.

Fixed return capital (FRC) is preference shares, debentures and loans.

	Company A	*Company B*
10% debentures	–	120
7% preference shares	45	45
	–	–
FRC	45	165
Ordinary shares	155	35
Capital gearing ratio	$\frac{45}{155} \times 100 = 29\%$	$\frac{165}{35} \times 100 = 470\%$
	Low geared	**High geared**

Company A is low geared (Table S54.1) and may be able to declare a dividend to ordinary shareholders even at low levels of profit. Company B is high geared (Table S54.2) and must therefore ensure that profits are stable and regular and above all sufficient to meet the interest due on debenture holdings. However, the highly geared company should be able to yield a higher rate of return to ordinary shareholders when profits are high.

Comments

Company A is able to pay the full dividend to preference shareholders at all times, even the ordinary shareholder receives payment at all the given levels of profit.

Company B is in a much more difficult position while profits are poor. In fact, for the first profit period a loan or bank overdraft must be negotiated in order to meet the interest charges on the debenture stock. The full dividend to preference shareholders cannot be met until profit level 4, when there is also a small amount available for distribution to ordinary shareholders. However, the high profits at levels 5 and 6 yield a far better return to the ordinary shareholders than in the low-geared company. This trend will continue for all increased levels of profit.

If the preference shares were issued at cumulative value, then in the case of the highly geared situation additional dividend would be paid out in the good years to make up for the lost years, but at the expense of the ordinary shareholder.

Table S54.1 Company A—low geared (£000).

	Item					
	1	2	3	4	5	6
Ordinary shares (50p)	155	155	155	155	155	155
Preference shares	45	45	45	45	45	45
Debentures		–	–	–	–	–
Total capital employed (£000)	200	200	200	200	200	200
Profit before tax	10	12	18	20	25	30
Corporation tax at 50%	5	6	9	10	12.5	15
Profit available as dividend	5	6	9	10	12.5	15
Dividend on pref. shares	3.15	3.15	3.15	3.15	3.15	5.15
Profit available to ordinary shareholders	1.85	2.85	5.85	6.85	9.35	11.85
Profit per ordinary share	0.6p	0.92p	1.89%	2.2p	3p	3.8p
Return on ordinary share capital	1.2%	1.84%	2.78%	4.4%	6%	7.6%
Return on total share capital	2.5%	3%	4.5%	5%	6.25%	7.5%

Table S54.2 Company B—high geared (£000).

	Item					
	1	2	3	4	5	6
Ordinary shares (50p)	35	35	35	35	35	35
7% Preference shares	45	45	45	45	45	45
10% Debentures	120	120	120	120	120	120
Total capital employed (£000)	200	200	200	200	200	200
Profit before tax	10	12	18	20	25	30
Interest on 10% debentures	12	12	12	12	12	12
Profit available	−2	Nil	6	8	13	18
Corporation tax at 50%	Nil	Nil	3	4	6.5	9
Profit available as dividend	Nil	Nil	3	4	6.5	9
Dividend to preference shareholders	Nil	Nil	3	3.15	3.15	3.15
Profit available to ordinary shareholders	Nil	Nil	Nil	0.85p	3.35p	5.85
Profit per ordinary share	Nil	Nil	Nil	1.2p	4.8p	8.35p
Return on ordinary share capital	Nil	Nil	Nil	2.4%	9.6%	16.7%
Return on total share capital	Nil	Nil	3.75%	5.0%	8.12%	11.25%

Solution 55 Working capital

Working capital requirements

	Time factor
Materials (40% of turnover)	
Materials held on sites	3 months
Work in progress	2 months
	5 months
Credit from materials suppliers	1 month
	4 months
Plant (15% of turnover)	
Work in progress	2 months
Credit from hire company	1 month
	1 month
Labour (20% of turnover)	
Work in progress	2 months
Overheads (15% of turnover)	
Work in progress	2 months

Let £X be the annual turnover required, then the proportion of plant materials, labour and overheads in the turnover is as follows:

$$\text{Plant}(X_p) = \frac{15}{100} \times X \quad \text{Materials}(X_m) = \frac{40}{100} \times X$$

$$\text{Labour}(X_L) = \frac{20}{100} \times X \quad \text{Overheads}(X_o) = \frac{15}{100} \times X$$

Each of the items must have sufficient working capital available to meet the commitments until payment is received from the client. Therefore the working capital required is

$$\text{working capital} = \frac{\text{time factor}}{12 \text{ months}} \times X_p + \frac{\text{time factor}}{12 \text{ months}} \times X_m + \frac{\text{time factor}}{12 \text{ months}} \times X_L + \frac{\text{time factor}}{12 \text{ months}} \times X_o$$

The working capital available = £180 000.

Thus the above equation is solved as follows:

$$180\,000 = \frac{1}{12} \times X_p + \frac{4}{12} \times X_m + \frac{1}{6} \times X_L + \frac{1}{6} \times X_o$$

$$180\,000 = \frac{1}{12} \times \frac{15}{100} \times X + \frac{1}{3} \times \frac{40}{100} \times X + \frac{1}{6} \times \frac{20}{100} \times X + \frac{1}{6} \times \frac{15}{100} \times X$$

$$= 0.0125X + 0.1333X + 0.0333X + 0.0250X$$

$$180\,000 = 0.2041X$$

$$X = 881\,920$$

Acceptable turnover, say £882 000 + 10% profit.

Summary of capital requirements

	£
Long-term capital for establishment	45 000
Working capital to be provided by private source or short-term loan	180 000
Total	225 000

Solution 56 Balance sheet

(1) Table S56.1 shows a tabulated revenue account, year ended 31 December.
(2) Table S56.2 shows a tabulated balance sheet at 31 December.

Table S56.1 Tabulated profit and loss statement, year ended 31 December.

	£000	£000	£000
Work certified		695	
Work in progress		10	
Value of production		705	705
Materials used			
Opening stock	15		
Delivered to site	300		
Closing stock	10		
	305	305	
Wages incurred		250	
Salaries		100	
Cost of production		655	655
Production profit			50
Depreciation on plant and equipment	20% of 100	20	
Net profit before tax			30
Corporation tax	50% of 30	15	
Profit after tax			15
Proposed dividend		7.5	
Balance carried forward			7.5

Table S56.2 Tabulated balance sheet at 31 December.

Employment of capital	£000	£000	£000
Fixed assets			
Land and buildings		50	
Plant and equipment	100		
Depreciation	20	80	130
Current assets			
Cash at bank		40	
Stock of materials		10	
Work in progress		10	
Debtors—work certified but not paid		25	85
			215
Current liabilities			
Creditors	300		
Materials	295		
	5		
Plant	100		
	95		
	5		
Wages and salaries	350		
	345		
	5		
	15	15.0	
Proposed dividend		7.5	
Corporation tax due		15.0	
		37.5	
		177.5	
Capital employed			
Share capital authorised, issued and paid up			
150 000 ordinary shares at £1 each		150.0	
Reserves			
General		20.0	
Capital		7.5	
Profit and loss account		177.5	177.5

Solution 57 Financial ratios

Table S57.1 shows the important financial ratios for Question 56.

Table S57.1 Important financial ratios for Question 56.

Type	Formula	Value	Comments
Sales (earnings) ratio %	$\frac{\text{Net profit before tax}}{\text{Sales}}$	$\frac{30000}{705000} \times 100 = 4.25\%$	Average for construction work
Turnover ratio	$\frac{\text{Sales}}{\text{Capital employed}}$	$\frac{705000}{177500} = 4$	Fair; should be between 4 and 8
Primary ratio %	$\frac{\text{Net profit before tax interest}}{\text{Current liabilities}}$	$\frac{15000}{177500} \times 100 = 8.45\%$	Fair
Current ratio	$\frac{\text{Current assets}}{\text{Current liabilities}}$	$\frac{85\,000}{37\,500} = 2.27$	Good; 2 : 1 usually accepted as sound
Liquidity ratio (acid test)	$\frac{\text{Cash + debtors}}{\text{Current liabilities}}$	$\frac{40\,000 + 25\,000}{37\,500} = 1.73$	High; 1 : 1 usually sufficient
	$\frac{\text{Debtors}}{\text{Sales}} \times 12$	$\frac{25000}{705000} \times 12 =$ 0.43 months	Excellent. Debtors are paying up quickly; usually three months is acceptable
	$\frac{\text{Creditors}}{\text{Purchases}} \times 12$ (materials and plant)	$\frac{15000}{300000} + 100000 \times 12 =$ 0.45 months	Company is paying bills far too quickly; 3–6 months is normal

Solution 58 Interpretation of the balance sheet

(1) Profit and loss account at 31 December

	(£)
Net profit	67 500
Interest paid to debenture holders	6 000
	61 500
Tax paid	30 000
	31 500
Dividend paid to ordinary shareholders	18 000
Balance carried forward	13 500

(2) Source and application of funds

Source of funds

Net profit before tax	67 500
Increase in bank overdraft	10 500
Depreciation on plant and equipment	33 000
Increase in creditors	45 000
Decrease in work in progress	7 500
Additional share capital: 10 000 × 1.25	12 500
	176 000

Application of funds

Plant and equipment purchased	100 000
Increase in debtors	30 000
Increase in stock of materials	15 000
Dividend paid to shareholders	18 000
Interest paid to debenture holders	6 000
Redemption of 7500 debentures	7 500
Tax paid to Inland Revenue	30 000
	207 000
Reduction in cash at bank	31 000

(3) Current and liquidity ratios

(a)

$$\text{Current ratio (CR)} = \frac{\text{current assets}}{\text{current liabilities}}$$

Previous year *This year*

$$CR = \frac{300}{90} = 3.33 \qquad CR = \frac{306.5}{145.5} = 2.11$$

(b)

$$\text{Liquidity ratio (acid test, AT)} = \frac{\text{cash} + \text{debtors}}{\text{current liabilities}}$$

Previous year *This year*

$$AT = \frac{31+112}{90} = 1.59 \qquad AT = \frac{142}{145.5} = 0.985$$

Generally, the measures taken during this year in delaying payments to suppliers and thereby increasing creditors were sensible and produced an acceptable working capital ratio. However, the cash balances were depleted and debtors were allowed additional time to pay. A better policy would be to reverse this arrangement. A tabulated balance sheet at 31 December is given in Table S58.1.

Table S58.1 Tabulated balance sheet at 31 December.

Employment of capital	£	£	£
Fixed assets			
Land and buildings		220 500	
Plant and equipment	229 500		
	100 500		
	330 000	297 000	
Less depreciation	33 000		
			517 500
Current assets			
Debtors		142 000	
Cash at bank		—	
Stock of materials		77 000	
Work in progress		87 500	
			306 500
			824 000
Current liabilities			
Creditors		135 000	
Bank overdraft		10 500	
			145 500
			678 500
Capital employed			
460 000 ordinary shares at £1 each		460 000	
Share premium account		2 500	
67 500 8% debentures at £1 each		67 500	
General reserve		135 000	
Profit and loss account		13 500	
			678 500

Solution 59 Mass haul problem

The last column of Table Q59.1 is expressed graphically in Fig. S59.1. If the cut material is used to supply the fill, then the volumes of earth moved over haul distances 200 to 500m can be read directly off the graph for each haul.

200 m haul

- Cut and cart away $= 132\,000 + (37\,000 - 19\,000) = 150\,000\ \text{m}^3$

$$\text{Cost} = 150\,000 \times 0.50 = £75\,000$$

- Fill from borrow pit $= (132\,000 - 19\,000) + (37\,000 - 26\,000) = 124\,000\ \text{m}^3$

$$\text{Cost} = 124\,000 \times 0.75 = £93\,000$$

- Cut and fill up to 200 m haul $= (140\,000 - 132\,000) + (19\,000 - 8000) + (48\,000 - 37\,000)$

$$= 30\,000\ \text{m}^3$$

$$\text{Cost} = 30\,000 \times 0.3 = £9000$$

Total cost $= 75\,000 + 93\,000 + 9000 = £177\,000$

300 m haul

- Cut and cart away $= 126\,000 + (32\,000 - 24\,000) = 34\,000\,\text{m}^3$

$$\text{Cost} = 134\,000 \times 0.5 = £67\,000$$

- Fill from borrow pit $= (126\,000 - 24\,000) + (32\,000 - 26\,000) = 108\,000\,\text{m}^3$

$$\text{Cost} = 108\,000 \times 0.75 = £81\,000$$

- Cut and fill up to 300m haul $= (140\,000 - 126\,000) + (24\,000 - 8\,000) + (48\,000 - 32\,000)$

$$= 46\,000\,\text{m}^3$$

$$\text{Cost} = 30\,000 \times 0.3 + (46\,000 - 30\,000) \times 0.8 = 9000 \times 12\,800 = £21\,800$$

$$\text{Total cost} = 67\,000 + 81\,000 + 21\,800 = £169\,800$$

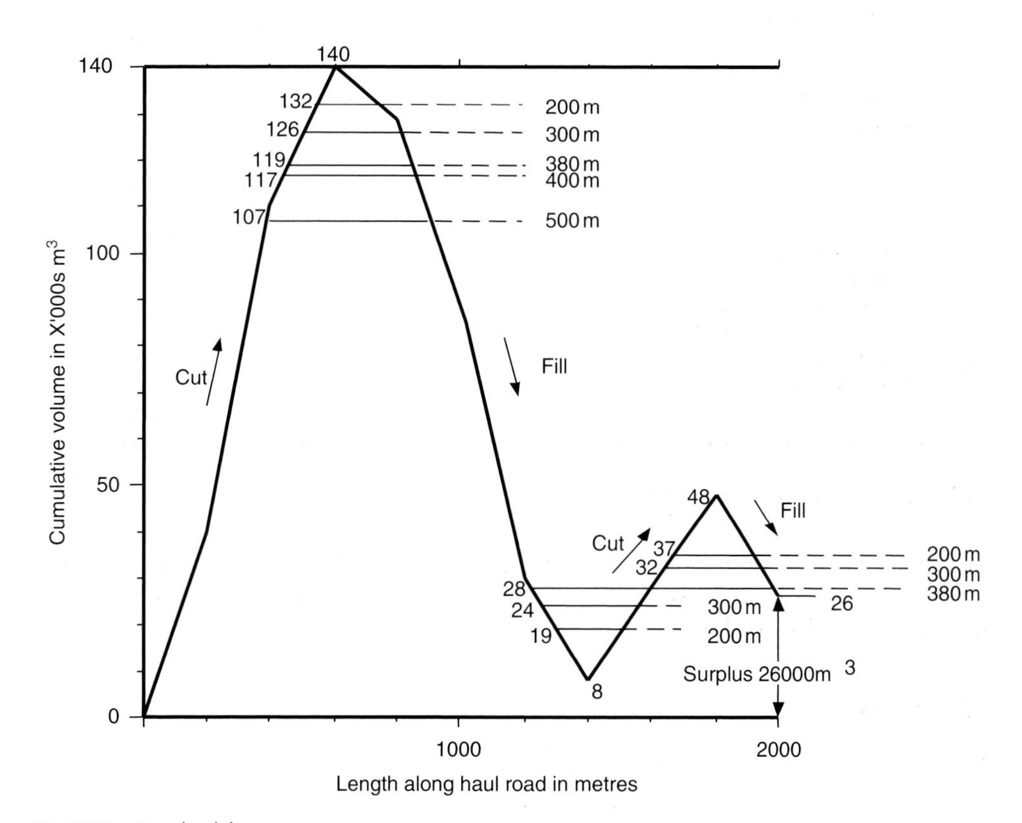

Fig. S59.1 Mass haul diagram.

380 m haul

- Cut and cart away $= 119\,000\ \text{m}^3$

$$\text{Cost} = 119\,000 \times 0.5 = £59\,500$$

- Fill from borrow pit $= (119\,000 - 28\,000) + (28\,000 - 26\,000) = 93\,000\ \text{m}^3$

$$\text{Cost} = 93\,000 \times 0.75 = £69\,750$$

- Cut and fill up to 380 m haul $= (14\,0000 - 119\,000) + (28\,000 - 8000) + (48\,000 - 28\,000)$

$$= 21\,000 + 20\,000 + 20\,000 = 61\,000\ \text{m}^3$$

$$\begin{aligned}\text{Cost} &= 30\,000 \times 0.3 + (46\,000 - 30\,000) \times 0.8 + (61\,000 - 46\,000) \times 1.6\\ &= 9\,000 + 12\,800 + 24\,000 = £45\,800\end{aligned}$$

$$\begin{aligned}\text{Total cost} &= 59\,500 + 69\,750 + 45\,800\\ &= £175\,050\end{aligned}$$

500 m haul

Clearly from a position 1200m along the road there can be no advantage in adopting a 500m haul, but it may well be more economic to use a 500 m haul up to this point.

- Cut and cart away $= 107\,000\ \text{m}^3$

$$\text{Cost} = 107\,000 \times 0.5 = £53\,500$$

- Fill from borrow pit $= (107\,000 - 28\,000) + (28\,000 - 26\,000) = 81\,000\ \text{m}^3$

$$\text{Cost} = 81\,000 \times 0.75 = £60\,750$$

- Cut and fill up to 380 m haul $= £41\,000$

$$\text{Cut and fill } 380 - 400\ \text{m haul} = (119\,000 - 117\,000) = 2000\ \text{m}^3$$

$$\text{Cost} = 2000 \times 1.6 = £3200$$

$$\text{Cut and fill } 400 - 500\ \text{haul} = (117\,000 - 107\,000) = 10\,000\ \text{m}^3$$

$$\text{Cost} = 10\,000 \times 2.8 = £28\,000$$

$$\text{Total cost} = 53\,500 + 60\,750 + 45\,800 + 3200 + 28\,000 = £191\,250$$

Figure S59.2 shows the costs plotted against haul distance. The economic haul is 326 m and the cost £169 600.

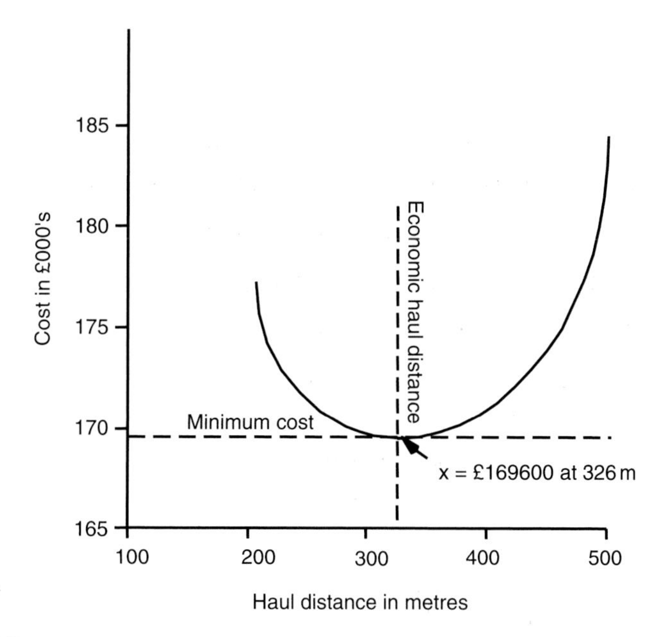

Fig. S59.2 Haul distance.

Solution 60 Development budget

(1) *Gross Development Value (GDV)*

Gross floor area	$1000\,m^2$
Utilisation	0.8
Net lettable space	$800\,m^2$
Rent	£50
	£40 000
Less management costs 10%	£4 000
Net income p.a.	£36 000

$$\text{GDV} = \frac{\text{Net income} \times 100}{\text{Yield}} = \frac{36\,000 \times 100}{5} = £720\,000$$

i.e. represents £720 000 capital investment

Cost of development

Construction	1000 × £400	= £400 000
Design fees	10%	£40 000
		£440 000
Finance (i.e. const. costs × av. period × interest)	440 000 × 1.5 × 0.05	£33 000
Development profit	£720 000 × 15%	£108 000
		£581000

Sum available for purchase of land		£720 000
	less	£581 000
		£139 000

Price of land at compound interest of 10% equals sum available, i.e.

$$\text{Bid price} = \frac{£139\,000}{1.2099} = £114\,885$$

(2) If the development suffered a lettable void of, say, 1 year then interest would be charged on the total cost including land as follows:

$$£720\,000 \text{ at } 10\% = £72\,000$$

Hence, the GDV increases to £792 000, with a drop in the yield to

$$\frac{36\,000}{792\,000} = 4.54\%$$

Consequently, rents will have to be correspondingly higher or developer's profit lower or a combination of both if the project is to provide the original yield.

Solution 61 Effects of a lease on development budget

(1) The cost of the development can only be recovered over the leasing period, and therefore similarly to the case of a lettable void, rents will need to rise by the equivalent sum to restore the equilibrium calculation against the costs.
Broadly, the cash flows as a DCF-type problem are:

Year	*Cash out*	*Income*	*Total*
0	£720 000	0	−£720 000
1	0	0	0
2	0	x	x
3	0	x	x
etc.			
50	0	x	x

At a Present Worth of 5% yield the sum of the total column equates to zero, i.e. from interest tables 720 000 − (x × PW of 1 per period over 50 years) × (PW of 1 over 2 years) = 0.

$$720000 - (x \times 18.256) \times 0.907 = 0$$

x = £43 483, i.e. annual net income required to generate sufficient funds to replace the development after 52 years.

(2) Finally if the development were planned for only the given life of the building, then as in the above situation the original capital has to be regenerated but in this case with the exclusion of the cost of land since this would remain in ownership.

20
Questions and solutions – operational research (OR)

Summary

The Operational Research Society explains OR as development of a scientific model of a system incorporating measurements of factors such as chance and risk, with which to predict and compare the outcomes of alternative strategies or means of control. In the main, the method is suited to those logistical problems that are highly dependent on detailed and accurate records of performance data and which can broadly be analysed and solved by mathematical programs. Combined with readily available computer software OR can positively assist in better construction planning, resources allocation, economic analysis and performance evaluation to achieve marginal improvements in scheduling and control.

The following methods provide examples that demonstrate potential applications to construction management:

- Linear, integer, binary integer linear programming
- Simplex method, transportation and assignment
- Quadratic assignment programming
- Goal programming
- Non-linear programming
- Dynamic programming
- Decision trees and utility
- Markov processes
- Queuing theory, simulation
- Stock control, inventory under uncertainty
- Theory of games

Questions

Question 1 Linear programming illustrating maximisation

A building contractor produces two types of house for speculative building—detached and semi-detached houses. The customer is offered several choices of architectural design and layout for each type. The proportion of each type of design sold in the past is shown in Table Q1.1. The profit on a detached house and a semi-detached house is £1000 and £800, respectively.

Table Q1.1 Proportion of sales.

Choice of design	Detached	Semi-detached
Type A	0.1	0.33
Type B	0.4	0.67
Type C	0.5	

The builder has the capacity to build 400 houses per year. However, an estate of housing will not be allowed to contain more than 75% of the total housing as detached. Furthermore, because of the limited supply of bricks available for Type B designs, a 200-house limit with this design is imposed.

(1) Determine how many detached and semi-detached houses should be constructed in order to maximise profits and state the optimum profit.
(2) Calculate the extra profit generated by increasing the capacity above 400 houses per year. Is this a sensible policy if the cost of providing the additional resources is £1000 per house?

Question 2 Linear programming illustrating minimising

A workshop in servicing engines replaces two parts A and B in the proportions shown below at a cost of £10 and £8 each, respectively. At least 10 per day of type A need to be completed and the workshop has the capacity to handle 40 parts in total per day.

Part A requires 0.5 hours while B requires 0.2 hours and 10 hours is worked each day.

Part type	Proportion
A	1/3
B	2/3

(1) Determine how many parts of each type per day should be fitted in order to minimise production cost and state the optimum production cost.
(2) Is this a sensible policy if the workshop overtime costs are £30 per hour?

Question 3 Simplex method—constraints: 'equal to or less than'

A construction crew of five workers is provided on site with two concrete delivery systems type A and type B, and is paid respectively £50 and £60 per 100 m^3 of concrete placed. The equipment and labour resources production output rates are as follows:

	Dumpers	Crane	Crew
Type A	12.5 m^3/hr	25 m^3/hr	2 m^3/hr
Type B	25 m^3/hr	20 m^3/hr	0.13 m^3/hr

How should the delivery systems be organised to maximise income over the 40-hour working week?

Question 4 Simplex method—constraints: 'greater than or equal to'

Solve the following problem of the type:

$$2x+3y\geq 14$$

$$8x+5y\geq 40$$

$$x\geq 0$$

$$y\geq 0$$

Maximise $1.5x+y$.

Question 5 Simplex method—constraints: 'greater than or equal to'

Find the solution to Question 4 as a minimum.

Question 6 Integer linear programming

Solve Question 3 as an integer problem

Question 7 Binary integer linear programming

A further application of binary variables in the form of 0 and 1 values can also be very useful for modelling logical conditions. Solve Questions 3–6 with the following additional information:

- System A set up costs £50 with a maximum delivery capacity 175 m^3/week.
- System B set up costs £125 with a maximum delivery capacity 700 m^3/week.

Question 8 Transportation problem

A civil engineering contractor is engaged to carry out the earth moving work for the construction of a new section of motorway. Fill material can be supplied from three borrow pits l, 2 and 3 located near the works, up to a maximum of 60 000, 80 000 and 120 000 m^3 from each, respectively. The material is to be delivered to three locations, A, B and C along the road, and each requires 90 000, 50 000 and 40 000 m^3, respectively. The costs in pence per cubic metre for delivery of the material from the pits to each of the three sites are shown in Table Q8.1.

Determine the best arrangements, for supply and delivery of the material, if the objective is to minimise cost.

Question 9 Assignment problem

A contractor has been successful in obtaining five new projects. The projects, however, are different in value, type of work and complexity. As a result, the experience and qualities required of the project manager for each will be different. After careful considerations, five managers are selected and their skills assessed against each project. Each manager is scored on a points scale, with a maximum of 100 marks indicating that the manager is highly suitable, and zero mark unsuitable for the work. The individual assessments are shown in Table Q9.1.

Which managers should be allocated to which projects, if the company wishes to distribute them in the most effective way?

Table Q8.1 Supply and delivery of fill material.

	Quantity required at sites			Quantity supplied from pits
	A (90 000 m^3)	**B** (50 000 m^3)	**C** (40 000 m^3)	
1	70p	40p	100p	60 000 m^3
2	50p	30p	90p	80 000 m^3
3	60p	50p	90p	120 000 m^3

Table Q9.1 Individual assessments.

		Manager				
		(1)	(2)	(3)	(4)	(5)
Project	(A)	75	28	61	48	59
	(B)	78	71	51	35	19
	(C)	73	61	40	49	68
	(D)	55	50	52	48	63
	(E)	71	60	61	74	70

Question 10 Assignment when number of agents and tasks differ

Solve Question 9 if only four managers 1, 2, 3 and 4 are available.

Question 11 Goal programming

If in Question 1 the first priority is to build 200 houses of Type B design, and second priority to complete no more than 400 in total, determine the optimum mix of housing.

Question 12 Non-linear programming

Problems not having linear properties are called non-linear and comprise relationships with variables expressed to powers other than one, e.g. quadratic form (squared), multiplied or divided by each other or cubed, etc. This occurs when either the objective function or the constraints or both are non-linear. Finding optimal solutions requires sophisticated methods of calculation and may provide only local optimum results.

(a) Maximise $x^2 - 2x + y^2 - 6y$
(b) Minimise $x^2 - 2x + y^2 - 6y$

subject to the constraints

$$2x + y < 8$$

$$x \geq 0, y \geq 0$$

Question 13 Quadratic assignment problem

In order to avoid the quadratic terms in a non-linear problem, expressions can sometimes be re-arranged into the assignment-type linear form to provide the solution. For example:

a contractor's stores depot receives goods from two suppliers on lorries, which are directed to two bays, truck A into Bay 1 and truck B to Bay 2, and after an inspection operation part loads are transferred between the trucks for onward despatch to the various projects under construction in the region as follows.

Amount of items transferred:

	To truck	1	2
From	1	0	30
truck	2	20	0

Transfer time in minutes from a bay to another bay:

	To Bay	A	B
From	A	0	5
Bay	B	15	0

Assign lorries to bays to minimise total transfer time between lorries and bays.

Question 14 Dynamic programming—batch processing

A contractor in purchasing materials for a large contract decides to minimise the level of bank overdraft. The budgeted cost of purchases at the beginning of each month shown below are recovered in revenues at the end of the month or alternatively held in stock at a cost of 1% per month on the value of stock and there is an administration charge of £300 for each purchase transaction.

When should stocks be replenished and by what amounts if the contractor aims to minimise total charges?

Month	Purchases
Jan	£5 000
Feb	£15 000
Mar	£20 000
Apr	£25 000

Question 15 Resources routing—knapsack problem

The manager of a design office schedules staff to projects over the next ten days aiming to deliver drawings to site according to priority. The estimated completion time for each job, the number of jobs and the priority value are listed in the table below.

Schedule the jobs for the period.

Job category	Number of jobs	Production time days	Priority
1	2	2	2
2	1	4	6
3	2	6	10

Question 16 Decision tree problem

A construction company is confronted with planning its market strategy over the next five-year period. Forecasts for the national economy are good, and the company planners assess that there is 0.8 probability of high growth and 0.2 probability of low growth in the demand for construction work.

The ability of the company to increase turnover depends largely upon the state of the market, but its market can be improved by investing in more facilities. The latter situation is possible either by setting up new area divisions immediately or by a more gradual expansion of existing facilities to suit the market conditions.

The planners, after carrying out investment appraisal on alternative strategies, calculate the profitability for the company as follows:

	Probability of Growth		
	High (0.8)	Low (0.2)	
Use existing facilities only	12%	10%	profitability
Apply gradual expansion	16%	9%	profitability
Immediate expansion	18%	zero	profitability

Express the problem in the form of a decision tree and recommend the decisions which should be made.

Question 17 Utility

A company contemplates expansion alternatives in two regions A and B away from its home base and estimates profitability as follows.

	High Demand (0.8)	Low Demand (0.2)
a_1 No expansion	12%	10%
a_2 Expand in new region A	16%	5%
a_3 Expand in new region B	18%	0%

The company also carries out a utility assessment to augment the expected cost evaluation by generating the following information:

Profitability	Indifference value (p)	Utility value
18%	1	10
16%	0.95	9.5
12%	0.6	6.0
10%	0.4	4.0
5%	0.30	3.0
0%	0	0

Determine which outcome the company might prefer.

Question 18 Markov processes

A plant firm's workshop has been recording the running performance of a machine at one hour periods and determined that the probability state of the machine being in a running or down state relates to its state in the previous period as follows:

	To running	To down
From running	0.85	0.15
From down	0.2	0.8

During maintenance the cause of breakdown was revealed to be failure of a specific part, which if replaced is assessed to improve the transition probabilities as shown below:

	To running	To down
From running	0.9	0.1
From down	0.3	0.7

(1) Determine the steady-state probabilities of the running and downtime states.
(2) What are the steady-state probabilities after the failed part is replaced?
(3) If the cost of downtime for a one-hour period is £100, determine the break even price that can be paid for the new component (including its fitting) on a time-period basis.

Question 19 Queuing

Trucks arrive at an excavator, where observations over a 136-minute period show that the numbers of trucks arriving per minute (i.e. time period) are 0, 1, 2, 3, 4 or more and have corresponding frequencies 78, 41, 15, 2, 0. Thirty service (loading) times are also recorded as follows: 0.5, 1, 1.5, 2, 2.5, 3 minutes or less, and the corresponding frequencies are 9, 9, 7, 3, 1, 1.

Determine whether the observed data for arrival times and service time represent respectively Poisson and Exponential probability distributions.

Calculate the average time trucks spent in the system (i.e. waiting and loading).

Calculate the average number of trucks in the system:

(a) if the calling ie. arrivals population is infinitely large
(b) if there are only four trucks in the fleet

Determine the optimum number of excavators to minimise cost, when the hire costs of an excavator and truck are respectively £50 and £60 per hour and a maximum of 3 excavators are available.

Question 20 Simulation

Trucks arrive at a batching plant from distribution points in an area serviced by a ready-mixed concrete depot. The arrival time intervals of the trucks are observed and yield the following results:

Arrival time interval (min)	*Frequency (%)*
2	10
3	15
4	30
5	25
6	20

The times taken to load the trucks, which are either 3 or 6 m^3 capacity, are fairly constant at 3 and 5 min respectively, and both types are equally represented at the depot.

If the batching plant loads each truck immediately it arrives, in the order that it arrives, calculate the total time likely that the mixer and trucks will be waiting in any one period of two hours selected at random.

Question 21 Stock control—simple case

Stocks of cement on a construction site are allowed to run down to a level of 3 tonnes before being replenished. Cement is used steadily at 50 tonnes per week. Cement costs £15 per tonne and the cost of storage and deterioration per week is 10% of the cement price. Each time cement is ordered there is a cost of processing this order of £1.

(1) How much cement should be ordered each time?
(2) What is the cost of ordering and storing the cement?

Question 22 Stock control—continuous usage

A manufacturer of pre cast concrete units is required to supply 1000 pre cast floor beams each week to a large building contract. The building contractor has very little storage space on site and thus requires the beams to be delivered at the rate at which they can be placed in position. The manufacturer has the capacity to produce 2500 beams per week. The cost of storing a single beam per week is 1p and the cost of setting up the equipment for a production run is £50.

(1) What is the optimum number of units to produce in a production run?
(2) What is the total cost of producing and storing the contractor's requirements for beams?
(3) How frequently should production runs be made?

Question 23 Stock control—shortages

A building project calls for the steady supply of 50 000 bricks each week; the price of the bricks is £30 per thousand. The brick supplier usually keeps sufficient stocks in the yard to meet customer demand and the cost of holding 1000 bricks per week is 10% of their cost price. The cost to the supplier each time a new order is processed is £10. However, sometimes the delivery date cannot be met so to make up the backlog there are special deliveries to the customer as soon as the supplier is able to continue with the order. The extra cost incurred by the supplier in this situation is £10 per 1000 bricks.

(1) Calculate the economic order quantity for the supplier.
(2) Calculate the total cost per week to the supplier of stockholding and processing orders.
(3) Calculate the level to which stock on site is topped up.

Question 24 Stock control—discounts

(This question should not be attempted before trying Question 21.)

Aggregate is used on a construction site at the rate of 2000 tonnes per month. The cost of ordering is £20 and the cost of storing the material on site is 50% of its purchase cost. The cost per tonne depends on the total quantity ordered as follows:

(a)	(b)	(c)
Less than 500 tonnes £1.21 per tonne	500–999 tonnes £1.00 per tonne	1000 or more £0.81 per tonne

Calculate the optimum order quantity and the optimum total cost per month of purchasing, storing and ordering the material.

Question 25 Inventory under uncertainty

Demand for equipment maintenance can vary between 95 and 105 items per week, accordingly the workshop manager budgets midway for servicing 100 items per week (Q) at a cost of £200 each.

Each item serviced is charged to customers at £300. If demand exceeds the plan, during the week customers are turned away. Conversely, when demand is less than plan the workshop resources are under-utilised but the budgeted costs are still incurred.

Suppose the probabilities of demand are those shown below. Determine the Optimal Order Quantity Q.

Possible demand	Probability (P_r)
95	0.05
96	0.06
97	0.09
98	0.11
99	0.15
100	0.20
101	0.13
102	0.10
103	0.07
104	0.03
105	0.01
	1.00

Question 26 Theory of games

Mixed strategy

A listed company A has two operating divisions, plant and construction, which have come under takeover speculation from company B. Company A has sufficient resources to successfully resist a bid for one or the other division. Similarly, the predator firm B is capable of mounting a successful takeover of either plant or construction. The construction division is worth £3 m, the plant division £2 m, and nothing is lost if the divisions ultimately remain under the ownership of the parent A firm.

Determine the value of the game and the best strategies for A and B.

Solutions

Lindo software at www.Lindo.com, and similar proprietary products, which facilitate interactive use from the keyboard or customised subroutines linked directly to form an integrated program, have been adopted where appropriate in this text.

Solution 1 Linear programming illustrating maximisation

This type of problem can be readily solved using the linear programming mathematical technique developed to determine a maximum value such as profit or minimum cost, where availability of the associated resources is constrained.

The descriptor 'linear' implies proportionality while 'programming' refers to the mathematical process of determining an optimum allocation of resources.

Problem formulation

The steps needed to be undertaken in solving a linear programming problem are as follows;

(1) Define the decision variables, these being the controllable inputs in both the objective and constraint equations and which describe the number or amounts of each item to produce.
(2) Describe the objective function in terms of the decision variables, this being the subject to be optimised, e.g. profit, output, etc. expressed as a formula.
(3) Establish the constraint value for each resource that has an imposed limit and express in terms of the decision variables.
(4) Solve for the decision variables and substitute into the objective function and determine its quantity or amount.

In this first problem the decision variables can be established as follows:

Part (a)

Decision variables:

Let X = number of detached houses constructed
Y = number of semi-detached houses constructed

Row 1. Total profit objective function:

$$MAX = 1000 \times X + 800 \times Y$$

Rows 2 and 3. Constraints to prevent negative values of X and Y:

$$X > 0$$
$$Y > 0$$

Row 4. Constraint on the builder's capacity:

$$X + Y \leq 400$$

Row 5. Constraint on detached:

$$X \leq 0.75 \times (X + Y)$$

Row 6. Constraint on B type design:

$$0.4 \times X + 0.67 \times Y <= 200$$

The equations plotted graphically in Fig.S1.1 illustrate the area of feasible solutions and the optimum profit is the solution which maximises the objective function $1000X + 800Y$.

Clearly various combinations of X and Y will lead to the same profit. For example, a profit of £160 000 is obtained for all pairs of X and Y lying on the line $1000X + 800Y = 160\,000$.

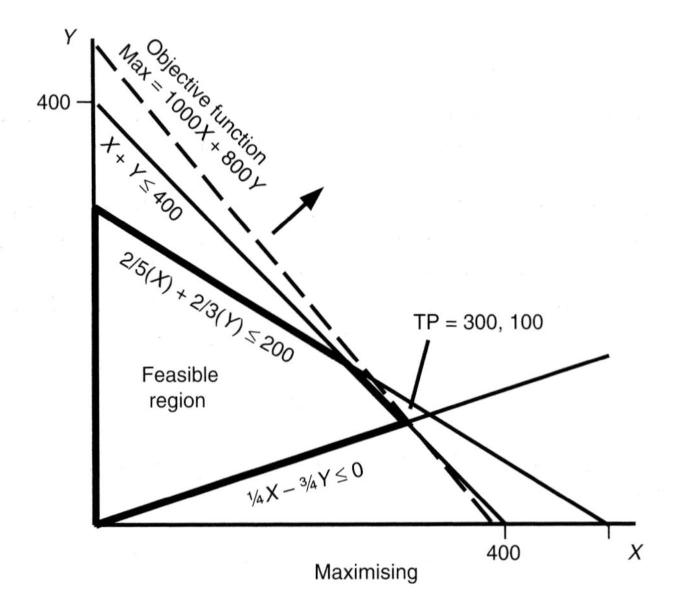

Fig. S1.1

The profit will continue to grow with any line drawn parallel to the objective function, moving away from the origin until the line is just tangential to the feasible area, the objective function then being at the tangential point.

The co-ordinates of X and Y at the tangent point (TP) are 300 and 100 respectively, which when substituted into the objective function yield a profit of £380 000. In other words, to optimise profit, 300 detached and 100 semi-detached is the best mix. The amount of each design is shown below.

Summary

Objective value:	£380 000
Variable	Value
X	300
Y	100
Row	Slack or surplus
1	380 000
2	300
3	100
4	0
5	0
6	13

The above summary illustrates that the constraints for all rows were not violated in not being negative values. However, for row 6 the constraint equation of $0.4X+0.67Y$ substituted with X, Y values of 300 and 100 respectively gives $200-0.4\times 300+0.67\times 1000=13$, i.e. produces a slack amount of 13 houses, meaning that this resource is under-utilised. A negative, i.e. surplus, value would have indicated insufficient availability of this resource.

Part (b) Shadow prices or duality

Clearly there would be little point in increasing the limit of 200 houses on design type B, as the constraints on rows 4 and 5 are presently fully utilised, i.e. no slack. However, it would be of interest to determine the effect on profit of increasing the limit on one or even both these resources, namely the builder's total and type B design capacities. Called the Shadow/Dual price, this is the amount the objective function would increase by raising the constraint by one unit and is shown in tabular format below for each constraint manipulated independently.

For example, row 4. Considering the constraint on total production capacity (row 4), increase the capacity to produce to 401 houses. The new constraint becomes

$$X+Y=401 \text{ (Figure S1.2)}$$

Again, the objective function $1000X+800Y=$ profit and the optimum solution co-ordinates are $X=300.75$ and $Y=100.25$

$$\text{New profit}=1000\times 300.75+800\times 100.25=£380\,950$$

The extra profit generated by adding 1 house is £950, i.e. £950/house.

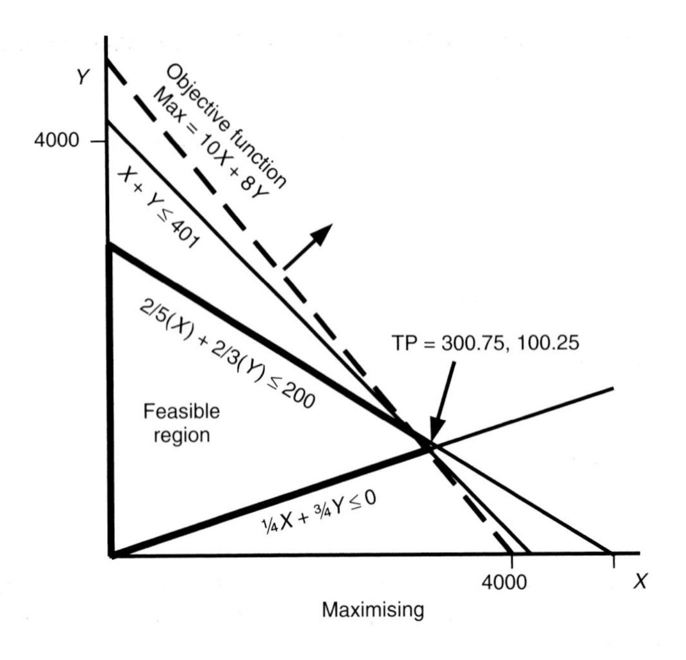

Fig. S1.2

This policy is *not* justified as the extra resources required cost £1000/house.

In a similar manner, the constraint of row 5 produces a shadow price of £200 if the 75% limit on detached is raised as summarised below for each constraint respectively:

Objective value:	£380 950.0	
Variable	Value	Reduced cost
X	300	750
Y	100	250
Row	Slack or surplus	Dual price
1	380 950	1
2	300.75	0
3	100.25	0
4	0	950
5	0	200
6	12.5325	0

It will be observed in the above summary that increasing production capacity 'improves' the solution by the dual price values.

Reduced cost

The result also shows the reduced cost, which is the amount that would have to be added to or subtracted from the variable coefficient in the objective function in order for that variable to feature in the solution, meaning an increase in the coefficient when maximising and a decrease when minimising. In the above example, X and Y are both used in the solution and therefore have zero reduced cost values. Nothing further could be gained, since no other variable is involved in the analysis.

Variable	Value	Reduced cost
X	300	0
Y	100	0

Alternative analysis by the canonical method
The primal arrangement of the problem is:

Row 1 $\text{Max} = 1000X + 800Y$
Row 2 $X + Y <= 400$
Row 3 $0.25X - 0.75Y <= 0$
Row 4 $0.4X + 0.67Y \leq 200$
Row 5 $X > 0$
Row 6 $X > 0$

The canonical form takes the coefficients of the primal equations to form new equations whose variables (A, B and C) represent the shadow or dual values and minimises if the primal is a maximising and vice versa, thus:

Canonical is:

$$\text{Min} = 400A + 200C$$
$$A + 0.25B + 0.4C \geq 1000$$
$$A - 0.75B + 0.67C \geq 800$$

Solution

Objective value:	£380 000.00
Variable	Value
A	950
B	0
C	200

These are the shadow prices and can be compared with the graphical solution shown above.

Solution 2 Linear programming illustrating minimising

Let XX = Number of parts type A completed per day;
Let YY = Number of parts type B completed per day;

Row 1. Production cost objective function:

$$\text{MIN} = 10\text{XX} + 8\text{YY}$$

Row 2. Constraint on type A proportion:

$$\text{XX} = (\text{XX} + \text{YY})^{1}/_{3}$$

Row 3. Constraint on shutter type B proportion:

$$YY = (XX + YY)^{2}/_{3}$$

Row 4. Constraint on total number of parts per day:

$$XX + YY \leq 40$$

Row 5. Constraint on number of part type A per day:

$$XX \geq 10$$

Row 6. Constraint on daily production time:

$$0.5XX + 0.2YY = 10$$

Rows 7 and 8. Constraints to prevent negative values of XX and YY:

$$XX > 0$$
$$YY > 0$$

Solving graphically

The equations plotted graphically below in Fig.S2.1 illustrate the area of feasible solutions and the optimum cost is the solution that minimises the objective function $10 \times XX + 8 \times YY$.

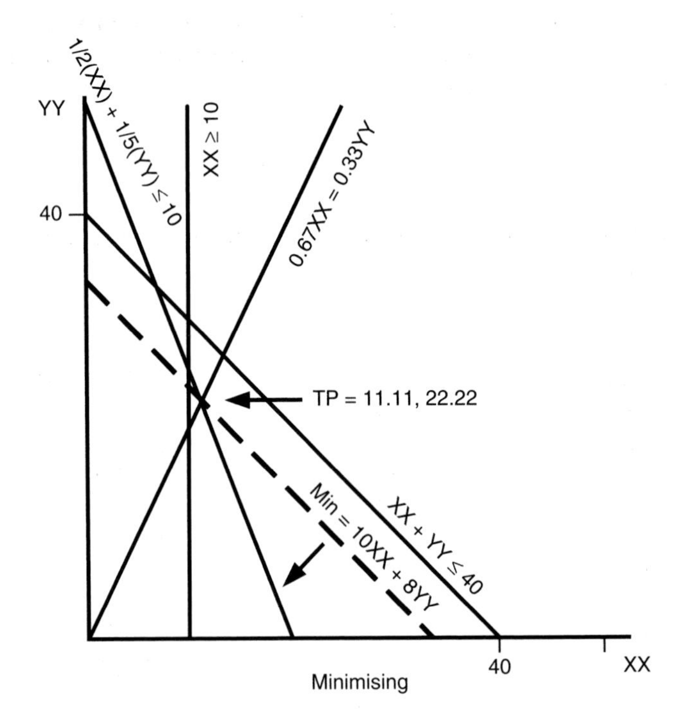

Fig. S2.1

The production cost will continue to reduce with any line drawn parallel to the objective function, moving towards the origin until the line is just tangential to the feasible area, the objective function then being at the tangential point.

The co-ordinates of XX and YY at the tangent point (TP) are 11.11 and 22.22, respectively. In other words, the minimum cost is £288.89 when approximately 11 type A and 22 type B are completed during the 8-hour working day.

Objective value:	£288.89	
Variable	Value	Reduced cost
XX	11	110
YY	22	220
Row	Slack/surplus	Dual price
1	288.89	1
2	0	67
3	0	0
4	6.67	0
5	1.11	0
6	0	–28.89
7	11.11	0
8	22.22	0

Shadow/dual quantities

Similarly to the maximising example, if XX and YY values are substituted into the constraint equations (rows 1 to 8) it will be evident that rows 2, 3 and 6 have no slack. However, if overtime is worked, the situation for row 6 becomes:

$$0.5 \times XX + 0.2 \times YY = 11$$

Once again the equations can be redrawn graphically and the objective function $MIN = 10 \times XX + 8 \times YY$ determined:

Objective value:	£317.78	
Variable	Value	Reduced cost
XX	12.22	0
YY	24.44	0
Row	Slack/surplus	Dual price
1	317.78	1
2	0	6.67
3	0	0
4	3.33	0
5	2.22	0
6	0	–28.89
7	12.22	0
8	24.44	0

In this case the optimum solution produces co-ordinates at the tangent point 12.22, 24.44 and objective value £317.78.

The additional income generated from the overtime working:

$$£317.78 - £288.89 = £28.89$$

This policy is not justified as the extra resources required cost £30.

Solution 3 Simplex method—constraints: 'equal to or less than'

Let x be hundreds of m^3 of concrete delivered each week in system A.
Let y be hundreds of m^3 of concrete delivered each week in system B.
Crew income objective function:

$$\text{Max}(M) = 50x + 60y$$

Constraint on dumper production time:

$$100/12.5x + 100/25y \leq 40$$

Constraint on crane production time:

$$100/25x + 100/20y \leq 40$$

Constraint on crew production time:

$$100/2x + 100 \times 0.13y \leq 40 \times 5$$

Constraints to prevent negative values of x and y

$$x \geq 0$$
$$y \geq 0$$

Solution by graphical or simultaneous equations is quite straightforward with two variables, and even three variables can be visualised in three dimensions, as shown in Figure S3.1, where the maximum is given by $x, y = {}^{12}/_3, {}^{62}/_3$. Max = £483.3.

However more sophisticated methods of analysis are required for a many-variable problem, the most prominent being the *Simplex method* developed by G. B. Dantzig in the USA. in 1947; although unknown elsewhere, L.V. Kantorovich was working on the problem in the USSR in 1939.

The equations need to be written in a particular form to remove the less-than-or-equal-to constraints; this is achieved by introducing 'slack' variables u, v, w.

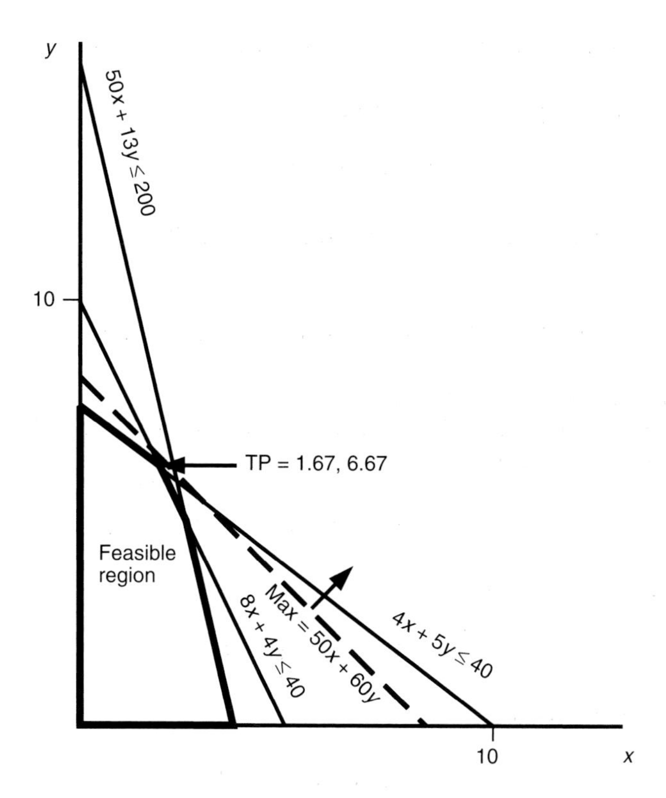

Fig. S3.1

x	y	u	v	w	M		
$8x$	$+4y$	$+u$				$=40$	...E1
$4x$	$+5y$		$+v$			$=40$	...E2
$50x$	$+13y$			$+w$		$=200$	...E3
$-50x-$	$60y$				$+M$	$=0$	...E4

$x \geq 0, y \geq 0, u \geq 0, v \geq 0, w \geq 0$ ...E5

All 'feasible' solutions will lie in the region given by E1, E2, E3, E5. The 'optimal' solution will be given by inclusion of the objective function, equation E4. The 'basic feasible' solution is defined as that given by $x = 0, y = 0$, etc.

Maximising

The procedure to maximise M is to use E1, E2, E3 **to modify E4 so as to convert the coefficients of *x*, *y* to zero or positive values.** As x, y cannot be negative (E5), this will mean that any valid values of these variables would start to reduce the value of M. That is, the conditions just reached will represent the optimum, i.e. maximum. The Simplex method is used as it leads to straightforward numerical calculation well suited to computers.

First, write the equations as a matrix of the coefficients (called the Simplex Table in the method of solution to be followed later).

The matrix represents a set of equations whereby:

(a) An equation is valid if divided or multiplied through by a constant.
(b) Subtraction or addition of two equations leads to another valid equation.
(c) Right-hand-side values must not be negative numbers.

x	y	u	v	w	M		
8	4	1	0	0	0	40	E1
4	5*	0	1	0	0	40	E2
50	13	0	0	1	0	200	E3
–50	–60	0	0	0	1	0	E4

Step 1. Operate on the objective function equation E4 and consider the most negative coefficient, –60 in column y, but it is equally possible to take any of the coefficients of E4. Take the positive coefficients of that column and divide into the column on the far right to produce the following quotients:

$$\frac{40}{4} = 10, \quad \frac{40}{5} = 8, \quad \frac{200}{13} = 15.38$$

Find the lowest quotient—in this case 8 above—and determine the position of the appropriate coefficient for this row i.e. the largest positive value—namely 5. This is called the 'pivot'.

Now divide E2 by 5 to reduce the pivot coefficient to 1, giving:

$^4/_5x$	$1y$	***0u***	$^1/_5v$	$0w$	$0M$	8 . . . **E6**

The matrix is now rewritten as:

x	y	u	v	w	M	
8	4	1	0	0	0	40 . . . E1
$^4/_5$	1	0	$^1/_5$	0	0	8 . . . E6
50	13	0	0	1	0	200 . . . E3
–50	–60	0	0	0	1	0 . . . E4

The coefficient of y is now removed from all equations by:

(i) Multiplying E6 × 4 and subtracting from E1 to give E7:

8	4	1	0	0	0	40 . . . E1
3 $^1/_5$	4	0	$^4/_5$	0	0	32 . . . E6 × 4
4 $^4/_5$	0	1	$-^4/_5$	0	0	8 . . . E7

(ii) Multiplying E6 × 13 and subtracting from E3:

50	13	0	0	1	0	200 . . . E3
10 2/5	13	0	2 3/5	0	0	104 . . . E6 × 13
39 3/5	0	0	−2 3/5	1	0	96 . . . E8

(iii) Multiplying E6 × 60 and adding to E4:

−50	−60	0	0	0	1	0 . . . E4
48	60	0	12	0	0	480 . . . E3 × 60
−2	0	0	12	0	1	480 . . . E9

The new matrix becomes:

4/5	1	0	1/5	0	0	8 . . . E6
4 4/5*	0	1	−4/5	0	0	8 . . . E7
39 3/5	0	0	−2 3/5	0	0	96 . . . E8
−2	0	0	12	0	1	480 . . . E9

Step 2. The new objective function equation is E9 and the most negative coefficient of E9 is −2. Dividing positive values of this column into the farthest right-hand column gives

$$\frac{5 \times 8}{4}, \frac{8 \times 5}{24}, \frac{5 \times 96}{198}$$

The smallest value of this manipulation above produces coefficient 4 4/5, which becomes the new pivot. Dividing E7 by 4 4/5 gives:

x	y	u	v	w	M	
1	0	24/5	−1	0	0	5/3 . . . E10

The matrix is now rewritten as:

1	0	24/5	−1/6	0	0	5/3 . . . E10
4/5	1	0	1/5	0	0	8 . . . E7
39 3/5	0	0	−2 3/5	1	0	96 . . . E8
−2	0	0	12	0	1	480 . . . E9

The coefficient of x is now removed from all equations by:

(i) Multiplying E10 by 2 and adding to E9:

2	0	$^{48}/_5$	$-^1/_3$	0	0	$^{10}/_3$ . . . E10×2
−2	0	0	12	0	1	480 . . . E9
0	0	$^{48}/_5$	$11\,^2/_3$	0	1	$483^1/_3$ . . . E13

(ii) Multiplying E10 × $^3/_5$ and subtracting from E7:

$^4/_5$	0	$^{96}/_{25}$	$-^4/_{30}$	0	0	$^4/_3$ . . . E10×4/5
$^4/_5$	1	0	$^1/_5$	0	0	8 . . . E7
0	1	$-^{96}/_{25}$	$^1/_3$	0	0	$^{62}/_3$ . . . E11

(iii) Multiplying E10 × 39 $^3/_5$ and subtracting from E8:

$39\,^3/_5$	0	$^{198}/_5 \times 2^4/_5$	$-^{198}/_5 \times {}^1/_6$	0	0	$^5/_3 \times {}^{198}/_5$ . . . E10×39 $^3/_5$
$39\,^3/_5$	0	0	$-2\,^3/_5$	1	0	96 . . . E8
0	0	$-^{198}/_5 \times 2^4/_5$	4	1	0	30 . . . E12

The new matrix is written as:

1	0	$^{24}/_5$	$-^1/_6$	0	0	$^5/_3$ . . . E10
0	1	$-^{96}/_{25}$	$-^1/_3$	0	0	$6^2/_3$ . . . E11
0	0	$^{198}/_5 \times {}^{24}/_5$	4	1	0	30 . . . E12
0	0	$^{48}/_5$	$11\,^2/_3$	0	1	$483\,^1/_3$ . . . E13

Step 3. It can be seen that the objective function equation E13 has reached the optimum condition whereby the coefficients of x and y in E13 are now 0,thus values for M, x, y and w may be read from the table as follows:

$$x = 1^2/_3, y = 6^2/_3, w = 30, M = 483^1/_3$$

$$x \geq 0, y \geq 0, u \geq 0, v \geq 0, w \geq 0$$

i.e. $x = 1.67, y = 6.67$, profit = £483.3 (compare with the graphical answer above).

Solution 4 Simplex method—constraints 'greater than or equal to'

The objective function to be maximised is $1.5x + y$.

$$\text{i.e. } M = 1.5x + y$$

Using a similar technique to that above, negative slack variables u and v are introduced as negative values together with artificial variables a and b in order to overcome the negativity elements. The equations may be written as:

$$2x+3y-u+a=14$$

$$8x+5y-v+b=40$$

$$u\geq 0, v\geq 0, x\geq 0, y\geq 0, a\geq 0, b\geq 0$$

The solution proceeds in the normal manner using the Simplex method. However, since artificial variables have been introduced, these will have to be eliminated before an optimum solution can be obtained. This is achieved by either adding very large positive numbers as the coefficients of a and b in the objective function or alternatively designating the coefficients as p to represent such a number. The maximising objective function is thus written as:

$$-1.5x-y+pa+pb+M=0$$

p is the large positive number and the maximising table becomes:

x	y	u	v	a	b	M	
2	3	−1	0	1	0	0	14 . . . E1
8	5	0	−1	0	1	0	40 . . . E2
−1.5	−1			p	p	1	0 . . . E3

$u\geq 0, v\geq 0, x\geq 0, y\geq 0, a\geq 0, b\geq 0$

It can be seen from the graph (Fig. S4.1) that a maximum solution to this problem is infinity as there is no limit on the upper boundaries of the constraints. A minimum value, however, is possible to calculate as shown in Solution 5.

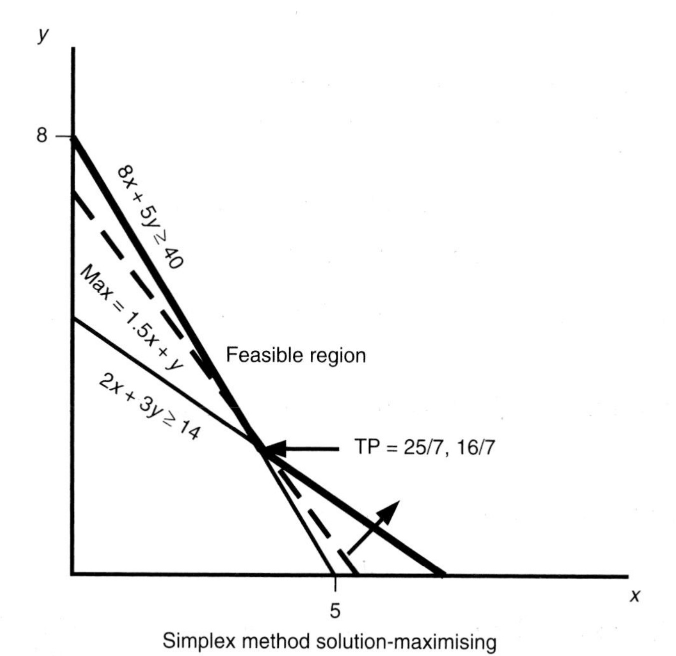

Fig. S4.1

Solution 5 Simplex method—constraints: 'greater than or equal to'

(This process can also be similarly applied in minimising Question 3.)
In minimising M, the procedure involves maximising $-M$, i.e. to minimise M merely maximise $-M$ or make M less negative.
Step 1. Determine the table involving the main constraints written as:

x	y	u	v	a	b	$-M$	
2	3	−1	0	1	0	0	14 . . . E1
8	5	0	−1	0	1	0	40 . . . E2
1.5	1	0	0	p	p	1	0 . . . E3

As in the maximising problem, the artificial coefficients a and b are introduced to counter the negative u, v coefficients of the 'greater than or equal to' constraints in E1 and E2.

Again p is introduced as the large positive number described previously and the theoretical basic feasible solution cannot reach a maximum value without a and b equal to zero. Thus, the first basic feasible table is provided with the a and b coefficients of E3 set to 0 to give E4 and E5,

$$\text{i.e. } pa + pb = 0$$

The initial maximising table thus becomes

x	y	u	v	a	b	$-M$	
2	3	−1	0	1	0	0	14 . . . E1
8	5	0	−1	0	1	0	40 . . . E2
1.5	1	0	0	0	0	1	0 . . . E4
p	p	0					0 . . . E5

$u \geq 0, v \geq 0, x \geq 0, y \geq 0, a \geq 0, b \geq 0$

From the above basic feasible solution, the Simplex method can now begin by first developing a new basis equation by subtraction or addition of two existing equations, which will lead to another valid equation. In this case an initial basis equation is obtained by multiplying each of E1 and E2 by $-p$ and adding to E5 to give E6. This produces the following table:

x	y	u	v	a	b	$-M$	
2	3	−1	0	1	0	0	14 . . . E1
8*	5	0	−1	0	1	0	40 . . . E2
1.5	1	0	0	0	0	1	0 . . . E4
$-10p$	$-8p$	$1p$	$1p$	0	0	0	$-54p$ . . . E6

Step 2. Select the most negative coefficient in the bottom line, $-10p$ is chosen, row E2 produces the pivot to give E7 and the transforming procedure is executed as in the previous example:

x	y	u	v	a	b	$-M$	
0	$^7/_4$*	-1	$^1/_4$	$-^1/_4$	0	4	. . . E8
1	$^5/_8$	0	$-^1/_8$	0	$^1/_8$	0	5 . . . E7
0	$^1/_{16}$	0	$^3/_{16}$	0	$-^3/_{16}$	0	$-^{15}/_2$ . . . E9
0	$-^7/_4p$	$1p$	$-^1/_4p$	0	0	1	$-4p$ . . . E10

Step 3. The most negative coefficient in the bottom line, $-7/4p$, is selected, row E8 produces the pivot in row E11 and the transforming procedure is continued to give the following table:

x	y	u	v	a	b	$-M$	
0	1	$-^4/_7$	$^1/_7$	$^4/_7$	$-^1/_7$	0	$^{16}/_7$ . . . E11
1	0	$^5/_{14}$	$-^3/_{14}$	$-^5/_{14}$	$^3/_8$	0	$^{25}/_7$ . . . E12
0	0	$-^1/_{28}$	$^5/_{28}$	$-^1/_{28}$	$-^5/_{28}$	1	$-^{107}/_{14}$ . . . E13
0	0	0	0	0	0	0	. . . E14

The test criteria for the objective equation (E14) are satisfied, coefficients are 0 and the function is now maximised with

$$x = {}^{25}/_7, y = {}^{16}/_7, -M = -{}^{107}/_{14}$$

i.e. minimum $M = +107/14 = 7.642$ (compare Fig S5.1 with graphical result in Fig S4.1 above).

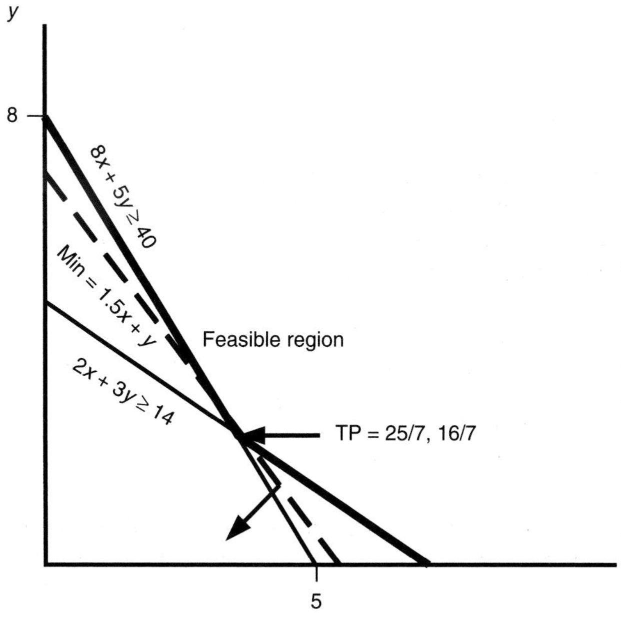

Fig. S5.1

Solution 6 Integer linear programming

General integers

The linear programming method previously described produces an optimal solution that may not be an integer value. However, in some practical situations only whole number values may be acceptable, for instance in the production of packaged quantities where an integer amount would be mandatory. While rounding to an integer solution by trial and error may be acceptable, the rounded up or down values need to be examined, to determine whether an 'optimum integer' solution has been found or is indeed feasible.

The non-integer (relaxed version) optimum solution is illustrated in Solution 3 and is reworked as an integer solution as follows:

Integer solution

It can be observed in Fig. S6.1 that, within the feasible area for a relaxed solution, instead of moving the objective function to the tangential point, the line is restricted to touching the last (i.e. extreme) whole value (integer) co-ordinates, irrespective of whether it is tangential or actually within the boundary area. The dots shown imposed upon the diagram demonstrate that the optimum integer solution occurs at point $X = 0$, $Y = 8$ to produce a smaller objective value £480.00. These X, Y values when substituted into the constraint equations produce the following summary:

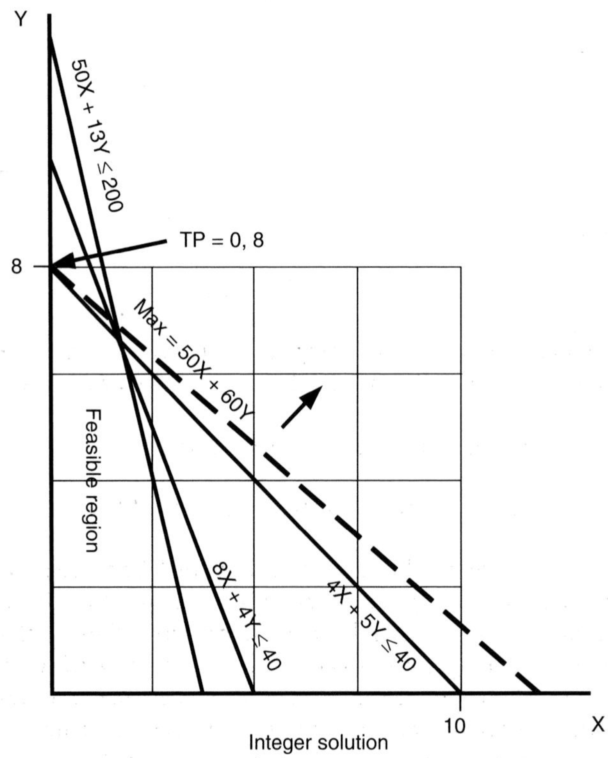

Fig. S6.1

Branch count:	2	
Variable	Value	Reduced cost
X	0	−50
Y	8	−60
Row	Slack or Surplus	Dual price
1	480.0	1.0
2	8.0	0.0
3	0.0	0.0
4	96.0	0.0
5	0.0	0.0
6	8.0	0.0

Branch and bound calculation of integer variable values

The branch count above is the number of times the integer values needed to be forced to an integer value in arriving at the optimum solution via the 'branch and bound' process of calculation, this being a specifically devised algorithm that enumerates all combinations of the integer variables (e.g. the whole numbers seen on the grid in the graph). The best feasible combination of variables is selected within the constraints permitted as defined in the linear programming problem. To illustrate the kind of difficulty that may arise, consider:

$$\text{Maximum} = Y$$

Subject to constraints:

$$Y + Z = 10.5$$

$$Y \geq Z$$

The relaxed solution is

$$Y = Z = 10.5/2 = 5.25.$$

In determining an integer solution by smoothing the graph manually and rounding up or down according to a guess might result in $Z = 6$ and $Y = 4.5$ which breaks the constraint $Y \geq Z$ and so the solution is infeasible.

The correct answer to comply with the constraints is clearly

$$Y = 5.5 \text{ and } Z = 5 \text{ and the Maximum} = 10.5$$

Obviously, when the number of variables exceeds three in total the problem lies outside a three-dimensional configuration and a graph cannot be drawn. The branch and bound algorithm carries out this calculation, first by using the relaxed solution values of the variables (determined by, for example, the Simplex method described earlier) to compute the best feasible combination of integers. Because of the tedium involved in such calculations, enumeration by computer is normally adopted. However, when a computer is not available, the most convenient approach is to 'eye in' the integer variables and select the combination that 'almost' produces the relaxed value of the solution.

Reduced costs

The table also shows reduced costs and that variable X is not used in the solution, thereby indicating that its coefficient in the objective function could be reduced by 50 without causing a change in income penalty. This is illustrated by superimposing $\text{MAX} = 0 \times X + 60 \times Y$ on the graph, which becomes tangential at point $X, Y\,(0, 8)$ when moving away from the origin.

Solution 7 Binary integer linear programming

(i)

The inclusion of binary integer constraints may be specified as:

$A = 1$ if system A is acceptable, $A = 0$ if not
$B = 1$ if system B is acceptable, $B = 0$ if not

and as before:

Let X = hundreds of m^3/week of concrete delivered with system A;
Let Y = hundreds of m^3/week of concrete delivered with system B.

Row 1. The objective function now includes the binary variables in the set up costs which will have 0,1 values applied or not depending on whether when the particular system A or B is operational:

$$\text{MAX} = 50 \times X + 60 \times Y - 50 \times A - 125 \times B.$$

Subject to the following constraints:

Row 2. Constraint on dumper production time:

$$8 \times X + 4 \times Y \leq 40$$

Row 3. Constraint on crane production time:

$$4 \times X + 5 \times Y \leq 40$$

Row 4. Constraint on crew production time:

$$50 \times X + 13 \times Y \leq 200$$

Rows 5 and 6. Constraints to prevent negative values of X and Y:

$$X \geq 0$$

$$Y \geq 0$$

Plus the following extra constraints:

Rows 7 and 8 are constraints that prevent concrete being produced by system A when production exceeds the 175 m^3/week permitted, i.e. $A = 0$, or by system B when production exceeds 700 m^3/week, i.e. $B = 0$:

$$X \leq 1.75 \times A$$

$$Y \leq 7.00 \times B$$

$$@BIN(A);@BIN(B);$$

Note: In a solution using Lindo the @BIN function denotes a shortcut setting the selected variables to be binary integers. A 0/1 binary integer variable is a special case of an integer variable that is required to be either zero or one and is used as a switch to model Yes/No decisions.

Computer solution

Objective value:	£308.3	
Branch count:	1	
Variable	Value	Reduced cost
X	1.67	0
Y	6.67	0
A	1	50
B	1	125

As can be seen from the above table, both A and B systems are needed and must operate within their respective capacities.

Clearly the new optimum is the same as for Example 1 minus the set-up costs i.e. £483.33 − £50 − £125 = £308.33.

Alternative equipment options

Other situations may be possible, for example changed set-up costs, which then become critical in determining the viability of a particular system for inclusion in the solution. Carrying out a number of trials for a range of set-up cost estimates enables evaluation of the alternative candidates.

(ii)

As an example, system B becomes critical ($B = 0$) at £396, but results in a correspondingly much reduced concrete output produced by system A ($A = 1$) alone, as follows:

With the set-up cost of system B = £396 the solution is:

Objective value:	£37.5	
Branch count:	1	
Variable	Value	Reduced cost
X	1.75	0
Y	0.00	0
A	1.00	−37.50
B	0.00	−24.00

The routine has tried

$$A=1 \text{ with } B=0$$
$$A=0 \text{ with } B=1$$
$$A=1 \text{ with } B=1$$

and found the optimum at $A=1$ combined with $B=0$.

(iii)

Similarly A becomes critical at a set-up cost of $A=£64$ where $A=0$ with $B=1$:

Objective value:	£295	
Branch count:	1	
Variable	Value	Reduced cost
X	0	0
Y	7	0
A	0	−23.50
B	1	−295.00

(iv)

Alternatively, the capacity limits on A or B may be evaluated where B becomes critical at Y limited to a capacity of 2.08 m^3/week, i.e. $A=1$ with $B=0$:

Objective value:	£37.50	
Branch count:	0	
Variable	Value	Reduced cost
X	1.75	0
Y	0	0
A	1	−37.5
B	0	0.2

(v)

Similarly A becomes critical at $X=1$ where $A=0$ with $B=1$ as follows:

Objective value:	£295	
Branch count:	0	
Variable	Value	Reduced cost
X	0	0
Y	7	0
A	0	0
B	1	−295

These examples illustrate that, by carrying out simulation routines, it is readily possible to determine which resources to adopt simply by attaching 0, 1 variables to the appropriate elements in the objective function.

Notably the original constraint equations were unaffected other than that the model was expanded by the inclusion of additional constraints to control the on/off parameter of the production systems A and B within their specified capacities.

Solution 8 Transportation problem

Expressed as a typical linear programming problem of maximising or minimising a supply/demand situation, the problem would be formulated as follows:

3 Suppliers		3 Consumers	
Capacity ($m^3 \times 1000$)		Requirements ($m^3 \times 1000$)	
60	S_1	C_1	90
80	S_2	C_2	50
120	S_3	C_3	40

Transport costs $S_1 - C_1 = 70$p per item $S_1 - C_2 = 40$p per item, etc.

Decision variables:

$$X_{i,j} = \text{amount sent from Supplier (row) } i \text{ to Consumer (column) } j$$

The objective function is to minimise the total cost of transport cost as follows (i.e. the sum of the amount multiplied by its cost for every cell in the matrix), thus:

$$\text{Minimise } 70X_{1,1} + 40X_{1,2} + \text{etc.} \ldots$$

Subject to constraints on supplier capacity and customers' requirements:
Supplier constraints (i.e. can supply less or equal to its capacity)

$$X_{1,1} + X_{1,2} + X_{1,3} \leq 60\,000$$

$$X_{2,1} + X_{2,2} + X_{2,3} \leq 80\,000$$

$$X_{3,1} + X_{3,2} + X_{3,3} \leq 120\,000$$

Customer constraints, i.e. requires the particular amount:

$$X_{1,1}+X_{2,1}+X_{3,1}=90\,000$$
$$X_{1,2}+X_{2,2}+X_{3,2}=50\,000$$
$$X_{1,3}+X_{2,3}+X_{3,3}=40\,000$$

and

$$X_{i,j}\geq 0$$

Alternative solution by the tabular method

Clearly to solve the problem by hand calculation using, for example, the Simplex method would be very time consuming and a tabular minimisation method is available to quickly derive a manual solution even for very large tables. (The table can also be maximised by transforming each cell into subtracted values from a base number, e.g. 100, and then minimising. The optimum results in the cells would thereafter be retransformed to enable calculation of the maximum.)

Solution

Total available for supply = 120 000 + 80 000 + 60 000 = 260 000
Total required to be delivered = 90 000 + 50 000 + 40 000 = 180 000
$80\,000\text{m}^3$

It is first necessary to get some idea of the most likely routes. This can be guessed simply from looking at the costs and routes involved sometimes. Alternatively, the following method has been shown to be very quick and in many instances gives the optimum solution also. The analysis is recommended for complex problems.

Preliminary stage

Step 1
The data are arranged in a suitable table, such as that shown in Fig. S8.1.

The route costs are indicated in the bottom right-hand corner of each square. In this instance the material available for supply exceeds that required, so a dummy site requiring the excess is

		Required				0 000s		
		A 9	B 5	C 4	Dummy 0			
	(40)	70	40	100	0	6	1	Available
	(30)	50	30	90	0	8	2	
Greatest difference →	(50)	60	50	90	8 0	4	3	
		(10)	(10)	(0)	(0)			

Fig. S8.1 Step 1.

invented in which the transport costs from supplier to site are zero. In other words, the material is never moved at all.

Now look along each row and deduct the lowest cost from the second lowest cost, repeat the procedure for each column. Allocate materials to the cheapest route along the row or column with the greatest difference in costs. The dummy site is fully supplied and so this column can now be eliminated. (All material quantities in 10 000 m^3.)

Step 2

Repeat Step 1 procedure with the routes remaining.

Site B is fully supplied and eliminated.

See Fig. S8.2.

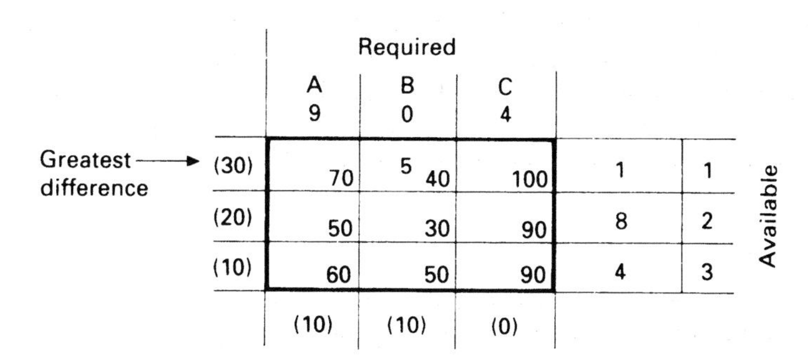

Fig. S8.2 Step 2.

Step 3

Supplier 2 is empty and may be eliminated, but Site A still requires 10 000 m^3 and Site B, 40 000 m^3. See Fig. S8.3.

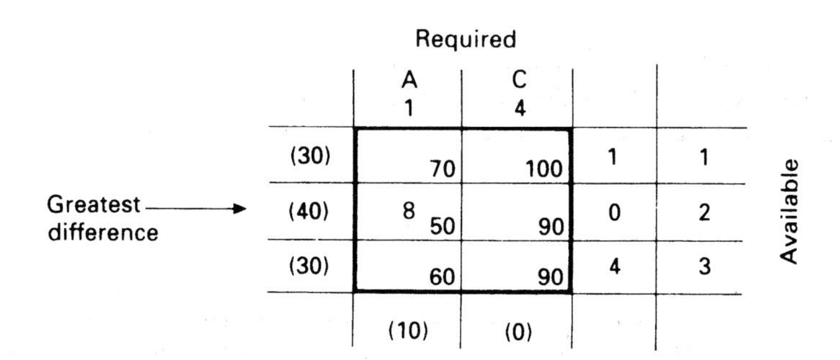

Fig. S8.3 Step 3.

Step 4

No clear choice is possible, so the material may be distributed along any combination of the available routes (Fig. S8.4)

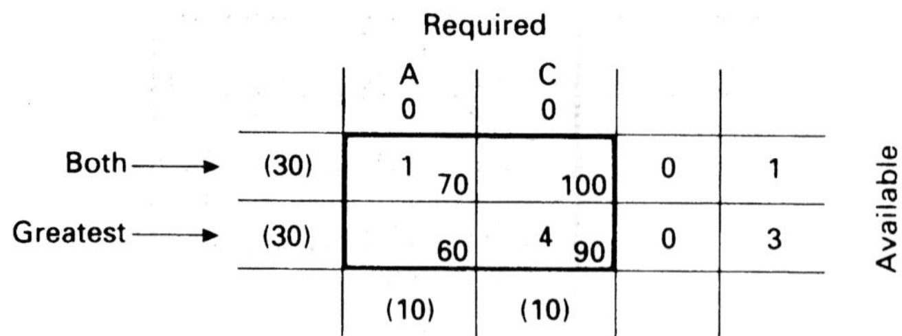

Fig. S8.4 Step 4.

Final guess

Note from Fig. S8.5 that the sum of the quantities in the routes used along a row or a column must be the amount available from a supplier or to the supplier respectively.

It is now necessary to test if the guess is in fact the optimum solution.

Required by						
A 9	B 5	C 4	Dummy 8	26 / 26		
1 70	5 40	100	0	6	1	Available from
8 50	30	90	0	8	2	
60	50	4 90	8 0	12	3	

Fig. S8.5 Final guess.

Final stage

Test for optimality

Consider the costs of the routes used in two components: (1) cost of despatch from the source of supply, (2) cost of reception at the destination. Start with one of the components arbitrarily chosen (usually zero). The remaining despatch and reception costs are determined by considering each used route in turn and testing the equation.

$$\text{Despatch cost} + \text{reception cost} = \text{actual cost along the route.}$$

Note that in the example (Fig. S8.6) the top left-hand corner is chosen. Unfortunately in this example the five routes used are less than the number of rows (m) plus columns (n) minus 1.

$$\begin{aligned} m+n-1 &= 3+4-1 \\ &= 6 \end{aligned}$$

It is therefore impossible to generate all the despatch and reception costs.

In order to overcome this difficulty a small quantity q is introduced into an unused route so that further row and column costs may be generated (Fig. S8.7).

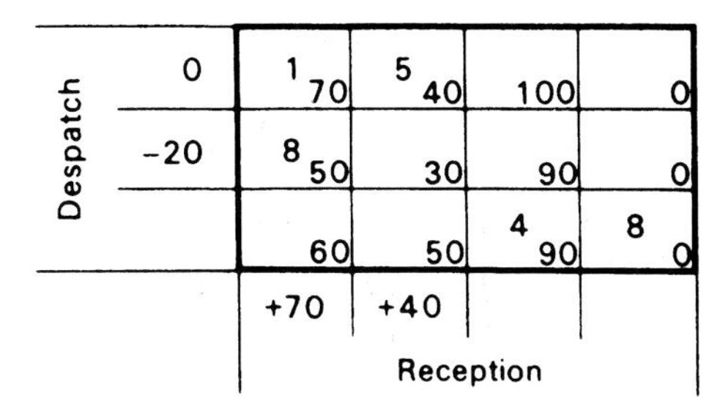

Fig. S8.6 Despatch cost + reception cost.

0	1 70	5 40	100	q 0
−20	8 50	30	90	0
0	60	50	4 90	8 0
	+70	+40	+90	0

Fig. S8.7 Introduction of q.

It is now possible to test for optimality. Compare for each unused route the actual costs with the sum of costs for its corresponding despatch and reception costs.

If the actual cost is *greater* than or *equal* to the sum for every unused route then the optimum has been found. If not, the solution can be improved by using that route (marked x) for which the actual cost is less than the sum (Fig. S8.8).

Clearly, we have a route which is cheaper and should be used. But because an amount x is introduced, the balance of supply and requirement is upset and thus it is necessary to deduct the quantity x from some of the routes (Fig. S8.9). The route containing q also contains a quantity x and so q can be dropped from further calculations as it is relatively small.

0	1 70	5 40	✓ 100	q 0
−20	8 50	✓ 30	✓ 90	✓ 0
0	x 60	✓ 50	4 90	8 0
	+70	+40	+90	0

Fig. S8.8 Route x.

Required	9	5	4	8^{+q}	Supplied
	1^{-x} 70	5 40	100	q $+x$ 0	6^{+q}
	8 50	30	90	0	8
	x 60	50	4 90	8^{-x} 0	12

Fig. S8.9 Balance of supply and requirement.

To reduce the solution to $m+n-1$ routes, it is necessary to remove a route containing x. Obviously by choosing x = minimum (1 or 8), i.e. $x=1$, the new route pattern becomes as shown in Fig. S8.10.

Note: $m+n-1=6$. This equals routes used and so the new despatch and reception costs may be generated.

The optimising test procedure is now repeated (Fig. S8.11).

The actual cost is greater than or equal to the sum of the despatch and reception costs for each unused route, therefore the optimum is found.

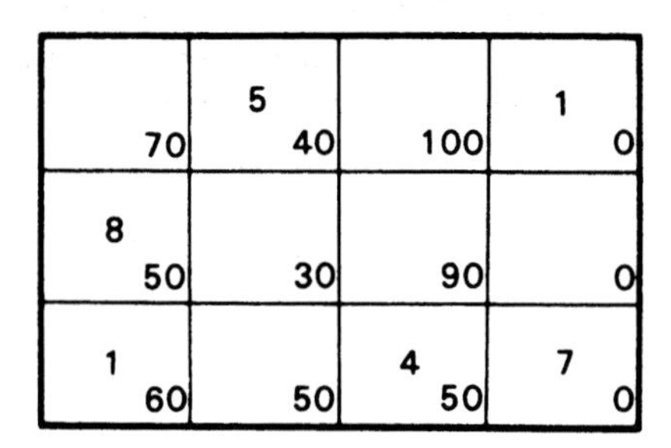

70	5 40	100	1 0
8 50	30	90	0
1 60	50	4 50	7 0

Fig. S8.10 New route pattern.

Despatch	A 9	B 5	C 4	D 8	Supplied from	
0	✓ 70	5 40	✓ 100	1 0	6	1
−10	8 50	✓ 30	✓ 90	✓ 0	8	2
0	1 60	✓ 50	4 90	7 0	12	3
Reception	+60	+40	+90	0		

(Required by: A, B, C, D)

Fig. S8.11 Repeat of optimising test procedure.

Summary

$$\begin{aligned} \text{Total cost} = 80\,000 \times 0.5 &= \quad 40\,000 \\ 10\,000 \times 0.6 &= \quad 6\,000 \\ 50\,000 \times 0.4 &= \quad 20\,000 \\ 40\,000 \times 0.9 &= \quad \underline{36\,000} \\ &\quad\ \pounds 102\,000 \end{aligned}$$

Minimum transport cost is £102 000 using routes A2, A3, B1 and C3.

Solution 9 Assignment problem

This is a problem characterised by the supplier only being able to assign a single task at each despatch location to a single receiver/consumer at its respective destination, which can also proceed as a linear program solved by the Simplex method as follows.
Decision variables:

$$X_{i,j} = \text{amount sent from Supplier (column) } i \text{ to Consumer (row) } j$$

$$\text{i.e. } X_{i,j} = \text{amount sent from Manager(column) } i \text{ to Project(row) } j$$

Manager performance rating $R_{i,j}$ out of 100 points is: $S_1 - C_1 = 75$ points $S_1 - C_2 = 78$ points, etc.
The objective function is to maximise the total amount or, where the problem requires, alternatively minimise the total amount as follows (i.e. the sum of the amount multiplied by its cell value for every cell in the matrix), thus the objective function is to maximise the total points for the six managers, written as:

$$\text{Maximise} = R_{1,1} \times X_{1,1} + R_{1,2} \times X_{1,2} + \text{etc.} + R_{1,5} \times X_{1,5} + R_{2,1} \times X_{2,1} + R_{2,2} \times X_{2,2} + \text{etc.} + R_{2,5} \times X_{2,2} + R_{3,1} \times X_{3,1} + \text{etc} \ldots$$

Subject to constraints:
Variable constraints (rows)

$$X_{1,1} + X_{1,2} + X_{1,3} + X_{1,4} + X_{1,5} \leq 1$$

$$X_{2,1} + X_{2,2} + X_{2,3} + X_{2,4} + X_{2,5} \leq 1$$

$$X_{3,1} + X_{3,2} + X_{3,3} + X_{3,4} + X_{3,5} \leq 1$$

$$X_{4,1} + X_{4,2} + X_{4,3} + X_{4,4} + X_{4,5} \leq 1$$

$$X_{5,1} + X_{5,2} + X_{5,3} + X_{5,4} + X_{5,5} \leq 1$$

$$X_{6,1} + X_{6,2} + X_{6,3} + X_{6,4} + X_{6,5} \leq 1$$

Project constraints (columns)

$$X_{1,1} + X_{1,2} + X_{1,3} + X_{1,4} + X_{1,5} = 1$$

$$X_{2,1} + X_{2,2} + X_{2,3} + X_{2,4} + X_{2,5} = 1$$

$$X_{3,1}+X_{3,2}+X_{3,3}+X_{3,4}+X_{3,5}=1$$
$$X_{4,1}+X_{4,2}+X_{4,3}+X_{4,4}+X_{4,5}=1$$
$$X_{5,1}+X_{5,2}+X_{5,3}+X_{5,4}+X_{5,5}=1$$
$$X_{6,1}+X_{6,2}+X_{6,3}+X_{6,4}+X_{6,5}=1$$

and $$X_{i,j}\geq 0$$

Alternative solution by the tabular assignment method

In a similar fashion to the transportation problem, a quick manual procedure involving matrix reduction has been devised to give solutions for this even more special purpose type of problem where supply and demand equals 1.

The transportation method may be used, but as only five routes are possible in the solution, and there are $m+n-1=5+5-1=9$ routes* needed to generate the 'despatch' and 'reception' costs, the procedure would be lengthy. Thus, the assignment method was developed to test for optimality. In this example, it is first necessary to subtract all the figures from 100 since it is required to maximise the point scores for the managers (i.e. minimise 100 minus point score).

* m = number of rows and n = number of columns in the matrix.

Initial stage
Subtract point scores from 100 (Fig. S9.1).

25	72	39	52	41
22	29	49	65	81
27	39	60	51	32
45	50	48	52	37
29	40	39	26	30

Fig. S9.1 Subtract point scores from 100.

Step 1
Subtract the lowest number in each row from every number in its row (Fig. S9.2).

0	47	14	27	16
0	7	27	43	59
0	12	33	24	5
8	13	11	15	0
3	14	13	0	4

Fig. S9.2 Step 1.

Step 2
Subtract the lowest number in each column from every number in its column (Fig. S9.3).

0	40	3	27	16
0	0	16	43	59
0	5	22	24	5
8	6	0	15	0
3	7	2	0	4

Fig. S9.3 Step 2.

Step 3
Test for an assignment.

An assignment is possible if the *minimum* number of horizontal and vertical lines drawn through the rows and columns needed to cover all the zeros *equals* the number of rows or columns in the table (Fig. S9.4). In this case a minimum four lines covers all zeros, thus an assignment is not possible.

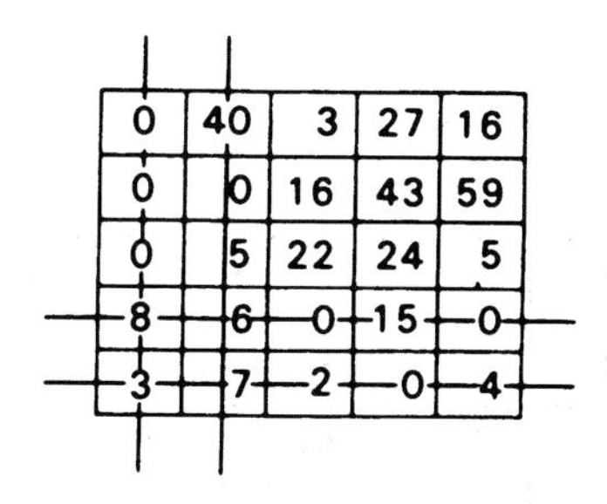

Fig. S9.4 Step 3.

Step 4
Find the smallest uncovered number remaining after Step 3 is undertaken, then (1) subtract this number from *every* number in the table (Fig. S9.5); (2) add the same number back to each

-3	37	0	24	13
-3	-3	13	40	56
-3	2	19	21	2
5	3	-3	12	-3
0	4	-1	-3	1

Fig. S9.5 Step 4.

number covered by Step 3 lines (i.e. a number covered by both a horizontal and vertical line will be added twice) (Fig. S9.6); (3) try the assignment test again.

(1) Smallest number is 3.

(2)

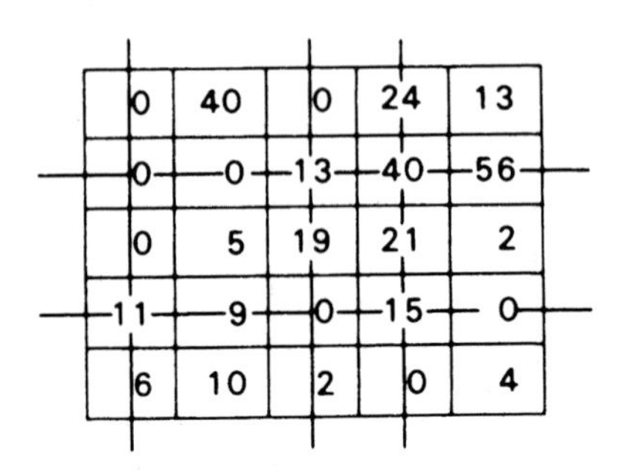

Fig. S9.6 Illustration of (2).

(3) The zeros cannot be covered by less than five lines now, therefore a zero assignment is possible. If not, repeat procedure from Step 4.

Step 5

In order to reach the optimal solution, it is only possible to use one route for each row and column (Fig. S9.7).

	(1)	(2)	(3)	(4)	(5)
(A)	■		0		
(B)	■	0			
(C)	0				
(D)			■		0
(E)				0	

Fig. S9.7 Step 5.

Columns 2, 4 and 5 each have a single zero, so these routes must be used. The extra zeros in Rows B and D can therefore be discarded. Hence the zero in Column 3 must be used and the zero in Column 1 now discarded, leaving the only possible route in Column 1 as the zero in Row C.

Final solution

This is shown in Fig. S9.8.

The final solution converted back to points is shown in Fig. S9.9.

Therefore, the best combination of manager to project is:

Manager	Project	Points
1	C	73
2	B	71
3	A	61
4	E	74
5	E	63
		342

		0		
	0			
0				
				0
			0	

Fig. S9.8 Final solution.

	(1)	(2)	(3)	(4)	(5)
(A)	75	28	(61)	48	59
(B)	78	(71)	51	35	19
(C)	(73)	61	40	49	68
(D)	55	50	52	48	(63)
(E)	71	60	61	(74)	70

Fig. S9.9 Final solution converted back to points.

Solution 10 Assignment when number of agents and tasks differ

The client needs to allocate the four managers best suited to the five projects available. The assignment method overcomes this difficulty by including a DUMMY Manager but who has ZERO points for each project as follows:

Five rows by five columns as in Q10.1

In Fig. S10.1 a DUMMY column (Manager) with zero cell rating values is introduced. Then each cell rating value is deducted from 100 (Fig S10.2) and minimised (Fig S10.3).
Using Lindo, the cell allocations are:

Managers		1	2	3	4
Project	Amount	1	1	1	1
A	1	75	28	61	48
B	1	78	71	51	35
C	1	73	61	40	49
D	1	55	50	52	48
E	1	71	60	61	74

Fig. S10.1 Assignments uneven.

Managers		1	2	3	4	Dummy
Project	Amount	1	1	1	1	1
A	1	75	28	61	48	0
B	1	78	71	51	35	0
C	1	73	61	40	49	0
D	1	55	50	52	48	0
E	1	71	60	61	74	0

Fig. S10.2

Managers		S1	S2	S3	S4	Dummy
Project	Amount	1	1	1	1	1
A	1	25	72	39	52	100
B	1	22	29	49	65	100
C	1	27	39	60	51	100
D	1	45	50	48	52	100
E	1	29	40	39	26	100

Fig. S10.3 Minimise.

$$X_{(1,3)}=1\ X_{(2,2)}=1\ X_{(3,1)}=1\ X_{(4,5)}=1\ X_{(5,4)}=1$$

$$R_{(1,3)}=39\ R_{(2,2)}=29\ R_{(3,1)}=27\ R_{(4,5)}=100\ R_{(5,4)}=26$$

Total = 221 or, when deducted from 500, maximum = 279 points.

Solution 11 Goal programming

Problems of the LP type can have multiple criteria with different levels of priority in making a decision, and can be solved by the goal programming technique. The procedure requires the following steps:

(1) Identify each goal and its target.
(2) Prioritise each goal in the order of importance, high to low, e.g. P1, P2, etc.
(3) Define the decision variables, these being the controllable inputs that describe the number or amounts of each item to produce, and formulate in the conventional LP manner.
(4) Develop an equation for each goal, involving the decision variables and goal deviation (amount α above or below the target value) on one side, and the goal target on the other.
(5) Write the objective function in terms of minimising the deviation value.

Thus, in the above problem the goals and priority levels are as follows:
Goal 1 and priority level 1: Build 200 houses with Type B designs.
Goal 2 and priority level 2: Complete no more than 400 houses.
The decision variables are:

$$X = \text{Number of detached houses to build}$$

$$Y = \text{Number of semi-detached houses to build}$$

Hence for Goal 1 the LP problem can be stated as:

$$\text{Minimise } \alpha_1$$

Constraint on proportion of detached houses:

$$X \leq 0.75(X+Y)$$

Goal 1 equation for Type B design limit:

$$0.4X + 0.67Y + \alpha_1 = 200$$

Constraints to prevent negative values of X, Y and α_1:

$$X \geq 0$$

$$Y \geq 0$$

$$\alpha_1 \geq 0$$

Drawn graphically (Fig. S11.1) the approach is similar to the LP approach except that each priority level is considered separately, thus beginning with priority level 1 as follows.
Substituting the G1 point = 321.4, 107.14 into the Goal 1 equation:

$$0.4X + 0.67Y + \alpha_1 = 200$$

$$0.4 \times 321.4 + 0.67 \times 107.14 = 200$$

$$\alpha_1 = 0, \text{ i.e. the goal can be achieved.}$$

Goal 2 is a secondary goal and must not degrade the above solution obtained with the priority level 1 conditions, hence the LP for the priority level 2 situation can be written as:

$$\text{Minimise } \alpha_1 + \alpha_2$$

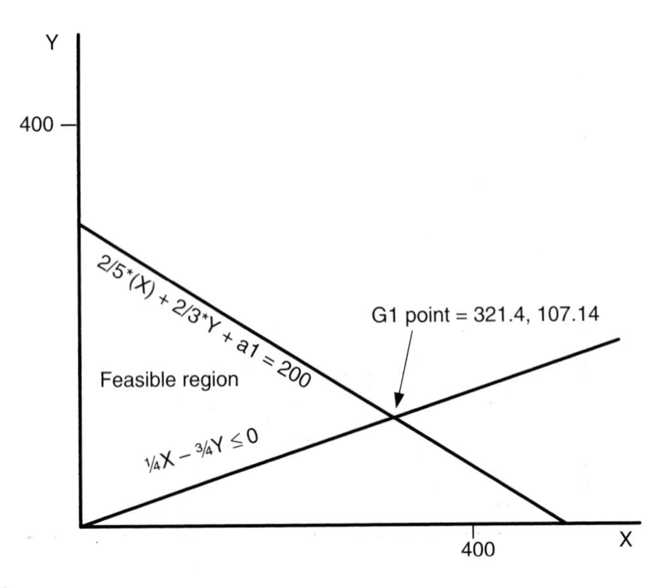

Fig. S11.1 Step 1.

Constraint on detached houses proportion:

$$X \leq 0.75(X+Y)$$

Goal 1 equation for type B design limit:

$$0.4X + 0.67Y + \alpha_1 = 200$$

Goal 2 equation for total housing limit:

$$X + Y + \alpha_2 = 400$$

Constraints to prevent negative values of X, Y, α_1 and α_2:

$$X \geq 0$$

$$Y \geq 0$$

$$\alpha_1 \geq 0$$

$$\alpha_2 \geq 0$$

The graphical interpretation is shown in Fig. S11.2 as follows:

In order for priority level 1 to be achieved, the G1 point cannot be violated, hence, substituting the G1 point value into the Goal 2 equation:

$$X + Y + \alpha_1 = 400$$

$$321.4 + 107.14 + \alpha_2 = 400$$

$$\alpha_2 = -28.54$$

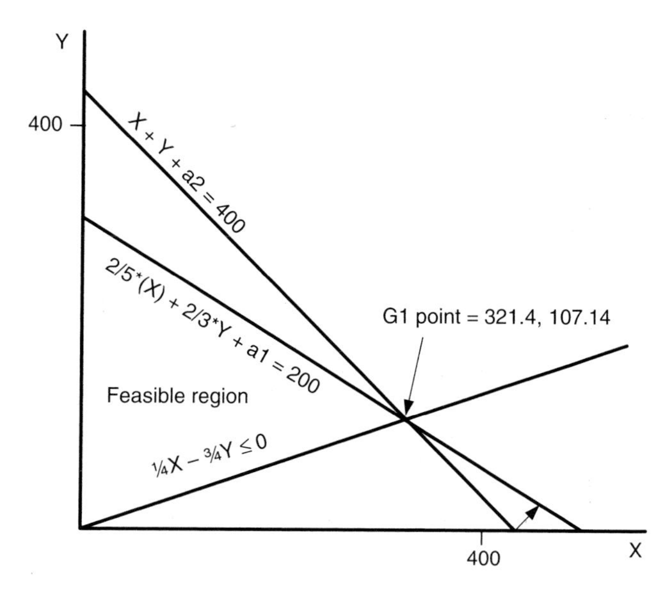

Fig. S11.2 Step 2.

Priority level 2 goal cannot be achieved, i.e. 428 houses will have to be built.
Notes:

(i) If the above priorities were switched, the solution would be different.
(ii) Several goals may have the same priority level as appropriate.
(iii) In more complex problems involving multiple decision variables, goals and priorities solutions a computer program becomes essential.

Solution 12 Non-linear programming

Equations of this form arise in a construction industry situation, for example where statistical data indicate that, say, the profit (in £s) achieved in groundworks per m^3 construction is governed by the above stated objective function variable relationship between excavation (x) and associated temporary works (y). A cubic metre of excavation takes 2 hours and fixing 1 m^2 of temporary works requires 1 hour, and an 8-hour day is available. The problem becomes one of determining the minimum and maximum profit expectations in a given period.

Graphical Solution (Fig. S12.1)

(a) Maximising

Let x be the amount of excavation in m^3. Let y be the amount of temporary works in m^2. The object is to maximise profit, thus, objective function is written as:

$$\text{Row 1 } \ \text{MAX}(\pounds) = x^2 - 2x + y^2 - 6y$$

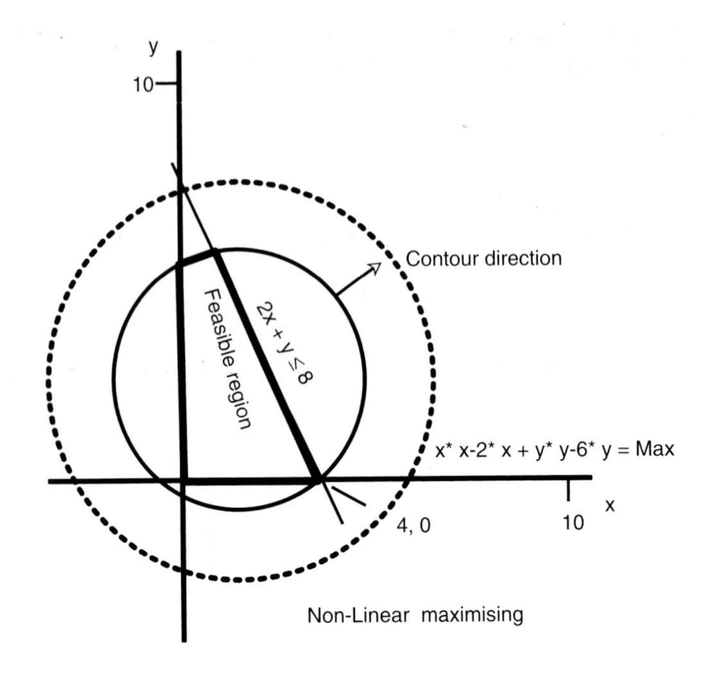

Fig. S12.1

Subject to constraint

$$\text{Row 2} \quad 2x + y \leq 8;$$

$$\text{Row 3} \quad x \geq 0;$$

$$\text{Row 4} \quad y \geq 0;$$

The objective function can be rearranged to represent a circle with the centre at (1, 3) and radius (r) as shown in Figure S12.1, i.e.:

$$(x-1)^2 + (y-3)^2 - 10 = \text{MAX}$$

where:

$$r^2 = (x-1)^2 + (y-3)^2$$

Hence

$$r^2 - 10 = \text{MAX}$$

In order to find the maximum the contours of the circle need to be moved away from the centre as shown in Fig S12.1. It can be observed that the last position of a contour of centre (1, 3) falling inside the constraint region coincides with it at point $x = 4, y = 0$.

Substituting x, y values in the objective function gives:

$$\text{MAX} = £8$$

Hence $r^2 = 18$, i.e. r = radius = 4.24 with the circle centre at $x = 1, y = 3$.

Note the alternative option of point $x=0, y=8$ would result in the larger contour falling outside the feasible region at point (4, 0) and would therefore be invalid.

Computer solution

Objective value:	£8.0	
Variable	Value	Reduced cost
x	4	0
y	0	9

The result shows that the best strategy is to work only on projects not requiring a temporary works element. Also:

Row	Slack or Surplus	Dual price
1	8	1
2	0	3
3	4	0
4	0	0

(b) Minimising

This time the objective is to find the minimum profit, hence the objective function is to minimise:

$$x^2-2x+y^2-6y$$

Again, this is the equation of a circle, which can be rearranged as:

$$(x-1)^2+(y-3)^2-10=\text{MIN}$$

with the centre at (1, 3) and radius (r) where:

$$r^2=(x-10)^2+(y-3)^2$$

Thus:

$$r^2-10=\text{MIN}$$

This time, as clearly shown in Fig. S12.2, for the minimum contour of the possible circles to fall wholly within the constraint region the condition must be that $x=1, y=3$, i.e. it coincides with the centre of the circle.

Substituting x, y values in the objective function gives:

$$\text{MIN}=-£10$$

Hence $r^2=0$, i.e. $r=$ radius $=0$ with the circle centre at $x=1, y=3$.

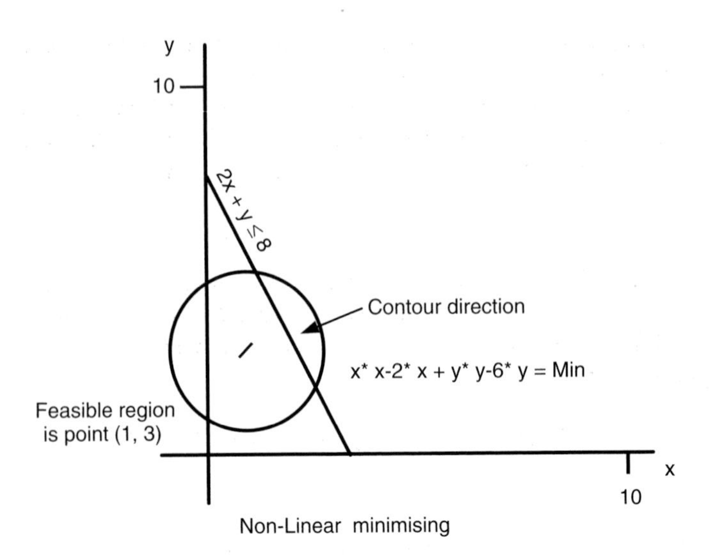

Fig. S12.2

Computer solution

In the same way as previously, the Lindo analysis produces corresponding results:

$$\text{MIN} = x^2 - 2x + y^2 - 6y$$
$$2x + y \leq 8$$
$$x \geq 0$$
$$y \geq 0$$

END

Objective value:	–£10.0	
Variable	Value	Reduced cost
x	1.0	0.0
y	3.0	–0.1184238E-07

This case shows that the project is unprofitable. Also:

Row	Slack or Surplus	Dual price
1	–10.0	1.0
2	3.0	–0.3552714E-07
3	1.0	0.0
4	3.0	0.0

Lagrangian multiplier and Kuhn–Tucker method

When the equations cannot be graphically interpreted, the analysis needs more sophisticated processes using, for example, the Lagrange multiplier method combined with the Kuhn–Tucker theorem, these being reasonably efficient developments to help find the saddle points, i.e. the turning points of the objective function curve.

The approach broadly builds on the principles described earlier in the basic LP, where it follows that the linear/non-linear problem for maximising $f(z)$ subject to constraints $g_i(z) = 0$ $(i = 1, \ldots, m)$ can be formulated as a Lagrangian function of the type:

$$L(z,u) = f(z) + \sum u_i g_i(z)$$

z being the variables, and u_i *unknown vectors* called the Lagrangian multipliers.

Solutions for z can be found among the extreme points of $L(z, u)$ by means of partial differential calculus, i.e. differentiating the Lagrangian function separately with respect to z and u to determine the gradients as follows:

$$\partial L(z,u)/\partial z = \partial f(z)\partial z + \sum u_i \partial g_i(z)/\partial z = \text{gradient}$$

and

$$\partial L(z,u)/\partial u = \sum g_i(z) = \text{gradient}$$

The optimum value (Kuhn–Tucker point) can be found at the stationary/turning point of the Lagrangian function where the gradient is zero, i.e. $\partial L(z, u)/\partial z = 0$, and $\partial L(z, u)/\partial u = 0$.

There are a variety of methods available, developed by different authors, for determining the Lagrangian Multiplier values (u)—see Lootsma. However, the equations can be quite intricate to manipulate in resolving the values for u in complex problems, and are more readily calculated with modern computer software packages.

By way of illustrating the Lagrangian principles for *maximising* the simple circle example, maximise:

$$x^2 - 2x + y^2 - 6y$$

subject to the constraints $2x + y \leq 8$ and $x \geq 0, y \geq 0$.

The Lagrangian function for the maximising situation gives the following formulation:

$$L(z, u) = x^2 - 2x + y^2 - 6y + u_1(-2x - y + 8) + u_2 x + u_3 y$$

Partially differentiating with respect to x and y gives:

$$\partial L/\partial x = 2x - 2 - 2u_1 + u_2 = 0$$

$$\partial L/\partial y = 2y - 6 - u_1 + u_3 = 0$$

and adding them together gives:

$$2x - 2 + 2y - 6 - 3u_1 + u_2 + u_1 = 0$$

Partially differentiating with respect to u gives:

$$\partial L/\partial u_1 = -2x - y + 8 = 0$$

$$\partial L/\partial u_2 = x = 0$$

$$\partial L/\partial u_3 = y = 0$$

The Kuhn–Tucker point is characterised by determining the appropriate positive Lagrangian multipliers to solve the partial differential equations and finding the stationary point (z, u) where the gradient is zero. In this example it can be found that for multipliers set at $u_1, u_2, u_3 = 0$, the maximum point $x, y = 4, 0$ satisfies all the above partial equations and therefore yields the $L(z, u)$ solution, Max = £8.

A similar calculation would demonstrate the minimising case:

$$L(z, u) = x^2 - 2x + y^2 - 6y - u_1(-2x - y + 8) - u_2x - u_3y$$

Local optimum solutions

While in the above simple case the global maximum or minimum values can be readily visualised, solutions to non-linear problems commonly produce local optima. Indeed, the available mathematical optimisation models generally operate by converging to a local optimum point. Hence, other and better solutions may thus exist.

For example, solutions of the objective function = $x\sin(\pi x)$ (i.e. non-linearly sinusoidal shaped) have different turning points or saddle points (i.e. local maximum and minimum) depending on the constraint limits chosen, as shown in Fig. S12.3. With constraint, say $x < 6$, a plot of this function produces positive turning points of 0.58, 2.52 and 4.51 at $x = 0.645$, $x = 2.54$ and $x = 4.52$ respectively, and a minimum value of -1.53 and -3.51 at $x = 1.56$ and $x = 3.52$. Other turning points exist for different limits on the constraint, thus clearly considerable trial and error would be necessary to establish the required result in analysing more complex problems. Indeed, the behaviours of the equations need to be understood and carefully considered in order to interpret the results meaningfully, as slight changes in the parameters can have marked effects.

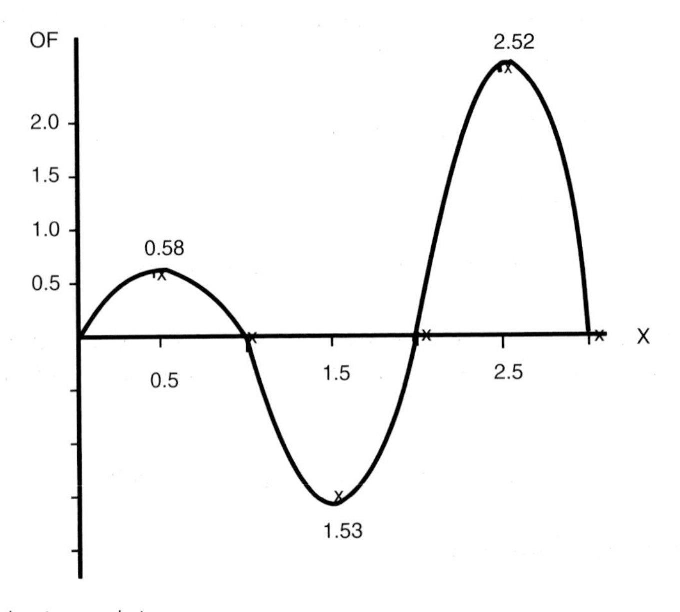

Fig. S12.3 Local optimum solutions.

Solution 13 Quadratic assignment problem

It will be observed that the problem consists of two variables, amount of items transferred from each lorry (variable L) and transfer time from a bay to another bay (variable B), thus, the objective function is quadratic i.e. multiplication of the two variables is involved and therefore not soluble by linear methods. However since all items in the table are already assigned the problem becomes merely calculation of the totals from which the minimum value is determined.

The problem is of the 0/1 binary integer variable type, where the variable can be either on or off (i.e. zero or one), which in this example would be according to the truck to bay designation and solved as follows:

Number of items transferred (variable L):

to truck	1	1	2
from	1	0	0
truck	2	20	0

Transfer time in minutes from one bay to another bay (variable B):

to Bay		A	B
from	A	0	5
Bay	B	15	0

By inspection in this simple case, the trucks directed as above would yield the minimum total transfer time calculated from the amount transferred from each truck to the other truck multiplied by the corresponding bay time as follows:

$$\text{Min} = L_{1,1} \times B_{1,1} + L_{1,2} \times B_{1,2} + L_{2,1} \times B_{2,1} + L_{2,2} \times B_{2,2}$$

and substituting values from the tables:

$$\text{Min} = 0 \times 0 + 30 \times 5 + 20 \times 15 + 0 \times 0 = 450 \text{ mins}$$

Obviously, however, if the trucks' arrivals were switched so that truck 1 arrived in Bay B and truck 2 arrived in Bay A and the amounts transferred were unchanged, the situation would now become:

$$\text{Min} = L_{1,1} \times B_{1,1} + L_{1,2} \times B_{2,1} + L_{2,1} \times B_{2,1} + L_{2,2} \times B_{2,2}$$

and substituting values from the tables:

$$\text{Min} = 0 \times 0 + 30 \times 15 + 20 \times 5 + 0 \times 0 = 550 \text{ mins}$$

Hence, the first case is preferable, but obviously as the size of the tables increases in a complex assignment problem, the manual method becomes impractical and is best tackled with a computer package.

Model formulation with Lindo software

The problem is to assign lorries to bays to minimise total transfer time between lorries and bays and can be generalised by introducing the 0/1 variable Y to represent whether a bay is being used or not. The calculations are similar the those described above, albeit in a more intricate algorithm, and can be seen in principle in the example available on the Lindo web site (www.Lindo.com).

Solution

Objective value: 450.0 minutes

Variable	Value	Reduced cost
$X(1,1)$	1	450
$X(1,2)$	0	0
$X(2,1)$	0	0
$X(2,2)$	1	450
$B(1,1)$	0	0
$B(1,2)$	5	0
$B(2,1)$	15	0
$B(2,2)$	0	0
$L(1,1)$	0	0
$L(1,2)$	30	0
$L(2,1)$	20	0
$L(2,2)$	0	0
$Y(1,1,2,2)$	1	0
$Y(1,2,2,1)$	0	550

Note: Substitution of reduced cost value = 550 for $Y(1,2,2,1)$ into the objective function produces $LBLB = 550$ for the opposite arrangement of lorry/bay positions.

Row	Slack or surplus	Dual price
1	0	0
2	0	0
3	0	0
4	0	0
5	0	–450
6	1	0
7	450	1

Solution 14 Dynamic programming—batch processing

The development of dynamic programming is closely associated with R.E. Bellman in the 1950s (see Bellman 2003) and provides solutions to linear and non-linear problems involving sequential decision-making, i.e. where the results of later decisions cannot affect earlier ones. The method of analysis is based on Belmopan's Principle of Optimality, which states that an optimal policy has the property that, whatever the initial state and initial decisions are, the remaining decisions must constitute an optimal policy with regard to the state resulting from the first decision.

Some management decisions are of this kind and critically depend on routing or scheduling in for example:

- Production and scheduling, e.g. make products now and keep in stock or make and use immediately.
- Resources scheduling, known as the knapsack problem, where the task is to determine how many items can be put into a knapsack of fixed capacity and handling weight limit according to an importance value for each item. Allocation of capital, prioritising transport deliveries, etc. are typical examples.

Dynamic programming solves these problems in stages, working backwards from the finish point objective and adopting the principle that:

> Irrespective of what decision was or might have been taken at the previous stage nearer the start point, the optimum decision at this stage (first stage in the back sequence) is the shortest route emanating from it to the finish point.
>
> Thus, beginning the analysis similarly at the next stage backwards, the optimum decision at this stage back from the finish point will again be the shortest route emanating from it, and so on repeated backwards in stages to the start point. The optimum solution is selected from the best result from the stages evaluation.

In this batch processing problem the evaluation procedure requires determination of the amounts of stock to purchase and when. The logical starting point begins with April purchases as Stage 1, working backwards from the finish date. Clearly, actions prior to Stage 1 would have already been executed and so could not affect the optimum decision remaining, i.e. what is the cheapest solution for the April purchase decision alone.

Thus there are *four stages* ($n = 1, 2, 3, 4$) January to April inclusive.

Let ${}_mD_n$ = decision variable, i.e. purchases in stage n; for decision options m, $m = 1, 2, \ldots, M$.

Let X_n = period in months remaining at beginning of stage n.

${}_mR_n$ is the return route costs from stage n = forward accumulated stocks $\times$ interest charge + administration charges.

In **bold** is the optimum (decision option) return route cost from stage n.

Stage 1—beginning of **April**
This is the first stage working back from the finish point

X_n	$_mD_n$	$_mR_n$	
X_1	$_1D_1$	$_1R_1$	
1 month	$_1D_1$	£300	**£300**

$_1D_1$ could be the decision to purchase in April for April. Thus, $_1R_1 = £300$ and $D_1F_1 = £300$. This is the only possible decision option.

Stage 2—beginning of **March**

X_2	$_mD_2$	$_mR_2$	
2 months	$_1D_2$	£600	–

$_1D_2$ could be the decision to purchase in each of March and April. Thus, costs are two administrative charges for purchases. Thus, $_1R_2 = 2 \times £300 = £300$, i.e. previous April optimum ($_1D_1$) plus March admin cost.

2 months	$_2D_2$	£550	**£550**

$_2D_2$ could be the decision to purchase in March for both March and April. Thus, costs are administrative charge plus interest charge for April stock.

Thus, $_2R_2 = £300 + 0.01 \times £25\,000 = £550$ and therefore $D_2F_2 = £550$ as being the smaller amount. Note that these are the only two decision options.

Stage 3—beginning of **February**

X_3	$_mD_3$	$_mR_3$	
3 months	$_1D_3$	£850	–

$_1D_3$ could be the decision to purchase in February for February plus purchases in March for March and April (previous optimum for March purchases is $_2D_2$. Thus, $_1R_3 = £300 + £550 = £850$.
Or

3 months	$_2D_3$	£800	**£800**

$_2D_3$ could be the decision to purchase in February for February and March, and purchase in April for April. Thus, $_2R_3 = £300 + 0.01 \times £20\,000 + £300 = £800$. $D_3F_3 = £800$ (i.e. this is the smallest value compared with the other cases in the decision options).
Or

3 months	$_3D_3$	£1000	–

$_3D_3$ could be the decision to purchase in February for February, March and April.

Thus, $_3R_3 = £300 + 0.01 \times £20\,000 + 2 \times 0.01 \times £25\,000 = £1000$.

Note that calculation for the fourth option ${}_1D_2$ has been avoided.

Stage 4—beginning of **January**

X_4	${}_mD_4$	${}_mR_3$	
4 months	${}_1D_4$	£1100	–

${}_1D_4$ could be the decision to purchase in January for January plus purchases in February for February and March, and purchase in April for April (Stage 3 optimum for February purchases i.e. ${}_2D_3$). Thus, ${}_1R_4$ = £300 + £800 = £1100.

Or

4 months	${}_2D_4$	£1000	**£1000**

${}_2D_4$ could be the decision to purchase in January for January and February, and purchase in March for March and April (Stage 2 optimum i.e. ${}_2D_2$). Thus, ${}_2R_4$ = £300 + 0.01 × £15 000 + £550 = £1000. Thus D_4F_4 = £1000 as the smallest amount.

Or

4 months	${}_3D_4$	£1150

${}_3D_4$ could be the decision to purchase in January for January, February and March and in April for April (Stage 1 optimum i.e. ${}_1D_1$). Thus, ${}_3R_4$ = £300 + 0.01 × £15 000 + 2 × 0.1 × £20 000 + £300 = £1150.

Or

4 months	${}_4D_4$	£1600	–

${}_4D_4$ could be the decision to purchase in January for January, February, March and April. Thus, ${}_4R_4$ = £300 + 0.01 × £15 000 + 2 × 0.01 × £20 000 3 × 0.01 × £25 000 = £1600.

Note that further options involving Stage 3 calculations have been avoided.

Summary of stage costs

${}_1D_1$	April	£300
${}_2D_2$	March	£550
${}_2D_3$	February	£800
${}_2D_4$	January	**£1000**

Comment

Clearly the best policy is to purchase in January and the rest in March, that is purchase in January for both January and February usage (£5000, £15 000) and purchase again in March for March and April usage (£20 000, £25 000).

Reconciliation

By way of comparison, all stock could be purchased in January as stock, so that, if a single purchase of £65 000 is made in January, the stock charges thereafter are as follows:

Month	Purchases	Stock charges
January	£5 000	no charge
February	£15 000	60 000 × 0.01
March	£20 000	45 000 × 0.01
April	£25 000	25 000 × 0.01
Total	£65 000	130 000 × 0.01

The total cost of a full stockholding strategy = £300 + 0.01 × 130 000 = £1600, i.e. this is ${}_4D_4 =$ £1600

Alternatively purchase at the beginning of each month for the month with no stock held over, giving 4 × £300 = £1200, which represents a route through the eliminated options.

The dynamic programming method required 10 calculations in the four stages, rather than $\sum 2^{n-1}$ over the $n = 1, 2, 3, 4$ stages, i.e. in this example $1 + 2 + 4 + 8 = 15$ calculations. Clearly, for a problem with many stages the number of avoided calculations becomes very significant.

Solution 15 Resources routing—the knapsack problem

Dynamic programming deals with this kind of problem in a similar manner to the previous example in so far as calculations are worked backwards from the delivery of drawings objective in stages from low (1) to highest priority (10). Scheduling which job type is the decision variable (D), production time is the return measure (R) in the available period (X) remaining.

Thus there are three stages ($n = 1, 2, 3$) *representing the delivery of jobs in order, namely: category 1, category 2 and category 3.*

Let ${}_mD_n$ = decision variable, i.e. number of jobs scheduled in stage n, for decision options m, $m = 1, 2, \ldots, M$. Let X_n = period in production days remaining available at beginning of stage n; ${}_mR_n$ is the return route weighting value. The optimum return route weighting is shown in **bold** type.

Stage 1

In the 10 days available the decision is to schedule jobs with the lowest priority i.e. job category 1 drawings, thus:

X_1	${}_mD_1$	${}_mR_1$	
0	0	0	**0**
1	0	0	**0**
2	1	2	**2**
3	1	2	**2**
4	2	4	**4**
5	2	4	**4**
6	3	6	**6**

7	3	6	**6**
8	4	8	**8**
9	4	8	**8**
10	5	10	**10**

Note: decision option $m = 1$ for each possibility.

Stage 2

The decision is to schedule jobs with the second lowest priority, i.e. job category 2, followed by any possible fitting in of job category 1 (second column). Thus in the 6 days (X_2) remaining after first allocating job category 2 the decision options are:

X_2	${}_mD_2$	${}_mR_2$	
4	1,0	6	**6**
5	1,0	6	**6**
6	1,0	6	
6	1,1	8	**8**
7	1,0	6	
7	1,1	8	**8**
8	2,0	12	**12**
8	1,1	8	
8	1,2	10	
9	2,0	12	**12**
9	1,1	8	
9	1,2	10	
10	2,0	12	
10	1,1	8	
10	1,2	10	
10	1,3	12	
10	2,1	14	**14**

Note that decision option m varies according to the combination of categories possible for each day.

Stage 3

The decision is to schedule jobs with the highest priority, i.e. job category 3, followed by any possible fitting in of job category 2 and category 1 in that order (second column). Thus in the 4 days (X_3) remaining after first allocating job category 3 the decision options are thus:

X_3	$_mD_3$	$_mR_3$	
6	1,0,0	10	**10**
7	1,0,0	10	**10**
8	1,0,0	10	
8	1,0,1	12	**12**
9	1,0,0	10	
9	1,0,1	12	**12**
10	1,0,0	10	
10	1,0,1	12	
10	1,0,2	14	
10	**1,1,0**	**16**	**16**

Note that decision option m varies according to the combination of categories possible for each day.

Result

Tracing back through the tables, the maximum total = 16 points illustrates the optimum result (stage 3 table), scheduled as follows:

Stage 3

(i) Schedule one category 3 job which occupies 6 days. There are 4 days left.
(ii) Schedule one category 2 job which occupies 4 days. There are 0 days left to schedule.
(iii) Schedule nil category 1 jobs.

The schedule would thus have to be repeated with additional periods of 10 days and recalculated for each to clear the rest of categories 3 and 2 in that order, finally delivering category 1.

Sensitivity analysis

Other alternatives could be considered. For example, it can be seen from stage 3 table that the effects of scheduling over a period as low as 6 days (i.e. the duration of the highest job category 3) can be quickly determined by simply recalculating for $X = 6$ days. Stage 3 then becomes:

X_3	$_mD_3 =$	$_mR_3$	
6	1,0,0	10	**10**

Stage 3 = 1 job
Stage 2 = 0 jobs
Stage 1 = 0 jobs
= 10 points

Alternatively, in scheduling over an 8-day period ($X = 8$), both stages 2 and 3 give 12 points, respectively:

Stage 3 = 1 job
Stage 2 = 0 jobs
Stage 1 = 1 job
Total = 12 points

Or

Stage 3 = 0 jobs
Stage 2 = 2 jobs
Stage 1 = 0 jobs
Total = 12 points

Computer software

While the general concept of finding the shortest route can be tackled through backwards stages of analysis, different problems need specific formulation and therefore bespoke software also commonly needs to be specific. However, suites of programs such as Lindo contain packages suitable for dynamic programming applications when suitably modified and tailored to the particular categories of routing, inventory and knapsack problems.

Solution 16 Decision tree problem

Decisions to be made: use existing facilities only, a_1; gradual expansion, a_2; immediate expansion, a_3.

If the growth in the demand for construction work can be guaranteed, the optimal act is to invest immediately and expand capacity, when the choice is simply between 'Do nothing' and 'Expand immediately', demonstrated as follows:

$$\text{Expected profitability } a_1 = 12 \times 0.8 + 10 \times 0.2 = 11.6\%$$
$$a_3 = 18 \times 0.8 + 0 \times 0.2 = 14.4\%$$

However, if the decision was to expand only to suit market conditions, an efficient method of evaluating the sequence of decisions is to express the alternative choices by a decision sequence diagram. Hence, let □ be points at which alternative decisions are possible. Let Ⓟ be points at which probability plays its part. The problem can thus be expressed in the way shown in Fig. S16.1.

Step 1 The company is confronted with decision 🅰, to do nothing, plan for gradual expansion or expand immediately.

Step 2 The probability situation Ⓟ, at the end of each act, may be high growth (0.8) or low growth (0.2) in the market.

Step 3 Travelling down the 'no expansion' path the expected profitability is 0.8 × 12% + 0.2 × 10% = 11.6%.

Step 4 Travelling down the 'expand immediately' path the expected profitability is 0.8 × 18% + 0.2 × 0 = 14.4%.

Step 5 Travelling down the 'expand gradually' path the company is faced with decisions B and C. Clearly the 12% 'do nothing' and the 9% 'expand' are unfavourable and are discarded. The structure of the tree can now be represented as shown in Fig. S16.2.

Step 6
Profitability:

$$a_1 = 0.8 \times 12 + 0.2 \times 10 = 11.6\%$$

$$a_2 = 0.8 \times 16 + 0.2 \times 10 = 14.8\%$$

$$a_3 = 0.8 \times 18 + 0.2 \times 18 = 14.4\%$$

Thus the final tree is as depicted in Fig. S16.3.

The best decision is to wait and see if the demand for construction work does improve and then put in facilities to meet the required demand.

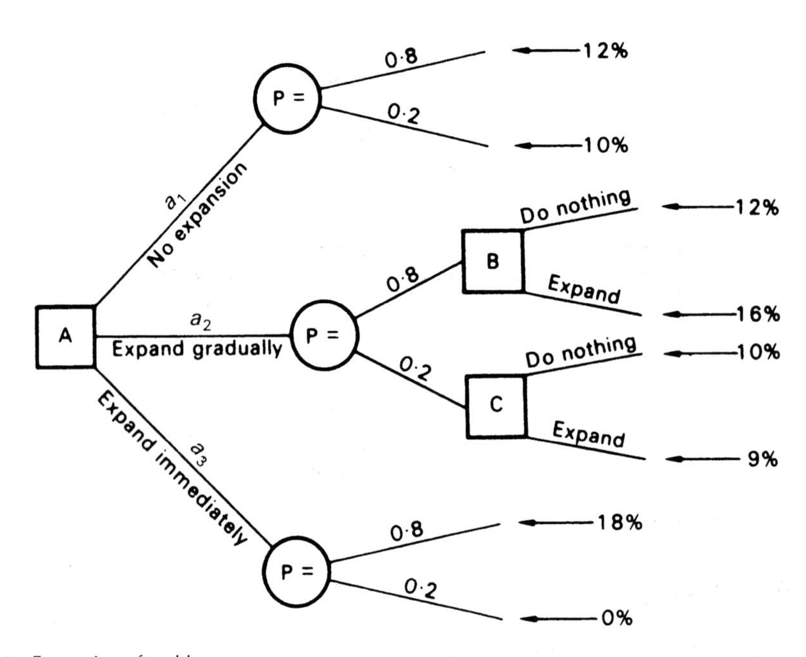

Fig. S16.1 Expression of problem.

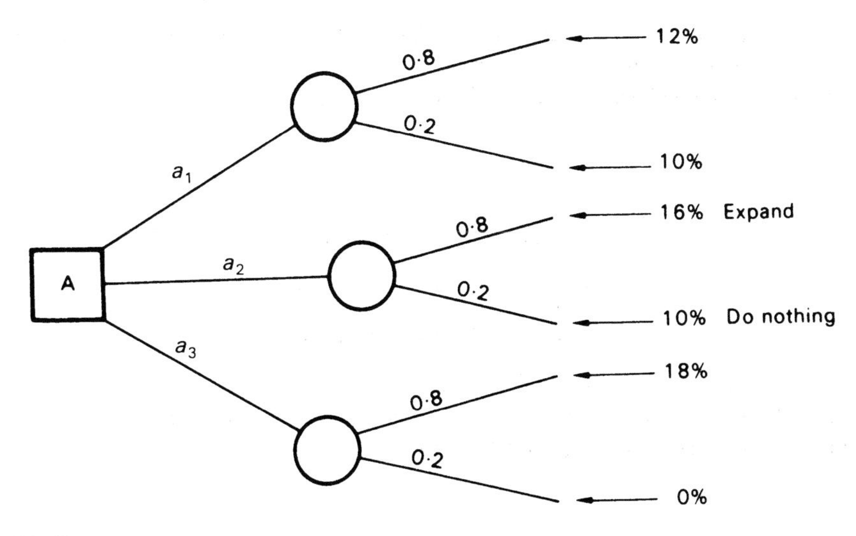

Fig. S16.2 Tree structure.

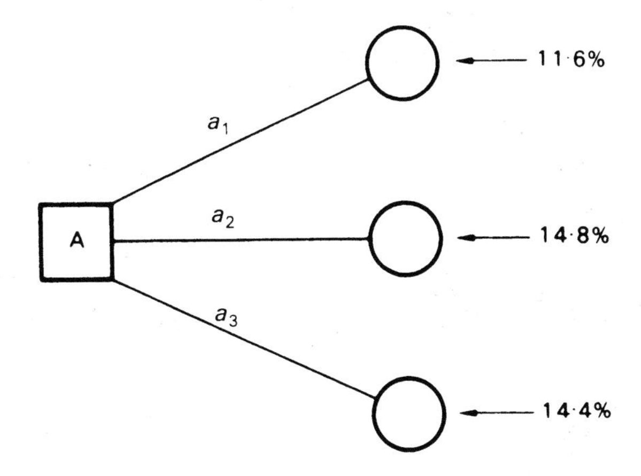

Fig. S16.3 Final tree structure.

Solution 17 Utility

The expected profitability calculations are;

$$a_1 = 0.8 \times 12 + 0.2 \times 10 = 11.6$$

$$a_2 = 0.8 \times 16 + 0.2 \times 5 = 13.8$$

$$a_3 = 0.8 \times 18 + 0.2 \times 0 = 14.4$$

The expected profitability analysis suggests a_3, but to avoid the possibility of incurring zero profitability if market conditions prove weak, option a_1 may be less risky. The utility table indicates worst and best case scenarios of 0% and 18% profitability respectively, which can be represented by utility values on a scale 0–10, for example:

$$\text{Utility } (\pounds 200) = 10$$

$$\text{Utility } (\pounds 2000) = 0$$

(i)

Consider determination of the utility of 16% profitability. If the indifference value (p) for this outcome were close to 1, say 0.95, then the company would most likely prefer to take the gamble between 18% and 0% of obtaining highest profitability with almost certainty compared to getting only the 16%. And conversely were p to be very close to 0, the company would want the 16% situation.

Given p is 95%, the utility for a 16% outcome is:

$$U(16\%) = pU(\pounds 18\%) + (1 - p)U(0\%)$$

$$U(16\%) = pU(18\%) + (1 - p)U(0\%) = 17.1\%$$

or

$$U(16\%) = 0.95(10) + 0.05(10) = 9.5$$

If the company prefers a guaranteed 16%, the difference between expected utility value of 17.1% and 16% is then in effect the premium incurred to avoid the 5% risk of zero profit.

In this manner, the utility values may be established for the rest of the profitability values.

(ii)

Suppose the company managing director is asked to state a preference for having 12% profitability rather than taking the gamble of getting the maximum profit of 18% if, say, $p = 0.9$. At this p level the MD might risk gambling on a 18% outcome. Repeated, the MD might be prepared to gamble until say $p = 0.6$, giving an ultimate utility value of:

$$U(12\%) = 0.6\,(18\%) + 0.4\,(0) = 10.8\%$$
$$= 0.6(10) + 0.4\,(0) = 6$$

i.e. the MD would just as soon accept 12% profit as gamble.

Rigorous questioning of the MD using this approach produced utility values for the different profitability outcome possibilities as stated in the question information as follows:

Profitability	Indifference value (p)	Utility Value
18%	1	10
16%	0.95	9.5
12%	0.6	6.0
10%	0.4	4.0
5%	0.30	3.0
0%	0	0

The expected utility for each of the cases a_1, a_2, a_3 can be calculated as:

$$a_1 = 0.8 \times 6 + 0.2 \times 4 = 5.6$$

$$a_2 = 0.8 \times 9.5 + 0.2 \times 3 = 8.2$$

$$a_3 = 0.8 \times 10 + 0.2 \times 0 = 8$$

	Expected profitability	Expected utility
$a_1 =$	11.6%	5.6
$a_2 =$	13.8%	8.2
$a_3 =$	14.4%	8.0

The optimal decision thus becomes a_2. The rationale for not accepting the highest expected profitability value of case a_3 being that at the demand probability of 0.2 profitability of zero was considered by the MD as too big a risk, not adequately reflected by the better expected profitability value.

However, it is observed that slight changes in indifference values would dramatically change the calculation outcomes. Hence the expected utility approach should be applied with extreme caution and only when the source of the advice for determining the p values can be confidently achieved.

Solution 18 Markov processes

A Markov process can be used to determine the probability that a situation will be in a particular state at any one period of time, depending only on the state in the immediate time period, called the transition probability. Clearly, such a technique has potential for ascertaining that equipment functioning in one period will still be working, or conversely break down in the next. Indeed a Markov process model can be developed to determine the 'steady-state' probabilities from a large number of previous transitions data to provide an estimate of the most appropriate time to make a replacement, carry out maintenance, etc.

The Markov process begins by assuming a given state, for example the equipment is initially running (period 0). Thus, as can be seen in Fig. S18.1 in period 1, there is a 0.85 transition probability of remaining as running and a 0.15 transition probability of being down. Similarly, the process continues into period 2 at the respective probabilities, and so on into period 3 and subsequent periods. The transition probabilities at period 2 can be calculated as illustrated, so that the probability of running in period 2 is 0.7225 + 0.03 = 0.7525 and the probability of being down is 0.1275 + 0.12 = 0.2475.

Continuing in this manner, eventually after enough periods the probability of a particular running or down situation can be shown to be independent of the initial state and these are called the 'steady-state' probabilities, denoted as running (π_1) and down (π_2), which are related by the following general formula for the particular state probabilities 0.85/0.15 and 0.2/0.8:

$$(\pi_1) = 0.85(\pi_1) + 0.2(\pi_2)$$

and

$$(\pi_2) = 0.15(\pi_1) + 0.8(\pi_2)$$

where

$$(\pi_1) + (\pi_2) = 1$$

(i) Solving (π_1) and (π_2) gives the following 'steady-state' probabilities:

$$(\pi_1) = 4/7, (\pi_2) = 3/7$$

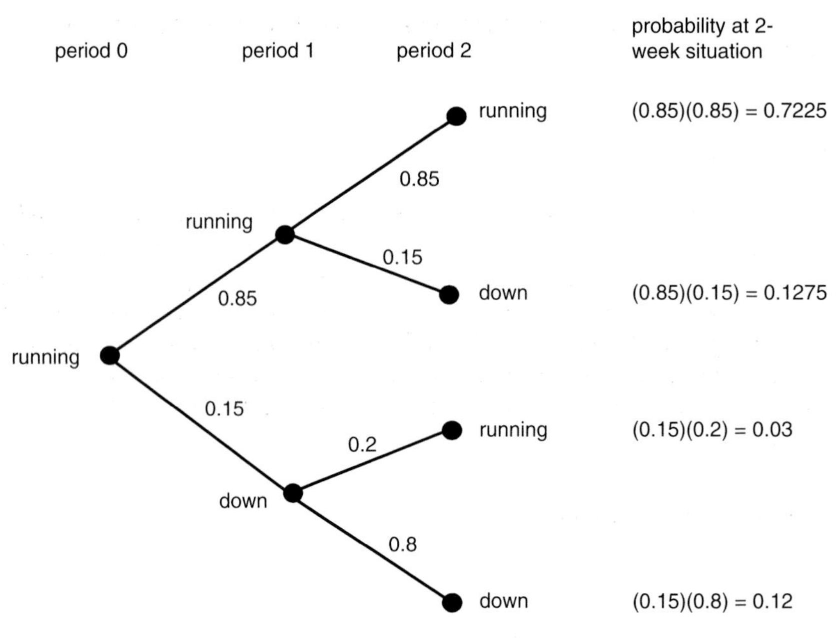

Fig. S18.1 Markov probabilities at 2 weeks.

This implies that eventually the equipment can be expected to be running 4/7 of the time and down 3/7 of the time.

(ii) The introduction of the replacement part changes the probability states to 0.9/0.1 and 0.3/0.7, allowing a new tree diagram to be constructed whereby:

$$(\pi_1) = 0.9(\pi_1) + 0.3(\pi_2)$$

and

$$(\pi_2) = 0.1(\pi_1) + 0.7(\pi_2)$$

Solving (π_1) and (π_2) for probability states 0.75/0.25 and 0.3/0.7 gives the following 'steady-state' probabilities

$$(\pi_1) = 3/4, (\pi_2) = 1/4$$

This implies that eventually the equipment can be expected to improve running time and reduce downtime.

(iii) Thus, in any one period of 100 hours downtime would reduce by $(3/7 - 1/4) \times 100 = 17.85$ hours or 0.1785 hours per hour, i.e. downtime losses would fall by $0.1785 \times £100 = £17.85$ per hour.

Hence, the breakeven cost of a replacement part including fitting is given by £17.85 × the chosen maintenance interval.

Solution 19 Queuing

Originally developed by A.K. Erlang (1917) in Denmark, the theory of queues was based on observed waiting and service time data and provides an alternative mathematical method to simulation using formulae. The underlying statistical analysis for a 'simple queue' requires that customers arrive randomly and independently, they occur within an infinite population, steady-state conditions apply and the number of arrivals per unit of time is a Poisson probability distribution with a constant mean and service time is an Exponential probability distribution. Otherwise, a different statistical model needs to be derived. The formulae for the two distributions are as follows.

Poisson probability distribution

$$P_p(x) = \frac{\lambda^x e^-}{x!} \text{ for } x = 0, 1, 2, \ldots$$

P_p = probability of x events (arrivals) in the time period;
x = number of events (arrivals) in the time period;
λ = mean number of events (arrivals) in the time period (i.e. mean arrival rate);
e = 2.71828.

Exponential probability distribution

$$P_e(\text{service time}) \leq t = 1 - e^{-\mu t}$$

P_e = probability that the service time will be equal to or less than time t;
μ = mean number of customers served per time period (i.e. mean service rate);
e = 2.71828.

Poisson distribution for rates of arrival

Arrival rate (trucks/min)	Frequency proportion	Poisson probability	
x	f	P_f	P_p
0	77	0.5662	0.5655
1	42	0.3088	0.3223
2	15	0.1103	0.0919
3	2	0.0147	0.0175
4 or more	0	0.0000	0.0028
Total	136	1.0000	1.0000

$$\text{Mean } x = \frac{0 \times 72 + 1 \times 42 + 2 \times 15 + 3 \times 2}{136} \text{ trucks per min}$$

$\lambda = 0.57$

Exponential distribution for service times

Service time t	Frequency f	Proportion Pf	Proportion $\leq t$	Exponential probability P_e
0.5	9	0.3	0.3	0.34
1	9	0.3	0.6	0.57
1.5	7	0.2333	0.8333	0.72
2	3	0.1000	0.9333	0.82
2.5	1	0.0333	0.9666	0.88
3	1	0.0333	1.0000	0.92
Total	30	1.0000		

$$\text{Mean } t = \frac{0.5\times9+1\times9+1.5\times7+2\times3+2.5\times1+3\times1}{30} = \frac{355}{300} = 1.18 \text{ mins}$$

$$\mu = 1/1.18 = 0.845$$

Data for the actual proportions must now be tested for goodness of fit with the theoretical probability distributions. The χ^2 test is appropriate for the Poisson distribution/arrival rate data, while the Kolmogorov–Smirnov procedure is more suited to comparing mean time to occurrence data with an exponential (P_e) probability distribution. The methods are described in statistical text books, which essentially require calculation of the χ^2 and KS values and finding their locations in appropriate probability tables. A value giving a location within the 95% confidence interval for the tabled distribution would be considered acceptable to confirm the null hypothesis (i.e. 95% likely), which for the above data indicated goodness of fit in each case.

(i) Queue with infinite population

The derived expressions for the 'simple queue' are as follows:

mean inter-arrival time = 1/mean arrival rate (λ)
mean service time = 1/mean service rate (μ)
probability of a customer having to wait for service = ρ = mean service time/mean inter-arrival time

$$= \frac{1/\mu}{1/\lambda} = \frac{\lambda}{\mu} < 1; \text{ otherwise the line will grow indefinitely}$$

probability of no customers in the system $= 1-\rho$
probability of n customers in the system $= (\rho)^n(1-\rho)$
average number of customers in the queue, including no queues $Q_q = \rho^2/(1-\rho)$
average number of customers in the queue, when a queue exists $= 1/(1-\rho)$
average number of customers in the system $Q = \rho/(1-\rho)$
average time customer in the queue $W_q = \rho/\mu(1-\rho)$
average time customer in the system $W = 1/\mu(1-\rho)$

Result

Manipulation of the data in the relevant formula gives an average 2.754 trucks in the queue when there is a queue, and a truck waits on average 1.95 minutes in the queue.

Variable	Value
ARV_RATE	0.57 trucks per min
SRV_TIME	1.18 mins
NO_SRVRS	1 crane
ρ	0.675
W	3.63 mins is average time a truck spends in the system
W_q	2.45 mins is average time a truck spends in the queue
Q	1.98 trucks on average in the system

(ii) Queue with finite population

Varying the number of servers and limiting the size of the arrivals population requires significant modification of the 'simple queue' formulae. Results are more conveniently achieved with, for example, the 'Machine Repairman model' within the Lindo suite of computer programs.

MODEL:

The mean time between arrivals;
MTBF = 1.754;
The mean time to service;
MTTR = 1.117;
The number of trucks;
NUSER = 4;
The number of servers/excavators;
NREPR = 1;
The mean number of trucks in the system;
NDOWN =
@PFS (MTTR × NUSER/MTBF, NREPR, NUSER);
The overall service rate, FR, must satisfy;
FR = (NUSER – NDOWN)/MTBF;
The mean time a truck is in the system, MTD, must satisfy;
NDOWN = FR × MTD;
END

Result

Manipulation of the data achieved with the Lindo 'Machine Repairman model' gives approximately an average 2.58 trucks in the system and a truck spends an average 3.2 mins in the system.

Variable	Value
MTBF	1.754 mins
MTTR	1.117 mins
NUSER	4 trucks
NREPR	1 crane
NDOWN	2.58 trucks waiting in the system on average
MTD	3.20 mins truck spends in the system on average

Cost optimisation

The cost trade-off between engaging more excavators and reducing truck waiting time can be developed in the Lindo 'Repairman model' as follows:

Total cost per hour of system = average number of trucks in the system × truck hourly hire rate per truck + number of excavators in the system × excavator hourly hire rate per excavator

MODEL:

The Lindo 'Repairman model' is as follows:

For each case of 1–3 excavators calculate expected number of trucks in the system, cost per hour of the trucks in the system and total cost per hour of the system. @PFS calculates the probability in a finite source, in this case expected number of trucks in the system.

SETS:

```
NREP/1.3/:! Consider 3 excavators;
NDOWN,! Expected number of trucks in the system;
CPERHR,! Cost/hour of the trucks in system;
TCOST;! Total expected cost/hour of the system;
ENDSETS
! The input data;
NMACH = 4;! Number of trucks;
RTIME = 0.0197;! Average service (loading) time in hours;
UPTIME = 0.0292;! Mean time between arrivals in hours;
CR = 50;! Hourly hire cost of excavator;
CM = 60;! Hourly cost of a truck;
@FOR(NREP(I):NDOWN(I) = @PFS(NMACH × RTIME/UPTIME, I, NMACH);
CPERHR(I) = CM × NDOWN(I);
TCOST(I) = CPERHR(I) + CR × I);
END
```

Result

The cheapest solution is to use one excavator, which produces a total system cost per hour of £205.2. The cost savings in reducing truck time in the system would be insufficient to cover the costs of additional excavators. It should also be noted that the costs of waiting periods for the service unit (excavator) are automatically taken into account in the calculation.

Variable	Value
NMACH	4
RTIME	0.0197
UPTIME	0.0292
CR	50
CM	60
NDOWN(1)	2.586

NDOWN(2)	1.818
NDOWN(3)	1.632
CPERHR(1)	155.2019
CPERHR(2)	109.1172
CPERHR(3)	97.93449
TCOST(1)	205.2019
TCOST(2)	209.1172
TCOST(3)	247.9345

Solution 20 Simulation

The trucks arrive at the mixer at varying times in an indistinguishable pattern. However, the percentage distributions of the arrival times are known, therefore it is possible to simulate the arrival time interval of a truck by drawing a random number lying in the range represented by the observed distribution. Thus:

Arrival time interval (min)	*Frequency (%)*	*Random number*
2	10	00–09
3	15	10–24
4	30	25–54
5	25	55–79
6	20	80–99

Table S20.1 simulates the arrival of 50 trucks, one after the other. The random numbers are selected from tables or generated from a hand-held calculator and the type of truck arriving is determined by tossing a coin. Truck Type A represents a 3 m^3 load. Truck Type B represents a 6 m^3 load.

A check on the distribution of arrival times simulated produces a reasonable fit, as shown below.

Arrival time interval (min)	*Observed frequency (%)*	*Simulated number of arrivals*	*Simulated frequency (%)*
2	10	4	8
3	15	8	16
4	30	17	34
5	25	11	22
6	20	10	20

It is now possible to simulate the whole loading process, since if the type of truck is known its loading time can also be determined. Table S20.2 shows the situation at the batching plant over a two-hour period.

The simulation should now be repeated several times, so that the results can be expressed with more confidence by the use of statistics.

Table S20.1 Simulated arrival times

Random number	Arrival time (min)	Type of truck	Clock time of arrival (min)	
			($3\,m^3$)	($6\,m^3$)
89	6	A	6	
29	4	B		10
73	5	A	15	
50	4	A	19	
47	4	B		23
51	4	A	27	
32	4	B		31
58	5	A	36	
07	2	A	38	
49	4	A	42	
14	3	A	45	
2	2	B		47
26	4	B		51
97	6	B		57
83	6	A	63	
47	4	B		67
51	4	A	71	
55	5	B		76
92	6	B		82
07	2	B		84
45	4	A	88	
85	6	A	94	
76	5	A	99	
15	3	A	102	
78	5	B		107
68	5	A	112	
83	6	B		118
33	4	A	122	
18	3	B		125
04	2	A	127	
11	3	B		130
98	6	A	136	
11	3	B		139
18	3	B		142
34	4	B		146
84	6	B		152
47	4	B		156
18	3	B		159
46	4	A	163	
71	5	A	168	
53	4	B		172
88	6	B		178
78	5	A	183	
30	4	A	187	
71	5	A	192	
35	4	B		196
96	6	A	202	
78	5	B		207
57	5	A	212	
19	3	B		215

Table S20.2 Simulated situation at batching plant

Arrival type truck	Clock time of arrival (min)	Starts loading (min)	Loading time (min)	Completes loading batcher	Waiting time truck (min)	Waiting time (min)
A	06	06	3	09	–	–
B	10	10	5	15	1	–
A	15	15	3	18	–	–
A	19	19	3	22	1	–
B	23	23	5	28	1	–
A	27	28	3	31	–	1
B	31	31	5	36	–	–
A	36	36	3	39	–	–
A	38	39	3	42	–	1
A	42	42	3	45	–	–
A	45	45	3	48	–	–
B	47	48	5	53	–	1
B	51	53	5	58	–	2
B	57	58	5	63	–	1
A	63	63	3	66	–	–
B	67	67	5	72	1	–
A	71	72	3	75	–	1
B	76	76	5	81	1	–
B	82	82	5	87	1	–
B	84	87	5	92	–	3
A	88	92	3	95	–	4
A	94	95	3	98	–	1
A	99	99	3	102	1	–
A	102	102	3	105	–	–
B	107	107	5	112	2	–
A	112	112	3	115	–	–
B	118	118	5	123	3	–
A	122	123	3	126	–	1
				Total	12 min	16 min

Waiting time mixer = 12 min in 2 hours.
Waiting time trucks = 16 min in 2 hours.

Solution 21 Stock control—simple case

Let B = minimum stock level, Q = tonnes of cement delivered with each order, D = rate of usage in tonnes per week, S = cost of processing an order, h = cost of storage and deterioration of cement per tonne per week as a percentage of cement price, P = cost of cement per tonne, t = time in weeks between orders.

The cycle of usage and replenishment is shown in Fig. S21.1.

Average amount of cement stored in time t is $\frac{1}{2}Q + B$. Therefore storing cost per cycle of length $t = \frac{1}{2}QthP + BthP$.

Total cost per cycle of length $t = \frac{1}{2}QhP + BthP + S$.

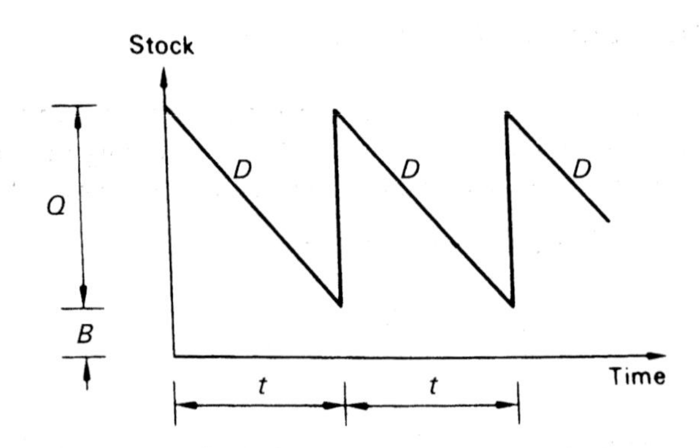

Fig. S21.1 $t = \frac{Q}{D}$.

Total cost per week = $QhP + BthP + \frac{S}{t}$

but
$$t = \frac{Q}{D}$$

$$\therefore \text{ Total cost (TC) per week} = \tfrac{1}{2}QhP + BhP + \frac{SD}{Q} \qquad (1)$$

To obtain optimum order size, differentiate w.r.t. Q.

$$\frac{\mathrm{dTC}}{\mathrm{d}Q} = \tfrac{1}{2}hP - \frac{SD}{Q^2}$$
$$= 0 \text{ for a maximum.}$$

Therefore $Q^2 = \frac{2SD}{hP}$.

$$Q = \sqrt{\frac{2SD}{hP}}$$
$$= \sqrt{\left[\frac{2 \times 1 \times 50}{1/10 \times 15}\right]}$$
$$= 8.16 \text{ tonnes.} \qquad (2)$$

Substituting Q in Equation (1):

$$\mathrm{TC} = hPB + \frac{Q^2hP + 2SD}{2Q}$$
$$= hPB + \frac{2SD + 2SD}{2 \times \sqrt{\frac{2SD}{hP}}}$$
$$= hPB + \sqrt{2SDhP}$$
$$= (\tfrac{1}{10}) \times 15 \times 3 + \sqrt{[2 \times 1 \times 50 \times \tfrac{1}{10} \times 15]}$$
$$= £16.75 \text{ per week.} \qquad (3)$$

Solution 22 Stock control—continuous usage

Q = number of beams made per production run, D = number of beams required by contractor each week, k = number of beams produced per week, H = cost of storing one beam per week, S = cost of setting up a production run, t = time interval in weeks between production runs.

From Fig. S22.1, length of the production run $t_1 = \dfrac{Q}{k}$.

Length of production and usage cycle $t = \dfrac{Q}{D}$, $AC = Q - Dt_1$,

$$\text{storage cost per cycle} = \tfrac{1}{2}(Q - Dt_1) \times H \times t = \tfrac{1}{2}Qt\left(1 - \frac{D}{k}\right)H,$$

$$\text{total cost per cycle} = \tfrac{1}{2}QtH\left(1 = \frac{D}{k}\right) + S.$$

$$\text{Therefore total cost (TC) per week} = \frac{S}{t} + \tfrac{1}{2}QH\left(1 - \frac{D}{K}\right).$$

$$\text{TC} = \frac{SD}{Q} + \tfrac{1}{2}QH\left(1 - \frac{D}{k}\right) \tag{1}$$

To obtain optimum run size differentiate with respect to Q.

$$\frac{\mathrm{dTC}}{\mathrm{d}Q} = \frac{SD}{Q^2} + \tfrac{1}{2}H\left(1 - \frac{D}{K}\right)$$

$$= 0 \text{ for a maximum.}$$

$$Q = \sqrt{\left\{\frac{2SD}{H[1-(D/k)]}\right\}} \text{ units.}$$

$$t = \sqrt{\left\{\frac{2S}{DH[1-(D/k)]}\right\}} \text{ weeks.}$$

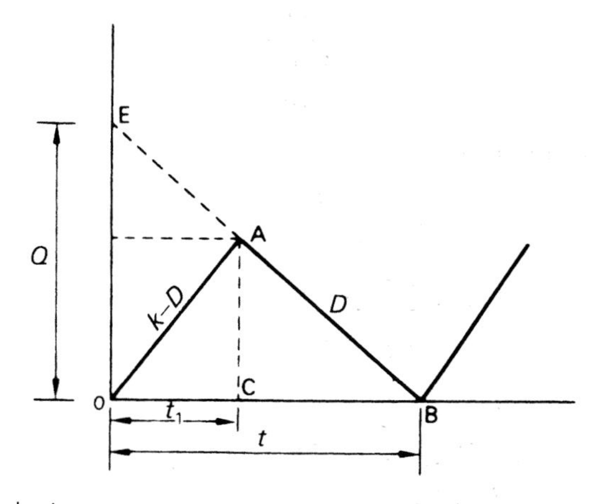

Fig. S22.1 Length of production run.

$$\mathrm{TC} = \frac{2SD + Q^2\mathrm{H}[1-(D/k)]}{2Q} \text{ per week.}$$

or substituting for Q

$$\mathrm{TC} = \sqrt{\{2SDH[1-(D/k)]\}} \text{ per week} \tag{2}$$

(a)

$$Q = \sqrt{\left\{\frac{2\times 50\times 100}{0.01\times[1-(1000/2500)]}\right\}}$$
$$= 4083 \text{ beams per run}$$

(b)

$$\mathrm{TC} = \sqrt{\{2\times 50\times 1000\times 0.01[1-(1000/2500)]\}}$$
$$= £24.50 \text{ per week}$$

(c)

$$t = \sqrt{\left\{\frac{2\times 50}{1000\times 0.01[1-(1000/2500)]}\right\}}$$

Solution 23 Stock control—shortages

Z = cost of shortage per 1000 bricks, Q = economic order quantity, A = top-up quantity, D = rate of usage of bricks per week, h = storage cost as a percentage of cost price of bricks, P = cost of bricks per 1000, S = cost of processing an order, t = time in weeks between supplies.

From Fig. S23.1:

(a) Storing cost = $\frac{1}{2} Dt_1 t_1 hP$ per cycle and $A = Dt_1$.

$$\text{Storing cost per week} = \frac{Dt_1^2 hP}{2t}$$
$$= \frac{At_1 hP}{2(t_1 + t_2)} \tag{1}$$

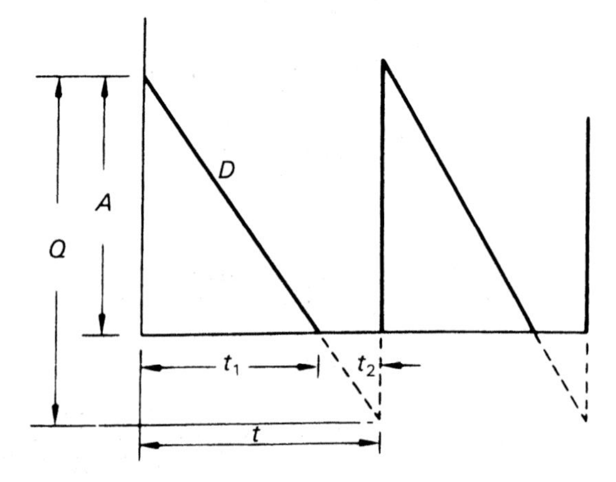

Fig. S23.1 Storing costs.

Now from similar triangles

$$\frac{Q}{t_1 + t_2} = \frac{A}{t_1}$$

Therefore

$$t_1 = \frac{At_2}{Q - A}$$

Substituting t_1 in Equation (1):

$$\text{Storing cost per week} = \frac{A^2hP}{2Q} \qquad (2)$$

(b) Cost of shortage per cycle $= \frac{1}{2} Dt_2^2 Z$

$$= \frac{Dt_2^2 Z}{2(t_1 + t_2)} \qquad (3)$$

Substituting t_1 in Equation (3):

$$\text{Shortage cost per week} = \frac{(Q - A)^2 Z}{2Q} \qquad (4)$$

(c) Total cost per week $= \dfrac{A^2hP}{2Q} + \dfrac{(Q-A)^2 Z}{2Q} + \dfrac{SD}{Q}$

To obtain optimum order size differentiate with respect to Q and A and maximise.

$$Q = \sqrt{\left(\frac{2SD}{hP}\right)} \times \sqrt{\left(\frac{Z + hP}{Z}\right)} \text{ units,}$$

$$A = \sqrt{\left(\frac{2SD}{hP}\right)} \times \sqrt{\left(\frac{Z}{Z + hP}\right)} \text{ units,}$$

$$\text{TC} = \sqrt{2SDhP} \times \sqrt{\left(\frac{Z}{Z + hP}\right)} \text{ per week.}$$

(i)

$$Q = \sqrt{\left[\frac{2 \times 10 \times 50}{(1/10) \times 30}\right]} \times \sqrt{\left[\frac{10 + (1/10) \times 30}{10}\right]}$$

$$= £48.04 \text{ per week}$$

(ii)

$$\text{TC} = \sqrt{[2 \times 10 \times 50 \times (1/10) \times 30]} \times \sqrt{\frac{10}{10 + (1/10) \times 30}}$$

$$= £48.04 \text{ per week}$$

(iii) $A = \sqrt{\left[\frac{2 \times 10 \times 50}{(1/10) \times 30}\right]} \times \sqrt{\left[\frac{10}{10 + (1/10) \times 30}\right]}$
$= 16.01$ thousand bricks.

Solution 24 Stock control—discounts

$$Q = \sqrt{\left(\frac{2SD}{hP}\right)}$$
$$= \sqrt{\left(\frac{2 \times 20 \times 2000}{P}\right)}$$
$$= \frac{400}{\sqrt{P}}$$

Discount situation

(1)

$$Q = \frac{400}{\sqrt{1.21}}$$
$$= \frac{400}{1.1}$$
$$= 363 < 500 \text{ tonnes.}$$

In range.

(2)

$$Q = \frac{400}{\sqrt{1}}$$
$$= 400$$

Outside range 500–999.

(3)

$$Q = \frac{400}{\sqrt{0.81}}$$
$$= \frac{400}{0.9}$$
$$= 444$$

Outside range 1000 or more.

On first inspection the optimum order quantity would be 363 tonnes. But the calculation so far does not take into account the cost of the material itself, which in this instance varies according to quantity ordered. Therefore looking at total costs:

$$\text{TP} = \text{cost of material} + \text{storage cost} + \text{order cost}$$

Therefore

$$\text{TC per month} = DP + \tfrac{1}{2}QhP + \frac{SD}{Q}.$$

Overall discount situation

(1)

(a) $$\text{TC} = 2000 \times 1.21 + \tfrac{1}{2} \times 363 \times 0.5 \times 1.21 + \frac{20 \times 2000}{363}$$

$= £2640$ per month.

(2)

(a) $$\text{TC} = 2000 \times 1.0 + \tfrac{1}{2} \times 500 \times 0.5 \times 1.0 + \frac{20 \times 2000}{500}$$

$= £2205$ per month.

(b) $$\text{TC} = 2000 \times 1.0 + \tfrac{1}{2} \times 999 \times 0.5 \times 1.0 + \frac{20 \times 2000}{999}$$

$= £2289$ per month.

(3)

(a) $$\text{TC} = 2000 \times 0.81 + \tfrac{1}{2} \times 1000 \times 0.5 \times 0.81 + \frac{20 \times 2000}{1000}$$

$= £1862$ per month.

(b) $$\text{TC} = 2000 \times 0.81 + \tfrac{1}{2} \times 2000 \times 0.5 \times 0.81 + \frac{20 \times 2000}{2000}$$

$= £2045$ per month.

The optimum order quantity is 1000 tonnes per month and the total monthly purchase, storage and ordering cost is £1862.

Solution 25 Inventory under uncertainty

Demands are commonly variable and subject to uncertainty and, thus, deterministic models are often of limited scope. Consequently, solutions embracing expected profits or costs by incorporating probability values can be developed to provide a range of solutions between maximum and minimum possibilities.

The objective of the workshop manager is to decide how many items Q^* should be serviced at the beginning of the week. The uncertain demand D expresses the number of items that customers will require during this period. Two types of outcome may occur. If demand is less than or equal to the order quantity, the number of equipments serviced will equal the quantity demanded (D); if demand is greater than the initial budgeted amount, the number serviced will equal the order quantity (Q).

Sales $$\begin{aligned} &= D \quad \text{if} \quad D \leq Q \\ &= Q \quad \text{if} \quad D > Q \end{aligned}$$

Thus let C_o be cost per item incurred for *overestimating the demand.* This represents the cost of budgeting for one extra serviced item of equipment and then finding that there is not a customer for it, i.e. £200 servicing cost for a machine.

Let C_u be the cost per item incurred for *underestimating demand.* This represents the opportunity loss of a sale in not planning for one extra serviced item of equipment, i.e. loss of profit = £300 − £200 = £100 per item.

Then for the following stated probabilities, the data can be expressed as follows:

(1)	(2)	(3)	(4)	(5)
Possible demand	Probability (*Pr*)	Demand × *Pr*	$C_o \times Pr$	$C_u \times Pr$
95	0.05	4.75	1000 × 0.05	–
96	0.06	5.76	800 × 0.06	–
97	0.09	8.73	600 × 0.09	–
98	0.11	10.78	400 × 0.11	–
99	0.15	14.85	200 × 0.15	–
100 = Q	0.20	20.00	0	0
101	0.13	13.13	–	100 × 0.13
102	0.10	10.2	–	200 × 0.10
103	0.07	7.21	–	300 × 0.07
104	0.03	3.12	–	400 × 0.03
105	0.01	1.05	–	5000 × 0.01
	1.00	μ = 99.58	226	71

Expected cost (Q = 100) = £226 + £71 = £297.

Clearly the workshop manager could have initially decided to budget for a different amount than Q = 100, for example Q = 95 or indeed any of the values between 95 and 105 wherein lies Q^*. The unit-by-unit calculation in finding the minimum expected cost ($Q = Q^*$) becomes quite tedious and can be more readily determined using marginal analysis (Fig. S25.1).

Expected cost (Q) is at a minimum value when the lowest level of Q is followed by an increase in expected cost, i.e.

$$EC(Q^* + 1) - EC(Q^*) = 0$$

which can be shown to reduce to the straightforward formula:

$$\text{Probability (Demand} \leq Q^*) = C_u/(C_u + C_o)$$

or in this example:

$$P(\text{demand} \leq Q^*) = 100/300 = 1/3$$

Since the range of Q is 95 to 105 = 10, then moving 1/3 × 10 from 95 towards 105 gives an optimal order quantity (Q^*) of $95 + 3^1/_3 = 98^1/_3$ items of equipment to budget for at the beginning of the week.

In practice the cost of servicing an item might change from £200 as a result of the revised plan and needs to be iteratively incorporated into the analysis, best achieved using a computer program.

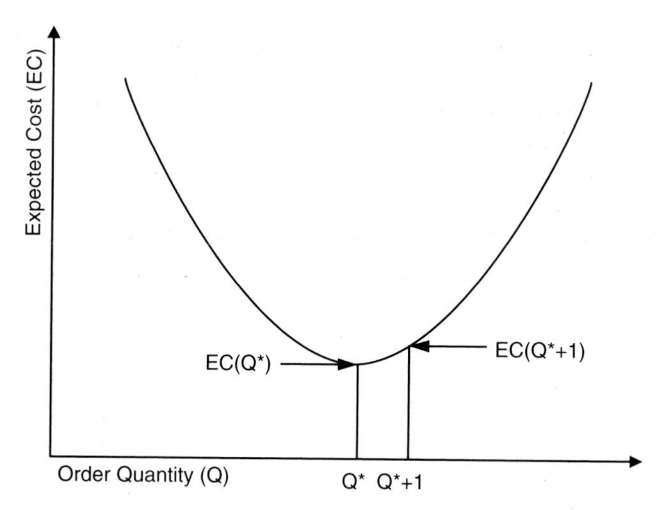

Fig. S25.1 Marginal or incremental analysis.

Solution 26 Theory of games

Gaming theory was largely developed during the 1940s by J. von Neumann (see Von Neumann & Morgenstern 2004) and with the advent of computer technology further improved to embrace business decisions.

In a two-person gaming situation the participants undertake a finite number of actions, which results in a gain to one and an identical negative gain (loss) to the other player, i.e. zero sum. The strategy is called 'pure' where the same course of action is always adopted. A 'mixed' strategy allows several courses of action, for example throwing a dice, which allows six potential courses of action each with 1/6 probability of being applied. The value of the game is the final expected gain for the successful player, which is the least bad outcome of all the potential strategies for that player. The game solution provides:

- The best strategies for the players
- The value of the game

In this question the two courses of action for each player are:

- A can decide to resist a bid for the plant division or alternatively resist a bid for the construction division.
- B can bid for the plant division or alternatively the construction division.
- The four potential gains for A are:
  —B bids for construction division, A successfully resists. Outcome for A = 0, i.e. nothing lost.
  —B bids for construction division, A does not resist. Outcome for A = −3.
  —B bids for plant division, A successfully resists. Outcome for A = 0.
  —B bids for plant division, A does not resist. Outcome for A = −2.

  The lower the number the worse the gain for A (or with the signs reversed the gain for B), which, represented as a matrix, produces the following situation:

- B strategy

Bid for plant or bid for construction

• A strategy	(I_1)	(II_2)
Resist plant bid and not construction (I)	0	−3
or		
Resist construction bid and not plant (II)	−2	0

Let A use course of action I with probability x
Let A use course of action II with probability $1-x$
Let B use course of action I_1 with probability y
Let B use course of action II_2 with probability $1-y$
If A uses a mixed strategy of I or II, then:

The expected gain for A when B always uses I_1 is $x(0)+(1-x)(-2)$ (i)
The expected gain for A when B always uses II_2 is $x(-3)+(1-x)(0)$ (ii)

However, if B also has a mixed strategy then:

$$\text{Expected gain for A} = y[x(0)+(1-x)(-2)]+(1-y)[x(-3)+(1-x)(0)] \quad \text{(iii)}$$

The best strategy for A is the x value giving the largest possible gain for A, which can be determined by substituting trial x values in equations (i) and (ii), and occurs for:

$$x(0)+(1-x)(-2)=x(-3)+(1-x)(0)$$

Hence $x=2/5$ and $1-x=3/5$.

Similarly for y (cell numbers deducted from 1):

The expected gain for B when A always uses I is $y(0)+(1-y)(+3)$ (iv)
The expected gain for B when A always uses II is $y(+2)+(1-y)(0)$ (v)

The best strategy for B is the y value giving the largest possible gain for B, which can be determined by substituting trial y values in equations (iv) and (v) and occurs for:

$$y(0)+(1-y)(+3)=y(+2)+(1-y)(0)$$

Hence $y=3/5$ and $1-y=2/5$.

Substituting x and y in equation (iii), the value of the game for A is:

$$v=3/5[2/5(0)+(1-2/5)(-2)]+(1-3/5)[2/5(-3)+(1-2/5)(0)]$$
$$v=-6/5$$

Conclusion

Since A does not know whether B will bid for construction or plant, and B does not know in advance which bid A will resist, the best strategy for A is to decide using random numbers, e.g. number 0 or 1 represent 2/5 probability, and 2, 3 or 4 represent 3/5 probability. The value (v) of the game to A is –£1.2 m.

Note: A game can have one or more *saddle* points, which happens when an element in the matrix is both the smallest number in its row and the largest number in its coincident column, for example:

		B strategy		
		(I_1)	(II_2)	(III_3)
A strategy	(I)	–2	–1	0
	(II)	4	5	**1**

The gain for A and least B coincide at **1** (shown in bold), i.e. A can be sure of at least a gain of 1, which cannot be prevented by B's strategy and vice versa. Thus, player A uses course of action II, and player B course of action III_3 throughout.

Games with saddle points are pure strategies, i.e. they do not have courses of action with probability. The mixed analysis method then does not hold.

Also when the courses of action and/or number of players exceed two, the analysis becomes complicated and is best undertaken with computer software.

References

Bellman, R. (2003) *Dynamic Programming*. Dover Publications, New York.

Lootsma, F.A. (ed.) (1972) Numerical Methods for Nonlinear Optimisation. Academic Press, London.

Von Neumann, J. & Morgenstern, O. (2004) *Theory of Games and Economic Behavior* (Commemorative Edition). Princeton University Press, Princeton, NJ.

Bibliography

Adeli, H. & Karim, A. (2001) *Construction Scheduling, Cost Optimization and Management.* Spon Press, London.

Akintoye, A., Beck, M. & Hardcastle, C. (2002) *Public Private Partnerships.* Blackwell Publishing, Oxford.

Anderson, D.R., Sweeney, D.J. & Williams, T.A. (2000) *An Introduction to Management Science,* South-Western College Publishing, Cincinnati, OH.

Apfelbaum, A. (2003) *Construction Cost Management: A Guide to Cost Engineering.* First Books Library.

Aqua Group (1999) *Tenders and Contracts for Building.* Blackwell Science, Oxford.

Armstrong, M. (2003) *A Handbook of Human Resource Management Practice.* Kogan Page, London.

Arnold, G. (2002) *Corporate Financial Management,* 2nd edn. FT Prentice Hall, Harlow.

Ashworth, A. (2002) *Pre-contract Studies: Development Economics, Tendering and Estimating.* Blackwell Science, Oxford.

Asmussen, S. (2003) *Applied Probability and Queues.* Springer-Verlag, London.

Atrill, P. & McLaney, E.J. (2003) *Accounting and Finance for Non-specialists.* FT Prentice Hall, Harlow.

Audit Commission (1990) *Preparing an Information Technology Strategy: Making it Happen.* HMSO, London.

Bahill, A.T. & Gissing, B. (1998) Re-evaluating systems engineering concepts using systems thinking. *IEEE Transaction on Systems, Man and Cybernetics,* **28**(4), 516–527.

Bakan, J. (2004) *The Corporation: The Pathological Pursuit of Power and Profit.* Constable and Robinson, London.

Baldwin, A., McCaffer, R. & Oteifa, S. (1995) *International bidding case study,* International construction management series No. 2, ILO, Geneva.

Barr, V. (1995) *Promotion Strategies for Design and Construction.* Wiley, Chichester.

Barrow, C. & Tracy, J.A. (2004) *Understanding Business Accounting for Dummies – UK Edition.* Wiley, Chichester.

Batty, J. (2003) *Cost and Management Accountancy for Students.* Technical Book Publishers, Midhurst.

Bazaraa, M.S., Shetty, C.M. & Sherali, H.D. (eds) (1993) *Non-linear Programming: Theory and Algorithms.* Wiley, Chichester.

Bennett, J. & Jayes, S. (1998) *Seven Pillars of Partnering.* Thomas Telford, London.

Bent, J.A. & Humphreys, K.K. (1996) *Effective Project Management Through Applied Cost and Schedule Control.* Marcel Dekker, New York.

Birkinshaw, J. & Piramal, G. (eds) (2005) *Sumantra Ghoshal on Management: A Force for Good.* FT Prentice Hall, Harlow.

Blackwell, M. (2001) *The PFI/PPP and Property: a Practical Guide.* Estates Gazette, London.

Blake, D. (1999) *Financial Market Analysis,* 2nd edn. Wiley, Chichester.

Blank, L.T. & Tarquin, A. (2005) *Engineering Economy,* 6th edn. McGraw-Hill, New York.

Bloomfield, B.P. (1997) *Information Technology and Organisations: Strategies, Networks, and Integration.* Oxford University Press, Oxford.

Boddy, D. and Boonstra, A. (2005) *Managing Information Systems: An Organisational Perspective.* FT Prentice Hall, Harlow.

Bratton, J. & Gold, J. (2003) *Human Resource Management: Theory and Practice*. Palgrave Macmillan, Basingstoke.

Braun, H. (1953) *Historical Architecture – the Development of Structure and Design*. Faber & Faber, London.

Bridges, A.H. (1996) *The Construction Net: Online Information Sources for the Construction Industry*. Spon, London.

Brook, M. (2004) *Estimating and Tendering for Construction Work*. Butterworth-Heinemann, Oxford.

Buchan, R.D., Fleming, F.W.E. & Grant, F.E.K. (2003) *Estimating for Builders and Surveyors*. Butterworth-Heinemann, Oxford.

Buchanan, D. & Huczynski, A. (2004) *Organisational Behaviour*. FT Prentice Hall, Harlow.

Burke, R. (2003) *Project Management: Planning and Control Techniques*. Wiley, Chichester.

Cahill, K. (2002) *Who Owns Britain*. Canongate Books, Edinburgh.

Cain, C.T. (2004) *Profitable Partnering for Lean Construction*. Blackwell, Oxford.

Camp, R.C. (1989) *Benchmarking: the Search for Industry Best Practices that Lead to Superior Performance*. Quality Resources, White Plains, NY.

Chaffey, D. & Wood, S. (2004) *Business Information Management: Improving Performance using Information Systems*. FT Prentice Hall, Harlow.

Chapman, C. & Ward, S. (2003) *Project Risk Management*. Wiley, Chichester.

Christensen, C.M. & Raynor, M.E. (2003) *The Innovator's Solution: Creating and Sustaining Successful Growth*. Harvard Business School Press, Cambridge, MA.

CIOB (1997) *Code of Estimating Practice*, 6th edn. Longman/CIOB, Harlow.

Clark, T. (2005) *Storage Virtualization: Technologies for Simplifying Data Storage and Management*. Addison-Wesley, London.

Cleland, D.I. & Ireland, L. R. (2002) *Project Management: Strategic Design and Implementation*. McGraw-Hill Professional, New York.

Clements-Croome, D. (2004) *Intelligent Buildings: Design Management and Operation*. Thomas Telford, London.

Clough, R.H., Sears, G.A. & Sears S.K. (2000) *Construction Project Management*. Wiley, Chichester.

Coakes, E. & Clarke, S.A. (2005) *Encyclopaedia of Communities of Practice in Information and Knowledge Management*. Idea Group, Harrisburg, PA.

Code of Practice for Project Management (2002) Blackwell Science, Oxford.

Cole, G.A. (2003) *Management Theory and Practice*. Thomson Learning, London.

Collins, J. (2004) *PFI: Meeting the Sustainability Challenge*. Green Alliance, London.

Collins, J.C. & Porras, J.I. (2004) *Built to Last: Successful Habits of Visionary Companies*. HarperBusiness, London.

Construction Industry Joint Council (2003) *Working Rule Agreement for the Construction Industry*. UCATT, London.

Cousins, D. (1980) *Book Keeping*. Hodder Arnold, London.

Cox, D. & Fardon, M. (2003) *Costing and Reports Tutorial AAT/NVQ Accounting*. Osborne Books, Worcester.

CSD, DOE & BT (1995) *Construct I.T.: Bridging the Gap: An Information Technology Strategy for the United Kingdom Construction Industry*. HMSO, London.

Currie, R.M. (1972) *Work Study*. Pitman, London.

Currie, W. (1995) *Management Strategy for IT: An International Perspective*. Pitman, London.

Curtis, F.W & Maines, P.W. (1974) Viewpoints: competitive bidding. *Operational Research Quarterly*, **25**(1), 179–181.

Curtis, G. & Cobham, D.P. (2005) *Business Information Systems: Analysis, Design, and Practice*. FT Prentice Hall, Harlow.

Dale, B. (2003) *Managing Quality*. Blackwell, Oxford.

Dantzig, G.B. (1999) *Linear Programming and Extensions*. Princeton University Press, Princeton, NJ.

Davis Langdon and Everest (2003) *A Survey of Contracts in Use during 2001*. RICS Construction Faculty, www.rics.org.

Dayananda, D., Irons, R., Harrison, S., Herbohn, J. & Rowland, P. (2002) *Capital Budgeting: Financial Appraisal of Investment Projects*. Cambridge University Press, Cambridge.

DeGarmo, E.P., Sullivan, W.G. & Bontadelli, J.A. (1997) *Engineering Economy*. Prentice Hall, Englewood Cliffs, NJ.

Deming, W.E. (1992) *The Deming Management Method*. Mercury Business Books, London.

Deming, W.E. (2000) *Out of the Crisis*. MIT Press, Cambridge, MA.

Dent, R.J. & Montague, K.N. (2004) *Benchmarking Knowledge Management Practice in Construction*. CIRIA, London.

Department for Transport (2002) *Design Manual for Roads and Bridges*. Vol.14. *Economic Assessment of Road Maintenance*. The Stationery Office, London.

Dewhurst, F. (2002) *Quantitative Methods for Business Management*. McGraw-Hill, New York.

Dixon, R. & Franks, R. (eds) (1992) *IT Management Handbook*. Butterworth-Heinemann, Oxford.

Drury, C. (2004) *Management and Cost Accounting*. Thomson Learning, London.

Dunn, C.L. & Cherrington, J.O. (2005) *Enterprise Information Systems: A Pattern-Based Approach*. McGraw-Hill/Irwin, New York.

Dutta, P.K. (1999) *Strategies and Games: Theory and Practice*. MIT Press, Cambridge, MA.

Dyson, J.R. (2003) *Accounting for Non-accounting Students*. FT Prentice Hall, Harlow.

ECI (1996) *Implementing TQ in the Construction Industry: a Practical Guide*. Thomas Telford, London.

Edwards, D.G., Harris, F.C. & McCaffer R. (2003) *Management of Off-highway Plant and Equipment*. Spon Press, London.

Egan, J. (2002) *Accelerating Change: A Report by the Strategic Forum for Construction*. Construction Industry Council, London.

Elliott, J. & Elliott, B. (2005) *Financial Accounting and Reporting*. Pearson Higher Education, London.

Erlande-Brandenburg, A. (1993) *Cathedrals and Castles – Building in the Middle Ages*. Abrams Discoveries, New York.

Flanagan, R. & Norman, G. (1993) *Risk Management and Construction*. Blackwell Science, Oxford.

Flanagan, R. & Tate, B. (1997) *Cost Control in Building Design*. Blackwell Science, Oxford.

Ford D., Gadde, L.E., Hakansson, H. and Snehota, I. (2003) *Managing Business Relationships*, 2nd edn. Wiley, Chichester.

Foster, N. (1995) *Construction Estimates: From Take-off to Bid*. McGraw-Hill, New York.

Franklin & Andrews (2005) *Hutchins UK Building Costs Blackbook*. Franklin & Andrews, Norwich.

Friend, G. & Zehle, S. (2004) *Guide to Business Planning*. Economist Books, London.

Fritz, M.B.W., Narasimhan, S. & Rhee, H-S. (1998) Communication and coordination in the virtual office. *Journal of Management Information Systems*, **14**(4), 7–28.

Gann, D.M. (2000) *Building Innovation: Complex Constructs in a Changing World*. Thomas Telford, London.

Gardiner, P.D. (2005) *Project Management: A Strategic Planning Approach*. Macmillan, Basingstoke.

Gardner, S.R. (1998) Building the data warehouse. *Communications of the ACM*, **41**(9), 52–60.

Goldratt, E.M., Cox, J. & Whitford, D. (2004) *The Goal: A Process of Ongoing Improvement*. North River Press, New York.

Gorman, G. E. (2004) *Metadata Applications and Management*. Facet, London.

Gottschalk, P. (2005) *Strategic Knowledge Management Technology*. Idea Group, London.

Gould, F.E. (2002) *Managing the Construction Process: Estimating, Scheduling, and Project Control*. Prentice Hall, Upper Saddle River, NJ.

Gray, C. & Hughes, W. (2000) *Building Design Management*, Butterworth-Heinemann, Oxford.

Green, R. (2001) *The Architect's Guide to Running a Job*. Architectural Press, Oxford.

GRI (2006) *Sustainability Reporting Guidelines*, G3. Global Reporting Initiative, Amsterdam.

Griffith, A. & Stephenson, D. (2000) *Management Systems for Construction*. Longman, Harlow.

Gympel, J. (1996) *The Story of Architecture*. Koenemann, Cologne.

Haag, S. & Cummings, M. (2004) *Management Information Systems for the Information Age*. McGraw-Hill, Boston.

Hackett, M. (2002) *Pre-contract Practice and Contract Administration for the Building Team.* Blackwell Science, Oxford.
Handy, C. (1995) *Gods of Management: The Changing Work of Organisations.* Arrow, London.
Handy, C.B. (1993) *Understanding Organizations.* Penguin Books, Harmondsworth.
Hardin, J.W. & Hilbe, J. (2003) *Generalized Estimating Equations.* Chapman & Hall/CRC, Boca Raton, FL.
Harrington, H.J. (1991) *Business Process Improvement: the Breakthrough Strategy for Quality, Productivity and Competitiveness.* McGraw-Hill Education, New York.
Harris, F. (1994) *Modern Construction and Ground Engineering Equipment and Methods.* Prentice Hall, London.
Harris, F. & McCaffer, R. (1986) *Construction Plant: Management and Investment Decisions.* Blackwell Scientific, Oxford.
Harris, F. & McCaffer, R. (1986) *Worked Examples in Construction Management,* 2nd edn. Blackwell Scientific, Oxford.
Havard, T. (2000) *Investment Property Valuation Today.* Chandos, Oxford.
Hayler, R. & Nichols, M. (2005) *What is Six Sigma Process Management?* McGraw-Hill, New York.
Haynes, D. (2004) *Metadata for Information Management and Retrieval.* Facet, London.
Health and Safety Executive (2001) *Managing Construction for Health and Safety.* HSE, Sudbury.
Heil, G., Bennis, W. & Stephens, D. (2000) *Douglas McGregor Revisited – Managing the Human Side of the Enterprise.* Wiley, Chichester.
Heller, R. & Hindle, T. (1998) *Essential Manager's Manual.* Dorling Kindersley, London.
Hill, R.M. (2001) *Construction Logistics (Digest).* Construction Research Communications Ltd, Peterborough.
Hillier, F.S. & Lieberman, G.J. *Introduction to Operations Research.* McGraw-Hill Higher Education, New York.
Hillier, F.S., Hillier, M. & Lieberman, G.J. (2002) *Introduction to Management Science.* McGraw-Hill, New York.
Hirschey, M. (2000) *Managerial Economics Revised Edition.* Harcourt Brace Jovanovich, Orlando, FL.
HM Treasury (2004) *Guidance on the Standardisation of PFI Contracts* (version 3). HMSO, London.
HMSO (1989) Companies Act 1989 c. 40. HMSO, London; see also www.opsi.gov.uk/acts/acts1989/Ukpga_19890040_en_1.htm.
HMSO (1998) *General Conditions of Contract for Building and Civil Engineering Works,* GC/Works/1(1998). HMSO, London.
Horner, M. & Duff, A.R. (2007) *Improving Productivity.* Blackwell, Oxford.
Hoskins, W.G. (1991) *The Making of the English Landscape.* Penguin Books, Harmondsworth.
Huczynski, A. & Buchanan, D. (2003) *Organizational Behaviour: An Introductory Text.* FT Prentice Hall, London.
Hughill, D. & Ross, A. (2007) *Financial Management in Construction Contracting.* Blackwell, Oxford.
Hunt, J. (1997) *Business and Commercial Aspects of Engineering.* Arnold, London.
Hutchins, T.V. (1987) *Hutchins' Priced Schedules,* 43rd edn. Technical Book Publishers, Midhurst.
Illingworth, J.R. (2000) *Construction Methods and Planning.* Spon Press, London.
Institution of Civil Engineers (1991) *Civil Engineering Standard Method of Measurement.* Thomas Telford, London.
Institution of Civil Engineers (1995) *The Engineering and Construction Contract: an NEC Document.* Thomas Telford, London.
Institution of Civil Engineers (1996) *Civil Engineering Procedure,* 4th edn. Thomas Telford, London.
Jaggar, D., Ross, A., Smith, J. & Love, P. (2002) *Building Design Cost Management.* Blackwell, Oxford.
Jayawardane, A.K.W., Olomolaiye, P.O. & Harris, F.C. (1998) *Construction Productivity Management.* Addison Wesley Longman, London.
Jobber, D. (2003) *Principles and Practice of Marketing.* McGraw-Hill Education, London.
Johnson, V.B. (2005) *Laxton's Building Price Book.* Butterworth-Heinemann, Oxford.
Johnston, H. & Mansfield, G.L. (2001) *Bidding and Estimating Procedures for Construction.* Prentice Hall, London.

Juran, J.M. (2003) *Juran on Leadership for Quality.* Free Press, New York.

Kaming, P., Olomolaiye, P., Holt, G.D. & Harris, F.C. (1998) What motivates the construction trades worker. *Building and Environment,* **33**(2–3), 131–141.

Kanji, G. & Asher, M. (1996) *100 Methods for Total Quality Management.* Sage, London.

Kelly, J., Male, S. & Drummond, G. (2001) *Value Management of Construction Projects.* Blackwell, Oxford.

Kelly, J., Morledge, R. & Wilkinson, S. (eds) (2002) *Best Value in Construction.* Blackwell, Oxford.

Kenley, R. (2003) *Financing Construction: Cash Flows and Cash Farming.* Spon Press, London.

Kerzner, H. (2003) *Project Management: A Systems Approach to Planning, Scheduling, and Controlling.* Wiley, Chichester.

Khan, F. & Parra, R. (2003) *Financing Large Projects: Using Project Finance Techniques and Practices.* Pearson Education Asia, Singapore.

Kibert, C. (2005) *Sustainable Construction: Green Building Design and Operation.* Wiley, New York.

King, B. A. (2004) *Performance Assurance for IT systems.* Auerbach Publications, Boca Raton, FL.

King, M. & Mercer, A. (1985) Problems in determining bidding strategies. *Journal of the Operational Research Society,* **36,** 915–923, 1985.

King, M. & Mercer, A. (1987) Note on a conflict of assumptions in bidding models. *European Journal of Operational Research,* **32,** 462–466.

King, M. & Mercer, A. (1990) The optimum mark-up when bidding with uncertain costs. *European Journal of Operational Research,* **47**(3), 348–363.

Kitchens, M. (1996) *Estimating and Project Management for Building Contractors.* American Society of Civil Engineers, Reston, VA.

Kloppenborg, T.J., Shriberg, A. & Venkatraman, J. (2003) *Project Leadership.* Management Concepts, Vienna, VA.

Knocke, J. (ed.) (1993) *Post Construction Liability and Insurance.* Spon Press, London.

Koontz, H., O'Donnell, C. & Weihrich, H. (eds) (1990) *Essentials of Management.* McGraw-Hill Education, New York.

Kotler, P., Armstrong, G., Saunders, J. & Wong, V. (2004) *Principles of Marketing: European Edition.* Prentice Hall, Harlow.

Lamb, B. & Merna, T. (2004) *A Guide to PFI.* Thomas Telford, London.

Lambert, D., Stock, J. & Ellram, L. (1997) *Fundamentals of Logistics Management.* Irwin/McGraw-Hill, Boston, MA.

Langdon, D. (2005) *Spon's Architects' and Builders' Price Book.* Spon Press, London.

Langdon D. (2005) *Spon's Civil Engineering and Highway Works Price Book.* Spon Press, London.

Langford, D. & Murray, M. (2004) *Architect's Handbook of Construction Project Management.* RIBA Enterprises, London.

Langmaid, J. (2003) *Estimating: Getting Value from Function.* BSRIA, Bracknell.

Lapin, L.L. (1994) *Quantitative Methods for Business Decisions with Cases,* 6th edn. Wadsworth, Belmont, CA.

Laudon, K.C. & Laudon, J.P. (2004) *Essentials of Management Information Systems: Managing the Digital Firm.* Prentice Hall, Upper Saddle River, NJ.

Lester, A. (2000) *Project Planning and Control.* Butterworth-Heinemann, Oxford.

Lewis, J.P. (2001) *Project Planning, Scheduling and Control: A Hands-on Guide to Bringing Projects in on Time and on Budget.* McGraw-Hill, London.

Limbachiya, M. & Roberts, J. (2004) *Sustainable Waste Management and Recycling.* Thomas Telford, London.

Lock, D. (2004) *Project Management in Construction.* Gower Publishing, Aldershot.

Lomborg, B. (2001) *The Sceptical Environmentalist.* Cambridge University Press, Cambridge.

Loosemore, M., Dainty, A. & Lingard, H. (2003) *Human Resource Management in Construction Projects: Strategic and Operational Approaches.* Spon Press, London.

Lucey, T. (2002) *Costing.* Thomson Learning, London.

Luenberger, D. (1998) *Investment Science.* Oxford University Press, Oxford.

Lumby, S. & Jones, C. (2003) *Corporate Finance: Theory and Practice*. Thomson Learning, London.
Magretta, J. (2003) *What Management Is*. Profile Books, London.
Makower, M.S. & Williamson, E. (1970) *Operational Research*. Teach Yourself Books, London.
Martin, E. W. (2004) *Managing Information Technology*. Prentice Hall, London.
Mathur, K.S., Betts, M.P. & Tham, K.W. (eds) (1993) *Management of Information Technology for Construction*. World Scientific Publishing Co., Singapore.
Mawby, W.D. (2004) *Decision Process Quality Management*. ASQ Quality Press, Milwaukee, WI.
Mawdesley, M., Askew, W. & O'Reilly, M. (1997) *Planning and Controlling Construction Projects: The Best Laid Plans*. Longman, Harlow.
Mawhinney, M. (2001) *International Construction*. Blackwell, Oxford.
McCabe, S. (2001) *Benchmarking in Construction*. Blackwell, Oxford.
McCaffer, R. & Baldwin, A.N. (1991) *Estimating and Tendering for Civil Engineering Works*, 2nd edn. Blackwell Scientific, Oxford.
McEntegart, R.C. (1980) *Costing and Budgetary Control*. Prentice Hall, Englewood Cliffs, NJ.
McGeorge, W.D. & Palmer, A. (2002) *Construction Management: New Directions*. Blackwell, Oxford.
McLeod, A. (2003) *Performance Coaching: the Handbook for Managers, HR Professionals and Coaches*. Crown House Publishing, Carmarthen.
Moore, D. (2002) *Project Management: Designing Effective Organisational Structures in Construction*. Blackwell, Oxford.
Morgan, D.B. (2005) *International Construction Contract Management*. RIBA Enterprises, London.
Murdoch, J. & Hughes, W. (2000) *Construction Contracts: Law and Management*. Spon Press, London.
Myers, D. (2004) *Construction Economics: A New Approach*. Spon Press, London.
Nah, R., Akmanligil, M., Hjelm, K., Sakaguchi, T. & Schultz, M. (1998) Electronic commerce and the internet: issues, problems, and perspectives. *International Journal of Information Management*, **18**(2), 91–101.
Nakayama, M. & Sutcliffe, N. (2004) *Managing IT Skills Portfolios: Planning, Acquisition and Performance Evaluation*. Idea Group, Hershey, PA.
National Joint Consultative Committee for Building (1994) *Code of Procedure for Single Stage Selective Tendering*. RIBA, London.
Navarrete, P.F. & Cole, W.C. (2001) *Planning, Estimating, and Control of Chemical Construction Projects*. Marcel Dekker, New York.
Ndekugri, I. (1986) *Construction Contract Information Management*. PhD thesis, Loughborough University of Technology.
Ndekugri, I. & McCaffer, R. (1986) Valuations – an interactive system linked to estimating. *Construction Computing*, July.
Ndekugri, I. & Rycroft, M. (2000) *The JCT98 Building Contract – Law and Administration*. Butterworth-Heinemann, Oxford.
Nelson, R. (2000) *Probability, Stochastic Processes and Queuing Theory*. Springer-Verlag, New York.
Nunnally, S.W. (1999) *Managing Construction Equipment*. Pearson US Imports, London.
Nuttgens, P. (1997) *The Story of Architecture*. Phaidon Press, London.
O'Brien, J. A. (2004) *Management Information Systems: Managing Information Technology in the Business Enterprise*. McGraw-Hill/Irwin, London.
O'Brien, J.J. (1994) *Preconstruction Estimating*. McGraw-Hill, London.
O'Brien, J.J. and Plotnick, F. L. (2006) *CPM in Construction Management*. McGraw-Hill, New York.
O'Brien, M.J. & Pantouvakis, J.P. (1993) A new approach to the development of computer-aided estimating for the construction industry. *Construction Management and Economics*, **11**(1), 30–44.
O'Leary, R. (ed.) (1996) *Construction Information Directory: A Guide to UK Resources*. RIBA Publications, London.
Oakland, J. (2004) *Total Quality Management*. Butterworth-Heinemann, Oxford.
Oakland, J.S. (2005) *Oakland on Quality*. Butterworth-Heinemann, Oxford.
Oakland, J.S. & Marosszeky, M. (2005) *Total Quality in the Construction Supply Chain*. Butterworth-Heinemann, Oxford.

OGC (2004) *Tendering for Government Contracts: A Guide for Small Businesses.* Business Link, HMSO, London.
Olomolaiye, P., Jayawardane, A. & Harris, F.C. (1998) *Construction Productivity Management.* Longman, Harlow.
Olomolaiye, P.O. & Ogunlana, S.O. (1988) A survey of construction operative motivation. *Building and Environment,* **23**(3), 179–186.
OPERC – Off-highway Plant and Equipment Research Centre. Various publications and guides relevant to plant and equipment, www.operc.com.
Orna, E. (2004) *Information strategy in practice.* Gower, Aldershot.
Orna, E. (2005) *Making Knowledge Visible: Communicating Knowledge through Information Products.* Gower, Aldershot.
Ostwald, P.F. & McLaren, T.S. (2003) *Cost Analysis and Estimating for Engineering and Management.* Pearson Education, Upper Saddle River, NJ.
Packard, V. (1985) *The Hidden Persuaders.* Pocket Books, New York.
Packer, A.D. (1996) *Building Measurement.* Longman, Harlow.
Palmer, J. & Platt, S. (2005) *Business Case for Knowledge Management in Construction.* CIRIA, London.
Pande, P. & Holpp, L. (2001) *What Is Six Sigma?* McGraw-Hill Education, New York.
Pansini, A.J. (1996) *Engineering Economic Analysis Guidebook.* Fairmont Press, Upper Saddle River, NJ.
Park, W.R. & Chapin, W.B. (1992) *Construction Bidding: Strategic Pricing for Profit,* 2nd edn. Wiley, Chichester.
Parker, C. S. and Case, T. (1993) *Management Information Systems: Strategy and Action,* 2nd edn. McGraw-Hill, London.
Parkinson, C. N. (2002) *Parkinson's Law.* Penguin Classics, London.
Parsley, L.L & Robinson, R. (1982) *The TRRL Road Investment Model for Developing Countries* (RT1 M2). Transport and Road Research Laboratory, Crowthorne.
Pascal, D. (2002) *Lean Production Simplified: a Plain-Language Guide to the World's Most Powerful Production System.* Productivity Press, Portland, OR.
Patrick, C. (2003) *Construction Project Planning and Scheduling.* Prentice Hall, Harlow.
Peppers, D. & Rogers, D. (eds) (2004) *Managing Customer Relationships: A Strategic Framework.* Wiley, Chichester.
Peters, T. & Waterman, R.H. (2004) *In Search of Excellence.* Profile Business, London.
Pettinger, R. (1998) *Construction Marketing: Strategies for Success.* Palgrave Macmillan, Basingstoke.
Pettinger, R. (2000) *Investment Appraisal: A Managerial Approach.* Macmillan, Basingstoke.
Peurifoy, R. L. (2005) *Construction Planning, Equipment, and Methods.* McGraw-Hill College, New York.
Peurifoy, R.L. & Oberlander, G.D. (2002) *Estimating Construction Costs.* McGraw-Hill, London.
Peurifoy, R.L. & Schexnayder, C.J. (2002) *Construction Planning, Equipment and Methods.* McGraw-Hill, New York.
Pilcher, R. (1994) *Project Cost Control in Construction.* Blackwell Scientific, Oxford.
Pinson, L. (2004) *Keeping the Books: Basic Record Keeping and Accounting for the Successful Small Business.* Dearborn Trade Publishing, Chicago, IL.
Pizzey, A. (1987) *Cost and Management Accounting for Students.* Paul Chapman, London.
PMI (2001) *Project Management Institute Practice Standard for Work Breakdown Structures.* Project Management Institute, Newtown Square, PA.
Porter, M.E. (2004) *Competitive Advantage.* Free Press, New York.
Porter, M.E. (2004) *Competitive Strategy: Techniques for Analyzing Industries and Competitors.* Free Press, New York.
Porter, M.E., Schwab, K., Sala-I-Martin, X. & Lopez-Carlos, A. (eds) (2004) *The Global Competitiveness Report 2004–2005.* Palgrave Macmillan, Basingstoke.
Potts, K.F. & Patchell, B. (1995) *Major Construction Works – Contractual and Financial Management.* Longman, Harlow.
Preece, C.N. (2001) Marketing and promotional strategies in construction. In: Langford D. & Male S. (eds) (2001) *Strategic Management in Construction,* 2nd Edn. Blackwell Science, Oxford, pp. 175–186.

Pryor, P. (2001) *Marketing Construction Services*. Industrial Press, New York.
Quinn, J.B., Mintzberg, H., James, R.M., Joseph, B., Lampel, J.B. & Ghoshal, S. (eds) (2003) *The Strategy Process*. Prentice Hall, Harlow.
Raisinghani, M.S. (2004) *Business Intelligence in the Digital Economy: Opportunities, Limitations and Risks*. Idea Group, London.
Rao, M. (2005) *Knowledge Management Tools and Techniques: Practitioners and Experts Evaluate KM Solutions*. Elsevier Butterworth-Heinemann, London.
Read, A. & Loose, P. (eds) (2000) *The Company Director: Powers and Duties*. Jordans, Bristol.
Resnik, M.D. (1987) *Choices: Introduction to Decision Theory*. University of Minnesota Press, Minneapolis, MN.
Retik, A. & Langford, D.A. (2001) *Computer Integrated Planning and Design for Construction*. Thomas Telford, London.
Rice, A. (2003) *Accounts Demystified: How to Understand Financial Accounting and Analysis*. Prentice Hall, Harlow.
Richardson, B. (1996) *Marketing for Architects & Engineers: A New Approach*. Spon Press, London.
Ritz, G.J. (1994) *Total Construction Project Management*. McGraw-Hill, New York.
Roberts, J. (2004) *The Modern Firm: Organizational Design for Performance and Growth*. Oxford University Press, Oxford.
Rowlinson, S. (ed.) (2003) *Construction Safety Management Systems*. Spon Press, London.
Royal Institute of British Architects (1998) *The JCT Standard Form of Building Contract (JCT 98)*. RIBA Publications, London.
Rutherford, B.A. (1998) Developing a framework for the analysis of published cash flow information. *ACCA Occasional Research Paper* 20. Certified Accountants Educational Trust, London.
Sampson, A. (2004) *Who Runs This Place?* John Murray, London.
Schaeffer, H.A. (2002) *Essentials of Cash Flow*. Wiley, Chichester.
Schniederjans, M. *Goal Programming: Methodology and Applications*. Kluwer Academic Publishers, Boston, MA.
Seabright, P. (2004) *The Company of Strangers*. Princeton University Press, Princeton, NJ.
Shank, J.K. & Govindarajan, V. (1993) *Strategic Cost Management: The New Tool for Competitive Advantage*. Free Press, New York.
Shapiro, A.C. (2002) *Multinational Financial Management*. Wiley, New York.
Shiller, R.J. (2001) *Irrational Exuberance*. Princeton University Press, Princeton, NJ.
Shriberg, A. (2003) *Practicing Leadership – Principles & Applications*. Wiley, Chichester.
Silk, D. J. (1991) *Planning IT: Creating an Information Management Strategy*. Butterworth-Heinemann, Oxford.
Silver, M. S. (1991) *Systems that Support Decision-makers: Description and Analysis*. Wiley, Chichester.
Sinclair-Desgagne, B. (ed.) (2004) *Corporate Strategies for Managing Environmental Risk*. Ashgate, Aldershot.
Singh, J. (2001) *Heavy Construction: Planning, Equipment and Methods*. A.A. Balkema, Leiden.
Skitmore, M. & Lo, H.P. (2002) A method for identifying high outliers in construction contract auctions. *Engineering Construction and Architectural Management*, **9**(2), 90–130.
Slack, N., Chambers, S. & Johnston, R. (2004) *Operations Management*, 4th edn. Prentice Hall, Harlow.
Smith, A.J. (1995) *Estimating, Tendering and Bidding for Construction: Theory and Practice*. Macmillan, Basingstoke.
Smith, N. (1998) *Managing Risks in Construction*. Blackwell, Oxford.
Smith, N. (ed.) (2002) *Engineering Project Management*. Blackwell, Oxford.
Smith, N.J. (ed.) (1995). *Project Cost Estimating*. Thomas Telford, London.
Snook, K. (1995) *CPI – Co-ordinated Project Information*. Chartered Institute of Building, Ascot.
Society for Marketing (2004) *Marketing Handbook for the Design and Construction Professionals*. Wiley, New York.
St John-Holt, A. (2001) *Principles of Construction Safety*. Blackwell Science, Oxford.

Statham, W. & Sargeant, M. (1969) Determining an optimum bid. *Building*, **216**(6573).
Stewart, R.D. (1991) *Cost Estimating*. Wiley, Chichester.
Strategic Forum for Construction (2004) *Accelerating Change, Constructing Excellence*. Strategic Forum for Construction, London.
Strook, D.W. (2004) *An Introduction to Markov Processes*. Springer-Verlag, Berlin.
Stubbington, D.T. (1998) Preliminaries: improving cash flow. *Construction Papers*, No. 92. Chartered Institute of Building, Ascot.
Sweeting, J. (1997) *Project Cost Estimating: Principles and Practice*. Institution of Chemical Engineers, London.
Tang, S.L., Ahmad, I.U., Ahmed, S.M. & Lu, M. *Quantitative Techniques for Decision Making in Construction*. Hong Kong University Press, Hong Kong.
Tatnall, A. (2005) *Web Portals: The New Gateways to internet Information and Services*. Idea Group, London.
Tennent, J. & Friend, G. (2005) *Guide to Business Modelling*. Economist Books, London.
Thorpe, A., Edum-Fotwe, F.T. & Mead, S. P. (1998) Managing construction projects within emerging information driven business environments. In Fahlstedt, K. (ed.) *Managing for Sustainability – Endurance Through Change*, CIB World Building Congress 98, Symposium D. Construction and the Environment, Sweden, 7–12 June, 1901–1910.
Thorpe, B. & Sumner, P. (2005) *Quality Management in Construction*, 3rd edn. Gower, Aldershot.
Thuesen, G.J. (2001) *Engineering Economy*, 9th edn. Prentice-Hall International, Upper Saddle River, NJ.
Tiffin, R.C. & Ellis H. (1992) *Practical Investment Appraisal*. Butterworths, London.
Tressell, R. (1993) *The Ragged Trousered Philanthropists*. Flamingo, London.
Tricker, R. (2000) *ISO 9001–2000 for Small Businesses: A Guide to Cost-effective Compliance*. Butterworth-Heinemann, Oxford.
Turban, E. & Rainer, R.K. (2005) *Introduction to Information Technology*. Wiley, New York.
Uff, J. (2002) *Construction Law*. Sweet and Maxwell, London.
Vose, D. (1999) *Quantitative Risk Analysis: Guide to Monte Carlo Simulation Modelling*. Wiley, Chichester.
Walker, A. (2002) *Project Management in Construction*. Blackwell Publishing, Oxford.
Walker, D. & Hampson, K. (eds) (2002) *Procurement Strategies: A Relationship-based Approach*. Blackwell, Oxford.
Walker, I. & Wilkie, B. (2002) *Commercial Management in Construction*. Blackwell Science, Oxford.
Waters, D. (2002) *Logistics: An Introduction to Supply Chain Management*. Palgrave Macmillan, Basingstoke.
Weatherhead, M., Owen, K. & Hall, C. (2005) *Integrating Value and Risk in Construction: C639*. CIRIA, London.
Weiser, M. & Morrison, J. (1998) Project memory: information management for project teams. *Journal of Management Information Systems*, **14**(4), 149–166.
Whitin, T.M. (1953) *Theory of Inventory Management*. Princeton University Press, Princeton, NJ.
Whitmore, J. (2002) *Coaching for Performance: Growing People, Performance and Purpose*. Nicholas Brealey Publishing, London.
Williams, J. (ed.) (1996) *Estimating for Building and Civil Engineering Works*, 9th edn. Butterworth-Heinemann, Oxford.
Williamson, D. (1996) *Cost and Management Accounting*. Prentice Hall, London.
Willis, J.A. & Chappell, D. (2005) *The Architect in Practice*. Blackwell, Oxford.
Wilson James Ltd (2004) *Construction Logistics: Models for Consolidation*. Constructing Excellence, London (see www.constructingexcellence.org.uk).
Winch G. (2002) *Managing Construction Projects*. Blackwell , Oxford.
Wisniewski, M. (2000) *Quantitative Methods for Decision Makers*. FT Prentice Hall, Harlow.
Womack, J.P. & Jones, D.T. (2003) *Lean Thinking*. Free Press, New York.
Wood, F. & Robinson, S.I. (2004) *Book-keeping and Accounts*. FT Prentice Hall, Harlow.
Wood, F. & Sangster, A. (eds) (2002) *Business Accounting*. Vol. 1. Prentice Hall, Harlow.
Woudhhuysen, J. & Abley, I. (2004) *Why is Construction so Backward?* Wiley-Academy, Chichester.

Wren, D.A. (2004) *The History of Management Thought*, 5th edn. Wiley, Chichester.
Yegge, W. (2002) *A Basic Guide for Valuing a Company*. Wiley, Chichester.
Yescombe, E. (2002) *Principles of Project Finance*. Academic Press, London.
Young, T.L. (2003) *The Handbook of Project Management: A Practical Guide to Effective Policies and Procedures*. Kogan Page, London.
Zakeri, M., Olomolaiye, P.O., Holt, G.D. & Harris, F.C. (1997) Factors affecting the motivation of Iranian construction operatives, *Building and Environment*, **32**(2), 161–6.

Appendix
Abbreviations and acronyms

AB	Awarding Body
ACE	Association of Consulting Engineers
ACoP	Approved Code of Practice
AG	Aktiengesellschaft (German: stock corporation)
AGM	Annual General Meeting
AIDB	Accountancy Investigation and Discipline Board
ASME	American Society of Mechanical Engineers
AT	Acid Test
BAA	British Airports Authority
BCIS	Building Cost Information Service (Royal Institution of Chartered Surveyors)
BOO	Build Own and Operate
BOOT	Build Own Operate and Transfer
BOQ	Bill of Quantities
BOT	Build Operate Transfer
BPO	Business Process Outsourcing
BSI	British Standards Institution
BTEC	Business and Technology Education Council
CAD	Computer-Aided Design
CAM	Computer-Aided Manufacture
CAT	Highways Agency's Capability Assessment Toolkit
CBA	Cost Benefit Analysis
CBPM	Condition Based Predictive Maintenance
CBPP	Construction Best Practice Programme
CDM	Clean Development Mechanism
CDM	Construction (Design and Management) Regulations
CE	Constructing Excellence
CEBE	Constructing Excellence in the Built Environment
CEO	Chief Executive Officer
CER	Certified Emissions Reduction
CERN	European Laboratory for Particle Physics
CESMM	Civil Engineering Standard Method of Measurement
CFO	Chief Finance Officer
CHSW	Construction (Health, Safety and Welfare) Regulations
CIRIA	Construction Industry Research and Information Association
CITB	Construction Industry Training Board

CJIC	Construction Industry Joint Council
CLAIT	Computer Literacy and Information Technology
COO	Chief Operating Officer
COSSH	Control of Substances Hazardous to Health
CPD	Continuous Professional Development
CPI	Co-ordinated Project Information
CR	Current Ratio
CRD	Capital Requirements Directive
CSCS	Construction Skills Certification Scheme
CSR	Corporate Social Responsibility
D&B	Design and Build
DBFL	Design Build Finance and Lease
DBFM	Design Build Finance and Maintain
DBFO	Design Build Finance and Operate
DCF	Discounted Cash Flow
DFS	Document Filing Systems
DMADV	Define, Measure, Analyse, Design, Verify (Six Sigma process)
DMAIC	Define, Measure, Analyse, Improve, Control (Six Sigma process)
DQP	Detailed Quality Plan
EAP	Engineers Against Poverty
EAR	European Agency for Reconstruction
EBRD	European Bank for Reconstruction and Development
EC	European Commission
ECC	Engineering Construction Contract
ECI	Early Contractor Involvement/European Construction Institute
ECTIB	Engineering Construction Industry Training Board
EDI	Electronic Data Interchange
EDMS	Electronic Document Management Systems
EFQM	European Foundation for Quality Management
EGM	Extraordinary General Meeting
EIS	Executive Information Support System
ENAA	Engineering Advancement Association of Japan
EPC	Engineer Procure Contract
EPD	Expected Project Duration
EQF	Associated Qualifications Framework in Education
ETS	Emissions Trading Scheme
EU	European Union
EU-PHARE	European Union – *Pologne, Hongrie Assistance à la Reconstruction Economique* (Poland, Hungary Assistance in Economic Reconstruction)
EWF	Engineers Without Frontiers
FAR	Federal Acquisition Regulations (US)
FASB	US Financial Accounting Standards Board
FE	Further Education
FTTM	Fixed Time to Maintenance
FIDIC	International Federation of Consulting Engineers
FRB	Federal Reserve Bank/Board
FRC	Financial Reporting Council/Fixed Return Capital

FSA	Financial Services Authority
FTP	File Transfer Protocol
GC	Government Contract
GCE	General Certificate of Education
GCSE	General Certificate of Secondary Education
GD	Greatest Difference
GDP	Gross Domestic Product
GDV	Gross Development Value
GNP	Gross National Product
GRI	Global Reporting Initiative
H&S	Health and Safety
HE	Higher Education
HMSO	Her Majesty's Stationery Office
HO	Head Office
HP	Horse Power
HSE	Health and Safety Executive
IASB	International Accounting Standards Board
IBRD	International Bank for Reconstruction and Development
ICE	Institution of Civil Engineers
ICT	Information and Communications Technology
IDA	International Development Association
IFC	International Finance Corporation
IFRS	International Financial Reporting Standards
IOSCO	International Organization of Securities Commissions
IRC	internet Relay Chat
IS	Information System(s)
ISO	International Organization for Standardization
ITA	International Trade Association
JCT	Joint Contract(s) Tribunal
JIT	Just-in-Time
KPI	Key Performance Indicator
KTP	Knowledge Transfer Partnership
LA	Local Authority
LOSC	Labour-only Subcontractors
LP	Linear Programming
M&CM	Marketing and Communications Management
MAC	Managing Agent Contract
MBNQA	Malcolm Baldrige National Quality Award
MD	Managing Director
MTBF	Mean Time between Failures
MTTR	Mean Time to Repair
MVDA	Multivariate Discriminant Analysis
NEC	New Engineering Contract
NGO	Non-Governmental Organisation
NHS	National Health Service
NIC	National Insurance Contribution
NQF	National Qualifications Framework

NVQ	National Vocational Qualifications
OCR	Oxford Cambridge and Royal Examination Board
OJEU	*Official Journal of the European Union*
OR	Operational/Operations Research
PC	Prime Cost/Personal Computer
PCAOB	Public Company Accounting Oversight Board
PERT	Program Evaluation and Review Technique
PFI	Private Finance Initiative
PFP	Partial Factors of Production/Productivity
PLC	Public Liability Company
PM	Project Manager
PPC	Project Partnering Contract
PPM	Planned Preventive Maintenance
PPP	Public Private Partnership
PQP	Project Quality Plan
PSA	Public Service Agreement
PSB	Public Sector Borrowing
PSPC	Public Sector Partnering Contract
PW	Present Worth
QA	Quality Assurance
QCA	Qualification and Curriculum Authority
QS	Quantity Surveyor
R&D	Research and Development
RFID	Radio Frequency Identification
RIBA	Royal Institute of British Architects
RICS	Royal Institution of Chartered Surveyors
RM	Relationship Management/Risk Management
SCQF	Scottish Credit and Qualifications Framework
SDS	Supervisor Delay Survey
SE company	*Societas Europaea* company
SEC	Securities & Exchange Commission (US government)
SSC	Sector Skills Council
SSDE	Single-Stage/Double-Envelope (bidding procedure)
SSM	Soft Systems Methodology
SSM8	Shave Scheme Manual – 8
SSSE	Single-Stage/Single-Envelope (bidding procedure)
STEP	Standard for the Exchange of Product Model Data
SVQ	Scottish Vocational Qualifications
TA	Training Agreement
TCP/IP	Transmission Control Protocol/internet Protocol
TD	Target Duration
TFP	Total Factor Productivity
TP	Tangent Point
TQM	Total Quality Management
TRADA	Timber Research and Development Association
UK-DFID	UK Department for International Development
USAID	United States Agency for International Development

VM	Value Management
VR	Virtual Reality
WLAN	Wireless Local Area Network
WTO	World Trade Organisation
WWW	World Wide Web

Index